THE CHEMICAL STATICS
AND KINETICS OF SOLUTIONS

Physical Chemistry

A Series of Monographs

Ernest M. Loebl, Editor

Department of Chemistry, Polytechnic Institute of Brooklyn,
Brooklyn, New York

THE CHEMICAL STATICS AND KINETICS OF SOLUTIONS

E. A. MOELWYN HUGHES

Department of Physical Chemistry
University of Cambridge
England

1971

ACADEMIC PRESS
London and New York

ACADEMIC PRESS INC. (LONDON) LTD
24–28 Oval Road,
London NW1

U.S. Edition published by
ACADEMIC PRESS INC.
111 Fifth Avenue
New York, New York 10003

Library of Congress Catalog Card Number: 78-185196
ISBN: 0-12-503550-0

PRINTED IN GREAT BRITAIN BY
J. W. ARROWSMITH LTD, BRISTOL

PREFACE

Coherence of the diversified phenomena of chemical kinetics first became possible in terms of the collision theory, according to which the velocity coefficient of a simple bimolecular reaction can be regarded as the product of a standard collision frequency and a Boltzmann-like fraction, $\exp(-E/kT)$, where E is termed the energy of activation. In this simple form, the six bimolecular reactions known in the homogeneous gas phase in 1926 and the one hundred or so bimolecular reactions known in dilute solutions in 1932 appeared to conform with theoretical requirements. Reaction rates falling short of theoretical requirements could be ascribed to de-activations, and reaction rates exceeding theoretical requirements could be attributed to the operation of a chain mechanism, whereby activated molecules pass on their energy to other energizable molecules or, alternatively, to a more generous distribution of the activation energy than is contemplated in any simple theory. It is hardly surprising that considerable extensions should have been found necessary before the collision theory, especially in its application to solutions, could claim anything like generality.

In the first place, it has proved necessary to identify the true energy of activation as a free energy rather than a total energy. In the second place, a very precise interpretation is to be given to the almost axiomatic expression: energy of activation equals the average free energy of the molecules that react, less the average free energy of all the molecules in the ground (or unreactive) state. Reflection on the way the subject has developed shows that excessive attention has been paid to conjectures concerning the critically activated states, and that insufficient attention has been focussed on the properties of molecules in their ground states. The latter topic forms the whole subject of chemical statics and must precede the subject of chemical kinetics. It is based on a knowledge of the many types of intimate inter-actions between solute and solvent molecules. In aqueous solution, with which we shall be more particularly concerned, this is a recondite problem, still only partly understood despite the extensive information provided by modern experimental techniques of widely different types.

The logical application of the collision theory to the study of reaction kinetics in solution has afforded, among other benefits, insight into the mechanism of the transmission of electrostatic energy in aromatic chemistry, an independent method of evaluating certain interatomic distances, and a quantitatively satisfactory interpretation of steric hindrance in terms of interatomic forces. But its chief value is that it provides a standard of molecular behaviour against which actual behaviour may be systematically compared.

There are, of course, other standards, in particular that particular standard which is based, not on the collision frequency of classical dynamics, but on the partly quantal frequency kT/h first derived by Herzfeld and freely adapted by Eyring in a form generalized to express the rate constants of reactions of all kinetic orders. We shall discuss the experimental data as impartially as possible in terms of both theories, recalling that accurately measured, reproducible experimental results have an intrinsic value irrespective of the form in which they are summarized.

Darwin College,   E. A. Moelwyn-Hughes
Cambridge.
1971

CONTENTS

Chapter 4

Fundamentals of Chemical Kinetics

Chapter 5

Processes Controlled by Diffusion

Chapter 6

The Kinetic Course of Some Simple Reactions

Chapter 7

Ionic Reactions

Chapter 8

Substitution at the Saturated Carbon Atom

Chapter 9

Ions and Polar Molecules; Aromatic Substitution

Chapter 10

Unimolecular Reactions

Chapter 11

Catalyzed Reactions

Contents

Chapter 12

Hydrolysis and Other Solvolyses

Chapter 13

Pressure Effects

Chapter 14

Fast Reactions and Relaxation Phenomena

Chapter 15

Correlations

Chapter 16

Reactions Between Polar Molecules

Appendix 1

The Enthalpy of Cavity Formation

Appendix 2

Heat of Expansion of the Solvent

Appendix 3

The Temperature Variation of the Viscosity of Water

Appendix 4

Kinetic Formulation of Isotopic Exchange Reactions, with Simultaneous Allowance For Solvolysis and Radioactive Decay

Author Index

Subject Index

1

INTRODUCTION

Chemical statics deals with the equilibrium state of matter; chemical kinetics with the mechanism and velocity with which that state is attained or disturbed. The two subjects, although ostensibly separate, are inextricably bound together. Moreover, static data are generally required for the full interpretation of kinetic results. Thus, for example, the kinetics of the conversion of ortho-hydrogen to parahydrogen cannot be elucidated without a knowledge of the equilibrium established between the two stable forms of the hydrogen molecule; and the kinetics of the reaction between hydrogen and bromine can be interpreted only when the equilibrium constant governing the dissociation of hydrogen molecules into atoms is known. Reactions in solution are no exceptions to this requirement. In fact, more static data are necessary for their interpretation, for we must allow for equilibria between the solute molecules and the solvent molecules that surround them. Failure to supplement kinetic data with independently determined static data has, more than any other omission, impeded real advancement in this field, and is responsible for much of the present confusion in the subject of the kinetics of reactions in solution.

The Status of Dissolved Molecules

Let us consider the dissolution (i.e. the process of dissolving) of the halogen molecules in water at room temperature. The dissolution of chlorine is attended by an evolution of heat, which is near to the heat evolved when chlorine gas liquefies. Bromine dissolves in water with negligible heat change. When iodine dissolves, it absorbs a quantity of heat not very different in magnitude from the heat of fusion. Judged on this evidence, and supplemented by further thermal data, the status of a dissolved molecule resembles that of a liquid.

The Liquid State

Liquids have many properties intermediate in magnitude between those of solids and vapours under the same conditions, and may therefore be regarded as slightly released solids or highly condensed vapours. Near the freezing point, the densities of liquids, excepting bismuth and water, are slightly less than the densities of the corresponding crystals. This fact suggests that liquids contain holes, like untenanted graves formed by the conversion

FIG. 1. Shows a hole in an otherwise regular pattern.

of perfect crystals, wherein each lattice site is occupied, into imperfect structures wherein there are unoccupied sites (Fig. 1). The results of experiments on the diffraction of X-rays by liquids (Debye and Mencke, 1930; Kirkwood, 1935; Gringrich, 1943; Zernike and Prins, 1947; Kruh, 1962) confirm, in some measure, this view. They indicate that the co-ordination number, c (that is, the mean number of nearest neighbours surrounding a molecule) is less in a liquid than in the corresponding crystal, and decreases as the temperature is raised, while the distance between nearest neighbours is unchanged. In terms of this quasi-lattice model of a liquid, which is not to be taken too literally, we can imagine each liquid molecule as having at least one unoccupied site among the c sites that surround it. Mobility is thus ensured. The hole theory of liquids leads to numerous other consequences (Altar, 1936; Fürth, 1941; T. S. Ree, T. Ree, H. Eyring and R. Perkins, 1965).

Heat capacity measurements also throw light on the nature of liquids (Bernal, 1937; Eucken, 1948; Staveley, Tupman and Hart, 1955; Harrison and Moelwyn-Hughes, 1957). The molar heat capacity of monatomic liquids and the translational component of the molar heat capacity of complicated liquids approximate to 3R, indicating that the motion of the centre of gravity of liquid molecules resembles that of a classical three-dimensional oscillator. The magnitude of the rotational component of the heat capacity of liquid chloroform compared with that for carbon tetrachloride suggests that rotation about one axis, presumably that which is perpendicular to the figure axis, is strongly hindered—a conclusion which is confirmed by the contours of its infra-red absorption bands (W. J. Jones and N. Sheppard, 1960).

A Simple Model of a Liquid Molecule

A crude but interesting model of a liquid molecule may be constructed by regarding it as an incompressible sphere of mass m and diameter σ symmetrically surrounded by c identical molecules at a common distance a apart (Fig. 2). The reduced mass μ of such a molecule moving with respect to its neighbours is $[c/(c + 1)]m$, but if we regard it as moving with respect to all the other molecules in the liquid, we may safely take μ as equal to m. Collisions are said to occur whenever a equals σ. According to the kinetic theory of gases, the number of collisions made per second by a single molecule of an ideal gas on unit area of any surface is

$$n\left(\frac{2}{\pi}\right)^{1/2}\left(\frac{m}{kT}\right)^{3/2}\int_0^\infty \exp\left(-\frac{1}{2}\frac{mv^2}{kT}\right)v^3\,dv\int_0^{\pi/2}\frac{1}{2}\sin\theta\,d\theta\cos\theta,$$

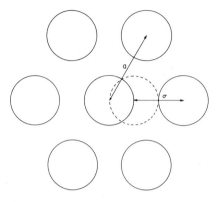

FIG. 2. a is the average distance apart of the centres of neighbouring molecules. σ is the minimum value of a, and is the effective diameter. The maximum amplitude is $a - \sigma$.

where n is the molecular concentration defined in the usual way as the quotient of the total number of molecules and the total volume of the system. v is the relative velocity, and θ is angle subtended between the direction of the relative velocity and that of the line joining the centres of the colliding molecules. On performing the integration, this expression yields the familiar result

$$n\left(\frac{kT}{2\pi m}\right)^{1/2}.$$

What is required in a general statistical formulation is not the concentration as conventionally defined but the concentration of point centres per unit free volume, which is defined as $\iiint dx\,dy\,dz$ integrated over the volume accessible to the motion of the centre of gravity. In the model under consideration (Fig. 3), the free volume is seen to be $(4/3)\pi r^3$, where r is the amplitude of its motion, as illustrated by the shaded sphere. The correct expression for n is therefore $1/(4/3)\pi r^3$, and the frequency of collisions on unit area is

$$\frac{3}{4\pi r^3}\left(\frac{kT}{2\pi m}\right)^{1/2}.$$

The area which the point centre must hit in making collisions with its warder molecules is $4\pi r^2$. Hence the number of collisions made per second by a caged molecule on the molecules that surround it is

$$Z_c = \frac{3}{4\pi r^3}\cdot 4\pi r^2 \cdot\left(\frac{kT}{2\pi m}\right)^{1/2} = \frac{3}{r}\left(\frac{kT}{2\pi m}\right)^{1/2}.$$

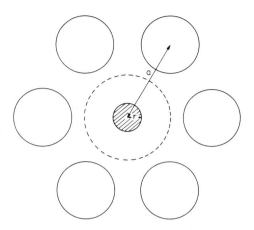

FIG. 3. *a* is the average intermolecular distance. *r* is the amplitude. The shaded portion represents the free volume, i.e. the volume accessible to the centre of gravity of the central molecule. The sphere within the dotted region is the total volume ascribable to one vibrating molecule.

The maximum value of r is clearly $a - \sigma$, and its average value, $\bar{r}$, is $(1/2)$ $(a - \sigma)$. The mean collision frequency is thus

$$Z_c = \frac{6}{(a - \sigma)} \left(\frac{kT}{2\pi m} \right)^{1/2},$$

as derived by various authors and applied to kinetic problems (Lamble and McC. Lewis, 1914; R. S. Bradley, 1935; T. S. Wheeler, 1938; Doss, 1936; R. H. Fowler and N. B. Slater, 1937).

The average vibration frequency may be obtained by dividing the average velocity in a radial direction, which is $(kT/2\pi m)^{1/2}$ by $4\bar{r}$, which is $2(a - \sigma)$:

$$v = \frac{1}{2(a - \sigma)} \left(\frac{kT}{2\pi m} \right)^{1/2}. \tag{2}$$

Consequently

$$Z_c = 12v. \tag{3}$$

Numerical values found in this way for v, by assuming that a may be identified with the average distance apart of near neighbours in the most condensed state of the liquid are correct with regard to order of magnitude. With carbon tetrachloride at 25°, for example, v is found to be 6.11×10^{11} sec^{-1}, which is about one-half of its average vibration frequency (1.31×10^{12}) in the crystal at low temperatures. In other respects, the model is inadequate. It gives, for example, a molar heat capacity of $(3/2)R$. The reason is that molecules are not incompressible, and that their actual motion is not in a force-free field. Before we can derive more reasonable expressions for v and C_V, we must deal with the forces and energies exerted between real molecules in the condensed states of matter.

Intermolecular Energies

Forces of repulsion and attraction are exerted between every pair of atoms, molecules and ions. They vary widely in their nature and magnitude, which depends upon specific properties of each particle, upon their distance, a, apart, and upon the angles at which their axes are inclined to the line joining their centres. Forces which depend in magnitude on specific molecular properties and the intermolecular separation, a, *only* are said to be spherically symmetrical; and it is with forces and energies of pairs of spherically symmetrical particles that we shall first deal. Experiment shows that the interaction energy, ϕ, of an isolated pair can then be summarized by the empirical equation (G. Mie, 1903):

$$\phi = Aa^{-n} - Ba^{-m}, \tag{4}$$

where A and B are specific constants, and n and m are integers, of which n is the larger. The first term represents an energy of repulsion, and the second an energy of attraction. The force exerted between the partners is

$$X = -\frac{d\phi}{da} = nAa^{-(n+1)} - mBa^{-(m+1)}, \tag{5}$$

which is zero when they have reached the separation a_e corresponding to the minimum value, ϕ_e, of the potential energy. By eliminating A and B, we can therefore write

$$\phi = \phi_e \frac{1}{(m-n)} \left[m\left(\frac{a_e}{a}\right)^n - n\left(\frac{a_e}{a}\right)^m \right]. \tag{6}$$

The general form of this equation is shown in Fig. 4. When the system is displaced, in either direction, from the separation a_e to a separation a, there is a gain in energy of $\phi - \phi_e$ which, for small displacements, is found by expansion to vary as the square of the displacement $a - a_e$. The resulting motion is therefore a simple harmonic one, with a vibration frequency

$$v_e = \frac{1}{2\pi a_e} \sqrt{\frac{mnD_e}{\mu}}, \tag{7}$$

where $D_e = -\phi_e$, and μ is the reduced mass of the pair (G. B. B. M. Sutherland, 1938).

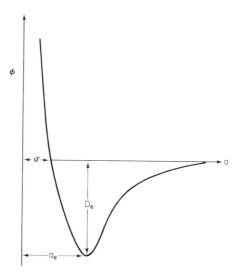

FIG. 4. $\phi = Aa^{-n} - Ba^{-m}$. Interaction energy of an isolated pair of molecules subject to spherically symmetrical forces of repulsion and attraction.

The separation of the pair when it has no potential energy is denoted by σ, and is frequently referred to as the molecular diameter. Its relation to a_e is clearly given by the equation

$$m\left(\frac{a_e}{\sigma}\right)^n = n\left(\frac{a_e}{\sigma}\right)^m$$

or

$$\frac{\sigma}{a_e} = \left(\frac{m}{n}\right)^{1/(n-m)}. \tag{8}$$

On eliminating a_e from this equation and equation (6), we obtain a third form of the general energy equation

$$\phi = \frac{D_e}{n-m}\left(\frac{n^n}{m^m}\right)^{1/(n-m)}\left[\left(\frac{\sigma}{a}\right)^n - \left(\frac{\sigma}{a}\right)^m\right], \tag{9}$$

which is often used in the particular form which it assumes when $n = 12$ and $m = 6$ (Lennard-Jones, 1937), and when, consequently

$$\phi = 4D_e\left[\left(\frac{\sigma}{a}\right)^{12} - \left(\frac{\sigma}{a}\right)^6\right]. \tag{10}$$

In order to apply these equations to the condensed states of matter, it may first be assumed that most of the interaction energy of the system arises from interaction energies of pairs of neighbouring molecules. On this assumption, the total potential energy of a system consisting of N identical molecules each surrounded by c equidistant neighbours would be

$$U = \tfrac{1}{2}Nc\phi, \tag{11}$$

since we must not count each interaction twice. The average potential energy per molecule would then be simply

$$u = \tfrac{1}{2}c\phi = \tfrac{1}{2}c(Aa^{-n} - Ba^{-m}). \tag{12}$$

Allowance must next be made for the interaction of each molecule with molecules in the second co-ordination shell and finally for its interaction with all the molecules in the system. This can be achieved with precision for all perfect crystal (Lennard-Jones and Ingham, 1925).

$$u = \tfrac{1}{2}c(s_n Aa^{-n} - s_m Ba^{-m}), \tag{13}$$

where the s terms are dimensionless summation constants which depend on the lattice type and on the values of n and m. If we now treat u as we treated ϕ, we find that the average potential energy per molecule can be expressed in the form

$$u = \frac{u_s}{m-n}\left[m\left(\frac{a_s}{a}\right)^n - n\left(\frac{a_s}{a}\right)^m \right], \tag{14}$$

where u_s is the mean value of the potential energy per molecule when the system is in the hypothetically motionless state of lowest potential energy, and a_s is the corresponding intermolecular separation. Slight displacements of the molecule from its position of minimum energy again lead to a simple harmonic vibration, of frequency v_s given by the equation

$$v_s = \frac{1}{2\pi a_s}\left\{ \frac{2mn|-u_s|}{3\mu} \right\}^{1/2}, \tag{15}$$

which is the analogue of equation (7). To take subcooled liquid carbon tetrachloride again as an example, we have from its density $a_s = 5.66 \times 10^{-8}$ cm, from its compressibility $mn = 60$, and from its vapour pressure $u_s = -7.8 \times 10^{-13}$ erg per molecule. Hence v_s becomes 1.02×10^{12} sec^{-1}, corresponding to a wave-number of 34 cm^{-1}, which agrees with the maximum of its absorption spectrum in the far infra-red region of the spectrum (N. E. Hill, W. E. Vaughan, A. H. Price and Mansel Davies, 1969).

Escape from the Cell

In the foregoing treatments of the liquid molecule, we have made no allowance for the possibility that it may leave its cell. The probability per second that the molecule will escape is obviously proportional to its vibration frequency, v. It must also be proportional to the probability that the molecule possesses a component of kinetic energy, resolved in a radial direction, amounting to at least ε. This is given by the kinetic theory as $e^{-\varepsilon/kT}$, where k is Boltzmann's constant. We may denote by p the probability that any remaining condition of escape must be fulfilled. Then the probability per second that the molecule escapes becomes

$$v' = vp\, e^{-\varepsilon/kT}. \tag{16}$$

The average lifetime of the molecule in its cell is $\tau = 1/v'$. It may readily be shown that τ for particles undergoing Brownian motion is also the average time taken for a molecule to move from its cell to a neighbouring one, the centre of which is at a distance a away. If the time interval considered is not too small, the process envisaged is that of diffusion. Now the coefficient of linear diffusion is

$$D = a^2/2\tau \tag{17}$$

(Einstein, 1926). Then

$$D = \tfrac{1}{2}a^2 v' = \tfrac{1}{2}a^2 pv\, e^{-\varepsilon/kT}. \tag{18}$$

In its dependence upon temperature, this equation for the coefficient of diffusion has the same form as one of the basic equations of chemical kinetics (Jost, 1952).

The Near-Neighbour Approximation to the Potential Energy

When interaction energies beyond those in the first co-ordination shell are ignored, the average potential energy of a molecule surrounded equidistantly by c neighbours is $u = (1/2)c\phi$, where ϕ is the interaction energy of a single pair of molecules. We may use this approximation to estimate the average co-ordination number, c_{liq}, of a molecule in the pure liquid state. For the crystal and the liquid, respectively, we have

$$u_{cr} = (1/2)c_{cr}\phi$$

and

$$u_{liq} = (1/2)c_{liq}\phi.$$

Therefore, the gains in potential energy attending fusion and sublimation respectively are

$$\Delta u(\text{fusion}) = (1/2)(c_{\text{liq}} - c_{\text{cr}})\phi \qquad (19)$$

and

$$\Delta u(\text{sublimation}) = -(1/2)c_{\text{cr}}\phi. \qquad (20)$$

To a first approximation, $\Delta u(\text{fusion})$ may be identified with the heat of fusion (L_f calories/mole), and $\Delta u(\text{sublimation})$ with L_s, which is the sum of the heat of sublimation (L_0 calories/mole) at the absolute zero of temperature and the residual energy of $(3/2)h\nu_s$ per molecule, or $N_0 h\nu_s$ per mole, where N_0 is Avogadro's number and h is Planck's constant. Then

$$c_{\text{liq}} = c_{\text{cr}}\left(1 - \frac{L_f}{L_s}\right). \qquad (21)$$

Some estimates of co-ordination numbers in liquids obtained in this way are given in Table I. The calculated co-ordination numbers of the liquids are, on the whole, in agreement with those given by data on the diffraction of X-rays. With liquid mercury, for example, the observed value (Kruh, 1962) is 7·5. In terms of holes of atomic size, this indicates the presence of one hole for every 15 atoms.

TABLE I

THERMAL ESTIMATES OF THE CO-ORDINATION NUMBER IN LIQUIDS

Substance	c_{cr}	L_f calories/mole	$L_s(= L_0 + (3/2)N_0 h\nu_s)$ calories/mole	L_s/L_f	c_{liq}
Ne	12	80	588	7·35	10·37
A	12	280·8	2,030	7·23	10·34
Kr	12	390·7	2,741	7·02	10·29
Xe	12	548·5	3,926	7·16	10·32
CO	12	200·6	2,083	10·39	10·84
CH$_4$	12	225·5	2,629	11·66	10·97
CCl$_4$	12	577·2	11,410	19·76	11·40
CO$_2$	12	1,913	6,513	3·39	8·46
H$_2$O	4	1,435·7	11,325	7·89	3·49
Hg	8	551·5	15,327	27·78	7·71

The near-neighbour approximation may be extended to other problems. Comparison of the total energy of a molecule in the surface of a liquid, E_σ, with its value in the bulk liquid, for example, shows that c_σ/c_{liq} ranges from

$\frac{1}{2}$ to $\frac{3}{4}$. Similarly, the ratio of the energy required by a molecule to escape from its cell to that required for sublimation or vaporization indicates that, at the point of escape, it has already rid itself of some of its neighbours.

Finally, we shall apply the near-neighbour approximation to a binary liquid solution formed of randomly mixed spherical molecules of the same size and co-ordination number. Let the solution consist of N_1 molecules of the solvent and N_2 molecules of the solute, where N_2 is so much less than N_1 that direct contacts of the solute–solute type do not occur. The total number of near-neighbour contacts is $(1/2)c(N_1 + N_2)$, of which cN_2 are of the solute–solvent type. The number of solvent–solvent contacts is thus $(1/2)c(N_1 - N_2)$ and the total interaction energy of the system is

$$U_{12} = N_2 c \phi_{12} + (1/2)c(N_1 - N_2)\phi_{11},$$

where ϕ_{11} and ϕ_{12} denote the interaction energies of like and unlike pairs. The potential energies of the separate liquids, before mixing, is

$$U_{11} + U_{22} = (1/2)N_1 c \phi_{11} + (1/2)N_2 c \phi_{22}.$$

The gain in potential energy on making the solution is thus

$$\Delta U = N_2[c(\phi_{12} - \tfrac{1}{2}\phi_{11} - \tfrac{1}{2}\phi_{22})] = N_2 \Delta u^0. \tag{22}$$

The term inside the square brackets is known as the interchange energy. It can readily be shown that the gain in potential energy attending the formation of solutions of any composition under the conditions stipulated is

$$\Delta U = \frac{N_1 N_2}{N_1 + N_2} \Delta u^0. \tag{23}$$

Solutions which obey this equation have been termed regular solutions by Hildebrand (J. H. Hildebrand and R. L. Scott, 1962). They cease to remain homogeneous, and separate into two layers when Δu^0 exceeds $2kT$. Hildebrand's theory has been modified in various ways. Even as it stands, it has proved to be qualitatively applicable to binary liquid mixtures of non-polar molecules, even when they are not spherical. Its validity diminishes with increasing polarity of either type of molecule, and vanishes almost completely for most aqueous solutions. The near-neighbour approximation to determine the potential energy of a liquid has proved to be totally misleading in the case of water, as we shall now show.

Enthalpy of Cavity Formation in Liquids

It has long been argued that the energy expended in creating a hole of molecular size in a liquid should have the same value as the energy attending its vaporization (Dupré, 1869; W. C. McC. Lewis, 1911). The argument as more recently advanced (Eyring, 1936) is given as follows: in terms of the co-ordination number, c, and the average energy of interaction of one pair, ϕ, corrected for the influence of distant molecules by including the summation constant, s. Then the energy required to remove a molecule of liquid, leaving a hole of molecular size, is $-cs\phi = 2\lambda$; the energy required to remove a molecule of liquid without leaving a hole, which is the heat of vaporization, is $-(1/2)cs\phi = \lambda$; therefore the energy required to form a hole of molecular size in the liquid is $-(1/2)cs\phi = \lambda$. The soundness of the argument may be judged by measuring the energy in two ways. (1) The increase in heat content associated with the vaporization of one gram-mole of liquid is given approximately by the Clausius–Clapeyron equation in terms of the saturation vapour pressure, p^0:

$$L_T = RT^2\left(\frac{d \ln p^0}{dT}\right). \tag{24}$$

(2) The increase in heat content attending the formation, at constant temperature and pressure, of a cavity of molecular size in a liquid is given approximately in terms of the molar volume, V, and the coefficients, α and β, of isobaric expansion and isothermal compressibility, respectively:

$$\Delta H_{T,P} = \frac{\alpha TV}{\beta}. \tag{25}$$

Table II gives a comparison of L_T and $\alpha TV/\beta$ at various temperatures for carbon tetrachloride, cyclohexane and water. For the two non-polar liquids there is satisfactory agreement. For water, however, there is no agreement, in magnitude or sign. The energy increase attending cavity formation is relatively small, and at the temperature of maximum density is nearly zero. At this temperature, no absorption of heat is necessary to form a cavity, however large, in liquid water. The conclusion is not exact, because it is based on the assumed temperature independence of α/β. It nevertheless explains why, at relatively low temperatures, liquid water can dissolve large quantities of massive molecules like those of haemoglobin and starch, with inappreciable absorption of heat (W. C. McC. Lewis, 1925; Eley, 1939). The derivation of equation 25 is given in Appendix 1.

The Slightly Anharmonic Oscillator

Only in a perfect crystal at the absolute zero of temperature is the vibration of an atom or molecule strictly harmonic. Deviations from pure harmonicity have long been formulated by expressing the displacement energy, w, in the form

$$w = \tfrac{1}{2}Kx^2 - gx^3, \tag{26}$$

where x is the displacement of the particle from the position of minimum potential energy, and w is the increase in potential energy attending the displacement. K is clearly the restoring force constant for the harmonic motion,

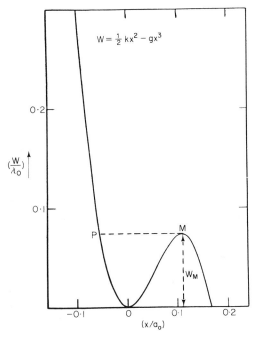

FIG. 5. Displacement energy for slight displacement according to Equation 26.

and g is an empirical term. The force acting on the particle is $-dw/dx = -Kx + 3gx^2$, which is zero when x is zero and when $x = K/3g = x_m$. The maximum energy is seen from Fig. 5 to be

$$
\begin{aligned}
w_m &= \tfrac{1}{2}Kx_m^2 - g\left(\frac{K}{3g}\right)^3 \\
&= \tfrac{1}{2}K\left(\frac{K}{3g}\right)^2 - g\left(\frac{K}{3g}\right)^3 \\
&= \frac{K^3}{54g^2}.
\end{aligned}
\tag{27}
$$

TABLE II

THE ENERGY OF CAVITY FORMATION (CALS/MOLE) FOR CARBON TETRACHLORIDE, CYCLOHEXANE AND WATER

Temperature	V (ccs/mole)	$\alpha \times 10^3$ (deg^{-1})	$\beta \times 10^6$ (atm^{-1})	$T\alpha V/\beta$	L_T	$\partial\left(\dfrac{T\alpha V/\beta}{\partial T}\right)$
			Carbon tetrachloride,* CCl$_4$			
250·2°K	91·77	1·136	77·3	8,168	8,402	—
270	93·89	1·171	87·3	8,233	8,188	− 10·0
290	96·14	1·204	101·9	7,989	7,952	− 13·7
310	98·51	1·233	118·8	7,670	7,727	− 15·8
330	101·00	1·260	138·3	7,358	7,502	− 16·8
349·9	103·59	1·284	160·6	7,017	7,278	− 17·4
			Cyclohexane,† (CH$_2$)$_6$			
20°C	108·08	1·224	111·6	8,414	8,076	− 13·0
30	109·41	1·239	120·4	8,262	8,020	− 17·0
40	110·79	1·254	130·4	8,076	7,909	− 20·2
50	112·20	1·270	141·6	7,874	7,797	− 22·1
60	113·64	1·287	154·3	7,645	7,686	− 24·4
			Water,‡ H$_2$O			
0°C	18·018	−0·678	51·3	− 157·5	(10,749)	37·8
20	18·048	+2·068	46·44	+ 569·7	10,545	32·8
40	18·157	3·846	44·75	1,183	10,341	28·5
60	18·323	5·229	45·00	1,716	10,141	25·0
80	18·538	6·419	46·6	2,184	9,929	21·9
100	18·798	7·496	49·1	2,593	9,705	19·3

REFERENCES

* Harrison and Moelwyn-Hughes, *Proc. Roy. Soc.*, **A239**, 230 (1957).
† G. Scratchard, S. E. Wood and T. M. Mochel, *J. Amer. Chem. Soc.*, **61**, 3206 (1939); Moelwyn-Hughes and Thorpe, *Proc. Roy. Soc.*, **A278**, 574 (1964).
‡ Hennig, *Ann. Physik*, **29**, 441 (1909).

Displacements of magnitude less than x_m result in stable but anharmonic motions. x_m may thus be regarded as the limiting value of the displacement for stable motion, and w_m as the corresponding limit to the displacement energy. Figure 5 represents the general shape of equation (26), where, however, the displacement energy, w, is shown as a fraction of λ_0, the average energy required to remove the molecule from the most condensed state in the crystal to the infinitely dilute gaseous state, and the displacement, x, is shown as a fraction of the equilibrium separation, a_0, of the molecules in the condensed state of lowest potential energy.

Expressions for K and g derived from Intermolecular Force Theory

Equation (14), applied to the crystalline state, may be written as follows:

$$u = \frac{\lambda_0}{n - m}\left[m\left(\frac{a_0}{a}\right)^n - n\left(\frac{a_0}{a}\right)^m\right] \tag{28}$$

where λ_0 is the energy of sublimation at the absolute zero of temperature, and a_0 is the corresponding average intermolecular separation. The displacement energy is consequently

$$w = u - u_0 = \frac{\lambda_0}{n - m}\left\{n\left[1 - \left(\frac{a_0}{a}\right)^m\right] - m\left[1 - \left(\frac{a_0}{a}\right)^n\right]\right\}.$$

On expanding, and taking the average value, we find, as an approximate expression for the displacement energy:

$$w = \tfrac{2}{3}mn\lambda_0\left[\frac{1}{2}\frac{x^2}{a_0^2} - \frac{1}{6}(n + m + 3)\frac{x^3}{a_0^3}\right]. \tag{29}$$

Hence,

$$K = \frac{2}{3}\frac{mn\lambda_0}{a_0^2}, \tag{30}$$

$$g = \tfrac{1}{9}(n + m + 3)\frac{mn\lambda_0}{a_0^3}, \tag{31}$$

and

$$\frac{K}{g} = \frac{6}{(n + m + 3)} \cdot a_0. \tag{32}$$

The maximum value of the oscillatory energy in a stable system is, from equation (27),

$$w_m = \frac{4mn\lambda_0}{9(n + m + 3)^2}. \tag{33}$$

With $m = 6$, we thus have

$$\frac{\lambda_0}{w_m} = 13 \cdot 5 \ (n = 9) \quad \text{or} \quad 13 \cdot 8 \ (n = 12).$$

These ratios are near to the observed ratios listed in Table III, of the latent heat of sublimation to kT_m, where T_m is the temperature of the melting point. If the treatment under discussion is correct, it would seem that the maximum vibrational energy, w_m, corresponds roughly with kT_m.

We have seen that the maximum value of the displacement, x_m, is $K/3g$. The space-average of the displacement is thus

$$\bar{x} = \frac{a_0}{n + m + 3}. \tag{34}$$

Furthermore, if we identify the average intermolecular distance at the melting point with $a_0 + \bar{x}$, we have

$$\frac{a_m}{a_0} = \frac{a_0 + \bar{x}}{a_0} = \frac{n + m + 4}{n + m + 3}. \tag{35}$$

It has long been argued (W. Sutherland, *Phil. Mag.*, **30**, 318 (1890); F. A. Lindemann, *Physikal. Z.*, **11**, 609 (1910)) that the atomic displacements of simple solids at the melting point are a constant fraction of the interatomic separations.

The ratio of the molecular volumes is

$$\frac{v_m}{v_0} = \left(\frac{n + m + 4}{n + m + 3}\right)^3. \tag{36}$$

Again, taking m as 6, we find $v_m/v_0 = 1.18$ ($n = 9$) or 1.15 ($n = 12$) which lie near to, but are on the whole somewhat higher than, the experimental ratios (Table III).

TABLE III

RATIOS OF MOLAR VOLUMES AT THE MELTING POINT TO THOSE AT THE ABSOLUTE ZERO, AND RATIOS OF THE LATTICE ENERGIES TO RT_m.

Crystal	V_m/V_0	$L_0 + 3/2\,N_0hv_0$ (cal/mole)	T_m (°K)	$\dfrac{L_0 + 3/2N_0hv_0}{RT_m}$
Ne	1·111	591	24·57	12·11
Ar	1·088	2,040	83·78	12·26
Kr	1·108	2,741	115·95	11·90
Xe	1·152	3,930	161·36	12·26

The elementary treatment given above (Moelwyn-Hughes, 1957) differs significantly from the classical derivation of Born and Debye, (1909–1926), whose treatment necessitates the introduction of a conjectural external force, and leads finally to the following expression for the total vibrational energy of the slightly displaced oscillator:

$$\varepsilon = \tfrac{1}{2}\mu\omega^2(\bar{x})^2 - g(\bar{x})^3 + \tfrac{1}{2}\mu(\dot{\xi})^2 + (\tfrac{1}{2}\mu\omega^2 - 3g\bar{x})\xi^2 - g\xi^3,$$

where μ is the reduced mass, ω is $2\pi v$, $\bar{x}$ is the time average of the displacement x, and $\xi = x - \bar{x}$. $\dot{\xi} = (\partial\xi/\partial t)$.

Modern interest and emphasis have, however, shifted from the study of representative single molecules to the investigation of large groups of molecules capable of forming planes of cleavage. (See, for example, Kittel, 1956; Mansel Davies and J. O. Thomas, 1960.) New Techniques have revealed a variety of hitherto undetected properties in most solids. They directly influence the kinetics of heterogeneous reactions in the condensed state of matter. The study of such reactions falls outside the scope of this work.

REFERENCES

Altar, *J. Chem. Physics*, **4**, 577 (1936).

Bernal, *Trans. Faraday Soc.*, **33**, 27 (1937).

Born and Debye, *Encyklopädie der mathematichen Wissenschaften*, *Vol. V*, p. 527, Teubner, Leipzig (1909–1926).

Bradley, R. S., *ibid.*, 1910 (1935).

Debye and Mencke, *Physikel. Z.*, **31**, 797 (1930).

Doss, *Proc. Indian Acad. Sci.*, **4**, 291 (1936).

Dupré, *Theorie Mécanique de la Chaleur*, Paris, 1869; W. C. McC. Lewis, *Trans. Faraday Soc.*, **7**, 94 (1911).

Einstein, *Investigations on the Theory of the Brownian Movement*, Eng. trans. by Cowper, Methuen, London (1926).

Eley, *Trans. Faraday Soc.*, **35**, 1281 (1939).

Eucken, *Z. Elekrochem.*, **52**, 255 (1948).

Eyring, *J. Chem. Physics*, **4**, 283 (1936).

Fowler, R. H. and N. B. Slater, *Faraday Soc. Discussion*, p. 81 (1937).

Fürth, *Proc. Camb. Phil. Soc.*, **37**, 252 (1941).

Gringrich, *Rev. Mod. Phys.*, **15**, 90 (1943).

Harrison and Moelwyn-Hughes, *Proc. Roy. Soc.*, **A239**, 230 (1957).

Hildebrand and R. L. Scott, *The Solubility of Non-electrolytes*, 4th ed., Reinhold, New York (1962).

Hill, N. E., W. E. Vaughan, A. H. Price and M. Davies, *Dielectric Properties and Molecular Behaviour*, Van Nostrand–Reinhold, London (1969).

Jones, W. J. and N. Sheppard, *Trans. Faraday Soc.*, **56**, 625 (1960).

Jost, *Diffusion*, Academic Press, New York (1952).

Kirkwood, *J. Chem. Physics*, **3**, 300 (1935).

Kittel, *Solid State Physics*, New York, 1956.

Kruh, *Chem. Rev.*, **62**, 319 (1962).

Lamble and McC. Lewis, *Trans. Chem. Soc.*, **105**, 2330 (1914).

Lennard-Jones, *Physica*, **IV**, 10, 947 (1937).

Lennard-Jones and Ingham, *Proc. Roy. Soc.*, **A107**, 636 (1925).

Lewis, W. C. McC., *A System of Physical Chemistry*, 4th ed., Vol. II, p. 30, Longmans (1925).

Mansel Davis and J. O. Thomas, *Trans. Faraday Soc.*, **56**, 185 (1960).

Mie, G., *Ann. Physik*, **11**, 657 (1903).

Moelwyn-Hughes, E. A., *Academia Nazionale dei Lincei*, Varenna, 1957.

Ree, T. S., Ree, T., Eyring, H., and Perkins, R., *J. Phys. Chem.*, **69**, 3222 (1965).

Staveley, Tupman and Hart, *Trans. Faraday Soc.*, **51**, 323 (1955).

Sutherland, G. B. B. M., *Proc. Indian Acad. Sci.*, **8**, 341 (1938).

Wheeler, T. S., *Trans. Nat. Inst., Sci., India*, **1**, 333 (1938).

Zernike and Prins, *Z. Physik.*, **41**, 184 (1947).

2

FUNDAMENTALS OF CHEMICAL STATICS

The term chemical statics aptly describes the study of chemical systems in their static or equilibrium states. Such states have been defined in various ways. Mathematically, the static or equilibrium state of a system is that for which the absolute probability of its existence is a maximum. Thermodynamically, there are many equivalent definitions. For a closed system (i.e. one of fixed amount) in the absence of gravitational, electrical, magnetic and surface forces, the static state is that for which the Gibbs free energy, G, is minimal. Equilibria in homogeneous and heterogeneous systems are most conveniently dealt with in terms of the chemical potential, μ, for each component. μ, in turn, is defined for each component in each phase by the partial differential equations

$$\mu_i = \left(\frac{\partial E}{\partial N_i}\right)_{S,V,N_j\ldots} = \left(\frac{\partial A}{\partial N_i}\right)_{T,V,N_j\ldots} = \left(\frac{\partial H}{\partial N_i}\right)_{S,P,N_j\ldots} = \left(\frac{\partial G}{\partial N_i}\right)_{T,P,N_j\ldots}$$

where E is the total energy, A the Helmholtz free energy, H the heat content or enthalpy, G the Gibbs free energy, S the entropy, V the volume, T the temperature, P the pressure, and $N_i, N_j, \ldots$ the numbers of the various types of molecules in the system. In a one-component system of two phases, the equilibrium condition is then

$$\mu_i \text{ (phase } \alpha) = \mu_i \text{ (phase } \beta). \tag{1}$$

This equation we shall frequently use to determine, for example, the distribution of a solute between two immiscible solvents, or between a liquid solution and the vapour phase.

The thermodynamic law governing the equilibrium between various chemical species, B, C, and L, say, in any particular phase, as symbolized by the chemical reaction

$$bB + cC \rightleftarrows lL$$

is

$$b\mu_B + c\mu_C = l\mu_L, \tag{2}$$

where b, c and l are the stoichiometric numbers, and the μ's are chemical potentials.

In a mixture of ideal gases, the chemical potential of component i is

$$\mu_i = \mu_i^0(T, P) + kT \ln p_i, \tag{3}$$

where p_i is its partial pressure, and $\mu_i^0(T, P)$, which is clearly the chemical potential at unit partial pressure, is a function of T and P only. Since the concentration, c_i, of this component is p_i/kT, we may also write

$$\mu_i = \mu_i^0 + kT \ln c_i, \tag{4}$$

where μ_i^0 is now the chemical potential of component i when its concentration is unity. The condition of equilibrium is therefore given by the relationship $b(\mu_B^0 + kT \ln c_B) + c(\mu_C^0 + kT \ln c_C) = l(\mu_L^0 + kT \ln c_L)$, which may be written in the form

$$\frac{c_L^l}{c_B^b c_C^c} = \exp\left(-\Delta\mu^0/kT\right), \tag{5}$$

where

$$\Delta\mu^0 = l\mu_L^0 - b\mu_B^0 - c\mu_C^0. \tag{6}$$

But the algebraic product of the concentrations of products and reactants is, by definition, the equilibrium constant, K. Then

$$K = \exp(-\Delta\mu^0/kT),$$

or

$$\Delta\mu^0 = -kT \ln K. \tag{7}$$

The equilibrium constant of a chemical reaction between ideal gases is thus determined by the temperature and pressure and by the algebraic sum (products minus reactants) of their chemical potentials in the reference state of unit pressure or unit concentration.

The chemical potential as hitherto used refers to one molecule. Its meaning may be gained by examining any of its four definitions. For example, μ_i is the gain in Gibbs free energy of a system when, at constant temperature, pressure and composition, one molecule of type i is added to it. To ensure that the composition is not affected by this addition, the numbers of the molecules of all types in it must be very large. Chemical potentials are often defined in molar rather than molecular terms. We may define the molar chemical potential as $G_i = N_0\mu_i$, where N_0 is Avogadro's number. With this notation,

$$\Delta G^0 = lG_L^0 - bG_B^0 - cG_C^0,$$

and

$$= -RT \ln K. \tag{7A}$$

By applying the Kelvin–Helmholtz equation

$$H = -T^2 \left[\frac{d(G/T)}{dT} \right]_P \tag{8}$$

to each participant in the equilibrium system, we obtain the following expression for the algebraic sum of the molar heat contents under the standard conditions of unit pressure:

$$\Delta H^0 = RT^2 \left(\frac{d \ln K}{dT} \right)_P. \tag{9}$$

This equation is known as the van't Hoff isochore.

The change in standard free energies is, by definition, related as follows to the changes in heat contents and entropies:

$$\Delta G^0 = \Delta H^0 - T \Delta S^0, \tag{10}$$

so that the equilibrium constant may be written as follows:

$$K = \exp (\Delta S^0/R) \exp (-\Delta H^0/RT). \tag{11}$$

ΔS^0 has the dimensions of R, the gas constant, multiplied by the logarithm of the pressure or concentration raised to the power $l - (b + c)$. The dimensions of K itself are those of pressure or concentration raised to the appropriate power.

The equations of this section are applicable to gaseous systems which obey the ideal gas laws. With but slight modifications, however, they can be applied to equilibria in dilute solutions.

The merit of the thermodynamic treatment of equilibria is that it is, in principle, applicable to systems comprising molecules of all complexities, since ΔH^0 and the various standard entropies, S^0, can be determined entirely from thermal sources. There is another approach to the equilibrium state which may be applied to systems made up of relatively simple molecules.

The Statistical Evaluation of Equilibrium Constants

The starting point here is the molecular partition function, f, defined for quantized systems in terms of the various occupied energy levels, ε_i, and their degeneracies, g_i, by the equation

$$f = g_0 \exp\left(-\varepsilon_0/kT\right) + g_1 \exp\left(-\varepsilon_1/kT\right) + \cdots = \sum_0^\infty g_i \exp\left(-\varepsilon_i/kT\right), \qquad (12)$$

and for classical systems by the equation

$$f = \frac{1}{h^s} \int_{-\infty}^\infty \cdots \int \exp\left(-\varepsilon/kT\right) dq_1\, dq_2 \ldots dq_s\, dp_1\, dp_2 \ldots dp_s, \qquad (13)$$

where h is Planck's constant, and ε is the total energy of the molecule, regarded as a continuous function of the s pairs of spatial and momentum co-ordinates, q_i and p_i. The chemical potential of a molecule of type i in dilute systems is related as follows to its partition function, f_i:

$$\mu_i = -kT\left[\ln f_i + \left(\frac{d\ln f_i}{d\ln N_i}\right)\right]_{T,V,N_j,\ldots}. \qquad (14)$$

It is convenient to introduce a term q_i, which has the dimensions of a concentration, and is related as follows to the molecular partition function:

$$f_i = q_i e/n_i = q_i eV/N_i. \qquad (15)$$

Here e is the base of the natural logarithm, and n is the molecular concentration expressed as molecules per cm^3. It follows that

$$\mu_i = -kT \ln \frac{q_i}{n_i}, \qquad (16)$$

and that the equilibrium constant for the reaction considered is

$$\frac{n_L^l}{n_B^b n_C^c} = \frac{q_L^l}{q_B^b q_C^c} = K_n. \qquad (17)$$

It has been often but wrongly stated that the equilibrium constant is the algebraic product of the partition functions, f_i, rather than of the concentrations, q_i.

Because the total energy which a molecule possesses can be constituted in a variety of ways, the exact evaluation of f may become difficult. If, however, the total energy can be resolved into components contributed independently by the various molecular motions, the total partition function may be resolved into the product of a number of simple partition functions, e.g. those due to translation, rotation, vibration, etc.:

$$f = f_{\text{trans}} f_{\text{rot}} f_{\text{vibr}} \cdots . \tag{18}$$

Partition functions for molecules exercising certain simple motions are listed in Table I. The symmetry number, σ, which appears in the rotational partition functions, is defined as the number of indistinguishable arrangements of the molecule which can be obtained by rotation of the molecule about its axes. Thus, for example, if the water molecule is rotated through 180° about an axis passing through the oxygen atom and midway between the two hydrogen atoms, there appears another setting for the molecule which is indistinguishable from the first setting: $\sigma = 2$. If the molecule CH_3X is rotated about the C–X axis through 60°, 120° and 180°, we realize that there are three indistinguishable permutations: $\sigma = 3$. In the vibrational component of the partition function for the angular triatomic molecule ABC, v_s denotes the frequency of the symmetrical vibration, when the wing atoms A and C move away from or towards the middle atom B synchronously; v_a is the frequency when atom A, for example, moves away while atoms B and C move towards each other, or when atom C moves away while atoms A and B move towards each other. This is the type of motion that can lead to decomposition of the molecule. The third vibration frequency, v_ϕ, is that which governs the rate of change of the valency angle.

In a molecule consisting of n atoms, the total number of modes of motion is $3n$. Whether a triatomic molecule is linear or bent, this number is 9. With the bent molecule 3 each are devoted to translation, rotation and vibration. The linear molecule, however, having only one moment of inertia, has only two modes of rotation. To compensate for the loss, it has 4 modes of internal vibration, namely the symmetrical and asymmetrical vibrations of the atoms along the lines joining them, and the wagging vibration, v_ϕ, which can take place in two dimensions, e.g. in the plane of the paper and perpendicular to it. For that reason, the wagging motion is said to be doubly degenerate, and its vibrational partition function appears twice. Another type of degeneracy, denoted by g, arises from the electronic levels of the atom or molecule. For diatomic molecules, g is usually unity. For the inert gases also, in their ground state, $g = 1$. For the alkali metal atoms $g = 2$. Generally g for atoms is the multiplicity of the terms required to reproduce their spectral lines.

Each partition function must, finally, be multiplied by a term $\exp(-u/kT)$ where u is the potential energy of the motionless atom or molecule. We then have, for atoms of mass m moving freely throughout a system of volume V at temperature T,

$$f = g \frac{(2\pi m k T)^{3/2} V e}{h^3 N} \exp(-u/kT). \tag{19}$$

TABLE I

Molecular motion	Molecular partition function f
Free linear translation of a molecule of mass m along a length l	$\dfrac{(2\pi mkT)^{1/2}\,l}{h}$
Free superficial translation over an area o, where $o = O/N = $ total area/total number of molecules.	$\dfrac{2\pi mkTo}{h^2}$
Free spatial motion confined to a cell of volume v, where $v = V/N = $ total volume/total number of molecules.	$\dfrac{(2\pi mkT)^{3/2}v}{h^3}$
Free spatial motion in a system of interchangeable cells, each of volume v. e is the base of the natural logarithm.	$\dfrac{(2\pi mkT)^{3/2}ve}{h^3}$
Free rotation of a linear molecule of moment of inertia, I, about its centre of gravity, when the rotational energy is less than kT. The symmetry number, σ, is discussed below.	$\dfrac{8\pi^2 IkT}{\sigma h^2}$
Free rotation of a rigid molecule with moments of inertia I_A, I_B and I_C about the axes passing through its centre of gravity.	$\dfrac{\pi^{1/2}(8\pi^2 kT)^{3/2}(I_A I_B I_C)^{1/2}}{\sigma h^3}$
Rotation of a rigid molecule of moment of inertia, I, possessing an electrical dipole of moment, μ, in an electrical field of strength, F.	$\dfrac{8\pi^2 IkT}{h^2}\cdot\dfrac{kT}{\mu F}\sinh\dfrac{\mu F}{kT}$
Linear harmonic oscillator with vibration frequency, v. This general expression reduces to kT/hv when hv is much less than kT.	$\left(2\sinh\dfrac{hv}{2kT}\right)^{-1}$
Isotropic three-dimensional harmonic oscillator.	$\left(2\sinh\dfrac{hv}{2kT}\right)^{-3}$
Harmonic internal vibrations of an angular triatomic molecule.	$\left(2\sinh\dfrac{hv_s}{2kT}\right)^{-1}\left(2\sinh\dfrac{hv_a}{2kT}\right)^{-1}$ $\left(2\sinh\dfrac{hv_\phi}{2kT}\right)^{-1}$
Harmonic internal vibrations of a linear triatomic molecule.	$\left(2\sinh\dfrac{hv_s}{2kT}\right)^{-1}\left(2\sinh\dfrac{hv_a}{2kT}\right)^{-1}$ $\left(2\sinh\dfrac{hv_\phi}{2kT}\right)^{-2}$

In dilute systems q, as shown, is always fN/eV; hence

$$q = g \cdot \frac{(2\pi mkT)^{3/2}}{h^3} \exp(-u/kT). \tag{20}$$

The partition function for a diatomic molecule consisting of atoms with masses m_A and m_B is

$$f = \frac{[2\pi(m_A + m_B)kT]^{3/2} Ve}{h^3 N} \cdot \frac{8\pi^2 IkT}{\sigma h^2} \left(2 \sinh \frac{hv}{2kT}\right)^{-1} \exp(-u/kT). \tag{21}$$

Here

$$q = \frac{[2\pi(m_A + m_B)kT]^{3/2}}{h^3} \cdot \frac{8\pi^2 IkT}{\sigma h^2} \left(2 \sinh \frac{hv}{2kT}\right)^{-1} \exp(-u/kT). \tag{22}$$

It is convenient to replace $[2 \sinh(hv/2kT)]^{-1}$ by $\exp(-hv/2kT)[1 - \exp(-hv/kT)]^{-1}$, so that

$$q = \frac{[2\pi(m_A + m_B)kT]^{3/2}}{h^3} \cdot \frac{8\pi^2 IkT}{\sigma h^2} [1 - \exp(-hv/kT)]^{-1}$$
$$\times \exp[-(u + \tfrac{1}{2}hv)/kT]. \tag{23}$$

In the same way, q for a linear triatomic molecule is seen to be

$$q = \frac{[2\pi(m_A + m_B + m_C)kT]^{3/2}}{h^3} \cdot \frac{8\pi^2 IkT}{\sigma h^2} [1 - \exp(-hv_s/kT)]^{-1}$$
$$\times [1 - \exp(-hv_a/kT)]^{-1} [1 - \exp(hv_\phi/kT)]^{-2}$$
$$\times \exp[-(\tfrac{1}{2}hv_a + \tfrac{1}{2}hv_s + hv_\phi + u)/kT]. \tag{24}$$

We are now in a position to derive theoretical expressions for various equilibrium constants.

1. Equilibria of the Type $AB \rightleftarrows (AB)^*$

Suppose a diatomic molecule AB acquires an additional energy, and thereby becomes activated. To express the equilibrium constant, which in this case is dimensionless, we require only the ratio of the partition functions for the normal and activated species. The translational partition functions are identical; hence, from equation (23), we have

$$K = \frac{n^*}{n} = \frac{q^*}{q} = \frac{I^*}{I} \frac{[1 - \exp(-hv/kT)]}{[1 - \exp(-hv^*/kT)]}$$
$$\times \exp(-\{[u^* + \tfrac{1}{2}hv^*] - [u + \tfrac{1}{2}hv]\}/kT). \tag{25}$$

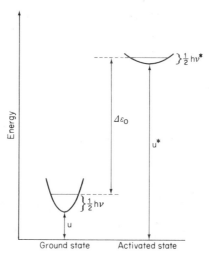

FIG. 1. Conversion of a diatomic molecule from the ground state to a stable activated state.

The u terms represent the potential energies of the static molecules at the absolute zero of temperature. Even at $T = 0$, molecules are never static, and their effective potential energy is thus $u_i + \frac{1}{2}h\nu_i$. The energy term in the exponential function is thus the observed gain in energy attending activation at the absolute zero of temperatures, and may be denoted by $\Delta\varepsilon_0$ (see Fig. 1). Since the moment of inertia varies as the square of the internuclear distance, we have

$$K = \frac{n^*}{n} = \left(\frac{r^*}{r}\right)^2 \frac{[1 - \exp(-h\nu/kT)]}{[1 - \exp(-h\nu^*/kT)]} \exp(-\Delta\varepsilon_0/kT). \qquad (26)$$

A simplified version of this equation results when the difference between r^* and r can be ignored, when $h\nu \gg kT$ and when $h\nu^* \ll kT$. Then

$$K = \frac{n^*}{n} = \frac{kT}{h\nu^*} \exp(-\Delta\varepsilon_0/kT). \qquad (27)$$

The assumptions leading to this equation must not be overlooked. Nor must it be forgotten that the potential energy difference $(u^* - u)$ includes not only the differences in energy associated with the distention of the molecule but the difference in the interaction energies of the molecule with other molecules surrounding it.

2. Metastable Equilibria

Let us suppose that the potential energy of the activated molecule in the preceding example has the form of an inverted parabola, as illustrated in

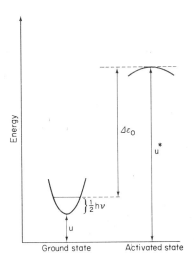

FIG. 2. Conversion of a diatomic molecule from the ground state to a metastable activated state.

Fig. 2. In a perfectly static system, a genuine but metastable equilibrium would then be attained. In a real system, however, such an equilibrium is always unstable. Equation (25), which refers to stable equilibria, may be written as follows, after ignoring the difference between r and r^*:

$$K = \frac{n^*}{n} = \frac{\sinh (h\nu/2kT)}{\sinh (h\nu^*/2kT)} \exp\left[-(u^* - u)/kT\right]. \tag{28}$$

Because of the curvature of the energy surface characterizing the unstable system, the vibration frequency ν^* has an imaginary value. Let $\nu^* = i\nu_a$, where ν_a is real. Then

$$K = \frac{n^*}{n} = \frac{\sinh (h\nu/2kT)}{i \sin (h\nu_a/2kT)} \exp\left[-(u^* - u)/kT\right]. \tag{29}$$

Although the equilibrium constant has only an imaginary value, the difference between the total energy of the activated and normal molecules has a real value, which is

$$\Delta \varepsilon_T = kT^2 \left(\frac{d \ln K}{dT}\right)_V$$

$$= (u^* - u) - (h\nu/2)\coth (h\nu/2kT) + (h\nu_a/2)\cot (h\nu_a/2kT). \tag{30}$$

When $h\nu \gg kT$ and $h\nu_a \ll kT$,

$$\Delta \varepsilon_T = u^* - u - \tfrac{1}{2}h\nu + kT \tag{31}$$

and therefore

$$\Delta \varepsilon_0 = u^* - (u + \tfrac{1}{2}hv). \tag{32}$$

as may be seen from Fig. 2. The equilibrium constant under these conditions is

$$K = \frac{kT}{hv^*} \exp(-\Delta \varepsilon_0/kT), \tag{33}$$

which is identical with equation (27), except that v^* is now imaginary.

3. Equilibria of the Type $A + B \rightleftarrows AB$

The equilibrium established between free atoms of type A and B and the stable diatomic molecule AB is governed by the relation

$$K = \frac{n_{AB}}{n_A n_B} = \frac{q_{AB}}{q_A q_B} \exp(-\Delta \varepsilon_0/kT)$$

$$= \frac{\dfrac{\{2\pi(m_A + m_B)kT\}^{3/2}}{h^3} \dfrac{8\pi^2 I_{AB}kT}{h^2} [1 - \exp(-hv_{AB}/kT)]^{-1}}{\dfrac{(2\pi m_A kT)^{3/2}}{h^3} \dfrac{(2\pi m_B kT)^{3/2}}{h^3}}$$

$$\times \exp(-\Delta \varepsilon_0/kT),$$

provided A and B are dissimilar and provided the vibration executed by the molecule is simple harmonic. Rearranging the terms, and noting that

$$I_{AB} = \frac{m_A m_B}{m_A + m_B} r_{AB}^2 = \mu_{AB} r_{AB}^2,$$

we have

$$K = h r_{AB}^2 \sqrt{\frac{8\pi}{kT\mu_{AB}}} [1 - \exp(-hv_{AB}/kT)]^{-1} \exp(-\Delta \varepsilon_0/kT).$$

When $hv \gg kT$, the equation reduces to

$$K = h r_{AB}^2 \sqrt{\frac{8\pi}{kT\mu_{AB}}} \exp(-\Delta \varepsilon_0/kT),$$

and when $hv \ll kT$, we have

$$K = \frac{r_{AB}^2}{v_{AB}} \sqrt{\frac{8\pi kT}{\mu_{AB}}} \exp(-\Delta \varepsilon_0/kT). \tag{34}$$

4. Equilibria of the Type $A + BC \rightleftarrows ABC$

Let us consider the equilibrium established between free atoms A and diatomic molecules BC on the one hand, and linear triatomic molecules ABC on the other hand. Making the usual assumption of harmonic vibrations and rigid rotations we obtain the equilibrium constant:

$$
K = \frac{n_{ABC}}{n_A n_{BC}} = \frac{q_{ABC}}{q_A q_{BC}} \exp\left(-\Delta\varepsilon_0/kT\right)
$$

$$
= \frac{\dfrac{[2\pi(m_A + m_B + m_C)kT]^{3/2}}{h^3} \dfrac{8\pi^2 I_{ABC}kT}{\sigma_{ABC}h^2} \prod\limits_{i=1}^{4} [1 - \exp(-h\nu_i/kT)]^{-1}}{\dfrac{[2\pi m_A kT]^{3/2}}{h^3} \dfrac{[2\pi(m_B + m_C)kT]^{3/2}}{h^3} \dfrac{8\pi^2 I_{BC}kT}{\sigma_{BC}h^2}[1 - \exp(-h\nu_{BC}/kT)]^{-1}}
$$

$$
\times \exp(-\Delta\varepsilon_0/kT),
$$

which, after cancelling and rearranging, becomes

$$
K = \frac{h^3}{(2\pi kT)^{3/2}} \left[\frac{m_A + m_B + m_C}{m_A(m_B + m_C)}\right]^{3/2} \frac{I_{ABC}}{I_{BC}} \frac{\sigma_{BC}}{\sigma_{ABC}} \frac{\prod\limits_{i=1}^{4}[1 - \exp(-h\nu_i/kT)]^{-1}}{[1 - \exp(-h\nu_{BC}/kT)]^{-1}}
$$

$$
\times \exp\left(-\Delta\varepsilon_0/kT\right).
$$

Of the four vibrations executed by the molecule ABC, let us suppose that the asymmetric one has a low frequency, ν_a, in which case we may write $(1 - e^{-h\nu_a/kT})^{-1} = kT/h\nu_a$. Then

$$
K = \frac{[ABC]}{[A][BC]} = \frac{1}{\nu_a}\left(\frac{h}{2\pi}\right)^2 \left(\frac{2\pi}{kT}\right)^{1/2}\left[\frac{m_A + m_B + m_C}{m_A(m_B + m_C)}\right]^{3/2} \frac{I_{ABC}}{I_{BC}} \frac{\sigma_{BC}}{\sigma_{ABC}}
$$

$$
\times \frac{[1 - \exp(-h\nu_s/kT)]^{-1}[1 - \exp(-h\nu_\phi/kT)]^{-2}}{[1 - \exp(-h\nu_{BC}/kT)]^{-1}} \exp(-\Delta\varepsilon_0/kT). \tag{35}
$$

$\Delta\varepsilon_0$, as in equation (25) is the difference between the total energy of the product molecule (ABC) and the energies of the atom (A) and reactant molecule (BC), both molecules being in their lowest vibrational states at the absolute zero. The increase, $\Delta\varepsilon_T$, in internal energy for the reaction, measured at any temperature, is given by the van't Hoff isochore as

$$
\Delta\varepsilon_T = \Delta\varepsilon_0 - \tfrac{1}{2}kT
$$

$$
- \tfrac{1}{2}h\{\nu_s(1 - \coth\beta\nu_s) + 2\nu_\phi(1 - \coth\beta\nu_\phi) - \nu_{BC}(1 - \coth\beta\nu_{BC})\}, \tag{36}
$$

where $\beta = h/2kT$.

The application of equation (25) to problems which are discussed later is most conveniently made when it is rewritten in a slightly modified form, so as to contain the difference, $\Delta\varepsilon_s$, between the energies of reactants and resultants when both are in the hypothetical static state: since

$$\Delta\varepsilon_0 + \tfrac{1}{2}h\nu_{BC} = \Delta\varepsilon_s + \tfrac{1}{2}h(\nu_s + 2\nu_\phi), \tag{37}$$

the equilibrium constant for the reaction $A + BC \rightleftarrows ABC$ becomes

$$K = \frac{1}{4\nu_a}\left(\frac{h}{2\pi}\right)^2\left(\frac{2\pi}{kT}\right)^{1/2}\left\{\frac{m_A + m_B + m_C}{m_A(m_B + m_C)}\right\}^{3/2}\frac{I_{ABC}}{I_{BC}}\frac{\sigma_{BC}}{\sigma_{ABC}}\frac{\sinh\beta\nu_{BC}}{\sinh\beta\nu_s\,\sinh^2\beta\nu_\phi}$$
$$\times\,\exp\left(-\Delta\varepsilon_s/kT\right). \tag{38}$$

By direct differentiation with respect to temperature, or by comparing equations (36) and (37), we then have

$$\Delta\varepsilon_T = \Delta\varepsilon_s - \tfrac{1}{2}kT + \tfrac{1}{2}h(\nu_s\coth\beta\nu_s + 2\nu_\phi\coth\beta\nu_\phi$$
$$-\nu_{BC}\coth\beta\nu_{BC}). \tag{39}$$

The Kinetic Treatment of the Equilibrium State

We have now dealt with the thermodynamic and statistical treatments of equilibria. There remains the kinetic treatment, according to which a system is said to be at equilibrium when the rate at which each type of molecule in the system is generated is equal to the rate at which it is destroyed. In other words, the net rate of change of the concentration of each species is zero. For obvious reasons, the number of instances in which this law can be directly verified is small.

It has been found that the rate of decomposition of nitrous acid in aqueous solution at 25°C is given by the empirical equation (Abel and Schmid, 1928)

$$-d[HNO_2]/dt = k_1[HNO_2]^4/[NO]^2,$$

where the square brackets denote concentrations. The rate at which it is formed is given by the equally empirical equation

$$+d[HNO_2]/dt = k_2[HNO_2][H^+][NO_3^-].$$

Since, at equilibrium, these rates must be equal, it follows that

$$\frac{[H^+][NO_3^-][NO]^2}{[HNO_2]^3} = \frac{k_1}{k_2}.$$

The term on the left-hand side is evidently the equilibrium constant for the reaction

$$3HNO_2 \underset{k_2}{\overset{k_1}{\rightleftharpoons}} H^+ + NO_3^- + 2NO + H_2O.$$

Hence

$$K = k_1/k_2. \tag{40}$$

The equilibrium constant is the ratio of the specific rate constants of the direct and reverse reactions. This important generalization will be frequently used in later sections of this work. We are here first concerned with its experimental verification. A second instance is provided by the system $H_3AsO_3 + I_3^- + H_2O \rightleftharpoons H_3AsO_4 + 2H^+ + 3I^-$, which has been examined by the analytical evaluation of K and the kinetic evaluation of k_1/k_2 separately. (Roebuck, *J. Physical Chem.*, **6**, 365 (1902); Bray, *ibid.*, **9**, 578 (1905); Liebhafsky, *ibid.*, **35**, 1648 (1931)). Further examples have been given by van't Hoff (1903) and by S. H. Maron and V. K. La Mer (1939).

The instantaneous rate of chemical change, dx/dt, may generally be formulated in terms of the instantaneous concentrations, $(a - x)$, $(b - x)$, etc. prevailing at time t after the start, when the initial concentrations were a, b, etc. On integrating the various expressions for dx/dt, we derive expressions for x, the extent of chemical change at time t. These, however complicated, invariably reduce when t becomes infinite to expressions for the equilibrium constant. Equilibrium constants may thus be regarded as the special forms taken by rate equations when the time becomes infinite. Numerous instances are worked out in detail in Chapter 6. In the meantime, we note that kinetic expressions generally are more complicated than equilibrium expressions, because they include time as an independent variable.

The immediate utility of the equation $K = k_1/k_2$ is that it affords an evaluation of any one of the constants in terms of the other two constants. Let us consider a few of its applications.

The rate constant, k_2, for the reaction between ethyl bromide and diethyl sulphide $C_2H_5Br + (C_2H_5)_2S \overset{k_2}{\rightarrow} (C_2H_5)_3Br$ has been evaluated from a knowledge of the equilibrium constant $K = [AB]/[A][B] = k_2/k_1$ (Corran, 1927) and the rate constant k_1 (Von Halban, 1909).

The rate constant for the ionization of water $H_2O \overset{k_1}{\rightarrow} H^+ + OH^-$ has been found by combining the equilibrium constant for the ionization of water, which, at 25°C, is 1.805×10^{-16} gram-ions per litre, with the observed rate constant for the union of the ions, which is 1.3×10^{11} litres per mole-second (Eigen and de Maeyer, 1955).

By comparing the observed rate constant for the reaction between hydrated electrons and water molecules at 25°C, $e_{aq}^- + H_2O \rightarrow H + OH^-$; $k_1 = 5 \times$

10^2 lit mole^{-1} sec^{-1} with the rate constant for the reverse reactions, $H + OH^- \rightarrow e_{aq}^- + H_2O$; $k_2 = 2 \times 10^7$ lit mole^{-1} sec^{-1}. The equilibrium constant

$$K = \frac{[H][OH^-]}{[e_{aq}^-][H_2O]} = \frac{k_1}{k_2}$$

is found to be $2 \cdot 5 \times 10^{-5}$, corresponding to the standard change in free energy of $-RT \ln K = 6 \cdot 3$ kcal mole^{-1} (Baxendale, 1964).

REFERENCES

Abel and Schmid, *Z. physikal. Chem.*, **132**, 56 (1928).

Baxendale, *Radiation Research Suppl.*, **4**, 139 (1964). Slightly different constants have been given by E. M. Fidden and E. J. Hart, *Trans. Faraday Soc.*, **63**, 2975 (1967), and a full account by E. J. Hart and M. Anbar, *The Hydrated Electron*, Wiley & Sons, Ltd, London (1970).

Corran, *Trans. Faraday Soc.*, **23**, 605 (1927).

Eigen and de Maeyer, *Naturwiss.*, **42**, 413 (1955),

Maron, S. H., and V. K. La Mer, *J. Amer. Chem. Soc.*, **61**, 2018 (1939).

van't Hoff, *Vorlesungen über theoretische und physikalische Chemie*, **I**, Braunschweig (1903).

von Halban, *Z. physikal Chem.*, **67**, 129 (1909).

3

EXPERIMENTAL DATA ON VARIOUS EQUILIBRIA

Equilibrium constants for homogeneous and heterogeneous systems are generally measured by direct analysis of their chemical composition, using any of the many available techniques, provided they are not such as to alter the position of the equilibrium during analysis. From the values of K thus found at various temperatures, the standard changes in free energy, heat content and entropy are derivable using equations (2.7), (2.9) and (2.10) respectively. Let us consider a few examples involving solutions.

Conformational Equilibria

α and β glucose are identical in all respects except for the positions of the hydrogen and hydroxyl groups attached to one carbon atom. An even simpler instance of equilibrium between conformationally different forms of a molecule is provided by orthochlorophenol (Fig. 1). The concentrations of the two forms under equilibrium conditions have been estimated by taking advantage of the fact that the intensity and position of the infra-red absorption band characterizing the hydroxyl group are modified by its

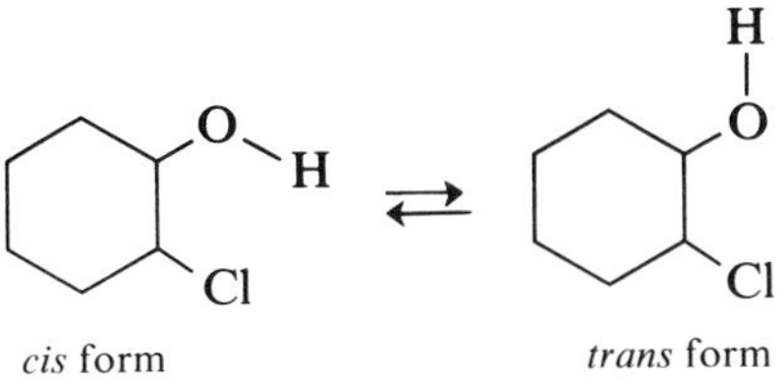

cis form   *trans* form

FIG. 1. Ortho-chlorophenol.

33

proximity to the chlorine atom. ΔH^0 for the conversion of the *cis* form to the *trans* form in carbon tetrachloride solution at 25°C is found to be 1·4 kcal/mole (M. M. Davies, 1938).

Dimerization Equilibria

Because of the mutual attraction of their positive and negative ends, polar molecules in all phases of matter are apt to form double molecules, or dimers, schematically represented in simple cases as $\overset{\rightarrow}{\leftarrow}$. Let us consider a monomer, denoted by A_1, to be distributed between the immiscible solvents B and W. From equation (2.1), we have the equilibrium

$$A_{1,W} \rightleftarrows A_{1,B}, \qquad K_1 = [A_{1,B}]/[A_{1,W}].$$

Partial dimerization is assumed to occur in solvent B only, according to the scheme

$$2A_{1,B} \rightleftarrows A_{2,B}, \qquad K_2 = [A_{2,B}]/[A_{1,B}]^2. \tag{2}$$

It follows that the ratio, c_W/c_B, of the concentrations in the two solvents, expressed as gram-*equivalents* per litre, is given by the law

$$c_B/c_W = K_1 + 2K_1^2 K_2 c_W, \tag{3}$$

from which both equilibrium constants may be found (P. Gross and K. Schwarz, 1930; Moelwyn-Hughes, 1940). The linear relationship between c_B/c_W and c_W has been established for many systems over wide ranges of concentrations and temperature. With acetic acid as solute and benzene and water as solvents, it is found that

$$\Delta S_1^0 = 7\cdot15 \text{ cal/mole-deg}; \qquad \Delta H_1^0 = 4{,}980 \text{ cal/mole},$$

$$\Delta S_2^0 = -22\cdot32 \text{ cal/mole-deg}; \qquad \Delta H_2^0 = -9{,}700 \text{ cal/mole}.$$

The spectroscopic estimate of ΔH_2^0 is $-9{,}300 \pm 1{,}000$ (M. M. Davies and G. B. B. M. Sutherland, 1938). ΔH^0 for the dimerization of aryl halides in the same solvent lies in the range $-1\cdot1$ to $-1\cdot7$ kcal, and for the dimerization of normal alcohols in normal hydrocarbons is about ten times as great (Sakurada, 1935; Wohl, Pahlke and Wehago, 1935; von Elbe, 1934). Nuclear magnetic resonance studies have revealed the existence of double molecules of alcohols in dilute solutions in non-polar solvents, and have provided an explanation of the departure of these solutions from regular behaviour (R. S. Scott and D. V. Fenby, 1969).

If there are no detectable monomers in solvent B, the second term in equation (3) far exceeds the first term, and we have

$$c_B/c_W^2 = 2K_1^2 K_2, \tag{4}$$

or

$$\sqrt{c_B/c_W} = \text{a constant}, \tag{5}$$

as derived by Nernst (1891). There is kinetic evidence that organic halides react simultaneously in nitrobenzene solution with monomeric and dimeric forms of alcohols (Bartlett and Nebel, 1940).

The Hydrogen Bond

The structure of the formic acid dimer as revealed by electron diffraction (L. Pauling and L. O. Brockway, 1934) is that of a plane hexagon, as shown in Fig. 2. The two links joining the molecules are identical, and each consists of a hydroxyl hydrogen atom placed between the hydroxyl oxygen atom in one molecule and the carbonyl oxygen atom in the other molecule. To such a link, Latimer and Rodebush (1920) gave the name of hydrogen bond. Similar bonds are formed between other polar groups ending in a hydrogen atom, such as —N—H, and other electronegative atoms (M. M. Davies, 1947; C. C. Pimental and A. L. McClellan, 1960), such as F. The hydrogen bond, at one time attributed to the phenomenon of resonance, is now regarded as due to electrostatic interactions (G. Briegleb, 1933; K. Berger, 1933, 1935; J. H. de Boer, 1936; M. Magat, 1936, 1937; E. A. Moelwyn-Hughes, 1938; H. Harms, 1939; M. M. Davies, 1940; E. J. W. Verwey, 1941; C. A. Coulson, 1957) and we shall seek to calculate its magnitude as found by Nernst, 1918; Fenton and Garner, 1930; Ritter and Simons, 1945; E. W. Johnson and Nash, 1950; and Coolidge, 1928 on that basis.

Denoting the C=O bond of one molecule by A and the O—H bond of the second molecule by B, and accepting the usual values for the bond lengths ($l_A = 1.16$, $l_B = 0.96$ Å) dipole moments ($\mu_A = 2.5$; $\mu_B = 1.6 \times 10^{-18}$ e.s.u), and O—C—O valency angle (125° 16'), we can show that, polarization

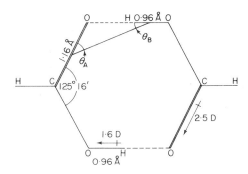

Fig. 2 The formic acid dimer.

effects being neglected, the distance (r_{AB}) of the lines joining the centres of the attracting dipoles, and the angles (θ_A and θ_B) which they make with it, have the following values:

r_{00} (Å)	2·85	2·55
r_{AB} (Å)	2·68$_7$	2·39$_3$
θ_A	51° 35′	50° 26′
θ_B	11° 3′	12° 26′
U (cal/g-mol)	$-4{,}310$	$-6{,}570$

The last row gives the contribution of the dipole–dipole interation to the energy of formation of the single bond, according to the equation

$$\phi = -2\mu_A\mu_B \cos\theta_A \cos\theta_B/r_{AB}^3. \tag{6}$$

If, however, there exists a repulsive energy varying as r_{AB}^{-9}, the minimum energy of the pair of molecules must be corrected by the factor 2/3. It follows that the estimated energy of the rupture of a single hydrogen bridge at $T = 0$ is 2·87 or 4·38 kcal/mole. The experimental value of ΔH^0 for the reaction $(RCOOH)_2 \rightleftarrows 2RCOOH$ in the gas phase at 400°C is 14·65 kcal. By subtracting RT and dividing by 2, the experimental value of ΔE_T^0 at this temperature is 6.76 ± 0.15 kcal, which is the gain in energy required to break a single bond. After giving equal weight to the two estimated values at $T = 0$, we conclude that ΔE^0 (cal/mole) $= 3{,}630 \pm 1{,}760 + (2.3 \pm 0.6)RT$. The breakdown of the bond is seen to be associated with a reasonable increase in the heat capacity, $d(\Delta E^0)/dT$.

ϕ is essentially a free energy, and the molar gain in free energy attending the dissociation of the dimer is consequently $\Delta G_g^0 = -N_0 2\phi$, where N_0 is the Avogadro number, and the subscript g is a reminder that the equation applies to equilibria in the gaseous phase. When the dimer dissociates in a solvent of permittivity D, we have

$$\Delta G_s^0 = \Delta G_g^0/D, \tag{7}$$

and, by means of the Kelvin–Helmholtz equation (2.8),

$$\Delta H_s^0 = \frac{(1 - LT)}{D} \Delta H_g^0, \tag{8}$$

where L $(= -d \ln D/dT)$ is a constant, independent of temperature, and is a specific property of the medium. It is more fully discussed in Chapter 7. Some doubt exists as to whether the macroscopic value of D is the right one to use in this connexion. Frank (1935) suggests replacing $1/D$ in the free energy equation by $(D + 2)/3D$. According to the simpler relationship, the observed increase in heat content for the dissociation of a given dimer in a variety of solvents should vary linearly with respect to the function

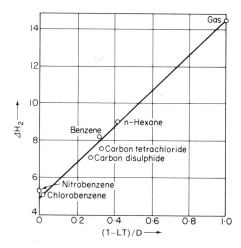

FIG. 3. Heats of dissociation of dimeric acetic acid in various media.

$(1 - LT)/D$. That this is true may be seen from the data for acetic acid in various solvents (Fig. 3 after M. M. Davies, P. Jones, D. Patnaik and E. A. Moelwyn-Hughes, 1951).

A more detailed calculation of ΔH_2 for the lower members of the mono-carboxylic acids in the gaseous phase has been given by T. Miyazawa and K. S. Pitzer (1959) and by L. Slutsky and S. H. Bauer (1954), whose form of ϕ is given as $\phi = \text{constant} [(r_0/r)^6 - 2(r_0/r)^3]$. Their values of n and m of equation (1.4) are thus 6 and 3 respectively, rather than 9 and 3 as here.

Further Studies on the Hydrogen Bond

The best known example of the hydrogen bond is that between chloroform and acetone $Cl_3CH \cdots OC(CH_3)_2$ for which ΔH was estimated at -4.1 kcal (A. Sherman and E. A. Moelwyn-Hughes, 1936). This value has been confirmed by Higgins, Pimemtel and Shorley (1955), whose work is based on measurements of the proton resonance of the binary liquid systems. Later extensions of their method have been made by D. L. Anderson, R. A. Smith, D. B. Myers, S. K. Alley, A. G. Williamson and R. L. Scott (1962) to binary liquid mixtures of 1-hydroperfluoroheptane and acetone. The investigators have also carried out accurate calorimetric measurements. The hydrogen bond in this case is $C_7F_{15}H \cdots OC(CH_3)_2$ which, except for the presence of fluorine atoms, resembles the chloroform–acetone bond. The observed values of the excess molar heat of mixing, ΔH^{ex}, derived from the temperature

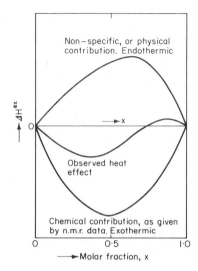

Fɪɢ. 3A. Detailed resolution of the total enthalpy of exchange in hydrogen bonding.

variation of ΔG^{ex}, vary with the composition as shown by the S-shaped curve in Fig. 3A. The lowest curve shows the symmetrical exothermic heats of complex formation, obtained from n.m.r. data. If the system contains N_A molecules of one kind, whether free or complexed, $N_A = N'_A + N_C$, where N'_A is the number of free or uncomplexed molecules, and N_C is the number of complexes. Similarly, $N_B = N'_B + N_C$, and the equilibrium constant for the reaction $A + B \rightleftarrows AB$ becomes

$$K = \frac{N_C(N_A + N_B - N_C)}{(N_A - N_C)(N_B - N_C)},$$

as in Dolezalek's (1910) treatment of binary mixtures. When values on the lowest curve are subtracted from those on the S-shaped curve, we find the uppermost, skew-shaped curve, which consequently represents the non-specific, or physical, interaction between the components of the mixture. This curve resembles that given by the van Laar–Hildebrand–Scatchard theory for binary liquid mixtures without hydrogen bonding, such as mixtures of $C_7F_{15}H$ with hydrocarbons. After thus resolving the total interaction energy into non-specific, or physical, and specific, or chemical, contributions, K may be evaluated. From its temperature variation between -80 and $+20°C$, the authors find for the hydrogen bond formation $\Delta H^{ex} = -2.5 \pm 0.5$ kcal. The basis of this complicated but entirely satisfactory theory is that n.m.r. data refer to the hydrogen-bonded hydrogen atom irrespective of its surroundings, i.e. whether it is in the complex or not, whereas the ordinarily determined physicochemical data refer to interaction of the hydrogen-bonded hydrogen atom with its partner in the complex and with its surroundings in general.

Emphasis in this work is centred more on dilute solutions, rather than on mixed liquids, which have been the subject of numerous authoritative treatises.

Hydrolytic Equilibria

As an example of hydrolytic equilibria, we shall study the well investigated hydrolysis of chlorine:

$$Cl_2 + H_2O \rightleftarrows H^+ + Cl^- + HOCl,$$

for which the hydrolytic constant in dilute solution is

$$K = \frac{[H^+][Cl^-][HOCl]}{[Cl_2]}.$$

From Jakowkin's (1899) data, Liebhafsky (1935) concluded that ΔH^0 decreased with a rise in temperature at a rate indicating that ΔC_P^0 was approximately -100 cal/mole-deg. His conclusion has been confirmed by numerous investigators, notably Connick and Yuan-tsan Chia (1959), whose results, which allow for the ionization of hypochlorous acid and for the formation of the trichloride ion, have been summarized by the equation

$$\log_{10} K = 136.406 - 46.8 \log_{10} T - 7,157/T.$$

It follows that ΔC_P^0 is -93 cal/mole-deg. This very marked temperature coefficient of the heat of reaction was at that time regarded as exceptional, but similar ΔC_P^0 values have since been found in a large number of equilibria in aqueous solution. With possible explanations for it, we shall deal later. In the meantime, we shall show that it can be independently confirmed. The partial molar heat capacity of aqueous chlorine (Glew and E. A. Moelwyn-Hughes, 1953) at 25°C is 91·9 cal/mole-deg. The partial molar heat capacity of hypochlorous acid in water is not known, but that of methanol is 38·8 at 25°C. Provided this can be taken as equal to the value for HOCl, ΔC_P^0 may be estimated as follows:

$$\Delta C_P^0 = -15.9(H^+) - 14.3(Cl^-) + 38.8(HOCl) - 91.9(Cl_2) - 18.0(H_2O)$$

$$= -101.3 \text{ cal/mole-deg,}$$

which is in satisfactory agreement with Jakowkin's result. Alternatively, we may accept the observed value of ΔC_P^0 and estimate C_P^0 for HOCl as 47 cal/mole-deg.

Equilibria in Terms of Activities and of Intermolecular Forces

The equilibrium expressions hitherto employed have been given in terms of concentrations, c, and are strictly applicable to dilute solutions only. With high concentrations, the valid equilibrium laws are given in terms of activities, a, which are related to the concentrations by the equation

$$a = c\gamma, \tag{9}$$

where γ is the activity coefficient. In most of our systems, a is defined so as to equal c when the solution is infinitely dilute. Under that condition, γ becomes unity. The activity of a solute may be measured in many ways, as, for example, by determining the depression in the freezing point of a solvent caused by the presence of the solute. The thermodynamic expression for the equilibrium constant of the reaction $A + B \rightleftarrows AB$ becomes

$$K = \frac{a_{AB}}{a_A a_B} = \frac{c_{AB}}{c_A c_B} \cdot \frac{\gamma_{AB}}{\gamma_A \gamma_B}. \tag{10}$$

Madgin and his collaborators (1933, 1937) have investigated cryoscopically several equilibria of the type p-chlorophenol (A) + pyridine (B) $\rightleftarrows$ complex (AB) at various temperatures and concentrations in benzene solution. With this representative reaction, $\Delta H^0 = -6.8$ kcal, and $\Delta S^0 = -13.8$ cal/mole-deg. Different kinds of forces are responsible for the formation of complexes of this type. We shall merely explore the possibility that the attractive energy in this instance arises from the linear interaction of the dipoles ($\mu_A = 2.25 \times 10^{18}$ e.s.u.; $\mu_B = 2.16 = 10^{-18}$ e.s.u.), in which m of equation (1.4) is 3, and the term B of that equation is

$$B = 2\mu_A \mu_B \left(\frac{D + 2}{3D} \right). \tag{11}$$

At 25°C, D for benzene is 2.272. Hence $B = 6.09 \times 10^{-36}$ erg-cm³. If the energy of repulsion varies inversely as the ninth power of the distance apart of the centres of the dipoles, equation (1.4) takes the form

$$\phi = Aa^{-9} - Ba^{-3}, \tag{12}$$

and consequently the distance apart, a_0, of the centres of the dipoles and the minimum energy of the pair, ϕ_0, are related as follows:

$$a_0^3 = -\frac{2B}{3\phi_0}.$$

Madgin's value of $-\phi_0$ is 4.7×10^{-13} erg. Hence $a_0 = 2.05$ Å, and $A = Ba_0^6/3$ becomes 1.51×10^{-82} erg-cm⁹. The two electronic shells responsible for the repulsion approximate to those of atoms of neighbouring elements in the first periodic series. It is therefore not surprising that the value of A thus estimated should be equal, within a factor of 3, to that computed by Lennard-Jones (1937) for the intrinsic repulsion constant of a pair of neon atoms in the gas phase.

The Influence of Solvents on Chemical Equilibria

When molecules of a given substance are distributed between a dilute solution and a dilute vapour phase, the ratio of their partial pressure to their concentration in solution is often found to be constant in a system at constant temperature (Henry, 1803): $r = p/c$. When chemical equilibrium of the type $A \rightleftarrows B + C$ is established in both phases, we have

$$K_g = p_B p_C / p_A$$

and

$$K_s = c_B c_C / c_A. \tag{14}$$

It follows that

$$\frac{K_g}{K_s} = \frac{(p_B/c_B)(p_C/c_C)}{(p_A/c_A)} = \frac{r_B r_C}{r_A}, \tag{15}$$

where r is Henry's constant for each solute. If, in addition, Raoult's law (1888) is obeyed, the partial pressure of each solute equals its vapour pressure multiplied by its molar fraction in solution:

$$p = p^0 x. \tag{16}$$

In the dilute solutions here considered,

$$x = c V_s / 1{,}000, \tag{17}$$

where c is the concentration expressed in moles per litre of solution, and V_s is the molar volume of the solvent. Consequently

$$\frac{p}{c} = \frac{p^0 V_s}{1{,}000}, \tag{18}$$

and

$$\frac{K_g}{K_s} = \frac{p_B^0 p_C^0}{p_A^0} \cdot \frac{V_s}{1{,}000}. \tag{19}$$

By applying the van't Hoff isochore (equation 2.9), we arrive at the following relationship between the increases in heat content for such a reaction in the gas phase and in dilute solution:

$$\Delta H_g - \Delta H_s = L_B^0 + L_C^0 - L_A^0 + RT^2\alpha, \tag{20}$$

where the L^0 terms are the heats of vaporization of the solutes, and α is the coefficient of expansion of the solvent. In a more general form, we may write for systems in which each component obeys Raoult's law

$$\Delta H_g - \Delta H_s = \Sigma L_j^0 - \Sigma L_i^0 + (\Sigma j - \Sigma i)RT^2\alpha. \tag{21}$$

Few solutions, however, obey Raoult's law, and we must revert to equation (15), for which, again using van't Hoff's isochore, we reach a similar conclusion. Each L term now is the gain in heat content when one mole of solute escapes from solution at unit concentration to the gas phase at unit pressure. Then, omitting the small term due to the expansion of the solvent, we have

$$\Delta H_g - \Delta H_s = \Sigma L_j - \Sigma L_i, \tag{22}$$

where the heat of escape for each solute is

$$L = RT^2\left[\frac{d\ln(p/c)}{dT}\right]_P = RT^2\left(\frac{d\ln r}{dT}\right)_P. \tag{23}$$

The effect of the solvent on the heat of reaction is seen to be a specific one, depending on its properties and on those of the solutes. When the solutes

TABLE I

INCREASES IN HEAT CONTENT ATTENDING CERTAIN REACTIONS IN THE GAS PHASE (ΔH_g) AND IN CARBON TETRACHLORIDE SOLUTION (CAL)

Reaction	ΔH_g	ΔH_s	$\Delta H_g - \Delta H_s$
$Br_2 + Cl_2 \rightleftarrows 2BrCl$	-600	-756	$+156*$
$N_2O_4 \rightleftarrows 2NO_2$	$13,693\dagger$	$18,840\ddagger$	$-5,150$
$C_2H_6 \rightleftarrows C_2H_4 + H_2$	$32,000$	$30,400$	$1,600\S$
$IC_2H_4I \rightleftarrows C_2H_4 + I_2$	$22,300\|$	$11,300\P$	$+11,000$

REFERENCES

* Blair and Yost, *J. Amer. Chem. Soc.*, **55**, 4489 (1933).
† Giauque and Kemp, *J. Chem. Physics*, **6**, 40 (1938); cf. 13,600 given by Schreber, *Z. physikal Chem.*, **24**, 651 (1897) and J. H. van't Hoff, *Studies in Chemical Dynamics*, p. 154 (1896).
‡ Derived from the data of Cundall, *Trans. Chem. Soc.*, **67**, 794 (1895).
§ Benford and Wassermann, *ibid.*, 367 (1939).
‖ Mooney and Ludlam, *Proc. Roy. Soc. Edin.*, **49**, 160 (1929).
¶ Polissar, *J. Amer. Chem. Soc.*, **52**, 956 (1930).

obey Raoult's law, only the latter properties matter. The results of some well-investigated equilibria examined in the vapour phase and in solution are given in Table I.

Dissolution Equilibria

If a general answer could be found to the question "What happens when a molecule dissolves?", many of the problems besetting solution kinetics could be solved. No such answer, however, can be given. The process of dissolution and the magnitudes of the energy and entropy changes attending it depend, as stated, on properties specific to the solutes and solvents. Consider, for example, the dissolution of hydrogen chloride from the dilute gaseous state into liquid carbon tetrachloride and water, respectively. In the former case, the solute is only slightly affected, as the very small changes in its internuclear vibration frequency and its dipole moment reveal. In the latter case, however, the character of the solute is completely changed, and the solution formed exhibits such a high electrical conductivity as to indicate that the solute has dissociated almost completely into ions.

The simplest type of dissolution is that of one liquid into another, resulting in a non-electrolytic solution (J. H. Hildebrand and R. L. Scott, 1962). Provided that both types of molecules are roughly of the same size and exert spherically symmetrical forces, and provided that they mix randomly without a volume change, a satisfactory theory can be advanced to account for the properties of the solution. One consequence of the theory is that the gain in energy, ΔE, associated with the formation of a solution formed from N_1 molecules of the first liquid and N_2 molecules of the second liquid becomes

$$\Delta E = \left(\frac{N_1 N_2}{N_1 + N_2} \right) \Delta u^0, \tag{1.23}$$

where Δu^0, an interchange energy, is given in terms of the common coordination number, c, and the three interaction energies ϕ_{11}, ϕ_{22} and ϕ_{12} of the corresponding molecular pairs:

$$\Delta u^0 = c(\phi_{12} - \tfrac{1}{2}\phi_{11} - \tfrac{1}{2}\phi_{22}), \tag{1.22}$$

or, in terms of the molecular heats of vaporization,

$$\Delta u^0 = c\phi_{12} + \lambda_1^0 + \lambda_2^0. \tag{24}$$

On multiplying by the Avogadro number we have $\Delta U^0 = N_0 c\phi_{12} + L_1^0 + L_2^0$, where L_1^0 and L_2^0 are the molar heats of vaporization of liquids 1 and 2. Examples typifying the results found with mixtures of non-hydroxylic solvents are shown in Table II. The upper limit of ΔU^0 consistent with

TABLE II

MOLECULAR INTERACTIONS IN MISCIBLE LIQUIDS AT 25°C

Liquids	ΔU^0	$L_1^0 + L_2^0$	$-N_0 c\phi_{12}$	$-\phi_{12} \times 10^{14}$ ($c = 12$)
		calories		ergs
$CH_3I + CH_2Cl_2$	90 ± 2	13,849	13,759	7·964
$CH_3I + CHCl_3$	-32 ± 3	14,339	14,371	8·316
$CH_3I + CCl_4$	204 ± 8	14,614	14,410	8·341

homogeneity is $2RT$. With the aid of the partial molar volume, V_1^∞, of liquid 1 at infinite dilution in liquid 2, and of the corresponding quantity V_2^∞, it is possible to compute the attractive energy constant B_{12} between the dissimilar pairs involved (E. A. Moelwyn-Hughes and R. W. Missen, 1956).

It is seen from equation (1.23) that the gain in energy of the solution when one molecule of liquid 2 dissolves in pure liquid 1 is $\Delta\varepsilon_2$(liquid 2 → infinitely dilute solution in liquid 1) $= \Delta u^0$. Now $\Delta\varepsilon_2$(liquid 2 → vapour 2) $= \lambda_2^0$. Therefore

$$\Delta\varepsilon_2(\text{vapour 2} \to \text{infinitely dilute solution in liquid 1}) = c\phi_{12} + \lambda_1^0. \quad (25)$$

Similarly, $\Delta\varepsilon_1$(vapour 1 → infinitely dilute solution in liquid 2) $= c\phi_{12} + \lambda_2^0$. The second terms in these equations may be regarded as the gains in energy due to the creation of cavities of molecular size in the solvents. The first term represents the gain in energy due to the new interactions resulting from the process of dissolution.

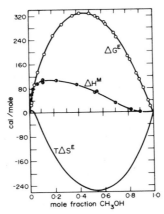

FIG. 4. Excess thermodynamic functions: CH_3OH—CCl_4 at 35°.

When one of the components in a binary liquid mixture contains a hydroxyl group, the theory as outlined above breaks down, and allowance must be made for the existence of dimers and higher polymers formed by hydrogen-bonding of the hydroxylic component. This is evident from Fig. 4, which shows that the heat of mixing of methanol with carbon tetrachloride does not vary symmetrically with respect to the molar fraction, as required by equation (1.23), and that the excess entropy, ΔS^{ex}, over that demanded by random mixing, which is $- N_1 kT \ln [N_1/(N_1 + N_2)] - N_2 kT \ln [N_2/(N_1 + N_2)]$, is far from zero. The properties of water are largely determined by hydrogen bonds, and we anticipate the behaviour of aqueous solution to depart so far from that of regular solutions as to make the concept irrelevant. Let us, therefore, renew our approach to the vast problem of aqueous solutions from the experimental angle, beginning with non-hydroxylic solutes.

Equilibria between Gases and Aqueous Solutions

The chemical potential of component 2 in a gaseous system obeying the ideal gas laws may be expressed as follows:

$$\mu_2 \text{ (gas)} = \mu_2^0 \text{ (gas)} + kT \ln p_2, \tag{2.3}$$

where p_2 is its partial pressure, and μ_2^0 is its chemical potential at unit pressure (say 1 mm). Similarly, the chemical potential of solute 2 in dilute solution may be written as follows:

$$\mu_2 \text{ (solution)} = \mu_2^0 \text{ (solution)} + kT \ln c, \tag{2.4}$$

where μ_2^0 is the chemical potential of the solute in a solution of unit concentration (say 1 g-mole/litre). Both standard potentials are functions of temperature, but are independent of the molecular compositions. The condition of equilibrium is that μ_2 shall be the same in the two phases [equation (2.1)]. Hence $kT \ln (p_2/c_2) = \mu_2^0(\text{solution}) - \mu_2^0(\text{gas}) = -\Delta\mu_2^0$, which is constant at constant temperature and pressure. This is the thermodynamic interpretation of Henry's law (equation (3.18)]. In molar notation,

$$RT \ln (p_2/c_2) = -\Delta G^0. \tag{26}$$

On using the Kelvin–Helmholtz relationship (2.8), we have

$$\Delta H_2^0 = RT^2 \left[\frac{d \ln (p_2/c_2)}{dT} \right]_P. \tag{27}$$

Experiments show that, to a good approximation,

$$(\Delta H_2^0)_T = (\Delta H_2^0)_0 + T\Delta C_P^0, \tag{28}$$

which implies that ΔC_P^0 is independent of temperature. The most accurate investigations, however, indicate that $d\Delta C_P^0/dT$ is real. On combining equations (27) and (28), and integrating, we find

$$\ln (p_2/c_2) = A + (\Delta C_P^0/R) \ln T - (\Delta H_2^0)_0/RT, \tag{29}$$

in agreement with the empirical equation involving three positive constants,

$$\log (p_2/c_2) = a - b \log T - c/T \tag{30}$$

established by Valentiner (1927) and Lannung (1930). Some representative data (Glew and Moelwyn-Hughes, 1953) are given in Table III. In terms of these constants, the molar increase in entropy attending the escape from a molar aqueous solution to the gas phase at a pressure of 1 mm is

$$(\Delta S_2^0)_T = RA + \Delta C_P(1 + \ln T). \tag{31}$$

The solubility of gases in liquids may be expressed equally well in terms of Ostwald's absorption constant, which is the ratio of the concentrations of the solute in the gas phase and in solution. We now have

$$\ln \left(\frac{c_{\text{gas}}}{c_{\text{sol}}} \right) = (A - \ln R) + \left(\frac{\Delta C_P^0}{R} - 1 \right) \ln T - \Delta H_0^0/RT, \tag{32}$$

where $\ln R = 4 \cdot 13294$. The standard gain in entropy on transferring one mole of solute from an aqueous solution at a given concentration to the gas phase at the same concentration is consequently

$$\Delta S_2^0(c_{\text{gas}}/c_{\text{sol}}) = R(A - \ln R) + (\Delta C_P^0 - R)(1 + \ln T). \tag{33}$$

TABLE III

SOME EXPERIMENTAL VALUES FOR THE CONSTANTS OF EQUATION (29) APPLIED TO
AQUEOUS SOLUTIONS (p_2 in mm Hg, c_2 in moles-litre^{-1})

Solute	$A/2 \cdot 303$	$-\Delta C_p$	ΔH_0	$\Delta H_{298 \cdot 16}$	$\Delta S_{298 \cdot 16}^0$ (p_2/c_2)	$\Delta S_{298 \cdot 16}^0$ $c(\text{gas})/c(\text{sol})$
CH_3F	58·97	35·26	14,937	4,405	33·8	12·3
H_2S	59·27	35·74	14,896	4,175	31·9	10·4
A	63·00	37·75	14,060	2,702	35·5	14·0
CO	68·39	41·33	15,009	2,690	35·5	14·0
CH_3Cl	71·01	43·02	18,496	5,670	37·0	15·5
CH_3Br	73·02	44·22	19,461	6,277	38·1	16·6
CH_4	176·62	55·0 (25° C)	29,687	3,173	37·0	15·5
CH_3I	133·25	85·35	31,812	6,364	38·3	16·8
CS_2	146·08	93·25	36,121	7,351	44·2	22·7
CCl_4	178·79	115·22	40,965	7,804	46·9	25·4

By comparing equations 31 and 33, we see that

$$\Delta S^0(c_{gas}/c_{sol}) = \Delta S^0(p/c) - R(1 + \ln RT). \tag{34}$$

The value of the last term at 25°C is 21·52 cal/mole-deg. Values thus obtained are given in the last column of Table III. We note that the escape of these solutes from aqueous solution to the gas phase is characterized by a large decrease in the partial molar heat capacity, and by a large increase in the partial molar entropy. Conversely, when the gases dissolve in water, there is a large increase in C_P and a large decrease in S. It will be shown in a later section that methyl chloride, which we shall consider as an example, is co-ordinated by 20 molecules of water. Hence, if these effects are divided equally among the water molecules forming the first solvation shell, the act of dissolving one molecule of methyl chloride in water at 25°C endows each water molecule in the first co-ordination shell with a heat capacity of $1·08k$ and deprives each water molecule of an entropy of $0·39k$ ($= k \ln 1·477$). The increase in C_P and the decrease in S attending dissolution are greater than can be ascribed to any changes in the methyl chloride molecule itself and therefore must arise from changes which it brings about in the motion and degree of organization of the water molecules surrounding it.

A Kinetic Treatment of the Escape of Solutes from Solution

The rate of escape of solute molecules from unit surface area of a solution may be equated to $v \cdot n_\sigma f(\varepsilon/T)$, where v is the frequency with which the solute molecule, of reduced mass μ, vibrates with respect to its c solvent neighbours at an average distance a from each solvent molecule; n_σ is the surface concentration of such clusters; $f(\varepsilon/T)$ is the probability that each solute molecule shall possess a total energy, ε, sufficient to ensure its escape. In terms of the integers m and n of Mie's equation (1.4), and of the average energy of interaction, ϕ_{12}, between a solute and solvent molecule, we have

$$v = \frac{1}{2\pi a}\left[\frac{mnc|-\phi_{12}|}{3\mu}\right]^{1/2}. \tag{1.15}$$

$n_\sigma = 2an_s$, where n_s is the bulk concentration of solute in solution. For $f(\varepsilon/T)$, we shall adopt Berthoud's expression.* The energy of escape is $\varepsilon = c\phi_{12}$. The rate at which solute molecules enter the unit area of the solution from the vapour phase, where the molecular concentration is n_g, is $n_g(kT/2\pi\mu)^{1/2}$. The ratio of the equilibrium concentrations in the two

---

* The derivation of Berthoud's equation is given in Chapter 10, equations (46)–(54).

phases is obtained by equating the rates of escape and dissolution:

$$\frac{n_g}{n_s} = \left(\frac{2mn\varepsilon}{3\pi kT}\right)^{1/2} . \exp\left(-\varepsilon/kT\right)\left(\frac{\varepsilon}{kT}\right)^{s-1} \frac{1}{(s-1)!}.$$

Taking $m = 6$ and $n = 9$, and expressing $\pi^{1/2}$ as $2(1/2)!$, we have

$$\frac{n_g}{n_s} = \frac{3}{(s-\frac{1}{2})!}\left(\frac{\varepsilon}{kT}\right)^{s-1/2} . \exp(-\varepsilon/kT), \tag{35}$$

from which we see that the gains in heat content and entropy attending the escape per molecule are

$$\Delta H^0 = \varepsilon - (s - \tfrac{1}{2})kT \tag{36}$$

and

$$\Delta S^0 = k\left[\ln 3 + (s - \tfrac{1}{2})\ln\frac{\varepsilon}{(s-\frac{1}{2})kT}\right]. \tag{37}$$

Equation (35) has the same form as equation (32), and will be referred to in connection with correlations between thermodynamic constants.

Gas Hydrates

When an aqueous solution of a gas in equilibrium with the vapour phase is cooled, the first solid to separate from solution is usually a crystalline hydrate of the solute, of definite chemical composition and crystallographic pattern. Measurement of the temperature variation of the pressure of the new system, hydrate plus aqueous solution plus vapour, affords the value of ΔH for a second phase change. So also does the p–T relationship for the system examined at still lower temperatures, when the phases in equilibrium are hydrate plus ice plus vapour. Values found by Glew (1962) for the methane–water system, adjusted to 0°C, are given in Table IV. Within the limits of error, heat effects (i) and (iv) are equal, indicating that the interaction energy of methane with its hydration shell in solution is the same as that of methane

TABLE IV

THERMAL DATA FOR THE METHANE–WATER SYSTEM (GLEW)

	Change	ΔH (cal/mole) at 0°C
(i)	CH_4 (aq.) $\rightleftarrows CH_4$ (gas)	$4{,}621 \pm 70$
(ii)	$CH_4 \cdot nH_2O$ (hydrate) $\rightleftarrows nH_2O$ (aq.) $+ CH_4$ (gas)	$12{,}830 \pm 140$
(iii)	$CH_4 \cdot nH_2O$ (hydrate) $\rightleftarrows nH_2O$ (ice) $+ CH_4$ (aq.)	$8{,}228 \pm 130$
(iv)	$CH_4 \cdot nH_2O$ (hydrate) $\rightleftarrows nH_2O$ (ice) $+ CH_4$ (gas)	$4{,}553 \pm 102$
(v)	nH_2O (ice) $\rightleftarrows nH_2O$ (aq.)	$n \times 1435.7$

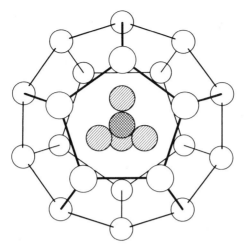

FIG. 5. Crystalline CH_4, 5·75 H_2O. Co-ordination number 20.

with its co-ordination shell in the crystalline hydrate. This is a conclusion of major importance in our understanding of the nature of the water molecules which surround an aqueous solute. $\Delta H(ii)-\Delta H(iv)$ is seen to be 8,277 cal, and this clearly corresponds to the heat absorbed when n moles of ice melt, according to equation (v). Hence $n = 5·765 \pm 0·120$, in agreement with the number 5·75 anticipated by Claussen and von Stackelberg (*vide infra*), whose established structure for the crystalline hydrate of methane is shown in Fig. 5. The organic molecule is seen to occupy a cavity formed by 20 water molecules, each retaining its normal co-ordination number of 4. Of the four bonds issuing from each water molecule, three are seen to be in the co-ordination shell, and one directed outwards from the central solute molecule. Without greatly straining the distances or angles which characterize pure ice, the figure formed is seen to be a regular dodecahedron, the surface of which consists of twelve equal plane pentagons. Crystalline hydrates are known for solutes as simple as helium and as complicated as haemoglobin, so that cavities of widely different sizes can evidently be formed in aqueous solution.

The Nature of Aqueous Solvation*

The tetrahedral nearest neighbour co-ordination of water exhibited in ices I, II, and III, cubic ice and in the solid gas hydrates (W. F. Claussen,

* This section is taken from a lecture delivered by D. N. Glew (*J. Phys. Chem.*, **66**, 605, (1962)) at a Symposium organized by the American Chemical Society to celebrate the 80th birthday of J. H. Hildebrand in 1961, and is reproduced with the consent of the author and the Society.

1951, L. Pauling and R. E. Marsh, 1952; M. V. Stackelberg and H. R. Müller, 1954) together with the small energy changes for the high pressure ice transitions, indicate that the water molecule invariably exerts a tetrahedrally directed force field in which the four hydrogen bonds with nearest neighbours do not rupture (R. L. McFarlan, 1936) and are not critically sensitive to angular distortions as large as $31°$ to $-39°$ from the regular tetrahedral angle. The persistence of general three-dimensional tetrahedral fourfold co-ordination of nearest neighbours in liquid water postulated by Bernal and Fowler (1953) has received support from later X-ray scattering (J. Morgan and B. E. Warren; C. L. van Panthaleon van Eck, H. Mendel and J. Fahrenfort, 1958) and Raman spectral measurements (M. Magat, 1937; E. F. Gross, 1959) which may be interpreted (J. A. Pople, 1951) in terms of complete tetrahedral fourfold co-ordination of nearest water neighbours at lower temperatures with thermal distortion and bending of hydrogen bonds $\pm 26°$ without significant rupture.

Eley (1939), presented an aqueous solubility theory for the inert gases, consistent with the Bernal and Fowler water model, in which the water at low temperatures was assumed to occupy quasi-lattice sites maintaining tetrahedral co-ordination and where the solute occupied interstitial sites or cavities formed at low energy expenditure within the hydrogen-bonded solvent water structure. H. S. Frank and M. W. Evans (1945) attributed the large gain in entropy and the large loss in heat capacity attending the escape of solute molecules from aqueous solution to iceberg formation of the water molecules next to the solute, and to the energy required at higher temperatures to break down the structures of the solution.

Independent evidence regarding the nature of aqueous solvation was deduced from reaction kinetics (D. N. Glew and E. A. Moelwyn-Hughes, 1952); and consequent solubility work (D. N. Glew and E. A. Moelwyn-Hughes, 1953) independently and simultaneously indicated (W. F. Claussen and M. F. Polglase, 1952) that the nature of the water solvent surrounding weakly interacting aqueous solutes should be likened geometrically to those co-ordination polyhedra experimentally observed (W. F. Claussen, 1951, L. Pauling and R. E. Marsh, 1952; M. v. Stackleberg and R. H. Müller, 1954) in the solid gas-hydrates. Such water polyhedra containing no ruptured hydrogen bonds provide a natural interpretation of the low energy cavities of Eley's solution theory (1939), consistent with a three-dimensionally hydrogen-bonded liquid water phase (J. D. Bernal and R. H. Fowler; 1933; J. A. Pople, 1951). The obvious co-operative alignment of the water molecules within such polyhedra or hydration shells likewise provides an interpretation of the relatively low entropies of aqueous solutes and of the reduced librational freedom (R. E. Powell and W. M. Latimer, 1951) of the water adjacent to the solute. The preferred regular tetrahedral hydrogen-bond forming directions of the water molecule (ice I, cubic ice) force an alignment and increased orientation of those water molecules adjacent to the solute in maintaining their three hydrogen bonds with nearest water neighbours also adjacent to the solute. The fourth bonds, radially directed away from the solute maintain bonding of the hydration shell with the external bulk water. The central attractive forces between the solute and the surrounding hydration shell further co-operatively stabilize the water members in a manner tending to promote intra-shell hydrogen bonding and the allied reduction of hydration water librational freedom. In no sense is it to be considered that the water molecules adjacent to the solute are permanently immobilized or rigid as in solid structures, but rather as being subject to greater orientational constraints permitting reduced hydrogen-bond bending as compared with those water molecules in the bulk liquid.

A Statistical Theory of Binary Solutions Formed from Molecules of Different Sizes

Hildebrand and Scatchard (1931) have independently shown that the volume fractions of the components in a binary liquid mixture are more significant than their molar fractions in determining its total potential energy. Their concept has been incorporated in several attempts at formulating a statistical theory of such solutions (I. Prigogine, 1957; Rowlinson, 1959). In the version which we shall here discuss,* the chemical potential of solute 2 in a dilute solution in solvent 1 is given as follows in terms of the partial molecular volumes, v_2 and v_1 (in cm^3 per molecule) and the concentration of solute, n_2, in molecules per cm^3:

$$\mu_2(\text{solution}) = \mu_2^0(\text{pure liquid}) + kT \ln n_2 v_2 + kT n_1 v_1 \left(1 - \frac{v_2}{v_1}\right)$$

$$+ (n_1 v_1)^2 \cdot \Delta u^\circ \tag{38}$$

Here Δu° is related as follows to the three interaction energies and the coordination numbers of the pure components:

$$\Delta u^\circ = \frac{1}{2}\left(c_1 \frac{v_2}{v_1} + c_2\right)\phi_{12} - \frac{1}{2}c_1 \frac{v_2}{v_1}\phi_{11} - \frac{1}{2}c_2\phi_{22}. \tag{39}$$

It is clear that, in a dilute gaseous system,

$$\mu_2(\text{gas}) = \mu_2^0(\text{sat vapour}) + kT \ln (p_2/p_2^0). \tag{40}$$

---

* Moelwyn-Hughes, *Physical Chemistry*, p. 827, equation (113), *2nd Ed.* (1965): The Δu^0 here is $(v_2/v_1)^{1/2}$ of the Δu^0 defined in the book.

Here p_2 is the partial pressure of component 2, and p_2^0 the vapour pressure of pure liquid 2. Now μ_2^0 for the pure liquid equals μ_2^0 for the saturated vapour at the same temperature. The condition of equilibrium between the solution and the vapour phase is that μ_2 must be the same for each. Hence, from equations (38) and (40),

$$\frac{p_2}{n_2} = p_2^0 v_2 \exp\left[n_1 v_1\left(1 - \frac{v_2}{v_1}\right)\right] \cdot \exp\left[(n_1 v_1)^2 \cdot \frac{\Delta u^0}{kT}\right], \tag{41}$$

in agreement with Henry's law. The gain in heat content attending the escape of one molecule of component 2 from a dilute solution in component 1 to the gas phase is $kT^2[d \ln (p_2/n_2)/dT]$, which is seen to be

$$\Delta H_2 = \lambda_2^0 + kT^2\alpha_2 - kT^2(v_2/v_1)(\alpha_2 - \alpha_1) - \Delta u^0$$
$$+ T(d\Delta u^0/dT), \tag{42}$$

where λ_2^0 is the molecular heat of vaporization of pure liquid 2, and the α terms are coefficients of cubical expansion. Equation (39) may be written in terms of the molecular heats of vaporization, λ^0:

$$\Delta u^0 = \frac{1}{2}\left(c_1\frac{v_2}{v_1} + c_2\right)\phi_{12} + \frac{v_2}{v_1}\lambda_1^0 + \lambda_2^0. \tag{43}$$

Hence, omitting the difference between the heats of expansion, which is small, and the term $T(d\Delta u^0/dT)$, which may well be considerable, we have

$$\Delta H_2 = -\frac{v_2}{v_1}\lambda_1^0 - \frac{1}{2}\left(c_1\frac{v_2}{v_1} + c_2\right)\phi_{12}. \tag{44}$$

For the reverse process, i.e. the dissolution of the vapour component 2 into a dilute solution in liquid component 1, the gain in heat content is

$$-\Delta H_2 = \left(\frac{v_2}{v_1}\right)\lambda_1^0 + \frac{1}{2}\left[c_1\left(\frac{v_2}{v_1}\right) + c_2\right]\phi_{12}. \tag{45}$$

Interaction energies between unlike molecules are seen, from the square-bracketed term, to be of two kinds, except when $c_1 = c_2$ and $v_1 = v_2$. Then, the interaction energy is simply $c\phi_{12}$. The first term in equation (45) represents, as shown in Table I.2, the energy expended in forming a cavity in liquid 1 capable of accommodating a molecule of vapour 2. Equation (45) is seen to be an elaborated form of equation (25).

Distribution Equilibria in the Methane–Water System

In his experimental investigation of the distribution of methane between the gaseous phase and aqueous solutions (Glew (1962)) finds that equation

(30) does not include sufficient terms to reproduce Henry's constant for this system with sufficient accuracy, and employs an equation of the form

$$\log_{10}(p_2/x_2) = a - b \log_{10} T - c/T + qT, \tag{46}$$

where p_2 is the partial pressure of the solute, in atmospheres, and x_2 is its molar fraction in solution. For the escape process under standard conditions, we thus have

$$\Delta H_2^0 = 2.303Rc - bRT + 2.303RqT^2, \tag{47}$$

$$\Delta C_P^0 = -bR + 4.606RqT, \tag{48}$$

and $$\Delta S_2^0 = \Delta C_P^0 + R(2.303a - b \ln_e T). \tag{49}$$

Glew's equation for the evolution of methane from aqueous solution is

$$\log_{10}(p_2/x_2) = 172.3632 - 61.99 \log_{10} T - 6501/T + 0.0249T. \tag{50}$$

Hence

$$\Delta H_2^0 = 29{,}746 - 123.18T + 0.114T^2 \tag{51}$$

$$\Delta C_P^0 = -123.18 + 0.228T, \tag{52}$$

$$\Delta S_2^0 = \Delta C_P^0 + R(342.51 - 61.99 \ln_e T). \tag{53}$$

From these equations, it is seen that p_2/x_2 at 25°C is 4.11×10^4 atm. At the same temperature, ΔH_2^0 and ΔC_P^0 have the values listed in Table III, and ΔS_2^0 becomes $+31.76$ cal/mole^{-1}-deg^{-1}. Before this value can be compared with the entropies of escape given in Table III, we must digress to consider the different standard states involved.

To express Henry's law in terms of molar fractions of solute is not only logical but is the only form applicable to such data as are given in Table III. On the other hand, for the kinetic object we have in mind, the law is more conveniently expressed in terms of the concentrations of the solutes. We note that in the methane–water system at 25°C and at one atmosphere, the molar fraction of solute is only 2.43×10^{-5}. With solutions as dilute as this, we may relate the molar fraction, x_2, to the solute concentration, c_2, expressed as moles per litre of solution as follows

$$x_2 = \frac{c_2}{1{,}000\rho_1/M_1}, \tag{17}$$

where ρ_1 is the density of the solvent and M_1 is its molar weight.

$$\frac{p_2 \, (\text{mm})}{c_2 \, (\text{mole-lit}^{-1})} = \frac{p_2 \, (\text{atm})}{x_2} \left(\frac{760M_1}{1{,}000\rho_1} \right). \tag{18}$$

The numerical factor for aqueous solutions at 25°C is 13.73, and the entropy

TABLE V

STANDARD ENTHALPY AND HEAT CAPACITY CHANGES FOR METHANE EVOLUTION
FROM AQUEOUS SOLUTION (D. N. Glew, 1962)

$t°C$	ΔH_2^0 cal/mole	ΔC_P^0 cal/mole-deg	$-\dfrac{d}{dT}\left[\dfrac{\alpha_1 T V_2}{\beta_1}\right]$
0	4,621	−60·7	−60·7
25	3,173	−55·0	−55·8
50	1,869	−49·3	−50·1
75	707	−43·6	−43·7
100	−312	−37·9	−36·5
		$\dfrac{d\Delta C_P^0}{dT} = 0·228$	$-\dfrac{d^2}{dT^2}\left[\dfrac{\alpha_1 T V_2}{\beta}\right] = 0·242 \pm 0·046$

change in the units required is greater than that in the atmosphere units by $R \ln_e 13·73 = 5·20\,\text{cal/mole}^{-1}\text{-deg}^{-1}$. The standard entropy change given in column 6 of Table III has been obtained in this way.

There remain slight corrections in ΔH_2^0 and ΔC_P since

$$\Delta H_2^0(p_2/c_2) = \Delta H^0(p_2/x_2) + RT^2\left(\frac{d \ln v_1}{dT}\right)_P = \Delta H^0(p_2/x_2) + RT^2\alpha, \quad (54)$$

where α is the coefficient of cubical expansion of the solvent. A rough estimate for water, in calories per mole, is given in Appendix 2:

$$RT^2\alpha = -551 + RT, \quad (55)$$

which necessitates a small adjustment in ΔH_0^0 and a difference of R in ΔC_P^0.

Values of ΔH_2^0 and ΔC_P^0 thus obtained are given in columns 2 and 3 of Table V. The last column gives ΔC_P derived from equation (1.25) (see Appendix 1). The agreement is within the limits of experimental error.

The process of dissolution of the gas is accompanied by two enthalpy changes, (i) the increase in enthalpy on forming the cavity, $\Delta H_{T,P}$, and (ii) the increase, $\Delta H_{1,2}$ due to the interaction of the solute with its solvent neighbours:

$$-\Delta H_2^0 = \Delta H_{T,P} + \Delta H_{1,2}. \quad (56)$$

Equation (1.25) may be written in the form

$$\Delta H_{T,P} = \frac{T\alpha_1 V_1}{\beta_1}\left(\frac{V_2}{V_1}\right). \quad (57)$$

Experimental values of the ratio of the molar volumes of methane and water are given in Table VI, the last column of which contains the enthalpy of molecular interaction between solute and solvent. According to Glew's

TABLE VI

ENTHALPY OF INTERACTION OF METHANE WITH WATER MOLECULES (CAL/MOLE)

$t°C$	(V_2/V_1)	$-\Delta H_2^0$	$\Delta H_{T,P}$	$\Delta H_{1,2}$
0	1·91	$-4,621$	-301	$-4,320$
20	1·96	$-3,437$	$+1,117$	$-4,554$
40	2·00	$-2,370$	$2,372$	$-4,742$
60	2·05	$-1,370$	$3,521$	$-4,891$
80	2·09	-480	$4,571$	$-5,051$
100	2·13	$+310$	$5,517$	$-5,207$

calculations, the temperature coefficient of $\Delta H_{1,2}$ is negligible, and we may accept an average value of about $-4,800$ cal for $\Delta H_{1,2}$, or -240 cal for each solute–solvent contact. Eley (1939) has applied intermolecular force theory to interpret $\Delta H_{1,2}$ data on aqueous solutions of the inert elements. With the methane–water system, if we ignore the difference between enthalpy and energy, we have an experimental interaction energy of $1·666 \times 10^{-14}$ erg/pair. The special form of Mie's equation

$$\phi = Aa^{-n} - Ba^{-m} \tag{1.4}$$

which has been widely applied by Lennard-Jones is

$$\phi = Aa^{-12} - Ba^{-6}. \tag{58}$$

When the interaction energy is at its minimum, we find

$$B = -2a_e^6\phi_e. \tag{59}$$

As a rough approximation, a_e may be identified with the sum of the radii of the methane and water molecules which, in Angstrom units, becomes $2·42 + 1·39 = 3·81$. The computed value of B is thus 102×10^{-60} erg-cm⁶. This compares favourably with the geometric mean of the B values for methane and water. In the units of 10^{-60}, we have $B_{1,2} = (B_{1,1}B_{2,2})^{1/2} = (226 \times 78)^{1/2} = 133$. The repulsion constant, $A_{1,2}$, for the methane–water pair is thus found to be 156×10^{-105} erg-cm¹² , which is near to Lennard-Jones' value of 162×10^{-105} for the argon–argon pair.

Several, but less successful, attempts have been made to estimate the free energy of cavity formation in terms of the surface tension, γ, and the cavity radius, r. Uhlig (1937), in particular, has shown that when simple gases such as hydrogen, nitrogen, oxygen and methane are dissolved in various solvents at the same temperature, there is a linear relationship between γ and the free energy of escape, according to the equation

$$kT \ln s = -4\pi r^2\gamma + E, \tag{60}$$

where Ostwald's absorption coefficient, s, is $n_{solution}/n_{gas}$ and E stands for the component of the free energy arising from solute–solvent interactions.

The Energy of Solvent Re-organization

Provided the solvent molecules surrounding a cavity in a liquid retain in full the properties of solvent molecules in the bulk liquid, we are justified in adopting Duprés's argument to calculate the energy expended in forming a cavity of volume V_2 in solvent 1. This we shall denote by ΔH_λ, where, as in equation 57,

$$\Delta H_\lambda = \left(\frac{V_2}{V_1}\right)L_1. \tag{61}$$

This term plus the energy ΔH_{or}, expended in re-organizing the solvent molecules surrounding the cavity must equal $\Delta H_{T,P}$, as given by equation 57. Hence the energy expended in the re-organization is

$$\Delta H_{or} = \Delta H_{T,P} - \Delta H_\lambda$$
$$= \left(\frac{T\alpha_1 V_1}{\beta_1} - L_1\right)\frac{V_2}{V_1}. \tag{62}$$

Estimates of the energy of solvent re-organization in the methane–water system are given in Table VII. If the loss in heat content attending the re-organization is identified as that due to the freezing of n molecules of solvent, n at $0°C$ is $20,830/1,435\cdot7 = 14\cdot5$. It may well be that the solvent molecules forming the shell are only partly frozen, say, for example, three-quarters frozen. This would bring the number concerned to $19\cdot3$, which is close to the co-ordination number of 20. Within the limits of accuracy of the volume ratios, ΔH_λ is constant. Since the increase in enthalpy attending

TABLE VII
ENTHALPY OF SOLVENT RE-ORGANIZATION IN THE METHANE–WATER SYSTEM
(CAL/MOLE)

$t°C$	L_1	ΔH_λ	$\Delta H_{T,P}$	$-\Delta H_{or}$
0	10,749	20,530	-300	20,830
20	10,545	20,670	$+1,117$	19,553
40	10,341	20,670	2,372	18,298
60	10,141	20,790	3,521	17,269
80	9,929	20,750	4,571	16,179
100	9,705	20,670	5,517	15,153

dissolution is

$$-\Delta H_2^0 = \Delta H_\lambda + \Delta H_{or} + N_0 c\phi_{12},$$

it follows that

$$\Delta C_p = \frac{d\Delta H_{or}}{dT} + N_0 \frac{d(c\phi_{12})}{dT}, \qquad (63)$$

and the explanation for the high positive value of ΔC_p is that when the temperature of the solution is raised, the heat taken in is utilized almost entirely, as Frank and Evans have argued, in breaking down the structure of the shell around the solute molecule.

Ionization Equilibria in Aqueous Solution

Most of our knowledge of ionic dissociation in aqueous solution refers to the ionization of bases and acids, with the second of which we are now to deal. The early data obtained electrometrically and conductimetrically on the ionization constants of acids have often been presented in empirical formulae showing the relation between K and various powers of t ($^\circ$ centigrade) or of $(t - 25)$. While such formulae suffice to summarize the data accurately, especially when K passes through a maximum or a minimum in the range of investigation, they are clearly not directly amenable to theoretical interpretation. Pitzer (1937) and Moelwyn-Hughes (1938) independently and for different reasons, examined the literature and recast some of the most reliable data into the form

$$\ln K = I + \left(\frac{\Delta C_P}{R}\right) \ln T - \frac{\Delta H_0^0}{RT} \qquad (64)$$

as suggested by Harned, MacInnes, and Harned and Owen, (1958). They found that, although ΔH^0 and ΔH_0^0 vary from one acid to another by as much as 17 kilocalories, ΔC_P^0 and ΔS^0 are not highly specific, having average values of about -40 and -21 cal/mole-deg respectively. Equation (64) implies that ΔC_P^0 is independent of temperature, which, as we shall see, is not exactly true. When $d(\Delta C_P^0)/dT$ is ignored, it is evident that

$$\Delta S^0 = RI + \Delta C_P^0(1 + \ln T). \qquad (65)$$

With acetic acid, for example, we have

$$\ln K = 103 \cdot 5 - \frac{34}{R} \ln T - \frac{10,020}{RT}$$

TABLE VIII

FIRST IONIZATION CONSTANTS IN WATER AT 25°C AND 1 ATMOSPHERE PRESSURE

Electrolyte	K molality or moles/litre	ΔH cal/mole	ΔC_P, cal/mole-deg	ΔS^0 cal/mole-deg
D_2O	3.524×10^{-17}	$+13,420$	-42.8	-30.2
H_2O	1.805×10^{-16}	$13,450$	-42.8	-26.8
H_2CO_3	4.310×10^{-7}	$2,075$	(-82)	-22.2
$H \cdot COOH$	1.772×10^{-4}	-13	-40.9	-17.2
$CH_3 \cdot COOH$	1.754×10^{-5}	-112	-33.9	-22.1
$nC_2H_5 \cdot COOH$	1.336×10^{-5}	-168	-37.7	-22.9
$nC_3H_7 \cdot COOH$	1.515×10^{-5}	-691	-46	-24.4
NH_4OH	1.810×10^{-5}	$+790$	-52.5	-19.0
I_2	1.840×10^{-5}	$-1,964$	$—$	-28.3
Violuric acid	2.730×10^{-5}	$+3,650$	-38.2	-17.8
$NH_2 \cdot CH_2 \cdot COOH$ (basic)	6.040×10^{-5}	$+2,765$	(-22.2)	(-10.0)
$HO \cdot CH_2 \cdot COOH$	1.480×10^{-4}	$+210$	-39.2	-16.8
$Cl \cdot CH_2 \cdot COOH$	1.378×10^{-3}	$-1,170$	-34.9	-14.3
$NH_2 \cdot CH_2 \cdot COOH$ (acidic)	4.470×10^{-3}	$+1,159$	-30.6	-15.7
H_3PO_4	7.520×10^{-3}	$-1,764$	-42.8	-21.6
H_2SO_3	1.300×10^{-2}	$-3,860$	$—$	$—$

or

$$\log K = 44.931 - 17.11 \log T - \frac{2,191}{T}. \tag{66}$$

The high negative values of ΔC_P^0 and ΔS^0 have been discussed from the standpoint of thermodynamics (Latimer, 1936; Pitzer, *loc. cit.*) electrostatics (Baughan, 1939; Everett and Coulson, 1940) and chemical kinetics (Moelwyn-Hughes, 1938). The iceberg hypothesis of Uhlich, affords an estimate of the number of water molecules which are thought to have become "frozen" in the process of ionization. ΔC_P^0 for the freezing of pure water at 25°C is -8.128 and ΔS^0 for the phase change at 0°C is -5.253. The ionization data for acetic acid yield the number 4.20 from the heat capacity and 4.21 from the entropy. These results are mutually consistent, and suggest that it is the hydrogen ion that is chiefly responsible. Moreover, its co-ordination number is known to be 4. As in the previous problem, it may be more reasonable to think of a higher number of water molecules partly frozen rather than of four molecules completely immobilized. The energy required to form cavities for the ions does not enter into this problem, as larger cavities have already been created to accommodate the unionized solute.

The apparent energy of activation for the hydrolysis of various solutes in water decreases as the temperature is raised, yielding values of ΔC_P^* which lie near to those which hold for ionizations. The principal difference between the ionization of an acid in water and a simple hydrolysis thus appears to be that only the former is successfully opposed by recombination of the ions.

Exact Formulation of the Dependence of Ionization Constants on Temperature

E. C. W. Clarke and D. N. Glew (1966) have devised a novel and exact method for deriving the thermodynamic functions of equilibrium (and velocity) constants in terms of experimental data. With the details of their method we shall not here be concerned. We shall merely quote, as an example, their conclusions relating to the ionization constants of cyanoacetic acid, as measured by Feates and Ives (1966). The simplest equation found adequately to reproduce the experimental ionization constants, K_p, at constant pressure has the form of the five-constant expression

$$\ln K_p = A + B/T + C \ln T + DT + ET^2,$$

from which the standard increase in enthalpy, per mole, is seen to be

$$\Delta H^0 = -RB + RTC + RT^2D + 2RET^3.$$

The standard increase in the molar heat capacity, and its temperature derivatives are clearly

$$\Delta C_P^0 = RC + 2RTD + 6RET^2,$$

$$d\Delta C_P^0/dT = \quad 2RD + 12RET,$$

$$d^2\Delta C_P^0/dT^2 = \quad 12RE,$$

$$d^3\Delta C_P^0/dT^3 = \quad 0.$$

The particular values afforded by these equations at 298·15°C and their limits of accuracy are as follows:

ΔG^0 (cal/mole)	$3369 \cdot 58 \pm 0 \cdot 16$
ΔH^0 (cal/mole)	-890 ± 5
ΔC_P^0 (cal/mole-deg)	$-36 \cdot 2 \pm 1 \cdot 3$
$d\Delta C_P^0/dT$ (cal/mole-deg^2)	$+0 \cdot 15 \pm 0 \cdot 09$
$d^2\Delta C_P^0/dT^2$ (cal/mole-deg^3)	$-0 \cdot 083 \pm 0 \cdot 036$
$d^3\Delta C_P^0/dT^3$ (cal/mole-deg^4)	$0 \qquad 0.$

Ion-pair Equilibria in Aqueous Solution

The chemical potential of a spherical ion, of radius r_i and charge $z_i\varepsilon$, in a medium of permittivity D at temperature T is given, at low concentrations, in terms of the ionic concentration, c_i, the ionic charge, $z_i\varepsilon$, and the standard chemical potential, μ_i^0:

$$\mu_i = \mu_i^0 + kT \ln c_i + \frac{z_i^2 \varepsilon^2}{2Dr_i} - \frac{z_i^2 \varepsilon^2 \kappa}{2D} \tag{67}$$

where κ is the variable appearing in the limiting equation of Milner, Debye and Hückel (equation 8.8a). In aqueous solutions the validity of the last term is limited to concentrations of about 10^{-3} molar for 1–1 electrolytes and about 10^{-5} for 2–2 electrolytes. The thermodynamic condition of equilibrium ($\mu_A + \mu_B = \mu_C$) may be applied to the ionic system

$$C \rightleftarrows A + B.$$

Since $z_C = z_A + z_B$, and $\mu_A^0 + \mu_B^0 - \mu_C^0 = \Delta\mu^0$, we find that

$$kT \ln \frac{c_A c_B}{c_C} = -\Delta\mu^0 + \frac{z_A z_B \varepsilon^2}{Dr_c}\left[1 - \frac{1}{2}\left(\frac{z_A}{z_B}\right)\left(\frac{r_C}{r_A} - 1\right) - \frac{1}{2}\left(\frac{z_B}{z_A}\right)\left(\frac{r_C}{r_B} - 1\right) \right]$$
$$- \frac{z_A z_B \varepsilon^2 \kappa}{D}. \tag{68}$$

The derivation of this relationship implies that when ions A and B unite to form ion C, the sum of their electrical charges is distributed uniformly over a sphere of radius r_c. The model underlying the mechanism is thus a highly improbable one, except when the sum of the charges on A and B is zero, i.e. when C is an electrically neutral solute.

The equilibrium constant in terms of concentration is defined as follows:

$$K_c = c_A c_B / c_C. \tag{69}$$

Then, since κ is zero at infinite dilution,

$$\ln K_c = \ln K^0 - \frac{z_A z_B \varepsilon^2 \kappa}{DkT}. \tag{70}$$

Using numerical values appropriate to water at 25°C, we have

$$\log_{10} K_c = \log_{10} K^0 - 1{\cdot}0124 z_A z_B \sqrt{I} \tag{71}$$

where I is the ionic strength of the solution and K^0 is the equilibrium constant extrapolated to solutions of zero concentration. When C is an uncharged solute molecule or a pair of ions with charges of equal magnitude but opposite sign, z_C is zero, and consequently $z_A = -z_B$. The observed ionization constant in the former case or the observed dissociation constant in the latter case is then given as

$$\log_{10} K_c = \log_{10} K^0 + 1.0124z^2\sqrt{I}. \tag{72}$$

Within the concentrations indicated above, this relationship has been experimentally verified for the ionization equilibria of acids and bases. We have here to deal more specifically with the dissociation equilibria of ion pairs, which have proved to be much more difficult to investigate (C. W. Davies, 1962; C. B. Monk, 1961).

Though potentiometric, polarographic and spectrophotometric methods have been employed to investigate ion-pair equilibria, we shall here describe the conductimetric method only. The first requirement is an expression for K_c which extends beyond the very narrow concentration range covered by equation (70). Davies achieves this by multiplying $\sqrt{I}$ of equation (71) by the factor $(1 + \sqrt{I})^{-1} - 0.30\sqrt{I}$, for which there are firm experimental and theoretical reasons. The revised equation,

$$\log_{10} K_c = \log_{10} K^0 - 1.0124z_Az_B\left[\left(\frac{\sqrt{I}}{1 + \sqrt{I}}\right) - 0.30I\right], \tag{73}$$

is found to hold up to concentrations of 10^{-1} molar. K_c written in terms of the total concentration, c, and the fraction, α, of the solute which is in the form of ions, is

$$K_c = c\alpha^2/(1 - \alpha) \tag{74}$$

as derived by Ostwald. If there were no un-ionized solutes or uncharged ion pairs, the equivalent conductivity, Λ, of the solution would be given by Kohlrausch's law $\Lambda = \Lambda_0 - A\sqrt{c}$, but when only a fraction α is in the ionic form, we have

$$\Lambda = \alpha(\Lambda_0 - A\sqrt{\alpha c}), \tag{75}$$

from which, with precise conductivity data, α may be found by successive approximations.

With the 3–3 electrolyte lanthanum ferricyanide, the decadic logarithm of the dissociation constant for the ion pair $[La^{+++}, Fe(CN)_6^{---}]^0$ when plotted against $\sqrt{I}$ gives a limiting gradient of 9.16, in favourable agreement with the theoretical limiting slope of 9.112 required by equation (72).

We next inquire into the magnitude of K^0, which, according to equation (68), can be expressed as follows:

$$kT \ln K^0 = kT \ln K^{00} + \frac{z_Az_B\varepsilon^2}{Dr_c}\left[1 - \frac{1}{2}\left(\frac{z_A}{z_B}\right)\left(\frac{r_c}{r_A} - 1\right) - \frac{1}{2}\left(\frac{z_B}{z_A}\right)\left(\frac{r_c}{r_B} - 1\right)\right]. \tag{76}$$

The first term on the right replaces $-\Delta\mu^0$; and K^{00} represents the hypothetical dissociation constant for a pair of uncharged spheres at infinite dilution. When there is no electrical charge on the ion pair, $z_A = -z_B$, and

$$kT \ln K^0 = kT \ln K^{00} - \frac{z^2 \varepsilon^2}{2D}\left(\frac{1}{r_A} + \frac{1}{r_B}\right). \tag{77}$$

Some values of the association constant, K_A, which is the reciprocal of the dissociation constant K^0, given by Monk (1961), are summarized in Table IX. The association constants for the 2–2 electrolytes are naturally greater than those for the 1–1 electrolytes. If we could select two of these

TABLE IX

$\log_{10} K_A^0$ AT ZERO IONIC STRENGTH IN AQUEOUS SOLUTION AT 25°C
(K_A IN LITRES/GRAM-ION)

Cation	Na^+	K^+	Mg^{++}	Ca^{++}	Sr^{++}	Ba^{++}
Anion						
IO_3^-	−0.47	−0.24	0.72	0.89	0.98	1.1
SO_4^{--}	0.72	0.96	2.3	2.31	—	—
$S_2O_3^{--}$	0.68	0.92	1.8	2.0	2.0	2.2

equilibria with common values of K^{00} and common values for the sum of the reciprocal radii, we would have

$$\log_{10}\frac{K^0(1:1)}{K^0(2:2)} = \frac{3\varepsilon^2}{2.303(2DkT)}\left(\frac{1}{r_A} + \frac{1}{r_B}\right). \tag{78}$$

An approximation to these conditions is furnished by $(Na^+, IO_3^-)^0$ and $(Ca^{++}, SO_4^{--})^0$, for which $\log_{10}(K^0(1:1) - \log_{10} K^0(2:2)$ is 2.78, suggesting ionic radii of about 3.3 Å. It is doubtful, however, whether K^{00} can be the same for the two pairs.

The electrostatic energy of two isolated ions of fixed charges and radii, separated from each other by a distance a in a medium of permittivity D, is

$$u = \frac{z_A^2 \varepsilon^2}{2Dr_A} + \frac{z_B^2 \varepsilon^2}{2Dr_B} + \frac{z_A z_B \varepsilon^2}{Da}. \tag{79}$$

The gain in energy when the ions are removed to infinity from their closest distance of approach is

$$\Delta u = u_{a=\infty} - u_{a=r_A+r_B} = -\frac{z_A z_B \varepsilon^2}{D(r_A + r_B)}. \tag{80}$$

Looked at in this way, the two ions forming the pair are still surrounded by solvent molecules, and each retains its electrostatic energy $z_i^2 \varepsilon^2 / 2Dr_i$. The presence of one solvent molecule between the two ions would satisfy these conditions, and in this treatment (Bjerrum and Larsson, 1927) no assumptions are necessary regarding the sphericity of the ion pair or the distribution of its charge. The square-bracketed term in equation (76) is now to be replaced by unity, and we have the simpler expression for the dissociation constant at zero ionic strength,

$$kT \ln K^0 = kT \ln K^{00} + \frac{z_A z_B \varepsilon^2}{Da}. \tag{81}$$

This equation is to be contrasted with equation (77). By plotting $\ln K^0$ against $1/D$ for a given electrolyte in various media, reasonable values have been obtained (C. W. Davies, 1934; W. F. K. Wynne-Jones, 1933) for the nearest distance of approach, a.

Let us apply this equation to the dissociation of some metal hydroxides in water: $(M^{zm}, mOH^-)^0 \rightleftarrows M^{zm} + mOH^-$. Thus, since $z_A = -1$ and $z_B = z_m$,

$$kT \ln K^0 = kT \ln K^{00} - \frac{z_m \varepsilon^2}{D(r_m + r_{OH})}. \tag{82}$$

Matlock (see Davies, 1934) finds agreement with experiment when r_m is taken as the crystal radius, and $\mathring{r}_{OH}$ is taken as 0·74 Å. Reasonable agreement is also found when $\mathring{r}_{OH}$ is taken as 0·474 Å (Table X). With the constants appropriate for water at 25°C, the theoretical equation becomes

$$\log_{10} K^0 = \log_{10} K^{00} - 3·097 \left(\frac{z_m}{\mathring{r}_m + \mathring{r}_{OH}} \right). \tag{83}$$

The experimental gradient from Fig. 6 is $-3·37$.

TABLE X

DISSOCIATION CONSTANTS OF METALLIC HYDROXIDES IN WATER AT 25°C
IN GRAM-IONS/LITRES ($\mathring{r}_{OH} = 0·474$ Å)

Cation	z_m	$\mathring{r}_m$	$z_m/(\mathring{r}_m + \mathring{r}_{OH})$	$4 + \log_{10} K^0$
Na^+	1	1·012	0·673	4·70
Li^+	1	0·758	0·812	4·08
Ba^{++}	2	1·395	1·070	3·36
Sr^{++}	2	1·175	1·213	3·04
Ca^{++}	2	1·051	1·311	2·70
Mg^{++}	2	0·78	1·595	1·42
La^{++}	3	1·14	1·859	0·70

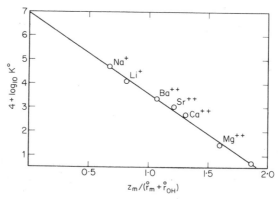

Fig. 6. Dissociation Constants of Metallic Hydroxides in Water at 25°C in Gram-ions/litre ($\mathring{r} = 0.474$ Å). (After Matlock).

The agreement on the whole is better than could be expected. The interionic distance in the ion pair, separated by a water molecule, should exceed the sum of the crystal radii, though not by as much as the diameter of a water molecule because the relatively strong forces of induction, which have not hitherto been mentioned, bring oppositely charged ions in an isolated pair closer together than they are in the crystal. The theory of ion pairs in aqueous solution has been refined by C. W. Davies.

Ionic Association in Non-aqueous Media

We shall here discuss the formation of ion pairs from univalent ions in organic solvents. Denison and Ramsey (1955) and Fuoss and Kraus (1957) have independently argued that the association constant, K_A, must be given, to a first approximation, by the expression

$$K_A = K_A^0 . \exp{(\varepsilon^2/DakT)}, \tag{84}$$

where K_A^0 can be interpreted as the association constant of a pair of uncharged particles. An approximate estimate for ions of equal radii (Moelwyn-Hughes, 1963) is

$$K_A^0 = \frac{N_0}{1,000} . \tfrac{7}{6}\pi a^3 . \tag{85}$$

If the ionic separation in the pair is a true constant, the plot of $\ln K$ against $1/D$ should be linear, since

$$\log_{10} K_A = \log_{10} K_A^0 + \frac{\varepsilon^2}{2.303akT} . \frac{1}{D}, \tag{86}$$

or

$$\log_{10} K_A = \log_{10} K_A^0 + B/D, \tag{87}$$

where B is an isothermal constant, whose value depends on the nature of the solute but not on that of the solvent. When applied to a temperature of $298 \cdot 16°$ K, we have

$$\log_{10} K_A = \log_{10} K_A^0 + 243/(åD), \tag{88}$$

where $å$ is the value of a expressed in Ångström units. A typical graph, due to Inami, Bodenseh and Ramsey (1961), is reproduced in Fig. 7. Further data, summarized in the form of equation (87), are given in Table XI. They differ from one another in their accuracy, and in the ranges of permittivities covered. The data for silver nitrate refer to solutions in water, methanol and ethanol: those for tetrabutylammonium picrate to pure solvents; and those for tetra*iso*amylammonium nitrate to dioxane–water mixtures of different compositions, with permittivities varying from 2 to 80. It will be observed that this extremely simple formulation of ion pair theory suffices to account for a change of about 10^7 in the association constant. Equation (88), applied to silver nitrate in the gaseous phase ($D = 1$), shows that it is completely associated, since $\log_{10} K_A^0$ becomes $82 \cdot 7 \pm 4 \cdot 5$. Fuoss and Kraus have

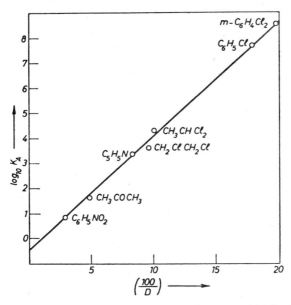

FIG. 7. Association constants of tetrabutylammoniumpicrate at 25°C (after Inami, Bodenseh and Ramsey[5]).

TABLE XI
TEMPERATURE = 25°C.
$$\log_{10} K_A \text{ (LITRES/MOLE)} = \log_{10} K_A^0 + B/D = \log_{10} K_A^0 + \varepsilon^2/(2{\cdot}303 Dak\,T)$$

Electrolyte	$\log_{10} K_A^0$	B	$a[\text{Å}]$		
			From B	From K_A^0	Mean
$AgNO_3$	$\bar{2}{\cdot}802$	$88{\cdot}5$	$2{\cdot}75$	$3{\cdot}06$	$2{\cdot}91 \pm 0{\cdot}16$
$(C_4H_9)_4NBr$	$\bar{1}{\cdot}041$	$56{\cdot}4$	$4{\cdot}31$	$3{\cdot}68$	$4{\cdot}00 \pm 0{\cdot}31$
$(C_4H_9)_4NI$	$\bar{1}{\cdot}892$	$43{\cdot}9$	$5{\cdot}55$	$7{\cdot}06$	$6{\cdot}31 \pm 0{\cdot}76$
$(C_4H_9)_4NClO_4$	$\bar{1}{\cdot}755$	$50{\cdot}2$	$4{\cdot}85$	$6{\cdot}37$	$5{\cdot}61 \pm 0{\cdot}76$
$(C_4H_9)_4NPic$	$\bar{1}{\cdot}543$	$46{\cdot}0$	$5{\cdot}29$	$5{\cdot}41$	$5{\cdot}35 \pm 0{\cdot}06$
$(isoC_5H_{11})_4N{\cdot}NO_3$	$\bar{1}{\cdot}549$	$41{\cdot}7$	$5{\cdot}83$	$5{\cdot}43$	$5{\cdot}63 \pm 0{\cdot}20$

pointed out that, when one of the ions is asymmetric and possesses a permanent dipole of moment μ, the Coulombic energy term must be augmented by an ion–dipole term, $\varepsilon\mu/Da^2$. Even with a pair of symmetrical ions, it is clear that forces other than Coulombic must be included, for Bodenseh and Ramsey (1965), working with tetra-propylammonium bromide in various solvents find a to be $4{\cdot}11$ Å, which is significantly less than the interionic separation of $4{\cdot}94$ Å in the crystal.

A statistical treatment of ion pair equilibria is given in the following section.

Ion-pair Equilibria Involving Repulsive Forces

If ions repel one another with a force varying as the inverse $(s + 1)$th power of their distance, a, apart, the energy of interaction of a pair of univalent ions with charges of opposite signs in a medium of permittivity, D, is

$$\phi = \frac{A}{a^s} - \frac{\varepsilon^2}{Da}. \tag{89}$$

The interaction energy has the following minimum value when $a = a_e$:

$$\phi_e = -\frac{\varepsilon^2}{Da_e}\left(1 - \frac{1}{s}\right). \tag{90}$$

Small displacements from the equilibrium separation are attended by a harmonic vibration of frequency

$$v_e = \frac{\varepsilon}{2\pi a_e}\left[\frac{s - 1}{m^* Da_e}\right]^{1/2}, \tag{91}$$

where m^* is the reduced mass of the pair. In deriving the partition functions for the ions, it is assumed that their internal motions are the same in the free and associated states, so that only the translational components of their total partition functions need be considered. In terms of the masses, m_+ and m_-, of the ions, whether free or solvated, of their numbers, N_+ and N_-, in a total volume V and of their average potential energies, u_+ and u_-, these are, according to equation (2.19),

$$f_+ = \frac{(2\pi m_+ kT)^{3/2} Ve}{h^2 N_+} \cdot \exp(-u_+/kT),$$

and

$$f_- = \frac{(2\pi m_- kT)^{3/2} Ve}{h^3 N_-} \cdot \exp(-u_-/kT).$$

The motions of the ion pair can be resolved into (i) translation throughout the total volume V, (ii) rotation of the pair, with moment of inertia $I = m^* a_e^2$, about its centre of gravity, and (iii) vibration along the line of centres. The magnitude of the vibration frequency v_e is such that hv_e is ordinarily much less than kT, so that the classical partition function may be used for this motion. The total partition function of the pair thus becomes

$$f_{+-} = \frac{[2\pi(m_+ + m_-)kT]^{3/2} Ve}{h^3 N_+} \cdot \frac{8\pi^2 IkT}{h^2} \cdot \frac{kT}{hv_e} \cdot \exp(-u_{+-}/kT). \quad (2.21)$$

The condition of equilibrium is $\mu_+ + \mu_- = \mu_{+-}$, where the chemical potential of each species is given by equation (2.14). It follows that

$$K_A = \frac{n_{+-}}{n_+ n_-} = 4\pi a_e^3 \left[\frac{2\pi Da_e kT}{(s-1)\varepsilon^2}\right]^{1/2} \cdot \exp\left\{-\frac{u_{+-} - u_+ - u_-}{kT}\right\}.$$

According to equation (90), the algebraic sum of the potential energies appearing in the exponent is $(\varepsilon^2/Da_e)(s-1)/s$. On converting the units of K_A from ccs per molecule to litres per gram-mole, the association constant becomes

$$K_A = \frac{N_0}{1,000} 4\pi a_e^3 \left[\frac{2\pi Da_e kT}{(s-1)\varepsilon^2}\right]^{1/2} \cdot \exp\left\{\frac{\varepsilon^2}{Da_e kT}\left(\frac{s-1}{s}\right)\right\}. \quad (92)$$

When applied to the data of Table XI, reasonable values of s are found, ranging from 5 for silver nitrate to 50 for tetr-*iso*-amyl-ammonium nitrate. Two items concerning this treatment call for comment.

No appeal has been made to the concept of ionic radius. The term a_e is the average distance apart of the charges in the ion pair when in its state of lowest potential energy. The root-mean-square displacement about the average separation is given by the equation

$$\frac{\bar{\bar{x}}}{a_e} = \left[\frac{Da_e kT}{(s-1)\varepsilon^2}\right]^{1/2}. \tag{93}$$

If $\log_{10} K_A$, rather than $\log_{10} (K_A/D^{1/2})$ is plotted against $1/D$, a constant gradient is not obtained since

$$\frac{d \log_{10} K_A}{d(1/D)} = \frac{1}{2\cdot303}\left[\frac{\varepsilon^2}{a_e kT}\left(\frac{s-1}{s}\right) - \frac{D}{2}\right], \tag{94}$$

which increases as $1/D$ increases. It is in this direction that deviations from linearity have been observed (Bodenseh and Ramsey, 1965).

Ionization in Non-hydroxylic Media

Because work must be applied to separate oppositely charged ions during ionization, it is, perhaps, natural to expect all ionizations to be endothermic, and this, in fact, is true when we examine the increase in enthalpy, ΔH_0, at the absolute zero of temperature, though not always when ΔH_T only is considered, as in Table VIII. According to equation (81), the ionization constant of a 1-1-electrolyte at infinite dilution may be expressed as follows

$$K^0 = K^\infty . \exp(-\varepsilon^2/DakT), \tag{95}$$

where K^∞ is the constant for the dissociation into uncharged products, and a is the charge separation. By the conductimetric method, Lichtin and Bartlett (1951) have measured the ionization constant of triphenylchlor-methane in sulphur dioxide solution, $(C_6H_5)_3Cl \rightleftarrows (C_2H_5)_3C^+ + Cl^-$. Their results may be summarized by the equation K^0 (gram ions/litre) $= 3 \times 10^{-12} . e^{+8,900/RT}$. At 25°C, $K^0 = 1\cdot00 \times 10^{-5}$, which has the same order of magnitude as the ionization constants of ammonia and acetic acid in water. By means of the van't Hoff isochore, we have $\Delta H^0 = \Delta H^\infty + \varepsilon^2/Da(1 - LT)$, where $L = -(d \ln D/dT)_P$, as previously explained. The numerical values of D and L are given in Table 7.IV, according to which $(1 - LT)$ for this solvent is $-2\cdot11$ at 25°C. It is impossible, without further evidence, to estimate the magnitude of ΔH^∞, which is the non-electrostatic contribution to the enthalpy of ionization. If it were zero, we would have $-8,900\,\text{cal} = -2\cdot11\varepsilon^2/Da$, and therefore $a = 3\cdot10\,\text{Å}$, which is consistent with the exo-thermicity of the ionization. The magnitude of the standard entropy of ionization, $\Delta S^0_{298\cdot15} = -52\cdot7\,\text{cal/mole-deg}$, cannot be explained in such a simple way. The decrease in entropy arising from the electrostatic effect is to be added to that attending the conversion of the triphenyl groups from the pyramidal to the planar configuration.

Entropy of Association

It is sometimes convenient to have a simple expression for the decrease in entropy attending the association of unlike spherical particles without regard to any energy exchanges. The approximate association constant, expressed in ccs per molecule, is given by Fuoss as

$$K_n = \frac{n_{AB}}{n_A n_B} = \tfrac{4}{3}\pi(r_A + r_B)^3. \tag{96}$$

The standard gain in molar entropy attending the association is

$$\Delta^0 S = R \ln K_c = R \ln \left[\frac{N_0}{1,000} \cdot \tfrac{4}{3}\pi(r_A + r_B)^3 \right]. \tag{97}$$

The reference state is now that of one mole per litre. Estimated entropy changes for spheres of various radii are given in Table XII. These rough

TABLE XII
ENTROPIES OF ASSOCIATION

$\mathring{r}_A + \mathring{r}_B$	$\Delta^0 S$ (cal/mole^{-1}-deg^{-1})
2	$-7 \cdot 8$
3	$-5 \cdot 3$
4	$-3 \cdot 6$
5	$-2 \cdot 3$
6	$-1 \cdot 2$

estimates agree reasonably well with many of the published data. For the association of various free radicals in non-polar solvents, $\Delta^0 S$ is about -5: for associations of the type $A + B \rightleftarrows AB$, also in non-polar solvents, $\Delta^0 S$ is about -6: and for the athermal dimerization of acetic acid in water, $\Delta^0 S = -6 \cdot 7$.

REFERENCES

Anderson, D. L., R. A. Smith, D. B. Myers, S. K. Alley, A. G. Williamson and R. L. Scott, *J. Amer. Chem. Soc.*, **66**, 621 (1962).
Bartlett and Nebel, *J. Amer. Chem. Soc.*, **62**, 1345 (1940).
Baughan, *J. Chem. Physics*, **7**, 951 (1939).
Benford and Wassermann, *Trans. Chem. Soc.*, **67**, 367 (1939).
Bernal, J. D. and R. H. Fowler, *ibid.*, **1**, 515 (1933).
Bjerrum, N. and Larsson, *Z. Physikal Chem.*, **127**, 358 (1927).
Blair and Yost, *J. Amer. Chem. Soc.*, **55**, 4489 (1933).
Bodenseh, K. K. and J. B. Ramsey, *J. Phys. Chem.*, **69**, 543 (1965).

Briegleb, G., *Z. physikal Chem. B*, **23**, 105 (1933); K. Berger, *ibid.*, *B*, **22**, 283 (1933); *B* **28**, 95; J. H. Boer, *Trans. Faraday Soc.*, **32**, 10 (1936); M. Magat, *ibid.*, **33**, 714, *Ann. Physique*, **6**, 108 (1936); E. A. Moelwyn-Hughes, *J.*, 1243 (1938); H. Harris, *Z. physikal Chem.*, *B.*, **43**, 257 (1939); M. M. Davies, *Trans. Faraday Soc.*, **36**, 333 (1940); E. J. W. Verwey, *Rec. Trav. chim.*, **60**, 887; C. A. Coulson, *Valence*, p. 202, Oxford (1957).

Clarke, E. C. W. and D. N. Glew, *Trans. Faraday Soc.*, **62**, 539 (1966).

Claussen, W. F., *J. Chem. Phys.*, **19**, 259, 662, 1425 (1951).

Claussen, W. F. and M. F. Polglase, *J. Am. Chem. Soc.*, **74**, 4817 (1952).

Cundall, *Trans. Chem. Soc.*, **67**, 794 (1895).

Connick and Yuan-tsan Chia, *J. Amer. Chem. Soc.*, **81**, 1280 (1959).

Davies, C. W., *Ion Association*, Butterworths, London (1962).

Davies, C. W., *The Conductivity of Electrolytes*, 2nd ed., Chapman & Hall, London (1934).

Davies, M. M., *Ann. Rep. Chem. Soc.*, **43**, 5 (1947).

Davies, M. M., *Trans. Faraday Soc.*, **34**, 1427 (1938).

Davies, M. M. and G. B. B. M. Sutherland, *J. Chem. Physics*, **6**, 755 (1938).

Davies, M. M., P. Jones, D. Patnaik and E. A. Moelwyn-Hughes, *Trans. Chem. Soc.*, 278 (1951).

Denison, J. J. and J. B. Ramsey, *J. Amer. Chem. Soc.*, **77**, 2615 (1955).

Dolezalek, *Z. physikal Chem.*, **71**, 191 (1910).

Eley, D. D., *Trans. Faraday Soc.*, **35**, 1281 (1939).

Everett and Coulson, *Trans. Faraday Soc.*, **36**, 633 (1940).

Feates and Ives, *Ibid.*, **62**, 539 (1966).

Frank, *Proc. Roy. Soc.*, A., **152**, 171 (1935).

Frank, H. S. and M. W. Evans, *J. Chem. Phys.*, **13**, 507 (1945).

Fuoss, R. M. and Kraus, *ibid.*, **79**, 3304 (1957).

Giauque and Kemp, *J. Chem. Physics*, **6**, 40 (1938); cf. 13,600 given by Schreber, *Z. physikal Chem.*, **24**, 651 (1897) and J. H. van't Hoff, *Studies in Chemical Dynamics*, p. 154 (1896).

Glew, *J. Phys. Chem.*, **66**, 605 (1962).

Glew, D. N. and E. A. Moelwyn-Hughes, *Faraday Soc. Discussion*, **15**, (1953).

Glew, D. N. and E. A. Moelwyn-Hughes, *Proc. Roy. Soc.* (London), **A211**, 254 (1952).

Glew, D. N. and E. A. Moelwyn-Hughes, *Trans. Faraday Soc. Discussion*, **15**, 150 (1953).

Glew, D. N. and E. A. Moelwyn-Hughes, *Discussions Faraday Soc.*, **15**, 150 (1953).

Goss, E. F., "Hydrogen Bonding", ed. D. Hadži, Pergamon Press, London, p. 203 (1959).

Gross, P. and K. Schwarz, *Monatsh.*, **55**, 287 (1930).

Harned and Owen, *The Physical Chemistry of Electrolytic Solutions*, 3rd ed., Reinhold, New York (1958).

Henry, *Phil. Trans.*, **29**, 274 (1803).

Higgins, Pimental and Shorley, *J. Chem. Physics*, **23**, 1246 (1955).

Hildebrand, J. H. and G. Scatchard, *Chem. Rev.*, **8**, 321 (1931).

Hildebrand, J. H. and R. L. Scott, *The Solubility of Non-electrolytes*, 4th ed., Reinhold, New York (1962).

Inami, Y. H., H. K. Bodenseh and J. B. Ramsey, *J. Amer. Chem. Soc.*, **83**, 4745 (1961).

Jakowkin, *Z. physikal Chem.*, **29**, 613 (1899).

Lannung, *J. Amer. Chem. Soc.*, **52**, 67 (1930).

Latimer, *Chem. Rev.*, **18**, 349 (1936).

Latimer and Rodebush, *J. Amer. Chem. Soc.*, **42**, 1419 (1920).
Lennard-Jones, *Physica*, IV, **10**, 947 (1937).
Lichtin and Bartlett, *J. Amer. Chem. Soc.*, **73**, 5350 (1951).
Liebhafsky, *Chem. Rev.*, **17**, 89 (1935).
McFarlan, R. L., *J. Chem. Phys.*, **4**, 60, 253 (1936).
Madgin, *Trans. Chem. Soc.*, **193** (1933); 606 (1937).
Magat, M., *Trans. Faraday Soc.*, **33**, 114 (1937).
Missen, R. W. Private communication.
Moelwyn-Hughes, E. A., *Trans. Chem. Soc.*, 850 (1940).
Moelwyn-Hughes, E. A., *Trans. Faraday Soc.*, **34**, 91 (1938).
Moelwyn-Hughes, E. A., *Z. Naturforschg*, **18a**, 202 (1963).
Moelwyn-Hughes, E. A., and Missen, *Trans. Faraday Soc.*, **53**, 607 (1956).
Mooney and Ludlam, *Proc. Roy. Soc. Edin.*, **49**, 160 (1929).
Monk, C. B., *Electrolytic Dissociation*, Academic Press, New York (1961).
Morgan, J. and B. E. Warren, *J. Chem. Physics*, **6**, 666 (1938).
Miyazawa, T. and K. S. Pitzer, *J. Amer. Chem. Soc.*, **81**, 74 (1959); L. Slutsky and
    S. H. Bauer, *ibid.*, **76**, 270 (1954).
Nernst, W., *Die theoretischen und experimentallen Grundlagen des neuen Wärmesatzes*,
    Halle (1918); Fenton and Garner, *Trans. Chem. Soc.*, 694 (1930); Ritter and
    Simons, *J. Amer. Chem. Soc.*, **67**, 757 (1945); E. W. Johnson and Nash, *ibid.*, **72**,
    547 (1950); Coolidge, *ibid.*, **50**, 2166 (1928).
Nernst, *Z. physikal Chem.*, **8**, 110 (1891).
Owen, B. B. See Harned and Owen (above).
Panthaleon van Eck, C. L. van, H. Mendel and J. Fahrenfort, *Proc. Roy. Soc.* (London),
    A247, 472 (1958).
Pauling, L. and L. O. Brockway, *Proc. Nat. Acad. Sci.*, **20**, 336 (1934).
Pauling, L. and R. E. Marsh, *Proc. Natl. Acad. Sci.*, **38**, 112 (1952).
Pimental, C. C. and A. L. McClellan, *The Hydrogen Bond*, W. F. Freeman & Co.,
    San Francisco (1960).
Pitzer, K., *J. Amer. Chem. Soc.*, **59**, 2365 (1937).
Polissar, *J. Amer. Chem. Soc.*, **52**, 956 (1930).
Pople, J. A., *Proc. Roy. Soc.* (London), **A205**, 163 (1951).
Powell, R. E. and W. M. Latimer, *ibid.*, **19**, 1139 (1951).
Prigogine, I., The Molecular Theory of Solutions, North Holland, Amsterdam (1957).
Raoult, F. M., *Z. physikal Chem.*, **2**, 353 (1888).
Rowlinson, *Liquids and Liquid Mixtures*, Butterworth, London (1959).
Sakurada, *Z. physikal Chem.*, **28B**, 104 (1935); Wohl, Pahlke and Wehago, *ibid.*,
    28B, 1 (1935); von Elbe, *J. Chem. Physics*, **2**, 73 (1934).
Scott, R. S. and D. V. Fenby, *Ann. Rev. Phys. Chem.*, **20**, 111 (1969).
Sherman, A. and E. A. Moelwyn-Hughes, *Trans. Chem. Soc.*, 101 (1936).
Stackelberg, M. von and H. R. Müller, *Z. Elektrochem.*, **58**, 25 (1954).
Uhlig, *J. Phys. Chem.*, **41**, 1215 (1937).
Valentiner, *Z. Physik*, **42**, 253 (1927).
Wynne-Jones, *Proc. Roy. Soc.*, **A140**, 440 (1933).
Yuan-tsan Chia, *J. Amer. Chem. Soc.*, **81**, 1280 (1959).

4

FUNDAMENTALS OF CHEMICAL KINETICS

The rate at which a molecule decomposes, or a pair of molecules react, is determined principally by intrinsic properties of the molecule or the reacting pair, and, with few exceptions cannot be anticipated theoretically but must be measured experimentally. Factors which, in the absence of light, are found to influence the instantaneous rate of any reaction in solution include the molecular concentration, c, at the instant of measurement, the pressure, P, and temperature, T, of the solution, and the nature of the solvent. It is assumed, of course, that the solution is free from impurities, of which the most insidious is dissolved air.

The Order of Reaction

The instantaneous rate of any isothermal process in dilute systems is found to be proportional to the product of the concentrations, $n_A, n_B, \ldots$, of the reacting species, raised, respectively, to small integral or fractional powers, $v_A, v_B, \ldots$. Thus if n_P denotes the concentration, at any time, of the products in the reaction $v_A A + v_B B + \ldots \rightarrow P$, the instantaneous rate of reaction is given by the equation

$$dn_P/dt = k_v n_A^{v_A} . n_B^{v_B} \ldots, \tag{1}$$

where k_v is the velocity constant, or the number of molecules of resultant produced in unit time and in unit volume when the concentration of each reactant is unity. The power to which the concentration of a reactant must be raised in order to reproduce the experimental rate of reaction is known as the order of reaction with respect to that reactant. The net, or over-all, order of reaction is the sum of the orders with respect to all the reactants:

$$v = v_A + v_B + \dots \tag{2}$$

Experiments show that v is usually 1 or 2 for reactions in solution.

The prevalence of bimolecular reactions is readily understood, for molecules must meet before they can react, and the rate at which they meet must be directly proportional to the product of their concentrations. Unimolecular reactions also are not difficult to understand, at least in the dilute gaseous phase, because the stability of molecules, especially complicated ones, is governed by changes in intramolecular separations and by exchanges of various forms of internal energies. These, being the concern of the molecule itself, do not require the presence of a partner.

The few known reactions of the third kinetic order all belong to the category $A + 2B \rightarrow P$, which suggests that the rate-determining step may not be a termolecular process $A + B + B \rightarrow P$, but a bimolecular reaction between molecules A and dimers B_2.

When a solute reacts with the solvent, as in hydrolysis and other solvolyses, it is found that v may be as high as 8, indicating that several molecules of solvent participate simultaneously in the process that governs the rate. This fact explains why certain esters which are readily hydrolysed in water may, at the same temperature and pressure, be distilled without decomposition in steam.

Order and Molecularity

If the generation of the final product of reaction entails a number of consecutive steps proceeding with noncommensurate velocities, the net rate of production is governed by the slowest step. The value of v for the participants in this rate-determining step is termed the molecularity of the reaction. The order of the reaction *para*-$H_2 \rightarrow$ *ortho*-H_2, determined by measuring the times of half-completion at various total pressures is 3/2 (Farkas, 1930), but the molecularity is 2, since the rate-determining step is the reaction between atoms of hydrogen and molecules of *para*-H_2: $H +$ *para*-$H_2 \rightarrow$ *ortho*-$H_2 + H$. The rapid steps which do not influence the net rate of reaction are the dissociation of molecules into atoms, and the reassociation of the atoms into molecules.

The most frequently occurring fractional order in solution is 3/2, as is to be expected if one of the participants in the rate-determining step is an ion or a radical formed from a slightly dissociated parent. In such a case, the concentration of ion or radical is nearly proportional to the square-root of the total concentration of the weakly dissociated solute.

The Mechanism of Reaction

When the principle of the slowest step is applied to empirically established kinetic expressions, we generally arrive at the mechanism of reaction, provided the slow step is not determined by diffusion. Let us consider, as our first example, the reaction between the thiosulphate ion and the molecules $C_2H_4I_2$, C_2H_4IBr and C_2H_4IBr in aqueous methanol. The instantaneous rate of the chemical change, $C_2H_4I_2 + S_2O_3^{--} \rightarrow C_2H_4IS_2O_3^- + I^-$, is proportional to the product of the concentrations of the organic molecule and the inorganic ion, suggesting that the substitution is a simple process whereby an iodine atom is ousted by a thiosulphate group in a binary encounter, e.g.,

$$d[I^-]/dt = k_2[C_2H_4I_2][S_2O_3^{--}].$$

When, however, the similar reaction $IC_2H_4Cl + S_2O_3^{--} \rightarrow C_2H_4ClS_2O_3^- + I^-$ is examined under the same conditions, it is found that the instantaneous rate of reaction is proportional to the concentration of the organic molecule only,

$$d[I^-]/dt = k_1[IC_2H_4Cl]$$

and is independent of the concentration of the inorganic ion (Slator, 1904, 1905). The explanation is evidently that the rate-determining step is a unimolecular change in the organic molecule, such as an ionization

$$IC_2H_4Cl \rightarrow C_2H_4Cl^+ + I^-,$$

followed by a rapid reaction between the organic cation and the thiosulphate ion. Similarly, the rate at which benzhydryl chloride reacts with various anions ($X = OH^-$, $C_2H_5O^-$ or $CHPg_2O^-$) in queous methanol $CHPh_2Cl + X^- \rightarrow CHPh_2X + Cl^-$ is independent of the concentration of X^- and is governed entirely by the concentration of benzhydryl chloride (A. M. Ward, 1927).

The same behaviour is met with in the oxidation of various disubstituted aminomethylsulphonic acids. A given acid is found to react with widely different oxidants at the same rate, the instantaneous value of which, dx/dt, is proportional to the concentration of the sulphonic acid but independent of the concentration of the oxidizing agent. With iodine, for example, the chemical change is $R_2\overset{+}{N}H\cdot CH_2\cdot SO_3^- + I_3^- + H_2O \rightarrow R_2\overset{+}{N}:CH_2 + SO_4^{--} + 3I^- + 3H^+$, but, denoting by a the initial concentration of the acid, the rate of reaction is found to obey the equation

$$-d(a - x)/dt = dx/dt = k_1(a - x),$$

which is seen to be independent of the concentration $(b - x)$ of the iodine.

To account for these facts, it has again been suggested (T. D. Stewart and W. E. Bradley, 1932) that the rate-determining step is the relatively slow ionization $R_2\overset{+}{N}H \cdot CH_2 \cdot SO_3^- \rightarrow R_2\overset{+}{N}:CH_2 + HSO_3^-$, followed by the rapid oxidation $HSO_3^- + I_3^- + H_2O \rightarrow SO_4^{--} + 3I^- + 3H^+$.

We next consider the reaction of acetone, in aqueous solutions of strong acids, with the halogens. The chemical change is $CH_3 \cdot CO \cdot CH_3 + X_2 \rightarrow CH_3 \cdot CO \cdot CH_2X + H^+ + X^-$, but the instantaneous rate of chemical reaction is directly proportional to the product of the concentrations of acetone and hydrogen ion, and is independent of the concentration of halogen. These facts have been explained (Lapworth, 1904; H. M. Dawson, 1909) by supposing that the rate-determining step is the reaction of hydrogen-ion with acetone, resulting in its enolization. This step may be represented here for clarity by using the deuteron instead of the proton: $CH_3 \cdot CO \cdot CH_3 + D^+ \rightarrow CH_3 \cdot C(OD):CH_2 + H^+$. There follows a rapid reaction of the enol with the halogen molecule $CH_3 \cdot C(OD):CH_2 + X_2 \rightarrow CH_3 \cdot CO \cdot CH_2X + D^+ + X^-$. The rate-determining step in these reactions is the bimolecular reaction between the proton or deuteron and the acetone molecule.

The reaction of the phenyldiazonium ion with water is represented stoichiometrically by the equation (Cain and Nicoll, 1902; Saunders, 1936; Moelwyn-Hughes and P. Johnson, 1940) $C_6H_5 \cdot N_2^+ + H_2O \rightarrow C_6H_5 \cdot OH + N_2 + H^+$. Here again, it is possible that the rate-determining step is the dissociation $C_6H_5 \cdot N_2^+ \rightarrow C_6H_5^+ + N_2$, followed by the rapid reaction of the phenyl ion with water. Such a mechanism is often postulated for hydrolyses. Without extra-kinetic evidence that such an ion as $C_6H_5^+$ exists, however, the alternative mechanism $C_6H_5N_2^+ + (n + 1)H_2O \rightarrow C_6H_5OH + N_2 + [H, nH_2O]^+$ cannot be dismissed. According to it, the rate-determining step is a reaction of order $n + 2$, in which one of the many water molecules that simultaneously attack the phenyldiazonium ion is split.

The Stationary State Hypothesis

Let us consider the conversion of reacting molecules A and B into product molecules, P, which, however, are not formed directly but *via* an intermediate complex, C, according to the scheme:

$$A + B \underset{k_1}{\overset{k_2}{\rightleftarrows}} C \overset{k_3}{\rightarrow} P.$$

We assume that the complex is formed bimolecularly and is decomposed unimolecularly into either the products P or its progenitors A and B. The rate at which reactant molecules decompose is

$$-\frac{dn_A}{dt} = -\frac{dn_B}{dt} = k_2 n_A n_B - k_1 n_C,$$

and the rate at which product molecules are formed is $+dn_P/dt = k_3 n_C$. Under steady conditions, i.e. after intermediates have ceased to accumulate, these rates must be equal. This can be true only when

$$n_C = \left(\frac{k_2}{k_1 + k_3}\right) n_A n_B.$$

This result may be obtained directly by equating to zero the rate of production of the intermediate complex, thus:

$$\frac{dn_C}{dt} = k_2 n_A n_B - k_1 n_C - k_3 n_C = 0. \tag{3}$$

After eliminating the term n_C, the rate of reaction is seen to be

$$\frac{dn_P}{dt} = \left(\frac{k_2 k_3}{k_1 + k_3}\right) n_A n_B. \tag{4}$$

The observed bimolecular constant is

$$k = \frac{1}{n_A n_B} \frac{dn_P}{dt} = \frac{k_2 k_3}{k_1 + k_3}. \tag{5}$$

Thus, although a given reaction may obey second-order kinetics throughout its course, and yield, at constant temperature and pressure, a bimolecular constant which is independent of initial concentrations, that constant is nevertheless a composite term.

When $k_3 \gg k_1$, $k = k_2$, and the rate of reaction is the rate at which the complex is formed. When $k_3 \ll k_1$, $k = (k_2/k_1)k_3$. The term in brackets represents the equilibrium constant governing complex formation in the absence of chemical change. These two interpretations are those given for ionic reactions by Bjerrum (1924) and Brönsted (1925) respectively.

The present application of the stationary state hypothesis to bimolecular reactions is merely a particular instance of the equation derived for determining stationary concentrations. Replacing the term n_C by a, we may express the general condition of attainment of the stationary state as

$$da/dt = 0. \tag{6}$$

One of the earliest applications is to enzyme-catalysed reactions (Michaelis and Menten, 1913). By allowing for the fact that in such systems the concentration of complex may be a significant fraction of the concentration, n_E, of enzyme, it is readily shown that the rate of change in the concentration of substrate is

$$-\frac{dn_S}{dt} = \frac{k_2 k_3 n_E n_S}{k_2 n_S + k_1 + k_2}. \tag{7}$$

At low concentration of substrate, the rate of reaction is proportional to its concentration. At high concentrations, however, it is independent of n_S, since

$$-dn_S/dt = k_3 n_E. \tag{8}$$

The order of reaction with respect to the substrate thus changes from 1 to 0 with increasing initial concentrations. k_3, the number of substrate molecules destroyed in unit volume and unit time by one enzyme molecule, has been termed the turnover number (Warburg, 1938).

There have been extensive applications of the stationary state hypothesis to photochemical reactions in gases (Bodenstein, 1913) and to thermal reactions exhibiting a change in kinetic order (F. A. Lindemann, 1922; C. N. Hinshelwood and H. W. Thompson, 1926).

To understand the latter phenomenon, the postulated intermediates are regarded as molecules of the reacting gas which have, in a system at constant temperature, sufficient energy above the average value to enable them to undergo chemical change. The concept of such active molecules is due to Arrhenius (1889). Now the redistribution of energy in gaseous systems, i.e. the gain and loss suffered by individual molecules, must depend on collisions, so that, unless light is absorbed or emitted, the rates at which molecules are activated and deactivated are bimolecular processes. If, in addition, each active molecule has a chance of undergoing chemical change, we may formulate the reaction scheme, in terms of n, the concentration of normal molecules, and of a, that of active molecules, follows:

$$A + A \underset{k_4}{\overset{k_2}{\rightleftarrows}} A + A^* \overset{k_3}{\to} P$$

$$n \quad n \quad n \quad a \quad p$$

When the concentration of active molecules has reached its stationary value

$$da/dt = k_2 n^2 - k_4 na - k_3 a = 0, \tag{9}$$

then

$$a = \frac{k_2 n^2}{k_4 n + k_3},$$

and

$$-\frac{dn}{dt} = \frac{dp}{dt} = k_3 a = \frac{k_3 k_2 n^2}{k_4 n + k_3}. \tag{10}$$

As with the enzyme system (equation 5), we again have a first-order reaction $-(dn/dt) = k_3(k_2/k_4)n$, at high concentrations, and a second-order reaction $-(dn/dt) = k_2 n^2$, at low concentrations. When there is no chemical change,

$k_3 = 0$ and

$$a/n = k_2/k_4. \tag{11}$$

This equation is consistent with Boltzmann's law, according to which the fractional number of molecules with a specified energy each is constant in a system at equilibrium.

The Equation of Arrhenius

The basic idea of Arrhenius' theory is that only energized or activated molecules undergo chemical change. To formulate it, we combine the experimental relationships

$$K = k_1/k_2 \tag{2.40}$$

and

$$\Delta H^0 = RT^2 \left(\frac{d \ln K}{dT} \right)_P, \tag{2.9}$$

obtaining

$$RT^2 \left(\frac{d \ln K}{DT} \right)_P = RT^2 \left(\frac{d \ln k_1}{dT} \right)_P - RT^2 \left(\frac{d \ln k_2}{dT} \right)_P. \tag{12}$$

The terms on the right are the apparent, or Arrhenius, activation energies, usually denoted by E_A. Then

$$\Delta H^0 = E_{A,1} - E_{A,2}. \tag{13}$$

We shall be content with one experimental verification of this equation. It refers to the unimolecular conversion of urea in aqueous solution to ammonium cyanate, and to the reverse bimolecular reaction. We have, in cal/mole:

$(NH_2)_2CO \rightarrow NH_4^+ + CNO^-$ $E_{A,1} = 31,300$ (Burrows and Fawsitt, 1914)

$NH_4^+ + CNO^- \rightarrow (NH_2)_2CO$ $E_{A,2} = 24,000$ (Walker and Hambly, 1895)

ΔH^0 (kinetic) $= E_{A,1} - E_{A,2}$ 7,300

ΔH^0 (from equation 2.9)   7,600 (Lineken and Burrows, 1929)

ΔH^0 (calorimetric)   7,400 (Walker, 1902).

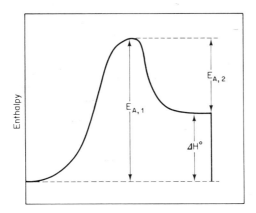

FIG. 1. The relationship between apparent energies of reaction and standard change in heat content.

The relationship between the standard change in heat content and the apparent energies of activation is shown schematically in Fig. 1. When the two energies of activation are equal, as in perfectly symmetrical systems, ΔH^0 is zero. When the energies are nearly equal, as in mutarotations, ΔH^0 is small. Finally, when $E_{A,2}$ is zero, the energy of activation for the forward reaction is equal to the increase in heat content. Equation (13) may be used to acquire a knowledge of any one of the terms from a knowledge of the other two.

By combining equations (12) and (13), we can write

$$\left[RT^2\left(\frac{d \ln K}{dT}\right)_P - \Delta H^0 \right] = \left[RT^2\left(\frac{d \ln k_1}{dT}\right)_P - E_{A,1} \right]$$
$$- \left[RT^2\left(\frac{d \ln k_2}{dT}\right)_P - E_{A,2} \right].$$

The term on the left hand side is zero by definition. Arrhenius argued that each of the bracketed terms on the right hand side is also zero, in which case we could write for the forward and reverse reactions separately two independent equations of the type

$$E_A = RT^2\left(\frac{d \ln k}{dT}\right)_P, \tag{14}$$

which is the differential and general form of the Arrhenius equation. We may also use its algebraic equivalent

$$E_A = -R\frac{d \ln k}{d(1/T)_P}, \tag{15}$$

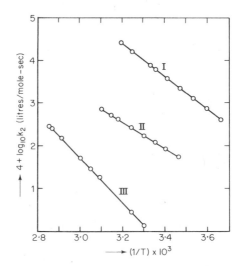

FIG. 2. Experimental determination of the apparent energy of activation.
I. $N_2CH \cdot COOC_2H_5 + H_3O^+$ in aqueous solution. $E_A = 17,500$ cal (P. Johnson and Moelwyn-Hughes, 1941).
II. $H_2O_2 + I^-$ in aqueous solution. $E_A = 13,400$ (Liebhafsky and Mohammad, 1933).
III. $CH_3Cl + CH_3O^-$ in methanol solution. $E_A = 23,690$ cal (Hurst and Moelwyn-Hughes, 1957).

which makes it clear that the apparent energy of activation of any reaction may be found by plotting the natural logarithm of the velocity constant against the reciprocal of the absolute temperature. Some examples are shown in Fig. 2. In these instances, the gradients are linear, and the apparent energies of activation are therefore constant within the temperature range explored. Within such limits, equation (14) may be integrated to give

$$k = A \cdot \exp(-E_A/RT), \qquad (16)$$

which is a much used but more restricted form of the Arrhenius equation. It was established by Hood (1878, 1885) and hotly contested by Harcourt and Esson (1895, 1913). Its form is reminiscent of earlier equations summarizing the effect of temperature on vapour pressure, thermionic current and other properties which depend on the presence of particles possessing energies above the average.

When E_A is found to be dependent on temperature, we may express its value at a temperature T as follows:

$$E_A = E_0 + \int_0^T \Delta C \, dT, \qquad (17)$$

where E_0 is the energy of activation extrapolated to the absolute zero, and

ΔC is the excess molar heat capacity of the active reactants over that of the normal reactants. If ΔC proved to be independent of temperature, integration of equation (14) leads to the more general equation

$$k = B \cdot T^{\Delta C/R} \cdot \exp\left(-E_0/RT\right), \tag{18}$$

as derived by Trautz (1909) and previously verified by Kooij (1893). We shall find that many reactions in solution conform to this equation.

Theories of Chemical Kinetics

The range of topics covered by chemical kinetics is so vast that they have naturally been treated in many different ways. The differences between them are seldom fundamental but arise from differences in emphases and, particularly, in notation. For purposes of discussion, we shall briefly examine: (1) the quasi-thermodynamic theory, (2) the statistical theory, (3) the transition-state theory, and (4) the collision theory.

(1) THE QUASI-THERMODYNAMIC FORMULATION OF KINETIC EXPRESSIONS

The basis of this formulation is that of Arrhenius, but here we recognize the difference between total energy, E, and free energy, G. Let the forward reaction proceed with velocity constant k_i and the reverse reaction with velocity constant k_j. Then the equilibrium constant is

$$K = k_i/k_j. \tag{2.40}$$

The van't Hoff isotherm is

$$\Delta G^0 = \Sigma G_j^0 - \Sigma G_i^0 = -RT \ln K. \tag{2.7a}$$

Here G^0 stands for the partial molar free energy of each reactant and resultant in a particular reference state, usually that of unit pressure or of unit concentration as a constant pressure, P, and temperature T. On combining these equations, we have

$$\Sigma G_j^0 - \Sigma G_i^0 = -RT \ln k_i + RT \ln k_j. \tag{19}$$

Molecules on changing from an initial state to a final state pass through an intermediate state of higher free energy (Fig. 3). Then let the partial molar free energy of the activated molecules in this state be denoted by $^0G^*$, which may be added to and subtracted from equation (19) to give $(^0G^* - \Sigma G_i^0) - (^0G^* - G_j^0) = -RT \ln k_i + RT \ln k_j$. As Arrhenius intuitively suggested, such an equation can be resolved as follows: $-RT \ln k_i = (^0G^* - \Sigma G_i^0) +$

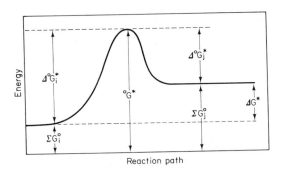

FIG. 3. Schematic representation of the standard free energies (ΣG_i^0 and ΣG_j^0) of reactants and resultants, the standard free energies of activation ($\Delta^0 G_i^*$ and $\Delta^0 G_j^*$) for the direct and reverse reactions, and the standard free energy change (ΔG^0) associated with the chemical change: $\Delta G^0 = \Sigma G_j^0 - \Sigma G_i^0 = \Delta^0 G_i^* - \Delta^0 G_j^*$.

$W = \Delta^0 G_i^* + W$ and $-RT \ln k_j = ({}^0 G^* - \Sigma G_j^0) + W = \Delta^0 G_j^* + W$, or generally for a reaction of kinetic order n,.

$$-RT \ln k_n = ({}^0 G^* - \Sigma G_{\text{reactants}}^0) + W = \Delta^0 G^* + W. \qquad (20)$$

The term in brackets is the standard free energy of activation, and was regarded by Arrhenius as independent of temperature. If that were true, free energies become replaceable by total energies, which were the terms employed by him. It was left to later workers to make the correction (P. Kohnstamm and F. E. C. Scheffer, 1911; La Mer, 1933; Soper, 1935).

There is no general agreement concerning the term W. It has been suggested that, for unimolecular reactions, it is a function of $\Delta^0 G^*$ (Dushman, 1924; Rodebush, 1923). Dynamical, as distinct from thermodynamical, reasoning indicates that for unimolecular reactions

$$W = -RT \ln \nu, \qquad (21)$$

where ν has the dimensions of frequency. Equation (20) then becomes

$$k_1 = \nu . \exp(-\Delta^0 G^*/RT)$$
$$= \nu . \exp(\Delta^0 S^*/R) . \exp(-\Delta^0 H^*/RT). \qquad (22)$$

Since the molar free energy of activation in unimolecular reactions is

$$\Delta^0 G^* = -RT \ln(k_1/\nu), \qquad (22)$$

the entropy and enthalpy of activation, respectively, are

$$\Delta^0 S^* = R \left\{ \ln(k_1/\nu) + \left[\frac{d \ln(k_1/\nu)}{d \ln T} \right]_P \right\} \qquad (23)$$

and

$$\Delta^0 H^* = RT^2 \left(\frac{d \ln k_1}{dT}\right)_P - RT^2 \left(\frac{d \ln v}{dT}\right)_P. \tag{24}$$

The heat of activation thus defined consists of two parts: (1) the excess standard heat content of the reactive molecules or complex over the sum of the standard heat contents of the reactants, and (2) the gain in heat content arising from the temperature dependence of the frequency term, v. Because v for reactions in solution may depend on many properties of the solvent, the latter contribution is seldom easy to disentangle from the total effect.

The Arrhenius energy of activation, from the general equation (14), is

$$E_A = \Delta^0 H^* + RT(d \ln v/d \ln T)_P. \tag{25}$$

The pre-exponential term in the integrated form of the Arrhenius equation (16) is thus given the following interpretation:

$$A = v . \exp(\Delta^0 S^*/R) . \exp[(d \ln v/d \ln T)_P]. \tag{26}$$

To obtain an expression for the temperature coefficient of the apparent energy of activation, expression (25) is to be differentiated with respect to T:

$$\frac{dE_A}{dT} = \Delta^0 C_P^* + \frac{d}{dT}\left[RT^2\left(\frac{d \ln v}{dT}\right)\right]_P. \tag{27}$$

Let us consider the possibility that v varies inversely as the viscosity, η. Then

$$\frac{dE_A}{dT} = \Delta^0 C_P^* + \left(\frac{dB}{dT}\right)_P, \tag{28}$$

where B is defined by equation (1.32). It is shown in Appendix 3 that the term dB/dT in aqueous solution at 25°C is -32.2 cal/mole^{-1}-deg^{-1}, which is a significant fraction of dE_A/dT.

(2) THE STATISTICAL THEORY OF CHEMICAL KINETICS

To calculate from first principles the absolute magnitude of the velocity of a chemical reaction is an ambitious project which, up to the present, has been achieved only for certain simple reactions, involving molecules containing not more than three or four atoms each undergoing chemical change in dilute gaseous systems. Statistical mechanics is applied to determine the equilibrium constant for association reactions of the type $A + B + C + \cdots \rightleftharpoons ABC \ldots$. The probability per second that the critically activated complex will break down is identified with one particular frequency, real or imaginary, of the various internal vibrations. The rate constant for the reaction then becomes the product of the equilibrium constant and the break-down frequency. Before the method can be called an absolute one, it is essential to calculate the energy of activation, which is rendered possible in simple systems by Eyring and Polanyi's (1931) adaptation of a wave-mechanical treatment due to London (1929) and by later methods (R. E. Weston, 1959; F. T. Wall and R. N. Porter, 1962).

We have seen that the equilibrium constant governing the formation of a stable complex $ABC\ldots$ from molecules or atoms, $A, B, C, \ldots$ in a system at constant volume is given by the expression

$$K = \frac{c_{ABC\ldots}}{c_A c_B c_C \cdots} = \frac{q_{ABC\ldots}}{q_A q_B q_C \cdots} \cdot \exp\left(-\Delta E_0/RT\right), \tag{2.17}$$

where the q term for each species is related as follows to its partition function, f, its equilibrium concentration, c, and the base of the natural logarithm, e:

$$q_i = c_i f_i / e. \tag{2.15}$$

ΔE_0 is the energy of the complex, less the sum of the energies of the reactants at the absolute zero of temperature. If the complex is unstable and the probability per second that it will split into products is v^*, the rate at which the products are formed is

$$\frac{dc_P}{dt} = -\frac{dc_{ABC\ldots}}{dt}$$

$$= c_{ABC\ldots} v^*$$

$$= c_A c_B c_C \ldots K v^*$$

but the velocity constant, k_n, for a reaction of the nth order is defined by the equation

$$\frac{dc_P}{dt} = k_n c_A c_B c_C \ldots . \tag{2.1}$$

Hence

$$k_n = K v^* = \frac{q_{ABC\ldots}}{q_A q_B q_C \cdots} \cdot \exp\left(-\Delta E_0/RT\right) \cdot v^*. \tag{29}$$

which is the desired relation.

The equilibrium constant governing activation in a single bond is given by equation (2.26), which reduces, when hv^*/kT is negligibly small, to

$$K = \frac{n^*}{n} = \left(\frac{r^*}{r}\right)^2 [1 - \exp\left(-hv/kT\right)] \frac{kT}{hv^*} \cdot \exp\left(-\Delta E_0/RT\right).$$

Hence the unimolecular constant is

$$k_1 = \left(\frac{r^*}{r}\right)^2 [1 - \exp(-hv/kT)]\frac{kT}{h} \cdot \exp(-\Delta E_0/RT), \tag{30}$$

or if hv/kT is large, and we ignore the difference between r and r^*,

$$k_1 = \frac{kT}{h} \cdot \exp(-\Delta E_0/RT). \tag{31}$$

These expressions were originally derived, in another way, by Herzfeld (1919). It follows that $E_A = E_0 + RT$, and that

$$k_1 = \frac{kTe}{h} \cdot \exp(-E_A/RT). \tag{32}$$

At 25°C, $kTe/h = 1\cdot689 \times 10^{13}$ sec^{-1}, which often tallies, as far as order of magnitude is concerned, with experimental values of the pre-exponential term in the Arrhenius equation

$$k_1 = A_1 \cdot \exp(-E_A/RT). \tag{16}$$

Herzfeld's equation has formed the starting point for more elaborate treatments of unimolecular reactions.

Similarly, for reactions between atoms and diatomic molecules, we have, from equation (2.34), the following expression for the bimolecular constant:

$$k_2 = \left(\frac{v^*}{v_{AB}}\right)r_{AB}^2\left(\frac{8\pi kT}{\mu_{AB}}\right)^{1/2} \cdot \exp(-\Delta E_0/RT), \tag{33}$$

which, if the break-down frequency v^* is identified with the vibration frequency of the molecule, is precisely the expression of the collision theory, as Eyring (1935) pointed out. Herzfeld, in fact, used equation (33) to derive equation (31).

We follow the same treatment to derive the theoretical expression for the bimolecular velocity constant governing the reaction between atoms, A, and diatomic molecules, BC. Using equation (2.38), we have, following Pelzer and Wigner (1932),

$$k_2 = \frac{1}{4}\left(\frac{h}{2\pi}\right)^2\left(\frac{2\pi}{kT}\right)^{1/2}\left[\frac{m_A + m_B + m_C}{m_A(m_B + m_C)}\right]^{3/2}\frac{I_{ABC}}{I_{AB}}\frac{\sigma_{BC}}{\sigma_{ABC}}$$

$$\times \frac{\sinh \beta v_{BC}}{\sinh \beta v_s \sinh^2 \beta v_\phi} \cdot \exp(-\Delta E_s/RT). \tag{34}$$

This equation is in complete agreement with all experimental rate constants of the many reactions of this type occurring in the gaseous phase. ΔE_s is the energy of the complex, less the sum of the energies of the atom and molecule,

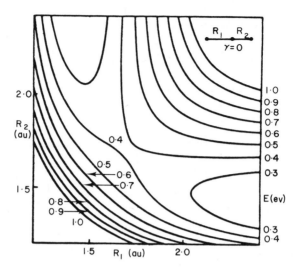

Fɪɢ. 4. Saddle-point detail for the linear configuration H···H···H (Porter and Marples, 1964). Energies in electron-volts.

each regarded as static (i.e. deprived of their residual energies) at the absolute zero of temperature.

In order to calculate ΔE_s by the wave-mechanical procedure mentioned, the potential energy of the linear triatomic system is plotted in a three-dimensional diagram in which two of the variables are the interatomic distances. There results the now familiar three-dimensional picture of an energy mountain traversed by a path which ascends as the complex is formed, and descends when it breaks up. The height of the col is identified as the energy of activation. Eyring and his collaborators (Hirschfelder, Topley and Eyring, 1936) find the mountain contour in the saddle region to have the shape of a shallow basin, suggesting the existence of a relatively stable critical complex. On the other hand, Farkas and Wigner (1936) and Moelwyn-Hughes (1936) independently conclude that the shape of the contour at the saddle point is that of an inverted saucer, indicating an imaginary vibration frequency, as in Fig. 2.2. Recent calculations (Porter and Marples, 1964) confirm this view, as may be seen from Fig. 4.

(3) Tʜᴇ Tʀᴀɴsɪᴛɪᴏɴ-sᴛᴀᴛᴇ Tʜᴇᴏʀʏ

Eyring (1935) has proposed the adoption of Herzfeld's expression, kT/h, as a general one for chemical reactions of all orders, to replace the specific break-down frequency, v^*, of the previous section. The thermodynamic-like terms in the resulting expression

$$k_n = \frac{kT}{h} \cdot \exp(-\Delta G^{\ddagger}/RT) = \frac{kT}{h} \cdot \exp(\Delta S^{\ddagger}/R) \cdot \exp(-\Delta H^{\ddagger}/RT) \qquad (35)$$

are referred to, respectively, as the free energy, entropy and enthalpy of activation. The last two variables are seen to be related as follows to the empirical parameters of the Arrhenius equation:

$$\Delta H^{\ddagger} = E_A - RT, \qquad (36)$$

$$\Delta S^{\ddagger} = R \ln [A/(kTe/h)]. \qquad (37)$$

These equations may be regarded as special forms of equations (22), (24) and (23), respectively, obtained by substituting kT/h for v. According to this convention we may derive, from the first empirical parameter of the Arrhenius equation, the gain in entropy attending the formation of an activated complex, and thus, given the absolute entropy of the reactants, obtain the defined entropy of the activated complex. This should sharpen our insight into many properties of molecules in the transitory state through which they pass during chemical reaction—the so-called transition state. M. G. Evans and M. Polanyi (1935), in a similar treatment, have dealt with the role of various thermodynamic factors in chemical kinetics. Wynne-Jones and Eyring (1935) have applied equation (35) to reactions of various orders in condensed systems. Doubts have been expressed concerning the validity of applying equation (35) directly to bimolecular reactions (Moelwyn-Hughes, 1936); see, however, H. Eyring and W. F. K. Wynne-Jones (1936), for which a simpler means of obtaining the entropy of activation is available. Equation

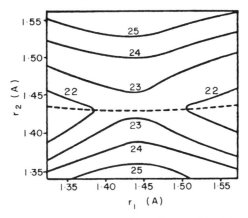

Fig. 5. Saddle-point detail for the linear configuration Br···H···H (Moelwyn-Hughes, 1936). Energies in kilocalories.

(35) has nevertheless been extensively used, especially by physical-organic chemists, to reactions of various orders. We shall in this work restrict its application to reactions of the first kinetic order.

The Entropy of Activation of Unimolecular Reactions

Herzfeld's equation for the unimolecular velocity constant,

$$k_1 = \frac{kT}{h} \cdot \exp\left(-\Delta E_0/RT\right) = \frac{kTe}{h} \cdot \exp\left(-E_A/RT\right), \qquad (32)$$

has long been accepted as providing a reasonable explanation of the magnitude of the term A_1 in the empirical equation

$$k_1 = A_1 \cdot \exp\left(-E_A/RT\right). \qquad (16)$$

At 25°C, $kTe/h = 1.689 \times 10^{13}$ sec^{-1}. Unimolecular reactions with values of A_1 lying near to this value may be regarded as proceeding with normal velocities. Deviations from equation (32) are naturally to be expected for all but the simplest reactions. Some idea of their origin can be gained from the form of equation (30), which was used in its derivation. The two terms there omitted, and all other terms responsible for departures from Herzfeld's simple equation have been expressed by Eyring in the form of $\exp\left(\Delta S^{\ddagger}/R\right)$, where $\Delta S^{\ddagger}$ is the entropy of activation as defined by equation (37). A few examples are given in Table I. Considering the undoubted complexity of many of these reactions, the simplicity of the theory, and a range of rate

TABLE I

ENTROPIES OF ACTIVATION FOR SOME UNIMOLECULAR REACTIONS

Reaction	Medium	A sec^{-1}	E_A kcal	$\Delta S^{\ddagger}$ cal/mole-deg
$ClCOOCCl_3 \rightarrow 2COCl_2$	gas	1.4×10^{13}	14.5	−0.04
	CCl_4	3.1×10^{13}	24.2	1.15
$N_2O_5 \rightarrow 2NO_2 + \frac{1}{2}O_2$	gas	3.7×10^{13}	24.7	2.01
	Br_2	1.9×10^{13}	24.0	0.28
Mutarotation of beryllium benzoyl camphor $\alpha \rightarrow \gamma$	CCl_4	1.1×10^{12}	19.0	−5.36
$\gamma \rightarrow \beta$	CCl_4	2.5×10^{14}	22.9	+5.36
$PhMe_3NI \rightarrow PhNMe_2 +$ MeI	$PhNO_2$	2.0×10^{15}	28.0	9.33
$(CH_3)_2O \rightarrow CH_4 +$ HCHO	gas	2.4×10^{13}	58.5	0.70

constants of about 10^{32}, the values of $\Delta S^{\ddagger}$ are remarkably near to zero, more especially for chemical changes in the gas phase and in non-polar solvents.

(4) THE COLLISION THEORY

Unlike the foregoing theories, the present theory applies chiefly to bimolecular reactions. It is based on the kinetic theory of gases, as applied to incompressible spheres of radii r_A and r_B, moving in a force-free field. If we define a collision as the appearance of the centre of one molecule within a distance $(r_A + r_B)$ of another molecule, the total number of collisions made per cc per second with relative velocities, resolved along the line of centres exceeding a value v_0 is found to be

$$_A Z_B = n_A n_B (r_A + r_B)^2 \left[8\pi k T \left(\frac{1}{m_A} + \frac{1}{m_B} \right) \right]^{1/2} . \exp\left(-E/RT \right),$$

where $E/N_0 = \frac{1}{2}mv_0^2$, n denotes molecular concentration, and m molecular mass. If the molecules react when they collide with an energy not less than E, the instantaneous rate of reaction is $-(dn_A/dt) = -(dn_B/dt) = {_A}Z_B$, and the bimolecular constant becomes (in the units of cc per mole-sec)

$$k_2 = (r_A + r_B)^2 (8\pi k T/\mu)^{1/2} . \exp\left(-E/RT \right).$$

Most bimolecular constants are given in the units of litres per mole-sec. In these units

$$k_2 = \frac{N_0}{1,000}(r_A + r_B)^2 \left(\frac{8\pi k T}{\mu} \right)^{1/2} . \exp\left(-E/RT \right)$$

$$= Z^0 . \exp\left(-E/RT \right) = Z^0 . \exp\left(1/2 \right) . \exp\left(-E_A/RT \right), \quad (38)$$

where Z^0 is referred to as the standard collision frequency. This equation was applied to data on bimolecular reactions in gases by M. Trautz (1916) and W. C. McC. Lewis (1924), who found the term A or the Arrhenius equation to be reproduced satisfactorily by the term Z^0 calculated using reasonable molecular radii. There are many reactions in solution to which the same remark applies. Some representative figures are given in Table II. We note that the mean value of A_2 is roughly 3.5×10^{10} litres/mole-sec for these reactions, and that $\log_{10} A$ lies in the range 10.5 ± 0.5. Its approximate constancy arises from the compensating effects of mass and diameter on the collision frequency. The empirical constants of the integrated form of the Arrhenius equation are seen to be:

$$E_A = E + (1/2)RT, \quad (39)$$

TABLE II
$$k_2 = A \cdot \exp(-E_A/RT)$$

Reaction	Medium	$A \times 10^{-10}$ (liter/ mol-sec)	E_A (cal/mol)	k_2 (298·16° K)
* $H + HBr \rightarrow H_2 + Br$	gas	1·35	1,090	$2·14 \times 10^9$
† $CO_2 + OH^- \rightarrow HCO_3^-$	H_2O	1·50	9,060	$3·41 \times 10^3$
‡ $2NOBr \rightarrow 2NO + Br_2$	gas	4·15	13,880	$2·74 \times 10^0$
§ $CH_3Br + I^- \rightarrow$ $CH_3I + Br^-$	$(CH_3)_2CO$	1·15	14,340	$3·55 \times 10^{-1}$
‖ $1:3:5Cl(NO_2)_2C_6H_3 +$ $I^- \rightarrow 1:3:5I(NO_2)_2C_6H_3 +$ Cl^-	$C_2H_4(OH)_2$	3·50	28,000	$1·03 \times 10^{-10}$
¶ $2HI \rightarrow H_2 + I_2$	gas	9·17	44,450	$2·30 \times 10^{-22}$

* Bodenstein and Lind, *Z. physikal. Chem.*, **57**, 168 (1907).
† Brinkman, Margaria and Roughton, *Phil. Trans.*, **232**, 65 (1933).
‡ Bodenstein and Krauss, *Z. physikal. Chem.*, **175**, 295 (1936).
§ Moelwyn-Hughes, *Trans. Faraday Soc.*, **45**, 167 (1949).
‖ Bennett and (Miss) Vernon, *Trans. Chem. Soc.*, 1783 (1938).
¶ Bodenstein, *Z. physikal. Chem.*, **29**, 295 (1889).

and

$$A_2 = Z^0 \cdot \exp(1/2). \tag{40}$$

Bimolecular reactions with rates given by the Trautz–Lewis equation are said to proceed with normal velocities. Accepting the logarithmic mean value given above, we may summarize as the general equation for normal bimolecular reactions:

$$k_2 \text{ (litres/mole-sec)} = 3·5 \times 10^{10·5 \pm 0·5} \cdot \exp(-E_A/RT). \tag{41}$$

Within the limits of experimental error, this equation correctly accounts for bimolecular reactions with velocity coefficients differing by a factor of 6×10^{31} under standard conditions. Thus the equation of Trautz and Lewis, despite its simplicity, affords a reasonable basis for discussions on bimolecular reactions. A statistical analysis of A_2 values available in 1933 for reactions in solution showed the most probable value of $\log_{10} A_2$ to lie in the region $11 \pm 0·5$ (Fig. 6). Equation (38), therefore, is not only quantitatively satisfactory for a wide range of many bimolecular reactions but is, in fact, the equation most likely to emerge for any bimolecular reaction in solution, chosen at random.

Numerous bimolecular reactions in the gas phase and in solution proceed with velocities differing by a factor P from those predicted by equation (38).

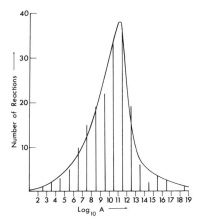

FIG. 6. Determination of the most probable value of the Arrhenius variable A for bimolecular reactions in solution (in 1947). From "The Kinetics of Reactions in Solutions, 2nd ed., by permission of The Clarendon Press, Oxford.

Empirically, therefore, we may write

$$k_2 = PZ^0 . \exp(1/2) . \exp(-E_A/RT). \tag{42}$$

When P exceeds or is less than unity, bimolecular reactions are said to be fast or slow, respectively. Some writers prefer to express this number in the form

$$P = \exp(\Delta S^*/R), \tag{43}$$

where ΔS^* is referred to as the entropy of activation. Equation (42) then becomes

$$k_2 = Z^0 . \exp(1/2) . \exp(\Delta S^*/R) . \exp(-\Delta E_A/RT). \tag{44}$$

Attempts have been made to resolve P on a dynamical basis into a factor p, intended to represent the probability of a correct phase of internal motion, a factor o, to allow for the need of a particular molecular orientation, and a factor s attributed to a steric hindrance occasioned by the presence of large groups adjacent to the point of attack. Then $P = pos$. Views differ concerning the order of magnitude of each factor. It is here contended that o is not likely to differ widely from unity, and that the steric effect manifests itself chiefly as an increase in the energy of activation, rather than a decrease in the effective collision frequency. The internal phase contribution may well prove to be important in some reactions, as suggested in Chapter 16.

To explain large values of P is not difficult. (1) If reaction proceeds by a chain mechanism, P may include the chain length, $(1 - \alpha)^{-1}$, where α allows for the fact that not every collision between active product and normal reactant molecules which leads to deactivation is necessarily fruitful in generating active molecules of reactants (Christiansen and Kramers, 1923). When α lies near to unity, the chain length $1/(1 - \alpha)$, becomes large, as in the oxidation of the sulphite ion by oxygen in aqueous solution, where it varies from 10^4 to 10^5 (Bäckström and Beatty, 1931). The operation of a chain mechanism is readily detected, and the fact that there are relatively few long-chain reactions in solution makes it evident that this cannot provide a general explanation for large values of P. (2) If the energy of activation is not restricted, as hitherto, to two quadratic terms, but must be expressed as the sum of $2s$ quadratic terms, the fraction of activated molecules is increased, and the bimolecular constant becomes

$$k_2 = Z^0 . \exp(-E/RT)\left[\frac{(E/RT)^{s-1}}{(s-1)!} + \frac{(E/RT)^{s-2}}{(s-2)!} + \cdots + 1\right]. \tag{45}$$

The expression in square brackets has been derived by Berthoud (1913). Simple derivations are given in Chapter 10. It is often sufficient to neglect all terms but the first term in the rapidly diminishing series shown inside the square brackets. Then

$$k_2 = Z^0 . \exp(-E/RT) . \frac{(E/RT)^{s-1}}{(s-1)!}. \tag{46}$$

It follows, since Z^0 is proportional to $T^{1/2}$, that

$$E_A = E - (s - \tfrac{3}{2})RT. \tag{47}$$

The Arrhenius energy of activation thus decreases linearly with respect to temperature, as is found in numerous reactions. For large values of s, we may approximate further, and write

$$k_2 = Z^0 . \exp(-E_A/RT) . \exp(1/2) . \left[\frac{E}{(s-1)RT}\right]^{s-1}, \tag{48}$$

from which we see that

$$A_2 = Z^0 . \exp(1/2) . \left[\frac{E}{(s-1)RT}\right]^{s-1}. \tag{49}$$

Let us consider a numerical example. For the hydrolysis of the disaccharide trehalose, catalysed by hydrogen ion in water, we find, in litres per mole-sec (Moelwyn-Hughes, 1929)

$$k_2 = 2.51 \times 10^{20} . \exp(-40,180/RT).$$

With any reasonable estimate of Z^0, it is clear that the term P, or $A_2/Z^0 e^{1/2}$, has the order of magnitude 10^{10}. The last term of equation (48), with

$E/RT = 100$, and $s = 12$, is 3.1×10^{10}. The entropy of activation is clearly

$$\Delta S^* = R(s - 1) \ln \left[\frac{E}{(s - 1)RT} \right]$$

$$= 227 \, \text{cal/mole-deg.} \tag{50}$$

In the second example, the bimolecular reaction between methyl iodide and trimethylamine at infinite dilution in carbon tetrachloride solution (Fahim and Moelwyn-Hughes, 1956) is found to be, in the same units

$$k_2 = 1\cdot33 \times 10^4 . \exp(-9,760/RT). \tag{51}$$

Clearly it is not necessary to appeal here for the participation of more than two quadratic terms, and we are left with an approximate estimate of 10^{-7} for P. This term is discussed in Chapter 16.

Christiansen included in the term P of equation (42) a factor to allow for the deactivation of activated solutes by molecules of the solvent and by various solute species. It was at one time thought that such deactivations were numerous and that, consequently, the rates of bimolecular reactions in solution would be far less than their rates in the gas phase. To settle the issue, several direct studies were carried out on the rates, in both phases, of reactions which, though not all uncomplicated bimolecular processes, seemed to have rates determined by binary collisions. They include the decomposition of chlorine monoxide (Hinshelwood and Prichard, 1923; Moelwyn-Hughes and Hinshelwood, 1931), the chlorine-sensitized decomposition of ozone (Bodenstein, Padelt and Schumacher, 1929; Bowen, Moelwyn-Hughes and Hinshelwood, 1931), the iodine-atom catalysed decomposition of ethylene diiodide (Cuthbertson and Kistiakowsky, 1935; Polissar, 1930), the ortho-para conversion of hydrogen catalysed by oxygen molecules (A. Farkas, 1935), numerous Diels–Alder reactions (Wassermann, 1936), the exchange reaction between bromine and bromtrichlormethane (A. A. Miller and J. E. Willard, 1949), and the quenching of the fluorescence of β-naphthylamine by carbon tetrachloride (H. G. Curme and G. K. Rollefson, 1952). All the investigators agree that A_2 is practically the same for a given reaction in all systems. On purely experimental grounds, therefore, equation 38 can be accepted as correct, in regard to order of magnitude, for collisions in solution.

The Entropy of Activation of Bimolecular Reactions

The entropy of activation of bimolecular reactions according to the collision theory is defined by the equation (Moelwyn-Hughes, 1934)

$$k_2 = Z^0 \exp(1/2) . \exp(\Delta S^*/R) . \exp(-E_A/RT) \tag{42}$$

and according to Eyring's formulation of the transition state theory by the equation

$$k_2 = \frac{kTe}{h} \cdot \exp\left(\Delta S^{\ddagger} R\right) \cdot \exp\left(-E_A/RT\right) \qquad (36, 37)$$

In terms of the empirical parameter of the equation $k_2 = A_2 \exp \cdot (-E_A/RT)$ the entropies of activation so defined are seen to be

$$\Delta S^* = R \ln\left(A_2/Z^0 \exp\left(1/2\right)\right) \qquad (52)$$

and

$$\Delta S^{\ddagger} = R \ln\left(A_2 h/kTe\right). \qquad (53)$$

Neither definition is wholly satisfactory. According to the collision theory, bimolecular reactions which take place with unit efficiency for each activating collision reveal no entropy of activation. According to the transition state theory, zero activation entropy is to be found when A_2, expressed in litres per mole-second, equals $kTe/h \, \sec^{-1}$. The difference is

$$\Delta S^{\ddagger} - \Delta S^* = R \ln\left[\frac{Z^0 \cdot \exp\left(1/2\right)}{(h/kTe)}\right]. \qquad (54)$$

Using the geometric mean value of the numerator, we have, at $298.16°\mathrm{K}$,

$$\Delta S^{\ddagger} - \Delta S^* = -11\cdot3 \, \text{cal/mole-deg}. \qquad (55)$$

It is instructive to derive and interpret the numerical value of the collision theory factor P which corresponds to a zero value of $\Delta S^{\ddagger}$. From equation (53), zero value of $\Delta S^{\ddagger}$ means the equality $A_2 h = kTe$, and from equation (42) we have $A_2 = PZ^0 e^{1/2}$. The required value of P is therefore

$$P = \frac{(kTe/h)}{Z^0 \cdot \exp\left(1/2\right)}$$

$$= \frac{(kT \cdot \exp\left(1/2\right)/h)}{(N_0/1{,}000)(r_A + r_B)^2(8\pi kT/\mu_{AB})^{1/2}} \qquad (56)$$

On inserting the geometric mean value of the denominator, we see that, at $25°\mathrm{C}$, the value of P, which makes $\Delta S^{\ddagger}$ zero is roughly 500. The value of P which makes ΔS^* zero is, of course, unity.

Kinetics and Equilibria in Simple Systems

Let us consider the simple system represented by the equations

$$AB \underset{k_2}{\overset{k_1}{\rightleftharpoons}} A + B$$

and

$$K = \left(\frac{n_A n_B}{n_{AB}}\right)_P = \frac{k_1}{k_2},$$

and apply to it the considerations of the preceding section, assuming the reversible reactions to be unimolecular and bimolecular.

According to the transition state theory, we have

$$k_1 = \frac{kTe}{h} \cdot \exp\left(\Delta S_1^\ddagger/R\right) . \exp\left(-E_{A,1}/RT\right),$$

and

$$k_2 = \frac{kTe}{h} \cdot \exp\left(\Delta S_2^\ddagger/R\right . \exp\left(-E_{A,2}/RT\right),$$

where E_A is the apparent energy of activation and equals $\Delta H^\ddagger + RT$. Then

$$K = \exp\left[(\Delta S_1^\ddagger - \Delta S_2^\ddagger)/R\right] . \exp\left[-(E_{A,1} - E_{A,2})/RT\right]$$
$$= \exp\left(\Delta S^\ddagger/R\right) . \exp\left(-\Delta^0 H/RT\right).$$

Basing our expressions for k_1 on Herzfeld's equation and for k_2 on the collision equation, we have

$$k_1 = \frac{kTe}{h} \cdot \exp\left(-E_{A,1}/RT\right),$$

and

$$k_2 = PZ^0 \exp(1/2) . \exp\left(-E_{A,2}/RT\right).$$

Then

$$K = \frac{kT \exp(1/2)}{hPZ^0} \cdot \exp\left[-(E_{A,1} - E_{A,2})/RT\right]$$

$$= \frac{kT \exp(1/2)}{hPZ^0} \cdot \exp\left(-\Delta H/RT\right),$$

and, since $\Delta H = \Delta H_0 + (1/2)RT$,

$$K = \frac{kT}{hPZ^0} \cdot \exp\left(-\Delta H_0/RT\right) = \exp\left(\Delta S^0/RT\right) . \exp\left(-\Delta H_0/RT\right)$$

where

$$\Delta S^0 = R \ln\left[\frac{1,000}{N^0} \cdot \frac{1}{hPr_{AB}^2}\left(\frac{kT\mu_{AB}}{8\pi}\right)^{1/2}\right],$$

in agreement with equation (2.34) when P is unity.

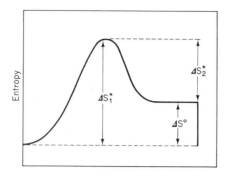

FIG. 7. The relationship between defined entropies of activation and the standard change in entropy.

Analogous with the diagram illustrating the relationship between apparent energies of activation and the standard change in heat content (Fig. 1), we have the entropy diagram drawn in Fig. 7, from which we see that, however the entropies of activation are defined, the experimental standard change in entropy of reaction is $\Delta S^0 = \Delta S_1^* - \Delta S_2^* = \Delta S_1^\ddagger - \Delta S_2^\ddagger$. Let us apply these equations to the reversible decomposition of phenyltrimethylammonium iodide in nitrobenzene solution, which is one of the few systems for which the kinetic data on the reverse reaction have been supplemented by equilibrium constants (Essex and Gelormini, 1920):

$$K = \frac{[C_6H_5N(CH_3)_2][CH_3I]}{[C_6H_5N(CH_3)_3I]} = \frac{k_1}{k_2} = 1.13 \times 10^8 . \exp(-15{,}070/RT)$$

$$k_2 = 3.94 \times 10^4 . \exp(-12{,}097/RT),$$

$$k_1 = Kk_2 = 4.45 \times 10^{12} . \exp(-28{,}040/RT).$$

Because there are two definitions of the entropy of activation for the bimolecular association reaction, we must accept two different values for the entropy of activation for the unimolecular decomposition. From equation (52), we have $\Delta S_2^* = -25.6$ cal/mole-deg, and, in the same units, $\Delta S_1^* = \Delta S^0 + \Delta S_2^* = 36.9 - 25.6 = 11.3$. From equation (53), on the other hand, $\Delta S_2^\ddagger = -36.9$, and $\Delta S_1^\ddagger = \Delta S^0 + \Delta S_2^\ddagger = 36.9 - 36.9 = 0$, in agreement with equation (37) (Eyring and Wynne-Jones, 1935). There is thus no difference between the entropy $S_1^\ddagger$ of the activated complex and the unactivated solute $C_6H_5N(CH_3)_3I$, since $\Delta S_1^\ddagger = S_1^\ddagger - S_1 = 0$. The discrepancy between this conclusion and that reached by taking account of the combination reaction arises from the different units of entropy afforded by the transition state theory and the collision theory in their different methods of expressing the velocity constants of bimolecular reactions.

REFERENCES

Arrhenius, S., *Z. physikal Chem.*, **4**, 226 (1889).
Bäckström and Beatty, *J. Phys. Chem.*, **35**, 2530 (1931).
Bennett and (Miss) Vernon, *Trans. Chem. Soc.*, 1783 (1938).
Berthoud, *Ann. Chim. Physique*, **11**, 580 (1913).
Bodenstein, *Z. physikal Chem.*, **29**, 295 (1889).
Bodenstein, *Z. physikal Chem.*, **85**, 329 (1913).
Bodenstein and Lind, *Z. physikal Chem.*, **57**, 168 (1907).
Bodenstein and Krauss, *Z. physikal Chem.*, **175**, 295 (1936).
Bodenstein, M., Padelt and Schumacher, *Z. physikal Chem.*, **B5**, 209 (1929).
Bjerrum, N., *Z. physikal Chem.*, **A.108**, 82 (1924).
Bowen, E. J., Moelwyn-Hughes and Hinshelwood, *Proc. Roy. Soc.*, **A134**, 211 (1931).
Brinkman, Margaria and Roughton, *Phil. Trans.*, **232**, 65 (1933).
Brönsted, *Z. physikal Chem.*, **102**, 169 (1922); **115**, 337 (1925).
Burrows and Fawsitt, *Trans. Chem. Soc.*, **105**, 609 (1914).
Cain and Nicoll, *Trans. Chem. Soc.*, **81**, 1412 (1902).
Christiansen, *Z. physikal Chem.*, **113**, 35 (1924).
Christiansen and Kramers, *Z. physikal Chem.*, **104**, 451 (1923).
Curme, H. G., and G. K. Rollefson, *J. Amer. Chem. Soc.*, **74**, 3766 (1952).
Cuthbertson and Kistiakowsky, *J. Chem. Physics*, **3**, 631 (1935).
Dawson, H. M., *Trans. Chem. Soc.*, **95**, 1806 (1909).
Dushman, *see* W. C. McC. Lewis, *A System of Physical Chemistry*, 3rd edn., Vol. III, p. 232, Longmans, 1924.
Essex and Gelormini, *J. Amer. Chem. Soc.*, **48**, 882 (1920).
Evans, M. G., and M. Polanyi, *Trans. Faraday Soc.*, **31**, 875 (1935).
Eyring, H., *J. Chem. Physics*, **3**, 107 (1935).
Eyring, H., and Polanyi, *Z. Physikal Chem.*, **B12**, 279 (1931).
Eyring, H., and W. F. K. Wynne-Jones, *J. Chem. Physics*, **3**, 492 (1935).
Eyring, H., and W. F. K. Wynne-Jones, **4**, 293 (1936).
Fahim and Moelwyn-Hughes, *Trans. Chem. Soc.*, 1034 (1956).
Farkas, A., *Light and Heavy Hydrogen*, p. 83, Cambridge (1935).
Farkas, A., *Z. Physikal Chem.*, **B.10**, 419 (1930).
Farkas, A., and Wigner, *Trans. Faraday Soc.*, **32**, 1 (1936).
Harcourt and Esson, *Phil. Trans.*, **186**, 187 (1895): *ibid.*, **212**, 187 (1913).
Herzfeld, *Ann. Physik*, **59**, 635 (1919).
Hinshelwood, C. N., and Prichard, *Trans. Chem. Soc.*, 2730 (1923).
Hinshelwood, C. N., and H. W. Thompson, *Proc. Roy. Soc.*, **A.113**, 221 (1926).
Hirschfelder, Topley and Eyring, *J. Chem. Physics*, **4**, 173 (1936).
Hood, *Phil. Mag.*, **6**, 371 (1878); *Phil. Mag.*, **20**, 323 (1885).
Hurst, R., and E. A. Moelwyn-Hughes, *Academia Nazionale dei Lincei*, p. 112 (1957).
Johnson, P., and E. A. Moelwyn-Hughes, *Trans. Faraday Soc.*, **37**, 282 (1941).
Kohnstamm, P., and F. E. C. Scheffer, *Verlag. Akad. Wetensch.*, Amsterdam, **18**, 879 (1911).
Kooij, *Z. physikal Chem.*, **12**, 155 (1893).
La Mer, *J. Chem. Physics*, **1**, 289 (1933).
Lapworth, *Trans. Chem. Soc.*, **85**, 30 (1904).
Lewis, W. C. McC., *Trans. Chem. Soc.*, **113**, 35 (1924).
Liebhafsky and Mohammad, *J. Amer. Chem. Soc.*, **55**, 3977 (1933).
Lindemann, F. A., *Trans. Faraday Soc.*, **17**, 599 (1922).
Lineken and Burrows, *J. Amer. Chem. Soc.*, **51**, 1106 (1929).

London, *Z. Elektro. Chem.*, **35**, 552 (1929).
Michaelis and Menten, *Biochem. Z.*, **49**, 333 (1913).
Miller, A. A., and J. E. Willard, *J. Chem. Physics*, **17**, 168 (1949).
Moelwyn-Hughes, E. A., *Trans. Faraday Soc.*, **32**, 1736 (1934).
Moelwyn-Hughes, *Trans. Faraday Soc.*, **45**, 167 (1949).
Moelwyn-Hughes, E. A., and C. N. Hinshelwood, *Proc. Roy. Soc.*, **A.131**, 177 (1931).
Moelwyn-Hughes, E. A., and P. Johnson, *Trans. Faraday Soc.*, **36**, 948 (1940).
Moelwyn-Hughes, E. A., *Trans. Faraday Soc.*, **25**, 81, 503 (1929).
Moelwyn-Hughes, E. A., *J. Chem. Phys.*, **4**, 292 (1936)
Moelwyn-Hughes, E. A., *Ann. Rep. Chem. Soc.*, **33**, 86 (1936).
Pelzer and Wigner, *Z. Physikal Chem.*, **B15**, 445 (1932).
Polissar, *J. Amer. Chem. Soc.*, **52**, 956 (1930).
Porter and Marples, *J. Chem. Physics*, **40**, 1105 (1964).
Rodebush, *J. Amer. Chem. Soc.*, **45**, 606 (1923).
Saunders, *The Diazo-Components and their Technical Applications*, Arnold (1936).
Slator, *Trans. Chem. Soc.*, **85**, 1286 (1904); **87**, 485 (1905).
Soper, *Trans. Chem. Soc.*, 1393 (1935).
Stewart, T. D. and W. E. Bradley, *J. Amer. Chem. Soc.*, **54**, 4183 (1932).
Stewart, T. D. and Fontana, *J. Amer. Chem. Soc.*, **62**, 3281 (1940).
Trautz, M., *Z. anorg. Chem.*, **96**, 1 (1916).
Trautz, M., *Z. physikal Chem.*, **66**, 496 (1909).
Walker, *Z. physikal Chem.*, **42**, 603 (1902).
Walker and Hambly, *Trans. Chem. Soc.*, **67**, 746 (1895).
Wall, F. T., and R. N. Porter, *J. Chem. Physics*, **36**, 3256 (1962).
Warburg, *Ergb. Enzymforsch.*, **7**, 210 (1938).
Ward, A. M., *Trans. Chem. Soc.*, 2285 (1927).
Wassermann, *Trans. Chem. Soc.*, 1027 (1936).
Weston, R. E., *J. Chem. Physics*, **31**, 892 (1959).
Wynne-Jones, W. F. K., and Eyring, H., *J. Chem. Physics*, **3**, 492 (1935).

5

PROCESSES CONTROLLED BY DIFFUSION

By diffusion is meant the movement of a particle due to random collisions with other particles. In the gas phase and in solution, the movement is relatively slow, and, since the slowest step in a sequence determines the net rate of change, diffusion may be answerable for the absolute, observable, rates of many chemical changes.

The Frequency of Encounters between Uncharged Spherical Solutes: Smoluchowski's Equation

According to Fick's law of linear diffusion, the number of molecules crossing an area O per second is proportional to that area and to the concentration gradient measured in the direction of flow:

$$dN/dt = -DO(\partial n_r/\partial r). \tag{1}$$

The constant D, termed the coefficient of diffusion, is positive, indicating that molecules move in a direction counter to that of the concentration gradient. There are more general forms of the law, but this simple version suffices for our purpose. Moreover, from the accurate solution of the differential equation governing spherical diffusion under steady conditions, the dependence of n on r takes the simple form

$$n_r = A + B/r, \tag{2}$$

where the constants A and B are to be determined by boundary conditions. We shall consider first the case where n_r equals the bulk concentration, n, when r is infinite. Then $A = n$. Let the concentration of solute particles be zero when r has a value r_0 or less. Then $B = -nr_0$, and consequently

$$n_r = n(1 - r_0/r),$$

and

$$\partial n_r/\partial r = nr_0/r^2.$$

The number of particles crossing the spherical area $4\pi r^2$ per second in the direction of increasing r is thus

$$dN/dt = -4\pi Dnr_0. \tag{3}$$

The number reaching the spherical area per second is evidently

$$dN/dt = 4\pi Dnr_0. \tag{4}$$

The number of molecules of type A which, by the process of diffusion, reach the surface of a sphere of radius $(r_A + r_B)$ per second is consequently $4\pi D_A(r_A + r_B)n_A$. This we can regard as the number of encounters made per second by all molecules of the A kind (radius r_A) with one molecule of the B kind (radius r_B). On this basis, the total number of encounters made per second between the two kinds of molecules is $_BZ_A = 4\pi D_A(r_A + r_B)n_An_B$.
This result would be, under the steady conditions envisaged, a correct one if only the A molecules were diffusing, and the B molecules were at rest. In real systems, both types of molecules diffuse simultaneously and independently; and we can make allowance for this by adding the number of B molecules which diffuse to the A molecules, assumed to be at rest. Then $_AZ_B = 4\pi D_B(r_A + r_B)n_An_B$. Because both processes are uncorrelated the total frequency of binary encounters is the sum of the two terms, i.e.

$$_AZ_B = 4\pi(D_A + D_B)(r_A + r_B)n_An_B, \tag{5}$$

which is one form of M. von Smoluchowski's (1918) equation. If chemical reaction occurs only when the molecules in their encounters possess between them an energy E per pair, expressible as the sum of two quadratic terms, the rate of reaction becomes

$$-dn_A/dt = 4\pi(D_A + D_B)(r_A + r_B)n_A^0 \cdot \exp(-E_A/RT)n_B^0 \cdot \exp(-E_B/RT),$$

and the bimolecular rate constant is

$$k_2 = 4\pi(D_A + D_B)(r_A + r_B) \cdot \exp(-E/RT), \tag{6}$$

where $E = E_A + E_B$. This equation has been applied to numerous bimolecular reactions in solution, with slight modifications, e.g., with the number 4 replaced by π (Ölander, 1929) or by 1/6 (R. S. Bradley, 1934). The treatment is doubtless valid if the rate of chemical reaction is the rate at which preactivated molecules collide by a diffusion process. If, however, E, is much greater than the energy of diffusion, the rate-determining process is that of chemical reaction, and is unaffected by diffusion. The region where the two mechanisms merge was first explored in connection with the dissolution of solids.

The Kinetics of Dissolution

Let us consider the dissolution of a homogeneous spherical particle. When the process has reached a steady state, we may apply equation (2), with the boundary conditions that $n_r = n_\sigma$ when r is equal to or less than σ, and that n_r equals the bulk concentration, n, when r is infinite. Then

$$n_r = n - (n - n_\sigma)(\sigma/r), \tag{7}$$

and

$$\frac{\partial n_r}{\partial r} = \frac{n - n_\sigma}{r^2}\sigma.$$

The number of molecules moving per second across the spherical surface $4\pi r^2$ in the direction of increasing r is thus

$$dN/dt = -D \cdot 4\pi r^2(\partial n_r/\partial r)$$

$$= 4\pi D(n_\sigma - n)\sigma, \tag{8}$$

which is the rate of dissolution of a single spherical particle by a diffusion mechanism. The rate of dissolution at the surface of the sphere may equally well be formulated as follows

$$dN/dt = 4\pi\sigma^2(n_s v - n_\sigma k_s) \tag{9}$$

where n_s is the number of molecules per unit area of the solid, and v is the average probability per second that a surface molecule shall dissolve. k_s is the average velocity with which solute molecules move towards and attach themselves to the surface. When the solution has become saturated, $n_s v = n_{sat} k_s$, which enables us to eliminate $n_s v$ and write

$$dN/dt = 4\pi\sigma^2 k_s(n_{sat} - n_\sigma). \tag{10}$$

In the steady state, the rate at which molecules become detached from the surface must equal the rate at which they diffuse away from it. By expressions (8) and (10), we have

$$n_\sigma = \frac{\sigma k_s n_{sat} + Dn}{\sigma k_s + D},$$

and, on eliminating n_σ from either equation (9) or (10), we obtain, for the number of molecules dissolving per second from the sphere

$$\frac{dN}{dt} = 4\pi D\sigma \cdot \frac{(n_{\text{sat}} - n)}{1 + (D/\sigma k_s)}. \tag{11}$$

When attachment to the surface is rapid,

$$dN/dt = 4\pi D\sigma(n_{\text{sat}} - n), \tag{12}$$

and the rate of dissolution is determined by diffusion. When, on the other hand, attachment to the surface is slow,

$$dN/dt = 4\pi\sigma^2 k_s(n_{\text{sat}} - n), \tag{13}$$

the rate of dissolution is determined by the rate of attachment at the solid surface, and is independent of diffusion. In both cases, the rate is directly proportional to the saturation deficiency. These are the principal features of the kinetics of dissolution, as formulated by Berthoud (1912) in his synthesis of the spontaneous escape mechanism of A. A. Noyes and Whitney (1897) and the diffusion layer theory of Nernst (1904).

Berthoud's method has been applied to the problems of dissolution from a plane surface (Roller, 1935; Moelwyn-Hughes, 1947; J. Crank, 1956) and of chemical reaction rates.

The Limiting Role of Diffusion in Chemical Reactions

The foregoing considerations are applicable to reactions taking place between spherical molecules in solution. According to equation (8), the number of molecules reaching the surface of a single sphere per second is $4\pi D(n - n_\sigma)\sigma$. The number destroyed by chemical reaction per second is $4\pi\sigma^2 n_\sigma k_s$, where, according to the kinetic theory,

$$k_s = \left(\frac{kT}{2\pi\mu}\right)^{1/2} \cdot \exp\left(-\varepsilon/kT\right).$$

μ is the relative mass, and ε the energy of activation. The rate of increase in the number of molecules in the interface is thus

$$dN_\sigma/dt = 4\pi D(n - n_\sigma)\sigma - 4\pi\sigma^2 n_\sigma k_s, \tag{14}$$

which is zero under steady conditions. Then

$$n_\sigma = Dn/(D + \sigma k_s).$$

The rate of reaction is found, by substituting this expression for n_σ in either of the rate terms of equation (14):

$$-\frac{dN}{dt} = 4\pi D\sigma n\left(\frac{\sigma k_s}{D + \sigma k_s}\right) = 4\pi\sigma^2 nk_s\left(\frac{D}{D + \sigma k_s}\right).$$

Let us suppose that molecules of type B diffuse towards the sphere A. Then

$$-\frac{dN_B}{dt} = 4\pi\sigma^2 n_B k_s\left(\frac{D_B}{D_B + \sigma k_s}\right).$$

If there are n_A spheres of type A per unit volume, the number of encounters in unit volume and unit time between dissimilar solutes due to this type of diffusion is

$$-\frac{dn_B}{dt} = 4\pi\sigma^2 n_A n_B k_s\left(\frac{D_B}{D_B + \sigma k_s}\right),$$

and the bimolecular velocity constant is

$$k_2 = 4\pi\sigma^2 k_s\left(\frac{D_B}{D_B + \sigma k_s}\right).$$

There is a complementary term due to diffusion of solutes of type A to be added, yielding for the observed bimolecular constant the expression

$$k_2 = 4\pi\sigma^2 k_s\left[\frac{D_A}{D_A + \sigma k_s} + \frac{D_B}{D_B + \sigma k_s}\right]. \tag{15}$$

When σk_s is small compared with either coefficient of diffusion, we recover the gas collision expression

$$k_2 = 8\pi\sigma^2 (kT/2\pi\mu)^{1/2} . \exp(-\varepsilon/kT). \tag{4.38}$$

When σk_s is large compared with D_A and D_B, we recover Smoluchowski's equation (5). (Recent discussions are given by F. C. Collins and G. E. Kimball, 1949; R. M. Noyes, 1961; S. R. Logan, 1967). $D/\sigma k_s$ is approximately unity for ordinary solutes in water at room temperatures, provided ε/kT also happens to be near to unity. When, however, ε/kT is about 7, $D/\sigma k_s$ is about 200, and the rate of reaction is effectively independent of the diffusion process.

The Stokes–Einstein Equation

By differentiating expression (2.4) with respect to r, the average force acting on one molecule due to Brownian motion is seen to be $-kT(\partial \ln n/\partial r)$. The force of resistance, assumed to be proportional to the steady velocity v is $-Cv$, where C is the resistance constant. Under steady conditions, the net force is zero:

$$-kT\,\frac{\partial \ln n}{\partial r} - Cv = 0.$$

Hence nv, which is the number of molecules passing unit area per second is

$$nv = -(kT/C)(\partial n/\partial r).$$

But, by the definition of the coefficient of diffusion $D = -nv/(\partial n/\partial r)$. Therefore

$$D = kT/C. \qquad (16)$$

For spherical particles of radius r in a medium of viscosity η, $C = 6\pi\eta r$:

$$D = kT/6\pi\eta r, \qquad (17)$$

which is the Stokes–Einstein equation (G. G. Stokes, 1856; A. Einstein, 1926):

von Smoluchowski's equation (5) may now be written in the alternative form

$$_AZ_B = \frac{2}{3}\frac{kT}{\eta}\frac{(r_A + r_B)^2}{r_A r_B}\cdot n_A n_B, \qquad (18)$$

or, for identical particles,

$$Z = \frac{8}{3}\frac{kT}{\eta}n^2. \qquad (19)$$

The numerical term is not exact.

The Kinetics of Coagulation

If colloid particles unite at every encounter, the bimolecular velocity constant governing the process should be $(8/3)(kT/\eta)$, which is 1.07×10^{-11} ccs/particle-sec for water, and 5.86×10^{-10} for air at 20°C. Some of the rates of coagulation of colloidal gold particles in water are given in Table I, which is due to Zsigmondy (1918) and to Wiegner and Tuorila (1926). The average value of k_2 is seen to agree with equation (19) to within a factor of 2, while the initial particle concentration varies by a factor of more than 10^4. The rate of coagulation of the colloid particles which form aerosols is also in agreement with equation (19). The experimental values of k_2 range from 5×10^{-10} for smokes formed from metallic oxides to 8×10^{-10} for smokes formed from resin and fatty acids (Whytlaw-Gray and Patterson, 1932).

TABLE I
THE COAGULATION COEFFICIENTS OF COLLOIDAL PARTICLES IN AQUEOUS SOLUTION

Colloid	r (Å)	n_0 (number of particles per cc at the start)	$t_{1/2}$ (time of half-completion, in seconds)	$k_2 \times 10^{12}$ (cc/particle-second)
Gold	970	$3 \cdot 60 \times 10^8$	507	$5 \cdot 48$
	—	$6 \cdot 20 \times 10^8$	254	$6 \cdot 35$
	320	$1 \cdot 06 \times 10^9$	134	$7 \cdot 04$
	512	$2 \cdot 02 \times 10^9$	79	$6 \cdot 27$
	$53 \cdot 7$	$5 \cdot 80 \times 10^{11}$	$0 \cdot 208$	$8 \cdot 29$
	$36 \cdot 9$	$8 \cdot 44 \times 10^{11}$	$0 \cdot 186$	$6 \cdot 36$
	$29 \cdot 1$	$2 \cdot 93 \times 10^{12}$	$0 \cdot 0424$	$8 \cdot 05$
	$29 \cdot 1$	$4 \cdot 87 \times 10^{12}$	$0 \cdot 0275$	$7 \cdot 47$
	$36 \cdot 9$	$5 \cdot 91 \times 10^{12}$	$0 \cdot 0269$	$6 \cdot 29$
Kaolin	3,040	$8 \cdot 90 \times 10^8$	211	$5 \cdot 33$

Again Smoluchowski's equation is obeyed. Finally, the effect of temperature on the coagulation coefficients is found to be the same as its effect on the fluidity, $1/\eta$ (P. J. R. Butler, 1930).

The Quenching of Fluorescence

A system is said to fluoresce when, simultaneous with the absorption of radiation of a given wave length, it emits radiation of longer wave length. Chlorophyll in water is the best-known example. The generally accepted theory of the phenomenon (Stern and Volmer, 1919; Förster, 1951) is based on the idea of radiationless transfers of energy between molecules or ions of the fluorescent substance and other molecular or ionic species in the system. We consider first the formation of electronically excited molecules, A^*, by the absorption of radiation at a rate proportional to the concentration, n_A, of ordinary unexcited molecules of the photosensitive species:

$A \rightarrow A^*$;   excitation by absorption of radiation; $k_3 n_A$.

The excited molecules may lose their energy by the spontaneous emission of radiation, at a rate proportional to their concentration, a;

$A^* \rightarrow A$;   spontaneous emission of fluorescent radiation; $k_1 a$.

k_1 is their decay constant, or the reciprocal of their mean life, which can be measured by a variety of methods which we need not discuss here. Typical values are given in Table II. The excited molecules, it is postulated, may also

TABLE II

Fluorescent substance	Average lifetime, τ (seconds)	Authority
Fluorescein	$4\cdot5 \times 10^{-9}$	Perrin, *Journal de Physique*, 7, 390 (1926)
Uranine (di-sodium salt)	$5\cdot0 \times 10^{-9}$	Gaviola, *Z. Physik.*, **42**, 862 (1927)
Quinine bisulphate	$(9\cdot5 \pm 2\cdot0) \times 10^{-8}$	Wawilow, *ibid.*, **53**, 665 (1929)

lose energy by radiationless transfers in binary collisions with molecules of their own kind and with other molecules:

$A^* + A \to 2A$;   collisions between excited and normal
molecules; rate $k_4 n_A a$

$A^* + B \to A + B$;   collisions between excited molecules
and normal molecules of species B;
rate $k_2 n_B a$

The rate of formation of the excited molecules is

$$+ \, da/dt = k_3 n_A - k_1 a - k_4 n_A a - k_2 n_B a,$$

which is zero under stationary conditions, when

$$a = \frac{k_3 n_A}{k_1 + k_2 n_B + k_4 n_A}.$$

The rate of emission of light is therefore

$$I = k_1 a = \frac{k_1 k_3 n_A}{k_1 + k_2 n_B + k_4 n_A}.$$

In the absence of added quenchers, this rate reaches its maximum value of

$$I_0 = \frac{k_1 k_3 n_A}{k_1 + k_4 n_A}.$$

Consequently

$$\frac{I_0}{I} = 1 + \left(\frac{k_2}{k_1 + k_4 n_A}\right) n_B. \tag{20}$$

This is the equation of Stern and Volmer, which has been extensively investigated by them in gaseous systems, and by Wawilow (1925, 1929) in solutions. The ratio $(I_0 - I)/I_0$ is known as the quenching constant, Q. Some of the results found by Jette and West (1928) on the quenching of the fluorescence of quinine by various salts in water are shown in Fig. 1. In order

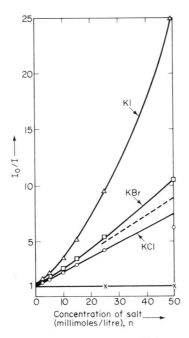

FIG. 1. Graphical test of the Stern–Volmer equation.

of diminishing quenching efficiency, Q follows the sequence $I^- > NCS^- > Br^- > Cl^- > NO_3^- > F^-$, which is also the order of diminishing polarizability (Fajans and Joos, 1924) and coagulation power, and of increasing electron-affinity. The departure from the linearity expected from equation 20 has been much discussed (Boaz and Rollefson, 1950; B. Williamson and La Mer, 1948: for quenching by xenon atoms, see Horrocks, Kearvell, Tickle and F. Wilkinson, 1966). It is in part attributable to the primary electolyte effect, which is dealt with in Chapter 7.

When self-quenching can be ignored, equation (20) becomes

$$\frac{I_0}{I} = 1 + \left(\frac{k_2}{k_1}\right) n_B, \tag{21}$$

from which the ratio of two rate constants can be found. In the present system, we see from Table II that $k_1 = 1\cdot05 \times 10^{-7} \ \text{sec}^{-1}$. The resulting values of the bimolecular velocity coefficients (Table III) are seen to lie close to the theoretical value consistent with one-hundred percent efficiency for deactivating collisions, which, according to equation (18) is

$$k_2 = \frac{N_0}{1{,}000} \cdot \frac{2kT}{3\eta} \frac{(r_A + r_B)^2}{r_A r_B}. \tag{22}$$

TABLE III
THE QUENCHING OF THE FLUORESCENCE OF QUININE IN WATER

Electrolyte	k_2/k_1 litres/mole	k_2 litres/mole-sec
KCl	150	$1\cdot58 \times 10^9$
KBr	178	$1\cdot87 \times 10^9$
KI	200	$2\cdot10 \times 10^9$

The calculated rate constant for solutes of equal radii in water at 20°C is $6\cdot5 \times 10^9$ litres/mole-sec. If we allow for the uncertainty in our knowledge of the ionic radii, we may conclude that each encounter between the fluorescent molecule and the iodine ion is successful in the deactivating sense. The other halide ions are slightly less efficient. The quenching efficiencies for many other systems have been evaluated by Doss (1937) and by Bäckström and Sandros (1958). For the nitrate ion, it is about 10^{-3}.

It may be concluded that the rate at which electronically excited solutes are deprived of their energy of excitation is determined by diffusion. Confirmation in the quinine sulphate iodide system has been provided by Wawilow who showed that Q measured between 18 and 95°C is directly proportional to T/η.

Interionic Encounters in Solution; Langevin's Treatment

Langevin (1903, 1908) has considered the mutual approach of oppositely charged ions when the forces due to Brownian motion can be neglected in comparison with the Coulombic force, which, for a uni-univalent pair is $-\varepsilon^2/Dr^2$, where ε is the charge on the proton, r is the interionic separation and D is the permittivity of the medium (see Nora Hill, W. E. Vaughan, A. H. Price and Mansel Davies, 1969). This force is common to both ions but the forces resisting their motions are different. When steady motion has been reached, the equations of motion are

$$-\frac{\varepsilon^2}{Dr^2} - C_+u = 0$$

and

$$-\frac{\varepsilon^2}{Dr^2} - C_-v = 0. \tag{23}$$

Both velocities are negative, in the direction of increasing r, which means that the ions are approaching each other. The relative velocity of motion in the direction of decreasing r is thus

$$u + v = \frac{\varepsilon^2}{Dr^2}\left(\frac{1}{C_+} + \frac{1}{C_-}\right).$$

The ionic mobilities, i.e. the velocities under unit field, are clearly $u^0 = \varepsilon/C_+$ and $v^0 = \varepsilon/C_-$. Hence

$$u + v = \frac{\varepsilon}{Dr^2}(u^0 + v^0).$$

In a system containing one ion of type A, the number of ions of type B which come within a distance r in one second and in unit volume is $4\pi n_B\varepsilon(u^0 + v^0)/D$, where n_B is the concentration of ions B. When there are n_A ions of the first type, the total number of dissimilar encounters per cc per sec is

$$_AZ_B = 4\pi n_A n_B(u^0 + v^0)(\varepsilon/D), \tag{24}$$

which is Langevin's equation (see also Önsager, 1934).

From equation (23), we have $\varepsilon(\varepsilon/Dr^2) = \varepsilon E$, where E is the electric field, and $\varepsilon E = C_- v$, so that $v^0 = v/E = \varepsilon/C_-$. But the coefficient of diffusion is $D_- = kT/C_-$ (equation 16). Hence $D_- = (kT/\varepsilon)v^0$. Equation (24) can now be expressed as follows

$$_AZ_B = 4\pi n_A n_B(D_A + D_B)\varepsilon^2/DkT. \tag{25}$$

D without a subscript is written for the permittivity of the solvent; D with a subscript denotes the coefficient of diffusion of the solute.

A More General Expression for Ionic Encounters: Debye's Treatment

Let us consider first two isolated spherical ions, with radii r_A and r_B, and electrical charges $z_A\varepsilon$ and $z_B\varepsilon$, where z denotes the electrovalency, in sign and magnitude, and ε is the charge on the proton. At a distance r apart, in a medium of uniform permittivity, D, their interaction energy is

$$\phi = \frac{z_A z_B \varepsilon^2}{Dr},$$

and the Coulombic force exerted between them is

$$-\frac{d\phi}{dr} = \frac{z_A z_B \varepsilon^2}{Dr^2}.$$

As we are concerned only with relative ionic motion, let us regard one ion, say A, to be at rest, and confine attention to the rate at which the other ion (B) approaches it. To the Coulombic force acting on B must be added the forces due to Brownian motion and to viscous resistance. The latter is assumed to be proportional to the velocity, dr/dt. Under steady motion, the total force is zero:

$$-\frac{d\phi}{dr} - \frac{kT}{n_B} \cdot \frac{dn_B}{dr} - C_B \frac{dr}{dt} = 0.$$

The velocity is therefore

$$\frac{dr}{dt} = -\frac{kT}{C_B}\left(\frac{1}{n_B} \cdot \frac{dn_B}{dr} + \frac{1}{kT} \cdot \frac{d\phi}{dr}\right)$$

$$= -D_B\left(\frac{1}{n_B} \cdot \frac{dn_B}{dr} + \frac{1}{kT} \cdot \frac{d\phi}{dr}\right),$$

where D_B is the coefficient of diffusion of ion B. The number of ions of type B crossing the surface of a sphere of radius r per second, in the direction of increasing r, is $4\pi r^2 n_B(dr/dt)$. Denoting this number by I, we have

$$I = -4\pi r^2 D_B\left(\frac{dn_B}{dr} + \frac{n_B}{kT} \cdot \frac{d\phi}{dr}\right).$$

If Boltzmann's distribution law is maintained during diffusion, $n_B = n_B^0 \times \exp(-\phi/kT)$, where n_B^0 is the bulk concentration at any time. It follows that

$$\frac{dn_B}{dr} = -\frac{n_B}{kT} \cdot \frac{d\phi}{dr} + \exp(-\phi/kT) \cdot \frac{dn_B^0}{dr}.$$

Consequently

$$I = -4\pi r^2 D_B \cdot \exp(-\phi/kT)\frac{dn_B^0}{dr},$$

and

$$dn_B^0 = -\frac{I}{4\pi D_B} \cdot \exp(+\phi/kT)\frac{dr}{r^2}.$$

Under steady conditions, I is constant, so that

$$n_B^0 = -\frac{I}{4\pi D_B} \int_{r=\sigma}^{r=\infty} \exp(+z_A z_B \varepsilon^2/DrkT)\frac{dr}{r^2},$$

where σ is the closest distance to which the charges can come. To perform the integration, let $y = 1/r$. Then $dy = -(1/r^2)\,dr$, and

$$n_B^0 = \frac{I}{4\pi D_B} \int_{y=1/\sigma}^{y=0} \exp\left[+(z_A z_B \varepsilon^2/DkT)y\right] dy$$

$$= \frac{I}{4\pi D_B} \cdot \frac{DkT}{z_A z_B \varepsilon^2}\left[1 - \exp(z_A z_B \varepsilon^2/D\sigma kT)\right].$$

The number of encounters made per sec per cc by all the B ions on one A ion is $-I$, which we can denote by $_1Z_B$.

$$_1Z_B = 4\pi D_B n_B^0 \frac{z_A z_B \varepsilon^2}{DkT}\left[\exp(z_A z_B \varepsilon^2/D\sigma kT) - 1\right]^{-1}$$

$$= 4\pi D_B \sigma n_B^0 \cdot \frac{z_A z_B \varepsilon^2}{D\sigma kT}\left[\exp(z_A z_B \varepsilon/D\sigma kT) - 1\right]^{-1}.$$

The total number of encounters per cc per sec due to diffusion of B ions towards all the A ions is greater by n_A. Simultaneously, there is an uncorrelated set of encounters due to diffusion of the A ions, giving in all

$$_AZ_B = 4\pi(D_A + D_B)\sigma n_A^0 n_B^0 \left[\frac{z_A z_B \varepsilon^2/D\sigma kT}{\exp(z_A z_B \varepsilon^2/D\sigma kT) - 1}\right], \qquad (26)$$

which is Debye's (1942) result (see also Umberger and La Mer, 1945). When the Coulombic energy is small, the equation reduces to that of Smoluchowski (equation (5), replacing σ by $r_A + r_B$). When the Coulombic energy is large and $z_A z_B = -1$, the equation reduces to that of Langevin (equation 25).

Equation (26), applied to encounters between univalent cations and anions ($z_A z_B = -1$), and expressed in terms of ionic mobilities ($u^0 = \varepsilon D_A/kT$; $v^0 = \varepsilon D_B/kT$), takes the form

$$_AZ_B = \frac{4\pi(u^0 + v^0)n_A^0 n_B^0 \varepsilon}{D} \cdot \frac{1}{1 - \exp(-\varepsilon^2/D\sigma kT)}, \qquad (27)$$

which, except for the last term, is Langevin's equation (24). If union of the ions takes place at each encounter, the bimolecular rate constant, expressed in the units of litres per mole-sec, is

$$k_2 = \frac{N_0}{1,000} \cdot \frac{4\pi(u^0 + v^0)\varepsilon}{D} \cdot \frac{1}{1 - \exp(-\varepsilon^2/D\sigma kT)}. \qquad (28)$$

By a variety of relaxation methods (see Chapter 14) Eigen (1954) and de Maeyer (1958) have measured the rates of many reactions of this type. Some of the results obtained by Eigen and Eyring (1962) are given in Table IV. With the exception of the first entry, the velocity constants are reproduced with sufficient accuracy by Langevin's equation. The rate of union of the

TABLE IV

BIMOLECULAR VELOCITY CONSTANTS FOR THE UNION OF
HYDROGEN IONS AND VARIOUS ANIONS IN WATER AT 25°C

Reacting ions	$k_2 \times 10^{-10}$ (l/mole-sec)
$H^+ + OH^-$	14
$H^+ + CH_3COO^-$	5·1
$H^+ + C_6H_5COO^-$	3·7
$H^+ + oNH_2 \cdot C_6H_4COO^-$	5·8
$H^+ + mNH_2 \cdot C_6H_4COO^-$	4·6
$H^+ + pNH_2 \cdot C_6H_4COO^-$	3·7

hydrogen and hydroxyl ions, however, exceeds that given by his equation. The mobilities and coefficients of diffusion of these ions in water are known with considerable accuracy. At 25°C, we have

$$u^0(H^+) = 1·087; \qquad v^0(OH^-) = 0·6166 \text{ cm-sec}^{-1} \text{ (e.s.u. of potential)}^{-1}$$

$$D^0(H^+) = 9·319 \times 10^{-5}; \qquad D^0(OH^-) = 5·285 \times 10^{-5} \text{ cm}^2\text{-sec}^{-1}.$$

The calculated value of the bimolecular constant is thus

$$k_2^0 = 8·08 \times 10^{10} \cdot \frac{1}{1 - \exp(-\varepsilon^2/D\sigma kT)} \text{ litres/mole-sec.} \qquad (29)$$

Comparison with the experimental value given in Table IV requires σ to be 8·5 Å, indicating that the ions unite when separated by a distance of about 3 molecular diameters. Eigen's work thus confirms the Grotthus (1806) mechanism involving rapid consecutive proton jumps transmitted through hydrogen-bonded water molecules. The rate constants determined at 5 temperatures (Ertl and Gerischer, 1962) between 5°C and 32°C are consistent with values of σ in the range 7 ± 1 Å.

If, as is probable in the case of large ions, the ionic mobilities vary inversely as the viscosity, the anticipated apparent energy of activation corresponding to equation (28) becomes

$$E_A = B + RT\left\{LT + \frac{\varepsilon^2}{D\sigma kT}(1 - LT)[\exp(\varepsilon^2/D\sigma kT) - 1]^{-1}\right\}, \qquad (27a)$$

where B is given by equation (5.32) and L by equation (7.26). With $r = 7$ Å and water as the solvent at 25°C, the term within the curly brackets is 1·16, indicating that the main effect of the temperature change is its effect on the viscosity. When E_A is greater than $B + RT$, the rate is no longer controlled by diffusion but by the probability that one of the ions in a pair can escape from its solvent sheath, as expressed in equation (1.16).

The Arrhenius Parameters for Diffusion-controlled Processes

When encounters alone are sufficient for the occurrence of chemical reaction, we have, from equations (6) and (22),

$$E_A = RT^2\left(\frac{d \ln D}{dT}\right) = E_D, \tag{30}$$

and

$$E_A = RT - RT^2\left(\frac{d \ln \eta}{dT}\right) = RT + B, \tag{31}$$

where

$$B = -RT^2\left(\frac{d \ln \eta}{dT}\right). \tag{32}$$

Since the energy terms E_D and B can be measured statically, a comparison with the kinetic term E_A can establish the relevance of the diffusion-control mechanism. As is well known, reactions such as the attack of various metals by the tri-iodide ion yield apparent energies of activation in agreement with the anticipated value, which is 4,612 cal/mole for aqueous solutions at 25°C. On the other hand, E_A for the catalytic reaction $2ClO^- \rightarrow 2Cl^- + O_2$ on the surface of suspended particles of cobalt chloride is 16,574 cal, and, like enzyme catalyses, is clearly not a diffusion-controlled process. In order to interpret the pre-exponential parameter, A, of the Arrhenius equation, we require the use of some empirical relationship such as that generally attributed to de Guzman (see also Andrade, 1936; Partington, 1951) to represent approximately the temperature variation of the viscosity of non-polar liquids, which is

$$\eta = be^{B/RT}, \tag{33}$$

where b and B are specific constants, the latter being defined by equation (32). Some numerical values are given in Table V. The bimolecular constants for reactions between solutes of equal radii are given by equation (22) as

$$k_2 = \frac{N_0}{1,000} \cdot \frac{8kT}{3\eta} \cdot \exp\left(-E/RT\right). \tag{34}$$

On substituting equation (33) for η, and recalling the definition of E_A, we see that

$$k_2 = \frac{N_0}{1,000} \cdot \frac{8kTe}{3b} \cdot \exp\left(-E_A/RT\right). \tag{35}$$

TABLE V

THE VISCOSITIES OF CERTAIN LIQUIDS

$$\eta = b e^{B/RT}$$

Liquid	$\eta_{25°C}$ (gm/cm-sec)	b (gm/cm-sec)	B (cal/gram-mole)
$CHCl_3$	5.50×10^{-3}	5.62×10^{-4}	$1,350 \pm 240$
$(CH_3)_2CO$	3.16×10^{-3}	2.12×10^{-4}	$1,600 \pm 60$
nC_6H_{14}	3.12×10^{-3}	1.218×10^{-4}	$1,660 \pm 5$
CCl_4	9.21×10^{-3}	1.664×10^{-4}	$2,375 \pm 5$
$C_2H_2Cl_4$	1.64×10^{-2}	1.047×10^{-4}	$2,990 \pm 30$
CH_3OH	5.56×10^{-3}	8.65×10^{-5}	$2,465 \pm 40$
C_2H_5OH	1.10×10^{-2}	4.65×10^{-5}	$3,225 \pm 130$
nC_3H_7OH	1.98×10^{-2}	1.46×10^{-5}	$4,270 \pm 280$
CH_3COOH	1.15×10^{-2}	1.305×10^{-4}	$2,650 \pm 40$
C_4H_6	6.06×10^{-3}	4.915×10^{-5}	$2,850 \pm 240$
$C_6H_5 \cdot CH_3$	5.54×10^{-3}	1.515×10^{-4}	$2,130 \pm 20$
$C_4H_5 \cdot COCH_3$	1.67×10^{-2}	1.09×10^{-4}	$2,980 \pm 640$
$C_6H_5 \cdot NO_2$	1.87×10^{-2}	1.10×10^{-4}	$3,040 \pm 170$
$C_6H_5 \cdot CH_2OH$	5.01×10^{-2}	7.16×10^{-6}	$5,240 \pm 440$
$C_6H_5 \cdot NH_2$	3.66×10^{-2}	6.56×10^{-6}	$5,330 \pm 10$
H_2O { (0°C)	1.79×10^{-2}	1.53×10^{-6}	$5,080$
(25°C)	8.94×10^{-3}	1.006×10^{-5}	$4,020$
(50°C)	5.50×10^{-3}	2.463×10^{-5}	$3,470$

The pre-exponential parameter of the Arrhenius equation is thus

$$A_2 = \frac{N_0}{1,000} \cdot \frac{8kTe}{3b}. \tag{36}$$

It is seen that A_2 for diffusion-controlled reactions in carbon tetrachloride at 25°C is 1.08×10^{12} litres/mole-sec, which is considerably higher than the A_2 term given by the kinetic theory.

It is clear from the contents of Table V that de Guzman's equation (33) is far from satisfactory for water, an empirical equation representing the temperature variation of the viscosity of which is given in Appendix 3.

Diffusion in Real Solutions

All the treatments of diffusion given in this chapter so far hold only for ideal, or infinitely dilute, solutions. When the activity coefficient of a solute at temperature T and concentration n_2 is γ_2, its chemical potential may be written in the form

$$\mu_2 = \mu_2^0 + kT \ln n_2 + kT \ln \gamma_2, \tag{37}$$

where μ_2^0 is its chemical potential at unit concentration and activity. The average force acting linearly on one molecule, due to Brownian motion, is now

$$X_2 = -\frac{\partial \mu_2}{\partial x} = -kT\left(\frac{\partial \ln n_2}{\partial x}\right)\left(1 - \frac{\partial \ln \gamma_2}{\partial \ln n_2}\right),$$

and the total force is $X_2 - Cv$, where C is the resistance factor and v the velocity. Under steady conditions, the total force is zero, and therefore

$$v = -\frac{kT}{C}\left(\frac{\partial \ln n_2}{\partial x}\right)\left(1 + \frac{\partial \ln \gamma_2}{\partial \ln n_2}\right).$$

The coefficient of diffusion is defined as

$$D = -nv\frac{\partial n}{\partial x} = \frac{kT}{C}\left(1 + \frac{\partial \ln \gamma_2}{\partial \ln n_2}\right), \tag{38}$$

which reduces to the Stokes–Einstein equation (16) when the activity coefficient is independent of the concentration (Crank, 1956).

Diffusion in Binary Mixtures

Hammond and Stokes (1955) have measured the coefficient of diffusion of carbon tetrachloride in a variety of hydrocarbon solvents at 25°C. They find that for every solvent, the coefficient of diffusion increases linearly with respect to the concentration, c_2, of solute, expressed in the units of grams of solute in every cc of solution, $D = D^0 + ac_2$, which enables them to obtain, by extrapolation, the value of the diffusion coefficient in an infinitely dilute solution. Some of the results are given in Table VI, which includes also the

TABLE VI
COEFFICIENTS OF DIFFUSION OF CARBON TETRACHLORIDE AT 25°C
AND AT INFINITE DILUTION

Solvent	$10^5 D^0$ cm^2/sec^{-1}	$10^2 \eta^0$ poise	$10^7 D^0 \eta^0$ dyne	V_1^0 cc/mole^{-1}
Benzene	2·00	0·602	1·20	89·4
Cyclohexane	1·49	0·900	1·34	108·7
Tetralin	0·735	2·03	1·49	137·1
Decalin	0·776	2·08	1·62	158·5

viscosities and the molar volumes, V_1^0, of the pure solvents. If equation (17) is applied to these data, it is clear that there will be found various values of r for the diffusion of a given solute in a variety of solvents:

$$D^0 = \frac{kT}{6\pi\eta^0 r}. \tag{17}$$

It is found that the product $D^0\eta^0$ increases linearly with respect to the molar volume of the solvent, according to the equation $D^0\eta^0 = 6.77 \times 10^{-8} + 6.0 \times 10^{-10}V_1^0$. While there is no apparent theoretical reason for this equation, the leading term in it can be thought of as the product $D^0\eta^0$ for diffusion in a medium the molecules of which have zero molecular volume, i.e. in the structureless continuum envisaged in the elementary derivation of Stokes' law. If the law is applied to data so obtained, i.e. if we use the equation

$$(D^0\eta^0)_{V_1^0 \to 0} = \frac{kT}{6\pi r},$$

we shall naturally obtain a value of r which is independent of the medium, having the value $r = 3.22 \times 10^{-8}$ cm, in satisfactory agreement with that given by X-ray diffraction experiments, which is 3.2×10^{-8} cm.

Self-diffusion

A more general expression than that of Stokes–Einstein gives the resistance factor as

$$C = 6\pi\eta r \left[\frac{1 + (2\eta/\beta r)}{1 + (3\eta/\beta r)} \right],$$

where β is a coefficient of sliding friction. When the radius of the solute greatly exceeds that of the solvent, the term in the square brackets tends towards unity. When the radius of the solute is relatively small, C becomes $4\pi\eta r$. For various reasons E. MacLaughlin considers that the equation

$$D = kT/4\pi\eta r$$

is the more appropriate expression for self-diffusion. Accurate measurements of the self-diffusion of liquid carbon tetrachloride have been made by A. F. Collings and R. Mills using a diaphragm-cell technique. Their results are shown in the second column of Table 5A. The last column shows that a constant value is found for the radius. It is greater than $(1/2)\sigma = 2.269A$ where σ is obtained from intermolecular force theory.

Empirically, the temperature dependence of D is given by the equation

$$D(\text{cm}^2 \text{ sec}^{-1}) = 2.386 \times 10^{-3} \exp(-3076/RT).$$

TABLE VII
THE SELF-DIFFUSION COEFFICIENTS OF CARBON TETRACHLORIDE

$T(°K)$	$D \times 10^5$ (cm^2 . sec^{-1})	$\eta \times 10^3$ (gr . cm^{-1} sec^{-1})	$(kT/4\pi\eta D) \times 10^8$ cm
288·2	1·092	10·67	2·722
298·2	1·296	9·21	2·744
303·2	1·428	8·50	2·746
313·2	1·683	7·51	2·721
323·2	1·937	6·70	2·745
		Average	2·736

This Arrhenius-type equation was obtained in the usual way by plotting $\ln D$ against $1/T$. E_D, the energy of diffusion, is seen to be 3,076 cals . mole^{-1}. Since $D = kT/4\pi\eta r$,

$$E_D = RT^2 \left(\frac{d \ln D}{dT}\right)_p = RT - RT^2 \left(\frac{d \ln \eta}{dT}\right)_p$$

$$= RT + B,$$

where B is the viscous enthalpy. The indirectly determined value of E_D is thus

$$E_D = 593 + 2,375 = 2,968 \text{ cal . mole}^{-1},$$

in satisfactory agreement with the directly measured value.

The Brownian Movement

In dealing with the limits of diffusion control, it will be helpful to consider the equation of linear motion of a particle of mass m, subject to a constant driving force f, and to a force of resistance which is proportional to the velocity:

$$m\frac{d^2x}{dt^2} = f - \frac{dx}{dt}. \tag{39}$$

The solution, averaged over all the particles in the system, is

$$x^2 = \frac{2kTt}{C}\left\{1 - \frac{m}{Ct}[1 - \exp(-Ct/m)]\right\}$$

$$= 2Dt\left\{1 - \frac{mD}{kTt}[1 - \exp(-kTt/mD)]\right\}, \tag{40}$$

where $D = kT/C$ (equation 16). After a short time interval, which for a particle of molar weight 50 in water at room temperatures is about 2×10^{-12} sec, $x^2/2Dt$ is within 90 per cent of its limiting value of 1. For longer times, equation (38) becomes

$$x^2 = 2Dt, \tag{41}$$

as has been abundantly verified. The motion of the particle is a zig-zag one through the system, due to randomly-directed collisions by solvent molecules and to the viscous resistance of the medium. When, however, time intervals considerably less than 10^{-12} sec are considered, expansion of equation (40) shows that the motion of the solute is independent of the viscous force. It is a to-and-fro motion, resembling a vibration of frequency,

$$v = \frac{1}{2\Delta}\left(\frac{kT}{m}\right)^{1/2}, \tag{42}$$

where $\Delta = \overline{(x^2)}^{1/2}$. The rate at which a solute molecule moves throughout the system and the rate at which it vibrates about a fixed position in it can both, as we have seen, be relevant to chemical kinetics.

Collisions Between Solute and Solvent Molecules

It has long been known that collisions between solute and solvent molecules present features distinguishing them from binary collisions between solute molecules. Jowett (1929) applied the kinetic theory to the problem, taking advantage of an expression derived by Jeans (1940) to allow for the tendency of velocity to persist after collision. The resulting expression for the number, Z, of collisions suffered per second by a single solute molecule with a solvent neighbour is

$$Z = 0.892 \times 3\pi\sigma\eta/\mu, \tag{43}$$

where σ is the diameter of the solute molecule, assumed to be spherical, η is the viscosity of the medium and μ the reduced mass of a solute–solvent pair. When the mass of the solute molecule greatly exceeds that of the solvent molecule, μ is effectively the mass, m, of the solute. Basing his argument on Einstein's theory of diffusion, Moelwyn-Hughes (1932) obtained a similar result, namely

$$Z = \frac{3\pi}{2} \cdot \frac{\sigma\eta}{m}. \tag{44}$$

The direct proportionality between collision frequency and viscosity follows from Maxwell's theory of the elasticity of compressible liquids, according to which

$$Z = \frac{3\kappa_1 a\eta}{\pi^2 m}. \tag{45}$$

Solute–solvent collision frequencies in solution, according to equation (44), decrease as the temperature is raised, necessitating a correction to the energy of activation for chemical reactions with rates proportional to Z. In a dilute solution containing n molecules of solute per cc, the number of solute–solvent collisions taking place in unit time and unit volume is Zn. The instantaneous rate of a chemical reaction governed by such collisions is

$$-\frac{dn}{dt} = Zn \cdot \exp\left(-E/RT\right), \tag{46}$$

and the first-order velocity constant is

$$k_1 = Z \cdot \exp\left(-E/RT\right) = \frac{3\pi}{2} \cdot \frac{\sigma\eta}{m} \cdot \exp\left(-E/RT\right). \tag{47}$$

The energy of activation is best found by plotting $\log\left(k_1/\eta\right)$ against $1/T$. In terms of the viscous enthalpy defined by equation (32), we see that

$$E = E_A + B. \tag{48}$$

On using the empirical parameter, b, of equation (33), we now write

$$k_1 = \frac{3\pi\sigma b}{2m} \cdot \exp\left(-E_A/RT\right). \tag{49}$$

These equations have been successfully applied to the kinetics of several simple uncatalysed hydrolyses in aqueous solution.

It follows from equation (48) that

$$\frac{dE}{dT} = \frac{dE_A}{dT} + \frac{dB}{dT}, \tag{50}$$

where dB/dT for water is a complicated function of T, which is discussed in Appendix 3.

The present treatment has been extended to bimolecular reactions in solution, and in particular to reactions catalysed by hydrogen ion and by hydroxyl ions in water. For the number of collisions per cc per sec between the substrate molecules and water, we have, from equation (44),

$$Z = \frac{3\pi\eta\sigma n_S}{2m_S} \cdot \exp\left(-E/RT\right), \tag{51}$$

where n_S is the concentration of substrate molecules. We next estimate crudely that the fraction of the water molecules (taking part in the collisions) which are already attached to the catalysing ion, say hydrogen ion, is the ratio of their concentrations, i.e. n_H/n_W. The catalytic coefficient thus becomes

$$k_c = -\frac{dn_S}{dt} \cdot \frac{1}{n_S n_H} = \frac{3\pi\sigma}{2m_S n_W} \cdot \exp(-E/RT). \tag{52}$$

A better though still crude estimate of the ratio would doubtless be given in terms of volume fractions. The equation as it stands accounts for the velocities of the hydrogen-ion catalysed hydrolyses of simple esters and amides, such as methyl acetate and acetamide.

Viscosity and Vibration Frequency in Liquids

According to equation (44), the viscosity, η, of a pure liquid consisting of incompressible spheres of diameter σ is directly proportional to the frequency, $v \,(= Z/2)$, with which the molecule vibrates about its mean position,

$$\eta = \frac{4}{3\pi} \cdot \frac{mv}{\sigma}. \tag{53}$$

Andrade (1936), treating viscosity as the transfer of momentum between parallel moving layers of harmonic oscillators, arrived at the equation

$$\eta = 2\left(\frac{\kappa_1}{\kappa_2}\right)\frac{mv}{a}, \tag{54}$$

where a is the average distance apart of the molecules, and κ_1 and κ_2 are structural factors relating the average volume, v, per molecule and the average relevant area, o, per molecule to the average separation, a:

$$v = \kappa_1 a^3, \qquad o = \kappa_2 a^2. \tag{55}$$

On account of the imperfection in the molecular arrangement of molecules in a liquid and of the ambiguity in the choice of the grazing surface, uncertainty attaches to the structural ratio. Its order of magnitude is unity. In a perfectly ordered face-centred cubic lattice, it is $2^{1/2}$:

$$\eta = 2^{1/2} \cdot \frac{mv}{a}. \tag{56}$$

Cheng (1947) uses 1·45 as the numerical factor (Partington, 1951). In terms of the density, ρ, the general relationship cited by Herzfeld and Litovitz (1959) is

$$\eta = \tfrac{1}{3}\rho a^2 v. \tag{57}$$

In terms of a, we have

$$\eta = \frac{1}{3\kappa_1} \cdot \frac{mv}{a}. \tag{58}$$

A critical review of most of the important theories has been given by these authors, to whom the reader is referred for particulars. We shall discuss only one of the theories here.

Jäger (1903) has related the viscosity of a liquid to its kinetic pressure, P_k, the average intermolecular distance, a, and the average molecular velocity, $\bar{c}$, as follows:

$$\eta = \frac{P_k}{\bar{c}} \cdot \frac{a}{2}. \tag{59}$$

The kinetic pressure is given, according to thermodynamical theory, in terms of the isothermal compressibility, β, and the coefficient of isobaric expansion, α:

$$P_k = \alpha T/\beta, \tag{60}$$

and $\bar{c}$ is given, according to the kinetic theory, as

$$\bar{c} = (8kT/\pi m)^{1/2}. \tag{61}$$

Liquid Viscosity in Terms of Intermolecular Force Theory

An approximate expression for the kinetic pressure of condensed monatomic systems at low temperatures is provided by the expression (Moelwyn-Hughes, 1951)

$$P_k = \tfrac{1}{2}(n' + m' + 1)\frac{kT}{v}, \tag{62}$$

where m' and n' are the integers of equation (1.4) and v is the molecular volume. By combining equations (59) and (62), we have

$$\eta = \frac{\tfrac{1}{2}(n' + m' + 1)\dfrac{kT}{v} \cdot \dfrac{a}{2}}{(8kT/\pi m)^{1/2}} = \frac{a}{v} \cdot \frac{(n' + m' + 1)}{16} \cdot (2\pi mkT)^{1/2} \tag{63}$$

The Viscosity of Liquid Mercury

Let us apply equation (63) to estimate the viscosity of liquid mercury at its melting point (234·34°K), at which temperature $a = 3·26$ Å. The integers

n' and m' are known to be 9 and 6 respectively (J. H. Hildebrand, H. R. R. Wakeham and R. N. Boyd, 1939; E. A. Moelwyn-Hughes, 1951; L. F. Epstein and M. P. Powers, 1953). Equation (63) now assumes the simple form

$$\eta = \frac{a}{v}(2\pi mkT)^{1/2}.\tag{64}$$

The numerical value at the melting temperature is $1{\cdot}95 \times 10^{-2}$ poise, which is within 4 per cent of the observed value of $2{\cdot}03 \times 10^{-2}$.

Eyring's Theory of the Viscosity of Liquids

H. Eyring has developed an interesting theory of the viscosity of liquids, the basic equation of which is

$$\eta = \frac{h}{v} \cdot \exp\left(E_0/kT\right),$$

where E_0 is an energy term, and h is Planck's (1936) constant. The appearance of h must not be construed as indicating that the motion of monatomic liquids at their melting temperatures obeys quantal rather than classical mechanics. On the whole, the evidence is to the contrary.

REFERENCES

Andrade, *Viscosity and Plasticity*, Heffer, Cambridge (1936).
Bäckström and Sandros, *Acta Chem. Scand.*, **15**, 823 (1958).
Berthoud, *J. Chem. Physique*, **10**, 633 (1912).
Boaz and Rollefson, *J. Amer. Chem. Soc.*, **72**, 3435 (1950).
Bradley, R. S., *Trans. Chem. Soc.*, 1910 (1934).
Butler, P. J. R., *Trans. Faraday Soc.*, **34**, 656 (1930).
Cheng, *Chinese Journal of Physics*, **7**, 56 (1947).
Collings, A. F. and R. Mills, *Trans. Faraday Soc.*, **66**, 2761 (1970).
Collins, F. C., and G. E. Kimball, *J. Colloid Science*, **4**, 425 (1949).
J. Crank, *The Mathematics of Diffusion*, Clarendon Press, Oxford (1956).
Debye, *Trans. Electrochem. Soc.*, **82**, 265 (1942).
de Maeyer, *Z. Electrochem.*, **59**, 986 (1955); *Proc. Roy. Soc.*, **247A**, 505 (1958).
Doss, *Proc. Indian Acad. Sci.*, **6**, 24 (1937).
Eigen, *Z. physikal Chem.*, **1**, 176 (1954).
Eigen and Eyring, *J. Amer. Chem. Soc.*, **84**, 3254 (1962).
Einstein, A., *Investigations on the Theory of the Brownian Movement* (*trans.* by Cooper),
    Methuen, London (1926).
Eptein, L. F., and M. D. Powers, *J. Phys. and Colloid Chem.*, **57**, 336 (1953).
Ertl and Gerischer, *Z. Electrochem.*, **66**, 560 (1962).
Fajans and Joos, *Z. Physik*, **23**, 1 (1924).
Förster, *Fluoreszenz Organischer Verbindungen*, Vanden-Hoek and Ruprecht, Göttingen
    (1951).

Grotthus, Th., *Ann. Chem.*, **58**, 54 (1806).

Hammond, B. R., and R. H. Stokes, *Trans. Faraday Soc.*, **51**, 1641 (1955).

Herzfeld, K. F. and Litovitz, *Absorption and Dispersion of Ultrasonic Waves*, p. 376, Academic Press (1959).

Hildebrand, J. H., H. R. R. Wakeham and R. N. Boyd, *J. Chem. Physics*, **7**, 1094 (1939).

Hill, Nora, W. E. Vaughan, A. H. Price and Mansel Davies, *Dielectric Properties and Molecular Behaviour*, van Nostrand, Reinhold Co., London (1969).

Horrocks, Kearvell, Tickle and F. Wilkinson, *Trans. Faraday Soc.*, **62**, 3393 (1966).

Jäger, *Ann. Physik*, **11**, 107 (1903).

Jeans, J. H., *The Kinetic Theory of Gases*, Cambridge Univ. Press (1940).

Jette and West, *Proc. Roy. Soc.*, **A.121**, 299 (1928).

Langevin, *Ann. Chim. Physique*, **28**, 28 (1903); *Comptes rendus*, **146**, 1011 (1908).

Logan, S. R., *Trans. Faraday Soc.*, **63**, 1713 (1967).

McLaughlin, E., *Trans. Faraday Soc.*, **55**, 29 (1959).

Moelwyn-Hughes, E. A., *J. Phys. Chem.*, **7**, 1246 (1951).

Moelwyn-Hughes, E. A., *J. Phys. and Colloid Chem.*, **55**, 1246 (1951).

Moelwyn-Hughes, E. A., *Kinetics of Reactions in Solution*, 2nd edn., p. 374, Oxford (1947).

Nernst, *Z. physikal Chem.*, **47**, 52 (1904).

Noyes, A. A., and Whitney, *Z. physikal Chem.*, **23**, 689 (1897).

Noyes, R. M., *Progress in Reaction Kinetics*, **1**, 129 (1961).

Ölander, *Z. physikal Chem.*, **144**, 118 (1929).

Onsager, *J. Chem. Physics*, **2**, 599 (1934).

Partington, *An Advanced Treatise of Physical Chemistry*, Vol. II, p. 95, Longmans, London (1951).

Partington, *An Advanced Treatise of Physical Chemistry*, Vol. II, p. 108.

Planck, *J. Chem. Physics*, **4**, 283 (1936).

Roller, *J. Phys. Chem.*, **39**, 221 (1935).

Smoluchowski, M. von, *Z. physikal Chem.*, **92**, 129 (1918).

Stern and Volmer, *Physikal Z.*, **20**, 183 (1919).

Stokes, G. G., *Proc. Camb. Phil. Soc.*, **9**, 5 (1856).

Umberger and La Mer, *J. Amer. Chem. Soc.*, **67**, 1099 (1945).

Wawilow, *Z. Physik*, **31**, 750 (1925); *Z. Physik*, **53**, 665 (1929).

Whytlaw-Gray and Patterson, *Smoke*, Arnold (1932).

Wiegner and Tuorila, *Kolloid-Zeitschrift*, **38**, 3 (1926).

Williamson, B., and La Mer, *J. Amer. Chem. Soc.*, **70**, 717 (1948).

Zsigmondy, *Z. physikal Chem.*, **92**, 600 (1918).

6

THE KINETIC COURSE OF
SOME SIMPLE REACTIONS

Although, in the ultimate analysis, most reactions are unimolecular or bimolecular, they are often concurrent, as if nature were reluctant to let any particular mechanism have things all its own way. This circumstance, coupled with the frequent need for allowing the attainment of equilibrium, renders it necessary to compare experimental data with mathematical equations that are sometimes complicated. A genuine disparity between observed rates and those reproduced by a kinetic equation based on a plausible mechanism demands investigation, and can lead to remarkable discoveries, of which ionization in solution is a notable example. On the other hand, some of the alleged drifts in velocity "constants" have proved to be spurious, due to faulty techniques, inaccurate analyses of the data, unreliable mechanisms, or failure to allow for equilibration, electrostatic effects and a zero time error.

Unopposed Unimolecular Reactions

If a denotes the initial concentration of reactant, and $a - x$ its concentration after a lapse of time, t, the instantaneous rate of reaction is

$$-\frac{d(a - x)}{dt} = \frac{dx}{dt} = k_1(a - x),\tag{1}$$

where the unimolecular constant, k_1, may be regarded as the fractional number of molecules decomposing in unit time, or as the average probability that any molecule taken at random shall decompose in unit time. It has the dimensions of frequency, or $(\text{time})^{-1}$. Integration* gives

$$k_1 = \frac{1}{t} \ln \frac{a}{a-x}. \tag{2}$$

The time, $t_{1/2}$, corresponding to half-completion, when $x = (1/2)a$, is seen to be

$$t_{1/2} = \frac{\ln 2}{k_1} = \frac{0.6933}{k_1},$$

which is independent of the initial concentration, a. The fractional extent of chemical change at a time t from the start is

$$x/a = 1 - \exp(-k_1 t). \tag{3}$$

Completion of reaction is therefore attained only after an infinite lapse of time. After a time, t, which equals $10 t_{1/2}$, however, the fractional change is 0·99902, and therefore, with an accuracy of 0·1 per cent, any unimolecular reaction can then be regarded as virtually complete.

According to equation (2),

$$\log_{10}(a - x) = \log_{10} a - (k_1/2.303)t. \tag{4}$$

The plot of $\log(a - x)$ against t should thus be linear as in Fig. 1, and a determination of the gradient by the method of least squares affords one of

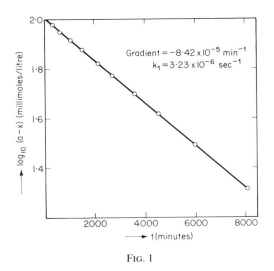

$$\text{Gradient} = -8.42 \times 10^{-5} \ \text{min}^{-1}$$
$$k_1 = 3.23 \times 10^{-6} \ \text{sec}^{-1}$$

Fig. 1

---

* Wilhelmy (1850). Detailed steps taken to integrate this and many other differential equations are shown in Moelwyn-Hughes, *Physical Chemistry*, p. 1298, 2nd Ed., Pergamon Press (1964).

the most reliable means of evaluating the velocity constant:

$$k_1 = -2 \cdot 303 d \log_{10} (a - x)/dt. \tag{5}$$

Because solutes do not dissolve or gases become heated instantaneously, it is difficult to say precisely when any reaction starts. In practice, a stop-clock or some other timing device is usually started half way through the operation of dissolving the solute or of admitting a gas into a reaction vessel. If the reaction does in fact start at that instant, the logarithmic plot, extrapolated to zero time, should correspond to $x = 0$. Otherwise, x at $t = 0$ is a small quantity, positive or negative, and the time axis is to be adjusted so as to yield the correct intercept as indicated by the chemical composition of the reactant solution and by the chemical analysis of the products.

After obtaining a first value of k_1 by a logarithmic plot, separate values should be calculated for each analytical datum, using equation (2) with the zero time correction, or the incremental equation

$$k_1 = \frac{1}{t_2 - t_1} \ln \frac{a - x_1}{a - x_2}. \tag{6}$$

Finally, x/a, calculated from equation (3) with the accepted mean value of k_1, should be compared with the experimental value to ensure that the difference is random and within the limits of analytical error.

The classical example of an unopposed unimolecular reaction in homogeneous systems is the decomposition of dinitrogen pentoxide

$$2N_2O_5 \rightarrow O_2 + 2N_2O_4$$
$$\updownarrow$$
$$4NO_2$$

in a variety of solvents (Lueck, 1922; Eyring and Daniels, 1930) and in the gaseous state at relatively high pressures (Daniels and E. H. Johnston, 1921; White and Tolman, 1925). The equilibrium between the nitrogen-containing products is known to be established rapidly compared with the rate of decomposition of the reactant (Selle, 1923; Brass and Tolman, 1932; W. T. Richards and J. A. Reid, 1933) and the relevant equilibrium constants are known (Natanson, 1886; Cundall, 1891).

The decomposition of ozone in carbon tetrachloride solution $2O_3 \rightarrow 3O_2$ is ostensibly an unopposed unimolecular reaction, though some slight attack on the solvent renders this instance somewhat ambiguous (Bowen, Moelwyn-Hughes and Hinshelwood, 1931).

The elimination of carbon dioxide from carboxylic acids $RCOOH \rightarrow RH + CO_2$ in a variety of solvents (Bredig and Balcom, 1908; Fajans, 1910; Wiig, 1928, 1930; Goldschmidt and Bräuer, 1906; Kappana, 1932; Moelwyn-Hughes and Hinshelwood, 1931; Dinglinger and Schröer, 1937; P. Johnson and Moelwyn-Hughes, 1940) as in the pure liquid phase (Barham and L. W. Clark, 1951), and of carbon disulphide from xanthogenic acid in various polar and non-polar media (von Halban and Kirsch, 1913) $C_2H_5OCSSH \rightarrow C_2H_5OH + CS_2$ are also unopposed unimolecular reactions. Certain cyclizations (Freundlich and M. B. Richards, 1912; Freundlich and Neumann, 1914; Freundlich and Kroepelin, 1926; Freundlich and Salomon, 1933; Salomon, 1936) follow a similar course,

$$X{-}(CH_2)_n{-}NH_2 \rightarrow (CH_2)_n\;\boxed{NH} + HX,$$

but are not invariably unopposed

Since k_1 has the dimensions of (time)$^{-1}$, its value is independent of the units of concentration employed, and we may replace the concentrations at time t and at infinity by the magnitude of any property known to be proportional to them. In the hydrolysis of the phenyldiazonium ion, for example, the extent, x, of reaction at time t is proportional to the pressure, p_t, which the system has developed at that time, and the total concentration, a, is proportional to p_∞. Equation (2) then takes the form

$$k_1 = \frac{1}{t} \ln \frac{p_\infty}{p_\infty - p_t}. \tag{7}$$

The results of a specimen run (E. A. Moelwyn-Hughes and P. Johnson, 1940) recorded in Table I illustrate the adequacy of this equation.

It is possible to determine the velocity constant of a reaction known to be unimolecular without following it to completion, by measuring the extent of the reaction, x, after fixed time intervals, each of extent Δt. From equation (3) we have $x_1/a = 1 - \exp(-k_1 t_1)$, and $x_2/a = 1 - \exp(-k_1 t_2)$. Therefore

$$\frac{x_2 - x_1}{a} = \exp(-k_1 t_1) - \exp(-k_1 t_2)$$

$$= \exp(-k_1 t_1)\{1 - \exp[-k_1(t_2 - t_1)]\}. \tag{8}$$

Similarly,

$$\frac{x_3 - x_2}{a} = \exp(-k_1 t_2)\{1 - \exp[-k_1(t_3 - t_2)]\}. \tag{9}$$

If measurements of x are made after equal time intervals, $\Delta t = t_2 - t_1 = t_3 - t_2$, the two terms in the curly brackets become equal. On dividing equation (8) by equation (9), we see that

$$\frac{x_2 - x_1}{x_3 - x_2} = \exp[k_1(t_2 - t_1)] = \exp(k_1 \Delta t),$$

$T = 313 \cdot 14 \pm 0 \cdot 03°\text{K}$. $k_1 = 4 \cdot 240 \times 10^{-4} \text{ sec}^{-4}$

t (sec)	$(p_\infty - p_t)_{obs}$	$(p_\infty - p_t)_{cal}$	t (sec)	$(p_\infty - p_t)_{obs}$	$(p_\infty - p_t)_{cal}$
0	22·62	22·68	1140	13·98	13·98
30	22·33	22·39	1200	13·62	13·63
60	22·08	22·10	1260	13·28	13·28
90	21·80	21·82	1320	12·96	12·95
120	21·55	21·55	1380	12·62	12·63
150	21·24	21·27	1440	12·30	12·31
180	20·97	21·00	1500	12·00	12·00
210	20·72	20·74	1560	11·74	11·70
240	20·47	20·47	1620	11·41	11·40
270	20·23	20·22	1680	11·10	11·12
300	19·97	19·96	1740	10·83	10·83
330	19·72	19·71	1800	10·54	10·57
360	19·45	19·46	1920	10·03	10·04
390	19·18	19·22	2040	9·53	9·54
420	18·97	18·97	2160	9·05	9·07
450	18·64	18·74	2280	8·61	8·62
480	18·48	18·49	2400	8·15	8·19
510	18·25	18·27	2520	7·76	7·79
540	18·05	18·03	2640	7·40	7·40
570	17·82	17·80	2760	7·00	7·03
600	17·60	17·58	2880	6·67	6·69
630	17·35	17·36	3000	6·34	6·35
660	17·12	17·14	3120	6·04	6·04
690	16·92	16·92	3240	5·77	5·74
720	16·73	16·70	3420	5·33	5·32
750	16·53	16·50	3600	4·88	4·88
780	16·32	16·29	3840	4·43	4·45
810	15·99	16·08	4080	4·02	4·02
840	15·90	15·88	4320	3·64	3·63
870	15·65	15·68	4560	3·32	3·28
900	15·49	15·47	4800	2·98	2·86
960	15·10	15·09	5100	2·62	2·61
1020	14·77	14·71	5400	2·32	2·30
1080	14·36	14·34	∞	0	0

so that

$$k_1 = \frac{1}{\Delta t} \ln \frac{x_2 - x_1}{x_3 - x_2}. \tag{10}$$

Mathematically, this method (E. A. Guggenheim, 1926) is unobjectionable, provided the unimolecularity of the reaction has already been undisputably established. Experimentally it is less informative than the other methods described.

Reversible Unimolecular Reactions

Let us consider the conversion of molecules of type A into molecules of type B and the reverse process, each obeying the unimolecular law:

$$A \underset{k_2}{\overset{k_1}{\rightleftarrows}} B$$
$$_{(a-x)} \qquad _{(b+x)}$$

The initial concentrations are a and b respectively, and x is the change in the concentration of either after a time t. Then

$$dx/dt = k_1(a - x) - k_2(b + x)$$
$$= (k_1 a - k_2 b) - (k_1 + k_2)x, \tag{11}$$

which gives, on integration (P. Henry, 1892; T. M. Lowry, 1899),

$$x = \left(\frac{k_1 a - k_2 b}{k_1 + k_2} \right) \{ 1 - \exp\left[-(k_1 + k_2)t \right] \}. \tag{12}$$

Equilibrium conditions are realized when the rate of reaction is zero, or the time infinite. From either equation (11) or (12), the equilibrium value of x is thus seen to be

$$x_e = \frac{k_1 a - k_2 b}{k_1 + k_2}. \tag{13}$$

This result can be rearranged to

$$\frac{b + x_e}{a - x_e} = \frac{k_1}{k_2}, \tag{14}$$

which is the kinetic derivation of the equilibrium law for this system:

$$\frac{[B]_e}{[A]_e} = K. \tag{15}$$

For convenience, equation (12) may be written as follows:

$$k_1 + k_2 = \frac{1}{t} \ln \frac{x_e}{x_e - x}. \tag{16}$$

Let us apply these equations to the mutarotation of glucose in water:

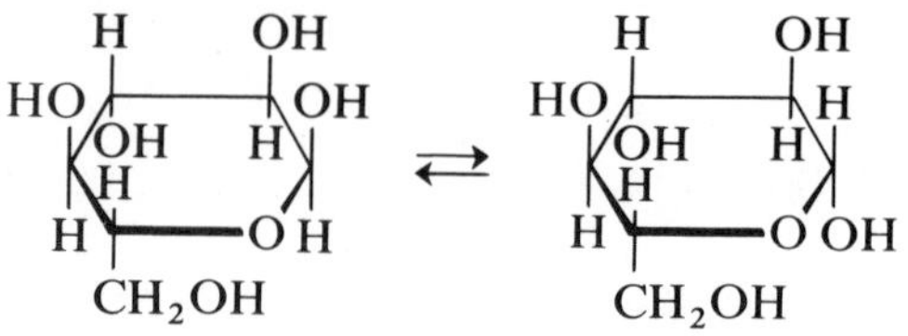

α-Glucose   β-Glucose

FIG. 2

The extent, x, of chemical change at time t is proportional to the change, $(\alpha_t - \alpha_0)$, in optical rotation; and the total change, x_e, is proportional to $(\alpha_\infty - \alpha_0)$, so that equation (16) becomes

$$k_1 + k_2 = \frac{1}{t} \ln \left(\frac{\alpha_0 - \alpha_\infty}{\alpha_t - \alpha_\infty} \right). \tag{17}$$

The mean value of $(k_1 + k_2)$ found using pure α-glucose in water at 278·26°K is $5\cdot13 \times 10^{-5}$ sec^{-1}. Changes in optical rotation reproduced by the equation

$$\alpha_t - \alpha_\infty = (\alpha_0 - \alpha_\infty) \exp \left[-(k_1 + k_2)t \right] \tag{18}$$

are compared in Table II with the observed values (J. C. Kendrew and E. A. Moelwyn-Hughes, 1940). The ratio k_1/k_2 at this temperature is 0·575, so that $k_1 = 1\cdot864 \times 10^{-5}$ sec^{-1}.

In racemizations, α_∞ is zero, and the rates of the direct and reverse reactions are equal, so that

$$k_1 = k_2 = \tfrac{1}{2}t \ln \frac{\alpha_0}{\alpha_t}. \tag{19}$$

An interesting class of racemizations is that of the substituted biphenyls, such as

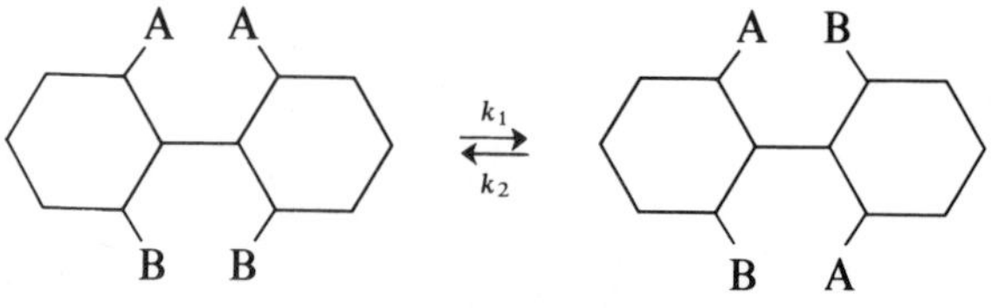

which have been studied in the gas phase (Kuhn and Albrecht, 1927; Li and Adams, 1935; Kistiakowsky and W. R. Smith, 1936), and more extensively in solution (M. M. Harris and K. R. Mitchell, 1960; A. S. Cook and Harris, 1963; C. C. K. Ling and Harris, 1964).

TABLE II
THE UNCATALYSED MUTAROTATION OF α-GLUCOSE IN WATER
AT 278·26°K

$a = 0.3193$ mole/litre. $k_1 + k_2 = 5.13 \times 10^{-5}$ sec^{-1}

| | $\alpha_t - \alpha_\infty$ | | |
t (min)	obs	calc	100 δ
0	13·01	13·01	0
20	12·23	12·23	0
30	11·86	11·86	0
40	11·50	11·50	0
50	11·17	11·15	+2
60	10·84	10·81	+3
75	10·35	10·32	+3
90	9·88	9·86	+2
105	9·43	9·41	+2
125	8·87	8·85	+2
154	8·10	8·10	0
161	7·90	7·92	−2
180	7·49	7·47	+2
190	7·27	7·25	+2
210	6·82	6·81	+1
220	6·61	6·61	0
240	6·22	6·21	+1
260	5·83	5·84	−1
280	5·49	5·49	0
300	5·16	5·16	0
330	4·72	4·71	+1
360	4·28	4·29	−1
390	3·90	3·91	−1
420	3·55	3·57	−2
450	3·23	3·25	−2
480	2·94	2·97	−3
∞	0	0	

The Relaxation Time in Reversible Unimolecular Reactions

On writing equation (11) in the form

$$-d(x - x_e)/dt = (k_1 + k_2)(x - x_e), \qquad (20)$$

it is seen that, when the composition of the system has been displaced from its equilibrium value to an extent $(x - x_e)$, the system reverts to its equilibrium state at a rate which is proportional to the extent of this displacement.

This is the basic law of relaxation phenomena. The relaxation time in this instance is defined as follows:

$$\tau = 1/(k_1 + k_2). \tag{21}$$

Hence

$$x_e = x + \tau(dx/dt), \tag{22}$$

which is another form of the same law. Finally, if we denote $x - x_e$ by Δx, we see that

$$-d(\Delta x)/dt = \Delta x/\tau, \tag{23}$$

which integrates to give a third form of the law:

$$\Delta x = \Delta x^0 . \exp(-t/\tau). \tag{24}$$

τ is thus the time taken for the magnitude of an initial displacement from the equilibrium state to be reduced by the factor e.

Pseudo-unimolecular Reactions

Let us suppose that, in the reaction $A + \nu B \to P$, the instantaneous rate of reaction is proportional to the product of the concentrations raised to the appropriate powers:

$$-dn_A/dt = kn_A n_B^\nu. \tag{25}$$

When the concentration of A is much less than that of B, n_B does not change sensibly, and the reaction appears to be one of the first order, with a velocity constant given as

$$k_1 = -\frac{1}{n_A} \cdot \frac{dn_A}{dt} = kn_B^\nu. \tag{26}$$

The reaction under these circumstances is said to be pseudo-unimolecular. The value of the integer ν may be found from the isothermal gradient

$$\frac{d \ln k_1}{d \ln n_B} = \nu. \tag{27}$$

Hydrolyses take this course, with ν equal to about 6. From a detailed analysis of the kinetic data of Lewis and his collaborators (C. M. Jones and W. C. McC. Lewis; Moran and W. C. McC. Lewis, 1922) on the hydrogen-ion catalysed hydrolysis of sucrose, G. Scatchard (1921) concluded that the order of reaction, with respect to the water molecules, is 6 ± 1. Kappanna (1932) has measured the rate of decomposition of the trichloracetate ion, over a concentration range from 0·25 to 5·00 molar, in aqueous solution at 80°C. The order of reaction, as given by equation (27), is 6 with respect to the water molecules. The value of v, found in the hydrolysis of tertiary butyl chloride is 5·7 from work on acetone–water mixtures (E. Tommila, M. Tiilikainer and A. Voipio, 1955) and 5·58 ± 0·18 from work on methanol–water mixtures (E. A. Moelwyn-Hughes, 1962). The reaction between ethyl iodide and di-methylaniline, which is bimolecular when carried out at low concentrations in an inert solvent, is pseudo-unimolecular when the latter reactant is used as a solvent (Hirniak, 1922). In the nitration of mesitylene in carbon tetra-chloride solution at 25°C, the order of reaction with respect to nitric acid is given by equation (27) as 5. There are thus present in the activated complex 5 molecules of HNO_3 and one of the hydrocarbon (R. G. Coombes, 1969), making a total order of 6.

Consecutive Unimolecular Reactions

When molecules of type A are converted into molecules of type B, which, in turn are transformed into molecules of type C, each by a unimolecular process, we have the simplest example of consecutive reactions:

$$A \overset{k_1}{\to} B \overset{k_2}{\to} C$$
$$\underset{a-x}{\overset{n_1}{}} \quad \underset{y}{\overset{n_2}{}} \quad \underset{z}{\overset{n_3}{}}$$

By integrating the differential equations

$$-dn_1/dt = k_1 n \tag{1}$$

and

$$-dn_2/dt = k_2 n_2 - k_1 n_1, \tag{28}$$

we find that

$$n_1 = a - x = n_0 \exp(-k_1 t), \tag{3}$$

$$n_2 = y = n_0 \frac{k_1}{k_2 - k_1}[(\exp(-k_1 t) - \exp(-k_2 t)], \tag{29}$$

and

$$n_3 = z = n_0 \left[1 - \left(\frac{k_2}{k_2 - k_1} \right) \exp(-k_1 t) + \left(\frac{k_1}{k_2 - k_1} \right) \exp(-k_2 t) \right], \tag{30}$$

where

$$n_0 = n_1 + n_2 + n_3,$$

and

$$x = y + z.$$

We note in the first place that when $k_1 > k_2$,

$$n_3 = n_0[1 - \exp(-k_2 t)],$$

and that when $k_2 > k_1$,

$$n_3 = n_0[1 - \exp(-k_1 t)],$$

which illustrates the general principle, enunciated in Chapter 1, that in a sequence of reactions the net rate is governed by that step which has the lowest velocity constant.

The concentration of reactant A decreases from the start of the reaction. The concentration of the intermediate reactant B, which is initially zero, at first increases, reaches a maximum when dn_2/dt is zero, and thereafter decreases exponentially with respect to time. When dn_2/dt is zero, the system is said to have reached a stationary state. At that instant and thereafter, we see from equations (29) and (3) that

$$n_2 = \frac{k_1}{k_2} n_1 = n_0 \frac{k_1}{k_2} . \exp(-k_1 t). \tag{31}$$

Thus the concentrations of A and B both diminish exponentially with respect to time while maintaining a constant ratio.

On differentiating equation (29), we obtain the following expression for the time, t_m, required for the attainment of the stationary state:

$$k_1 \exp(-k_1 t_m) = k_2 \exp(-k_2 t_m),$$

or

$$t_m = \frac{1}{k_1 - k_2} \ln \frac{k_1}{k_2}. \tag{32}$$

The maximum concentration of the intermediate compound, B, is seen to be:

$$n_2^{\max} = n_0 . e^{-k_2 t_m}. \tag{33}$$

The exceptionally accurate data of T. M. Lowry and R. Traill (1931) on the rate of change of the optical activity of aluminium benzoylcamphor in carbon tetrachloride solution at $293.16°K$ indicate that mutarotation of the solute takes place in two consecutive unimolecular stages, with $k_1 = 9.01 \times 10^{-3}$ sec^{-1} and $k_2 = 1.91 \times 10^{-3}$ sec^{-1}. According to equation (32), t_m is 218 sec. With their initial concentration of 10 millimoles per litre of solution, $n_2^{\max}$ becomes 6.59 in these units. The temporal course of n_2 is shown in Fig. 3. Both k_1 and k_2 are, of course, composite constants.

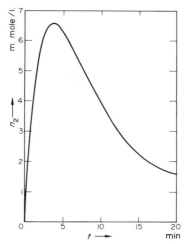

FIG. 3. Concentration of intermediate isomer in the uncatalysed mutarotation of aluminium benzoylcamphor (calculated from the data of Lowry and Traill).

Galactose in water mutarotates in consecutive unimolecular steps (G. F. Smith and T. M. Lowry, 1928). The best known example of such a mechanism is provided by the spontaneous disintegration of radioactive nuclei, such as that of uranium into radium, followed by the formation of radium emanation:

$$\underset{n_1}{U} \xrightarrow{k_1} \underset{n_2}{Ra} \xrightarrow{k_2} Rn.$$

Certain intermediate steps have here been omitted. The experimental value of k_2 is 1.381×10^{-11} sec^{-1}, and the ratio of the stationary concentrations, n_2/n_1 is 3.4×10^{-7}; hence $k_1 = 4.70 \times 10^{-18}$ sec^{-1}, which corresponds to a half-life of 4.4×10^9 years (Mme. Curie, 1935). Mathematical equations have been derived for systems undergoing any number of consecutive unimolecular changes (Bateman, 1910).

The decompositions of methylene oxide (Sickman, 1936; Fletcher and Rollefson, 1936) and dimethyl ether (Hinshelwood and Askey, 1927) in the gaseous phase have been found to follow the same kinetic sequence:

$$(CH_2)_2O \rightarrow (CH_3CHO)^* \rightarrow CH_4 + CO,$$

$$(CH_3)_2O \rightarrow CH_4 + HCHO \rightarrow H_2 + CO$$

Consecutive Pseudo-unimolecular Reactions

An electrometric determination of the rate of hydrolysis of β-β'-dichlordiethyl sulphide (mustard gas) in water showed that the change proceeded in two steps, which were presumed to be

$$\underset{\substack{CH_2CH_2Cl \\ \diagup \\ S \\ \diagdown \\ CH_2CH_2Cl}}{} \xrightarrow[-HCl]{+H_2O} \underset{\substack{CH_2CH_2OH \\ \diagup \\ S \\ \diagdown \\ CH_2CH_2Cl}}{} \xrightarrow[-HCl]{+H_2O} \underset{\substack{CH_2CH_2OH \\ \diagup \\ S \\ \diagdown \\ CH_2CH_2OH}}{} .$$

Thiodiglycol and hydrochloric acid are finally formed in quantitative yields, and the pseudo-unimolecular constants are 25°C were found to be 1.96×10^{-3} and $2.77 \times 10^{-3} \sec^{-1}$ respectively (K. Brookfield, 1950). The stable intermediate compound in this instance has been isolated bio-chemically (R. Peters), and its rate of hydrolysis found to be the same as that derived from the study of the composite reaction. Di-alkyl esters of the dicarboxylic acids, in the presence of excess alkali, follow a similar course (J. W. Mellor, 1904).

Unopposed Bimolecular Reactions

The instantaneous rate of an unopposed bimolecular reaction, such as

$$\underset{(a-x)}{A} + \underset{(b-x)}{B} \xrightarrow{k_2} \underset{x}{P}$$

is

$$dx/dt = k_2(a - x)(b - x), \tag{34}$$

where k_2, the velocity constant, is found by integration to be

$$k_2 = \frac{1}{t(a - b)} \ln \frac{b(a - x)}{a(b - x)}. \tag{35}$$

By re-writing equation (35) in the form

$$\frac{x}{a} = \frac{1 - \exp[k_2(a - b)t]}{1 - (a/b)\exp[k_2(a - b)t]}, \tag{36}$$

we see that, consistent with the irreversibility of the reaction, x at infinite time equals a or b, whichever is the less. The applicability of this equation to a typical reaction is shown in Fig. 4.

When $(a - b)$ is small, we may expand the exponential terms, obtaining

$$\frac{x}{a} = \frac{-k_2(a - b)t}{1 - (a/b)[1 + k_2(a - b)t]} = \frac{k_2bt}{1 + k_2at}.$$

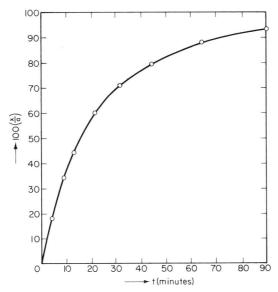

FIG. 4. Fractional change of an uncomplicated bimolecular reaction. Circles: observed values. Full line: equation 36.

But, when $(a - b)$ is small, a and b are effectively equal, so that

$$\frac{x}{a} = \frac{k_2 at}{1 + k_2 at},$$

or

$$k_2 = \frac{1}{ta} \cdot \frac{x}{a - x}. \tag{37}$$

This result is usually obtained by direct integration of the equation

$$dx/dt = k_2(a - x)^2. \tag{38}$$

The time for half-completion is clearly

$$t_{1/2} = 1/k_2 a. \tag{39}$$

In general, when equal concentrations are used, the time of half-completion of a reaction is inversely proportional to the initial concentration raised to the power of $v - 1$, where v is the order of reaction.

Equation (5) may be written in the form

$$\log_{10} \frac{b - x}{a - x} = \log_{10} \frac{b}{a} + \frac{k_2(b - a)}{2 \cdot 303} t, \tag{40}$$

from which we see that k_2 may be found from the gradient obtained when

$\log_{10}[(b - x)/(a - x)]$ is plotted against t, and the zero time error, positive or negative, from the difference between the intercept corresponding to zero time and the analytical value of $\log_{10}(b/a)$.

When the concentrations of the reactants are equal, both k_2 and the zero time error are found by plotting the reciprocal of the concentration of reactant as a function of t:

$$\frac{1}{a - x} = \frac{1}{a} + k_2 t. \tag{41}$$

The zero time error need not be considered when applying the incremental form of equation (37):

$$k_2 = \frac{1}{t_2 - t_1}\left(\frac{1}{a - x_2} - \frac{1}{a - x_1}\right). \tag{37a}$$

Most esters react with the hydroxyl ion in aqueous solution according to the mechanism of irreversible bimolecular reaction. The development of physical chemistry owes so much to the investigation of reactions in this category that they merit special discussion.

The Kinetics of the Saponification of Esters; The Origin of the Theory of Ionization in Solution

Before the idea of ions in solution had been conceived, the most rational way of defining a bimolecular velocity constant was in the units of litres per gram-mole-second. In the reaction between an ester and a base, for example, the velocity constant, k'_2, was defined by the equation

$$-\frac{d[\text{ester}]}{dt} = k'_2 [\text{ester}][\text{base}], \tag{42}$$

where the square brackets denote total molar concentration, and k'_2 is the stoichiometric second-order velocity constant. Data on the saponification of ethyl acetate were given in these terms by Warder (1881), Reicher (1886) and Arrhenius (1887). What seemed to be an anomaly appeared in Arrhenius' second publication on this subject (1887), leading to results and conclusions which may be summarized by reference to the following reactions:

1. $CH_3 \cdot COOC_2H_5 + NaOH \rightarrow CH_3 \cdot COONa + C_2H_5OH$

2. $CH_3 \cdot COOC_2H_5 + KOH \rightarrow CH_3 \cdot COOK + C_2H_5OH$

3. $2CH_3 \cdot COOC_2H_5 + Ba(OH)_2 \rightarrow (CH_3 \cdot COO)_2Ba + 2C_2H_5OH$

4. $CH_3 \cdot COOC_2H_5 + NH_4OH \rightarrow CH_3 \cdot COONH_4 + C_2H_5OH$.

k'_2 for reactions 1 and 2 were found to be identical. It therefore appeared that the real species reacting with the ester is the OH part of the base, and that the nature of the metallic part is irrelevant. k'_2 for reaction 3, as for reactions 1 and 2, obeyed equation (35) throughout each run, indicating that this reaction also is bimolecular, rather than termolecular, as suggested by the chemical equation. Moreover, the stoichiometric second-order rate constant obtained for reaction 3 was found to be twice as great as the value found for reactions 1 and 2. It thus seemed that the effective concentration of the inorganic reactant in $Ba(OH)_2$ solutions was twice the value for equimolar solutions of NaOH and KOH, and was, in short, an equivalent rather than a molar concentration. With ammonia as base, k'_2 was found to be anything but constant. Moreover, its initial value was lower by several orders of magnitude than k'_2 found for the other reactions. These facts Arrhenius interpreted by making the following hypotheses, each of which has since, in the main, been substantiated:

(a) the bases NaOH, KOH and $Ba(OH)_2$ are ionized almost completely in aqueous solutions, into Na^+ and OH^-, K^+ and OH^-, and Ba^{++} and $2OH^-$, respectively,

(b) the rate of chemical change is proportional to the product of the concentrations of ester and hydroxyl ion, so that the bimolecular velocity coefficient k_2, defined by the equation

$$-\frac{d\,[\text{ester}]}{dt} = k_2\,[\text{ester}][\text{OH}^-], \tag{43}$$

is a true constant; and, finally

(c) aqueous solutions of ammonia contain only a small fraction of the total concentration of base in the form of ions, indicating that its degree of ionization ($NH_4OH \rightleftarrows NH_4^+ + OH^-$) is small. It is now known that in a molar solution of ammonia in water at 25°C only a fraction 0·00424 exists in the form of free ions. Since the days when the foundation of the theory of ionization in solution was, in this way, laid down, the use of a stoichiometric second-order rate constant, k'_2, has, with few exceptions (e.g., E. D. Hughes, C. K. Ingold and Mackie, 1955; E. D. Hughes, C. K. Ingold and A. J. Parker, 1960) fallen into desuetude.

Concurrent Unimolecular and Bimolecular Reactions

Many types of organic molecules react with water and other hydroxylic solvents at a rate which is unimolecular with respect to the solute. They also frequently react simultaneously with ions in bimolecular reactions. Such is the case, for example, with the methyl halides in aqueous alkaline solution

(E. A. Moelwyn-Hughes, 1949). If x denotes the concentration of organic halide which has reacted in time t, according to the kinetic scheme:

$$CH_3X + H_2O \overset{k_1}{\to} CH_3OH + H^+ + X^-,$$
$$ {\scriptstyle (a-x)}$$

$$CH_2X + OH^- \overset{k_2}{\to} CH_3OH + X^-,$$
$${\scriptstyle (a-x)} \quad {\scriptstyle (b-x)}$$

the instantaneous rate of reaction is given by the differential equation

$$dx/dt = k_1(a - x) + k_2(a - x)(b - x), \tag{44}$$

and by the integrated equation

$$t = \frac{1}{k_1 + k_2(b - a)} \ln \left[\left(\frac{a}{a - x} \right)^{\frac{k_1 + k_2(b - x)}{k_1 + k_2 b}} \right]. \tag{45}$$

This expression naturally reduces to the ordinary laws of unimolecular and bimolecular processes when k_2 and k_1 are, respectively, zero. The fractional change at time t is seen to be

$$\frac{x}{a} = \frac{e^{\alpha t} - 1}{e^{\alpha t} - \beta}, \tag{46}$$

where

$$a = k_1 + k_2(b - a)$$

and

$$\beta = k_2 a/(k_1 + k_2 b).$$

Instances of the applicability of this equation are shown in Table III. When a and b are equal, the fractional change at time t is

$$\frac{x}{a} = \frac{1 - \exp(-k_1 t)}{1 - [k_2 a/(k_1 + k_2 a)] \exp(-k_1 t)} \tag{47}$$

When k_2 is zero, we recover equation (3). When k_1 is small compared with k_2, we expand the exponential term, obtaining

$$\frac{x}{a} = \frac{k_1 + k_2 a t}{1 + k_2 a t},$$

which, in the limit when k_1 is zero, reduces to equation (37). These results can also be obtained directly by integrating the equation

$$dx/dt = k_1(a - x) + k_2(a - x)^2. \tag{48}$$

Simultaneous unimolecular and bimolecular kinetics hold also for the cyclization of the ω-brompentadecanoate ion in aqueous solution (A. C.

TABLE III

FRACTIONAL CHANGES OF THE METHYL HALIDES IN AQUEOUS ALKALINE SOLUTION

(k_1 and α in $\sec^{-1}$)

t (min)	millimoles/l		$100(x/a)$	
	$(a - x)$	$(b - x)$	observed	calculated by equation (46)
CH$_3$Cl at 333·05°K : $k_1 = 2·16 \times 10^{-6}, k_2 = 4·88 \times 10^{-4}, \alpha = 3·50 \times 10^{-5}$, $\beta = 3·30 \times 10^{-1}$				
0	35·3	102·5	0	0
2	35·1	102·3	0·6	0·6
30	32·0	99·2	9·3	8·8
65	29·4	96·6	16·7	18·0
109	25·5	92·7	27·7	27·7
214	18·9	86·1	46·5	45·8
1160	2·1	69·3	94·0	93·9
∞	0	67·2	100·0	100·0
CH$_3$Br at 297·62°K : $k_1 = 3·21 \times 10^{-7}, k_2 = 1·373 \times 10^{-4}, \alpha = 6·88 \times 10^{-6}$, $\beta = 5·33 \times 10^{-1}$				
0	57·2	105·0	0	0
200	48·4	96·2	15·4	15·5
500	38·5	86·3	32·7	32·9
900	29·0	76·8	49·3	49·1
1400	21·1	68·9	63·1	62·7
2000	15·2	63·0	73·4	72·5
3000	8·9	56·7	84·4	84·4
∞	0	47·8	100·0	100·0
CH$_3$I at 342·80°K : $k_1 = 2·45 \times 10^{-5}, k_2 = 8·55 \times 10^{-3}, \alpha = 8·10 \times 10^{-4}$, $\beta = 1·105 \times 10^{-1}$				
0	11·76	103·60	0	0
3·5	9·79	101·63	16·8	17·2
8·5	7·52	99·36	36·0	36·5
13·5	5·57	97·41	50·7	51·0
19·5	4·20	96·04	64·3	64·0
26·5	2·92	94·76	75·2	74·7
37·5	1·75	93·59	85·1	85·3
∞	0	91·84	100·0	100·0

Davies, Mansel Davies and M. Stoll, 1954).

$$Br(CH_2)_{14}COO^- \rightarrow (CH_2)_{14} \begin{array}{c} C=O \\ | \\ O \end{array} + Br^-.$$

Opposing Bimolecular and Unimolecular Reactions

When amines and sulphides, aromatic or aliphatic, react with alkyl halides, salts are formed, such as tetramethylammonium chloride and triethylsulphonium bromide:

$$(CH_3)_3N + CH_3Cl \rightarrow (CH_3)_4NCl$$

$$(C_2H_5)_2S + C_2H_5Br \rightarrow (C_2H_5)_3SBr$$

An analogous reaction is the formation of diphenyliodonium iodide:

$$C_6H_5I + C_6H_5I \rightarrow [(C_6H_5)_2I]I$$

These reactions, however, are not complete, because the salts, though possibly insoluble in a pure solvent, are generally soluble to some extent in the presence of the reactants which form it. The simplest means of formulating the chemical kinetics of such a system, assuming that no salt is precipitated, is as follows:

$$\underset{(a-x)}{A} \quad \underset{(b-x)}{B} \underset{k_1}{\overset{k_2}{\rightleftarrows}} \underset{x}{C}$$

when there is no salt present initially. The differential equation

$$dx/dt = k_2(a - x)(b - x) - k_1 x \tag{49}$$

gives, on integration,

$$k_2 = \frac{1}{2\beta t} \ln \left\{ \frac{1 + x/[\beta - (1/2)(a + b + K)]}{1 - x/[\beta + (1/2)(a + b + K)]} \right\}, \tag{50}$$

where $K = k_1/k_2$, and $\beta^2 = (1/4)(a + b + K)^2 - ab$. The extent of change at time t is given by the equation

$$x = \frac{ab}{(1/2)(a + b + K) + \beta \coth (\beta k_2 t)}. \tag{51}$$

After an infinite lapse of time, when coth t becomes unity, we have

$$x_\infty = \frac{ab}{(1/2)(a + b + K) + \beta}.$$

Hence,

$$\frac{(a - x_\infty)(b - x_\infty)}{x_\infty} = x_\infty \left(\frac{a}{x_\infty} - 1\right)\left(\frac{b}{x_\infty} - 1\right) = K. \tag{52}$$

These equations have been found to apply to the reaction between dimethylaniline and methyl iodide in nitrobenzene and other solvents (Essex and Gelormini, 1926) and to the reaction between diethyl sulphide and ethyl bromide in benzyl alcohol and in mixtures of this solvent with toluene and glycerol (Corran, 1927).

A special case of the kinetic scheme described above is that when the concentrations of the two reactants are equal, i.e., $a = b$, as when aqueous ammonium cyanate, in the absence of other cyanates or ammonium salts, is converted into urea:

$$NH_4^+ + CNO^- \rightleftarrows (NH_2)_2CO.$$

From the scheme

$$\underset{(a-x)}{A} + \underset{(a-x)}{B} \underset{k_1}{\overset{k_1}{\rightleftarrows}} \underset{x}{C},$$

we now have the differential equation

$$dx/dt = k_2(a - x)^2 - k_1 x, \tag{53}$$

which gives, on integration,

$$k_2 = (1/2\beta t) \ln \frac{[1 + x/(\beta - a - K/2)]}{[1 - x/(\beta + a + K/2)]}, \tag{54}$$

where, as before, $K = k_1/k_2$, and β is now given by the equation $\beta^2 = (K^2/4) + Ka$. In this particular instance

$$x/a = [1 + K/2a + (\beta/a) \coth (\beta k_2 t)]^{-1}, \tag{55}$$

and

$$\frac{x_\infty}{a} = \left(1 + \frac{K}{2a} + \frac{\beta}{a}\right)^{-1}.$$

It is again easily verified that

$$\frac{(a - x_\infty)^2}{x_\infty} = x_\infty \left(\frac{a}{x_\infty} - 1\right)^2 = K. \tag{56}$$

When the reaction begins with the unimolecular step, and no products of decomposition are initially present,

$$\underset{(a-x)}{A} \underset{k_2}{\overset{k_1}{\rightleftarrows}} \underset{x}{B} + \underset{x}{C},$$

we have the differential equation

$$\frac{dx}{dt} = k_1(a - x) - k_2 x^2, \tag{57}$$

and the integrated equation

$$k_2 = \frac{1}{2\alpha t} \ln \left\{ \frac{a + x[(k_2/k_1)\alpha - 1/2]}{a - x[(k_2/k_1)\alpha + 1/2]} \right\}, \tag{58}$$

where

$$\alpha^2 = \frac{k_1}{k_2}a + \frac{1}{4}\left(\frac{k_1}{k_2}\right)^2. \tag{59}$$

The fractional change at time t is

$$\frac{x}{a} = \left[\frac{1}{2} + \frac{k_2}{k_1}\alpha \coth(k_2\alpha t) \right]^{-1}, \tag{60}$$

which at infinite time again gives the equilibrium law

$$\frac{x_\infty^2}{a - x_\infty} = \frac{k_1}{k_2} = K. \tag{61}$$

These equations may be cast in other forms, using x_∞ rather than α, where x_∞ is the equilibrium concentration of products. Equation (57) applied to equilibrium conditions is $k_1(a - x_\infty) - k_2 x_\infty^2 = 0$, from which we have the only positive root

$$x_\infty = \frac{k_1}{k_2}\left[\left(\frac{k_2}{k_1}a + \frac{1}{4}\right)^{1/2} - \frac{1}{2} \right].$$

The only positive root of equation (59) is

$$\alpha = \frac{k_1}{k_2}\left(\frac{k_2}{k_1}a + \frac{1}{4}\right)^{1/2},$$

from which it follows that

$$\alpha = x_\infty + \frac{1}{2}\frac{k_1}{k_2}.$$

With the aid of equation (61), the expression for the bimolecular constant (equation 58) can now be written as follows to yield the unimolecular constant:

$$k_1 = \frac{1}{t}\left(\frac{x_\infty}{2a - x_\infty}\right) \ln\left(\frac{1 + x/x_\infty - x/a}{1 - x/x_\infty}\right). \tag{58a}$$

When the products of decomposition do not combine, $x_\infty = a$, and we recover equation (2).

These relationships have been applied to the decomposition of tri-ethyl sulphonium bromide in a variety of solvents (von Halban, 1909), to the dissociation of dinitrogen tetroxide in the gaseous phase (H. Selle, 1923; Brass and R. C. Tolman, 1932; W. T. Richards and J. A. Reid, 1933), $N_2O_4 = 2NO_2$, and to ionizations in solution (E. A. Moelwyn-Hughes, 1938; Eigen, 1954).

If the extent of decomposition is very slight, $a - x$ may be taken as approximately a throughout the reaction. Then

$$\frac{dx}{dt} = k_1 a - k_2 x^2. \tag{62}$$

The integrated equation is now

$$k_2 = \frac{1}{2\beta t} \ln \frac{\beta + x}{\beta - x}, \tag{63}$$

where

$$\beta^2 = k_1 a / k_2. \tag{64}$$

The extent of change at time t is

$$x = \beta \tanh (k_2 \beta t). \tag{65}$$

These equations hold for the ionization of gaseous molecules by X-rays, and k_2 denotes the bimolecular constant governing the rate of combination of the ions (J. J. Thomson and G. P. Thomson, 1928).

The Relaxation Time in Opposing Bimolecular and Unimolecular Reactions

Let us re-formulate the kinetic scheme discussed above by expressing the concentration of each species in terms of the equilibrium concentration, n^0, and the extent, Δn, by which it differs from n^0. For the system

$$AB \underset{k_2}{\overset{k_1}{\rightleftarrows}} A + B$$

we then have

$$n_{AB} = n_{AB}^0 + \Delta n, \qquad n_A = n_A^0 - \Delta n, \qquad n_B = n_B^0 - \Delta n. \tag{66}$$

The rate of chemical change in the system displaced from its equilibrium state is

$$dn_{AB}/dt = k_2(n_A^0 - \Delta n)(n_B^0 - \Delta n) - k_1(n_{AB}^0 + \Delta n). \tag{67}$$

The rate of chemical change under equilibrium conditions is

$$dn_{AB}^0/dt = k_2 n_A^0 n_B^0 - k_1 n_{AB}^0. \tag{68}$$

But

$$d\Delta n/dt = dn_{AB}/dt - dn^0_{AB}/dt.$$

Hence

$$d\Delta n/dt = -[k_1 + k_2(n^0_A + n^0_B)]\,\Delta n, \tag{69}$$

and therefore

$$\Delta n = \text{constant} \times \exp\{-[k_1 + k_2(n^0_A + n^0_B)]t\}, \tag{70}$$

which is the standard equation of a relaxation process (equation 24), the relaxation time being now given by the equation

$$1/\tau = k_1 + k_2(n^0_A + n^0_B). \tag{71}$$

If, therefore, $1/\tau$ can be measured at various total concentrations of the reactants, there results a linear relationship, yielding k_2 from the gradient and k_1 from the intercept.

In terms of the equilibrium constant, $K = k_1/k_2$, we have

$$1/\tau = k_2(K + n^0_A + n^0_B), \tag{72}$$

and in terms of the degree of dissociation of AB

$$1/\tau = k_2\left[\left(\frac{n\alpha^2}{1-\alpha}\right) + 2n\alpha\right]$$
$$= k_2 n\left[\frac{\alpha(2-\alpha)}{1-\alpha}\right], \tag{73}$$

where n is the total formal concentration of AB, associated and dissociated.

In order to measure the relaxation time, it is necessary to subject the system to some perturbing influence, such as the passage through it of ultrasonic waves. The foregoing formulation has then to be modified because it is illogical to assume that the velocity constants, k_1 and k_2, are unaffected by the perturbation. Let us then write

$$k_1 = k^0_1 + \Delta k_1 \qquad \text{and} \qquad k_2 = k^0_2 + \Delta k_2. \tag{74}$$

We then find

$$d\Delta n/dt = -[k^0_1 + k^0_2(n^0_A + n^0_B)]\,\Delta n - k^0_2 n^0_A n^0_B\,\Delta \ln K. \tag{75}$$

This equation will be required in the treatment of fast reactions (Chapter 13).

Opposing Bimolecular Reactions

When both the direct and reverse reactions are bimolecular, we have the scheme

$$\underset{(a-x)}{A} + \underset{(b-x)}{B} \underset{k_4}{\overset{k_2}{\rightleftarrows}} \underset{(c+x)}{C} + \underset{(d+x)}{D}$$

where a, b, c and d are, respectively, the initial concentrations of the four components in the system. The differential rate equation is now

$$dx/dt = k_2(a - x)(b - x) - k_4(c + x)(d + x), \qquad (76)$$

which gives, on integration

$$k_2 = \frac{1}{2(1 - K)\beta t} \ln \left[\frac{1 - \left(\dfrac{x}{\alpha + \beta} \right)}{1 - \left(\dfrac{x}{\alpha - \beta} \right)} \right], \qquad (77)$$

where

$$K = k_4/k_2,$$

$$\alpha = \frac{(a + b) + K(c + d)}{2(1 - K)} = \frac{k_2(a + b) + k_4(c + d)}{2(k_2 - k_4)},$$

and

$$\beta^2 = \alpha^2 - \left(\frac{ab - Kcd}{1 - K} \right) = \alpha^2 - \left(\frac{k_2 ab - k_4 cd}{k_2 - k_4} \right).$$

The change in concentration occurring in time t is given by the equation

$$x = \left(\frac{ab - Kcd}{1 - K} \right) \{ \alpha + \beta \coth [(k_2 - k_4)\beta t] \}^{-1}. \qquad (78)$$

These relationships are usually applied to reacting systems when either a and b or c and d are zero. Among reactions found to obey the resulting equations are (Bodenstein, 1894, 1897, 1898; Bonner, Gore and Jost, 1936; Gross and Steiner, 1936):

$$H_2 + I_2 \rightleftarrows 2HI$$

$$H_2 + ICl \rightleftarrows HI + HCl,$$

$$D_2 + HCl \rightleftarrows DH + DCl,$$

in the gaseous state, and numerous reactions of the type

$$RCH_2X + Y^- \rightleftharpoons RCH_2Y + X^-$$

in solution (E. A. Moelwyn-Hughes, 1938; Bennett and (Miss) Vernon, 1938). When c and d are zero and $a = b$, we have the simpler relationships:

$$k_2 = \frac{1}{2at\sqrt{K}} \ln \left\{ \frac{(1 - x/a)(1 - \sqrt{K})}{(1 - x/a)(1 + \sqrt{K})} \right\} \tag{79}$$

and

$$\frac{x}{a} = \left\{ 1 + \sqrt{\frac{k_4}{k_2}} \coth [ta\sqrt{k_2 k_4}] \right\}^{-1}. \tag{80}$$

When A and B are chemically identical, a and b are necessarily equal, and we must allow for the fact that the increase in concentration of each resultant is only one half the decrease in the concentration of reactant. This becomes clear when we formulate the change as follows:

$$\underset{(a-x)}{A} + \underset{}{A} \underset{k_4}{\overset{k_2}{\rightleftharpoons}} \underset{(x/2)}{C} + \underset{(x/2)}{D}.$$

The differential equation is now $dx/dt = k_2(a - x)^2 - k_4(x/2)^2$, and equations (79) and (80) must be amended by replacing k_4 by $k_4/4$.

Unopposed Termolecular Reactions

There are very few reactions of the third kinetic order in solution or in the gaseous state, for the same reason that the number of accidents involving three aeroplanes is less than the number involving two. The known instances in both phases belong, as previously indicated, to the category

$$\underset{a-x}{A} + \underset{2(b-x)}{2B} \rightarrow \underset{x}{P},$$

where the reaction is unimolecular with respect to one reactant and bimolecular with respect to the other. The rate equation therefore becomes

$$dx/dt = 4k_3(a - x)(b - x)^2, \tag{81}$$

which gives, on integration,

$$k_3 = \frac{1}{4t(a - b)^2} \left[\frac{(a - b)x}{b(b - x)} + \ln \frac{a(b - x)}{b(a - x)} \right]. \tag{82}$$

This equation is obeyed by the cyanide-ion catalysis of the conversion of benzaldehyde into benzoin in aqueous solution (Bredig and Stern, 1904; Stern, 1905) at temperatures between 25°C and 60°C:

$$C_6H_5 \cdot CHO + C_6H_5 \cdot CHO + CN^- \rightarrow C_6H_5 \cdot CH(OH) \cdot CO \cdot C_6H_5 + CN^-,$$

and, modified to allow for non-ideality of the solutions, by the reaction between *p*-nitrobenzoyl chloride and various alcohols in anhydrous diethyl-ether (A. A. Ashdown, 1930):

$$pNO_2 \cdot C_6H_4 \cdot COCl + 2ROH \rightarrow pNO_2 \cdot C_6H_4 \cdot COOR + HCl + ROH.$$

Opposing Termolecular and Bimolecular Reactions

The reactions of chlorine, bromine, iodine and oxygen with nitric oxide in the gaseous phase (Bodenstein and Krauss, 1898; Trautz and Dalal, 1920) can be represented by the scheme

$$\underset{(a-x)}{X_2} + \underset{2(b-x)}{2NO} \underset{k_2}{\overset{k_3}{\rightleftarrows}} \underset{2x}{2NOX}$$

and the instantaneous rates by the equation

$$dx/dt = k_3(a-x)[2(b-x)]^2 - k_2(2x)^2$$
$$= 4[k^3(a-x)2(b-x)^2 - k_2x^2]. \tag{83}$$

The integrated form of the equation is too cumbrous to be of use. We may, however, take advantage of the equilibrium relationship

$$\frac{k_3}{k_2} = \left\{ \frac{[NOX]^2}{[NO]^2[X_2]} \right\}_\infty = \left\{ \frac{x^2}{(a-x)(b-x)^2} \right\}_\infty = \frac{1}{K} \tag{84}$$

and write

$$dx/dt = 4k_3[(a-x)(b-x)^2 - Kx^2]. \tag{85}$$

If the instantaneous rate of reaction, dx/dt, is now replaced by the incremental rate $\Delta x/\Delta t$ measured over a short time interval, and the subscript m is made to denote the average concentration during this interval, the termolecular constant becomes

$$k_3 = \frac{\Delta x}{\Delta t} \cdot \frac{1}{4[(a-x)_m(b-x)_m^2 - kx_m^2]}.$$

This method of measuring a rate constant is the one resorted to when the integrated form of the rate equation is unwieldy or unavailable. In the present instance (Bodenstein and Krauss, 1898; Trautz and Dalal, 1920), with

$X_2 = Br_2$, K at 273·7°K is $1·467 \times 10^{-4}$ gram-mole/litre, and k_3 is found to be $2·345 \times 10^{+3}$ (gram-mole/litre)$^{-2}$ sec^{-1}. Hence $k_2 = 0·344$ litre/mole-sec.

Reactions of Variable Order

There are not many established instances where the order of reaction changes during the course of a run. If the simple mechanism which led to equation (4.10) were to hold, we would obtain the integrated equation

$$t = \frac{1}{k_2} \cdot \frac{x}{a(a-x)} + \frac{k_4}{k_3 k_2} \ln \frac{a}{a-x}, \tag{86}$$

which is seen to contain bimolecular and unimolecular components. The real mechanism, however, is never quite so simple. Active molecules of reactants may be formed by collision between product molecules and normal molecules, according to the last term in the equation

$$da/dt = k_2 n^2 - k_3 a - k_4 na + k_6 np. \tag{87}$$

After applying the stationary state hypothesis, and integrating, there results a somewhat complicated expression (E. A. Moelwyn-Hughes, 1964) which allows for first order kinetics during a given run, where the first order constant is the product of a bimolecular constant and the initial concentration.

Catalysed Reactions

The unimolecular constant governing the decomposition of acetonedicarboxylic acid in water at 60°C is $k_0 = 5·48 \times 10^{-2}$ sec^{-1}. In the presence of 0·0165 gram-mole of aniline, the decomposition is again unimolecular with respect to the acid, but the constant is now $k = 11·9 \times 10^{-2}$. Moreover, the difference between k and k_0 for various concentrations of the base is proportional to the concentration (Wiig, 1928). These facts are typical of the simplest kind of catalysis, and are interpreted by assuming that the instantaneous rate of reaction consists of two components, the first due to the spontaneous reaction in the pure solvent, and the second due to catalysis. Denoting the reactant by A, the catalyst by C, and the product by P, the two steps are

$$\underset{a-x}{A} \overset{k_0}{\to} \underset{x}{P}$$

$$\underset{a-x}{A} + \underset{c}{C} \overset{k_c}{\to} \underset{x}{P} + \underset{c}{C}.$$

Then

$$dx/dt = k_0(a - x) + k_c c(a - x). \tag{88}$$

The observed first-order constant is thus

$$k = k_0 + k_c c. \tag{89}$$

In the instance cited, k_c, which is known as the catalytic coefficient, is seen to be 3·89 litres/mole-sec. There are numerous examples where the solvent constant k_0 is immeasurably small. In such cases

$$k_c = k/c. \tag{90}$$

There are also numerous instances of reactions simultaneously catalysed by a variety of solutes:

$$dx/dt = k_0(a - x) + k_1 c_1(a - x) + k_2 c_2(a - x) + \cdots.$$

A number of specific catalytic coefficients are to be found, according to the general equation for multiple catalysis (Dawson, 1913),

$$k = k_0 + \Sigma k_i c_i. \tag{91}$$

Autocatalysed Reactions

Acetic anhydride reacts with water to form acetic acid, and the hydrogen ion produced by partial ionization of the product acts as a catalyst to the hydrolysis:

$$\underset{a-x}{(CH_3CO)_2O} + H_2O \rightarrow \underset{2x}{2CH_3COOH} \rightleftarrows \underset{2x\alpha}{2CH_3COO^-} + \underset{2x\alpha}{2H^+}.$$

The instantaneous rate of reaction is accordingly

$$dx/dt = k_0(a - x) + k_H(a - x)2x\alpha, \tag{92}$$

where α is the degree of ionization of acetic acid. The observed velocity constant

$$k = k_0 + k_H 2x\alpha \tag{93}$$

thus increases as the reaction proceeds. Such changes are termed autocatalytic. A well investigated instance is the acetone–iodine reaction which is stoichiometrically

$$\underset{a-x}{CH_3COCH_3} + I_2 \rightarrow CH_3COCH_2I + H^+ + \underset{x}{I^-}$$

but kinetically

$$-\frac{d[CH_3COCH_3]}{dt} = k_2[CH_3COCH_3][H^+].$$

If the solution contains c gram-ions of hydrogen ions per litre initially, the rate law is

$$dx/dt = k_2(a - x)(c + x).$$

Integration gives us for the catalytic coefficient (Lapworth, 1904)

$$k_2 = \frac{1}{t(a + c)} \ln \frac{a(c + x)}{c(a - x)}, \tag{94}$$

and for the fractional change at time t

$$\frac{x}{a} = \frac{1 - \exp[-k_2(a + c)t]}{1 + (a/c)\exp[-k_2(a + c)t]}. \tag{95}$$

Isotopic Exchange Reactions

When an atom or ion displaces from a molecule another atom or ion with the same nuclear charge but a different atomic mass, the reaction is said to be an isotopic exchange. The simplest instance is the replacement of a hydrogen atom from a hydrogen molecule by a deuterium atom:

$$D + H_2 \rightleftarrows DH + H.$$

Accurate data on the kinetics of this reaction and of the numerous kindred interchanges involving the ortho and para states of H_2 and D_2 have played a vital part in developing the theory of chemical change (A. Farkas, 1935).

We shall here consider a very simple isotopic exchange reaction in solution, namely, the replacement of a normal iodine atom by a radioactive iodine atom, according to the scheme

$$\underset{a-x}{CH_3I} + \underset{b-x}{*I^-} \underset{k_2}{\overset{k_2}{\rightleftarrows}} \underset{x}{CH_3I^*} + \underset{c+x}{I^-},$$

where

$$[CH_3I]_0 = a \quad [*I^-]_0 = b \quad [I^-]_0 = c,$$

and x is the extent of change at time t. The difference between the velocity constants of the direct and reverse reactions can, in this instance, be ignored, so that

$$dx/dt = k_2[(a - x)(b - x) - x(c + x)] = k_2[ab - (a + b + c)x]. \tag{97}$$

Integration, taking x as zero when t is zero, gives the constant k_2 as

$$k_2 = \frac{1}{(a + b + c)t} \ln \frac{1}{1 - [(a + b + c)/ab]x}, \tag{98}$$

and the extent of change at time t as

$$x = \frac{ab}{(a + b + c)t}\{1 - \exp[-k_2(a + b + c)t]\}. \tag{99}$$

The maximum extent of change, x_∞, is given by the term preceding the curly bracket. Hence

$$x/x_\infty = 1 - \exp[-k_2(a + b + c)t], \tag{100}$$

which enables us to write equation (98) in the form (H. A. C. Mackay, 1938)

$$k_2 = \frac{1}{(a + b + c)t} \ln \frac{1}{1 - (x/x_\infty)}. \tag{101}$$

The ratio of the concentration of radioactive ion at time t to its initial concentration is

$$\frac{[*I^-]}{[I^{*-}]_0} = \frac{b - x}{b} = 1 - \frac{x}{b} = 1 - \frac{a}{(a + b + c)}\{1 - \exp[-k_2(a + b + c)t]\}. \tag{102}$$

By taking Geiger counts on the inorganic reactant after its separation from the organic reactant, this ratio is readily measured, and the rate constant for the exchange reaction evaluated. In practice, b can generally be ignored in comparison with a and c.

During these exchange reactions, the radioactive iodine, whether ionic or atomic, reverts to its normal state unimolecularly, and allowance must be made for its spontaneous decay. Let the decay constant (of $*I^-$ and CH_3*I) be λ sec^{-1}. We now have the additional changes:

$$\underset{b-x-y}{*I^-} \overset{\lambda}{\to} \underset{c+x+y}{I^-}$$

and

$$\underset{x-z}{CH_3*I} \overset{\lambda}{\to} \underset{a-x+z}{CH_3I}$$

where x retains its previous significance, and y denotes the decrease in concentration of $*I^-$ due to the decay, and z is the decrease in concentration of CH_3*I due to the same cause. The various concentrations at time t are now

$$[CH_3I] = a - (x - z)$$

$$[CH_3*I] = x - z$$

$$[I^-] = c + x + y$$

$$[*I^-] = b - (x + y).$$

Let us define new variables p and q as follows:

$$p = x + y \tag{103}$$

$$q = x - z. \tag{104}$$

Then

$$[CH_3I] = a - q$$

$$[CH_3*I] = q$$

$$[I^-] = c + p$$

$$[*I^-] = b - p.$$

We now have the rate equations

$$\frac{dq}{dt} = \frac{d[CH_3*I]}{dt} = k_2[CH_3I][*I^-] - k_2[CH_3*I][I^-] - \lambda[CH_3*I]$$

$$= k_2(a - q)(b - p) - k_2q(c + p) - \lambda q \tag{105}$$

and

$$-\frac{dp}{dt} = \frac{d[*I^-]}{dt} = -k_2[CH_3I][*I^-] + k_2[CH_3*I][I^-] - \lambda[*I^-]$$

$$= -k_2(a - q)(b - p) + k_2q(c + p) - \lambda(b - p). \tag{106}$$

By addition,

$$\frac{d(p - q)}{dt} = \lambda[b - (p - q)]$$

and by integration, noting that $p - q = 0$ when $t = 0$,

$$p - q = b[1 - \exp(-\lambda t)]$$

or

$$b - (p - q) = b \cdot \exp(-\lambda t). \tag{107}$$

But $p - q = y + z$. Hence equation (107) merely states that the rate of decay of the total radioactive iodine is unimolecular, i.e.

$$[*I^-] + [CH_3*I] = [*I^-]_0 \cdot \exp(-\lambda t). \tag{108}$$

By eliminating p from equations (105) and (107), we see that

$$dq/dt = -[k_2(a + b + c) + \lambda]q + k_2ab \cdot \exp(-\lambda t) \tag{109}$$

$$= -Aq + B \cdot \exp(-\lambda t) \tag{110}$$

where
$$A = k_2(a + b + c) + \lambda \tag{111}$$
and
$$B = k_2ab. \tag{112}$$

Equation (110) can be rearranged as follows:
$$dq + Aq\,dt = B.\exp(-\lambda t)\,dt.$$

Each term may be multiplied by $\exp(At)$:
$$\exp(At)\,dq + qA.\exp(At)\,dt = B.\exp[(A - \lambda)t]\,dt.$$

The integral on the left hand side is $q.\exp(At)$. Hence
$$q.\exp(At) = \frac{B}{A - \lambda}.\exp[(A - \lambda)t] + C,$$

where C is a constant. On multiplying throughout by $\exp(-At)$, we have
$$q = \frac{B}{A - \lambda}.\exp(-\lambda t) + C.\exp(-At). \tag{113}$$

It may be verified that differentiation of (113) reproduces (110) as required. This mathematical solution is valid when $A \neq \lambda$, which generally holds for systems of interest. Since q is zero when t is zero, $C = -B/(A - \lambda)$, and
$$q = \frac{B}{A - \lambda}[\exp(-\lambda t) - \exp(-At)]. \tag{114}$$

On substituting the expressions (111) and (112) for A and B, we have
$$q = \frac{ab}{(a + b + c)}\{\exp(-\lambda t) - \exp\{-[k_2(a + b + c) + \lambda]t\})$$
$$\therefore \quad [CH_3\text{*}I] = \frac{ab}{(a + b + c)}.\exp(-\lambda t)\{1 - \exp[-k_2(a + b + c)t]\}. \tag{115}$$

Now $[\text{*}I^-] = b - p$, and, from equation (107), $b - p = b.\exp(-\lambda t) - q$.
$$\therefore \quad [\text{*}I^-] = b.\exp(-\lambda t)\left(1 - \frac{a}{(a + b + c)}\{1 - \exp[-k_2(a + b + c)t]\}\right). \tag{116}$$

Since $[\text{*}I^-]_0 = b$, the ratio, R, of the concentration of radioactive inorganic iodide at time t to its initial value is
$$R = \frac{[\text{*}I^-]}{[\text{*}I^-]_0} = \exp(-\lambda t)\left(1 - \frac{a}{(a + b + c)}\{1 - \exp[-k_2(a + b + c)t]\}\right). \tag{117}$$

For a perfectly stable isotope ($\lambda = 0$), this equation reduces to equation (102).

In order to evaluate the bimolecular constant, k_2, from a knowledge of R, a, b, c and λ, equation (117) is rearranged as follows (Swart and Le Roux, 1956, 1957):

$$k_2 = \frac{-1}{(a + b + c)t} \ln \left\{ 1 - \left(\frac{a + b + c}{a}\right)[1 - R \cdot \exp(\lambda t)] \right\}.$$

A slightly more complex kinetic equation is found when, as is often necessary, allowance must be made for solvolysis. This is derived in Appendix 4.

Discussion

We have now derived a number of mathematical expressions which have been shown by experiment to summarize the temporal course of a variety of chemical reactions in solution. The processes selected as examples include decarboxylation, cyclization, hydrolysis, saponification, ionic substitution, mutarotation, racemization, radioactive decay, salt formation, isotopic exchange, ionization and catalysis. In the next chapter, we shall introduce the first theory which was found capable of treating such diversified phenomena in a unified way.

REFERENCES

Arrhenius, *Z. physikal Chem.*, **1**, 110 (1887).
Arrhenius, *Z. physikal Chem.*, **1**, 631 (1887).
Ashdown, A. A., *J. Amer. Chem. Soc.*, **52**, 268 (1930).
Barham and L. W. Clark, *J. Amer. Chem. Soc.*, **73**, 4638 (1951).
Bateman, *Proc. Camb. Phil. Soc.*, **15**, 423 (1910).
Bennett and Vernon (Miss), *Trans. Chem. Soc.*, 1783 (1938).
Bodenstein, *Z. physikal Chem.*, **13**, 56 (1894); **22**, 1 (1897); **29**, 295 (1898).
Bodenstein and Krauss, *Z. physikal Chem.*, **175**, 295 (1898).
Bonner, Gore and Jost, *J. Amer. Chem. Soc.*, **58**, 690 (1936).
Bowen, Moelwyn-Hughes and Hinshelwood, *Proc. Roy. Soc.*, **134A**, 211 (1931). For references to numerous papers on the gas reaction, *see* Laidler, *Chemical Kinetics*, McGraw-Hill (1965).
Brass and Tolman, *J. Amer. Chem. Soc.*, **54**, 1003 (1932).
Bredig and Balcom, *Berichte*, **41**, 740 (1908).
Bredig and Stern, *Z. Elektrochem.*, **10**, 582 (1904).
Brookfield, K. F., *see* E. A. Moelwyn-Hughes, *Physical Chemistry of the War Gases*, H.M. Stationery Office (1950).
Cook, A. S., and M. M. Harris, *Trans. Chem. Soc.*, 2365 (1963).
Corran, *Trans. Faraday Soc.*, **23**, 605 (1927).
Cundall, *Trans. Chem. Soc.*, **59**, 1076 (1891).
Curie, Mme., *Radioactivité*, p. 525, Hermann et Cie., Paris (1935).

Daniels and E. H. Johnston, *J. Amer. Chem. Soc.*, **43**, 53 (1921).
Davies, A. C., Mansel Davies, and M. Stoll, *Helv. Chim. Acta*, **37**, 1351 (1954).
Dawson, H. M., *Trans. Chem. Soc.*, **103**, 2135 (1913).
Dinglinger and Schröer, *Z. physikal Chem.*, **179A**, 401 (1937).
Eigen, *Z. physikal Chem., Neue Folge*, **1**, 176 (1954).
Essex and Gelormini, *J. Amer. Chem. Soc.*, **48**, 882 (1926).
Eyring and Daniels, *J. Amer. Chem. Soc.*, **52**, 1473 (1930).
Fajans, *Z. physikal Chem.*, **73**, 25 (1910).
Farkas, A., *Orthohydrogen, Parahydrogen and Heavy Hydrogen*, Cambridge (1935).
Fletcher and Rollefson, *J. Amer. Chem. Soc.*, **58**, 2129 (1936).
Freundlich and Kroepelin, *Z. physikal Chem.*, **122**, 39 (1926).
Freundlich and M. B. Richards, *Z. physikal Chem.*, **79**, 681 (1912).
Freundlich and Neumann, *Z. physikal Chem.*, **87**, 69 (1914).
Freundlich and Salomon, *Berichte*, **66**, 355 (1933).
Goldschmidt and Bräuer, *Berichte*, **39**, 109 (1906).
Gross and Steiner, *J. Chem. Physics*, **4**, 165 (1936).
Guggenheim, E. A., *Phil. Mag.*, **1**, 538 (1926).
Harris, M. M., and R. K. Mitchell, *Trans. Chem. Soc.*, 1905 (1960).
Henry, P., *Z. physikal Chem.*, **10**, 98 (1892).
Hirniak, *Tables Annuelles*, **2**, 508 (1922).
Hughes, E. D., C. K. Ingold and A. J. Parker, *Trans. Chem. Soc.*, 4400 (1960).
Hughes, E. D., C. K. Ingold and Mackie, *Trans. Chem. Soc.*, 3177 (1955).
Johnson, P., and E. A. Moelwyn-Hughes, *Proc. Roy. Soc.*, **175A**, 118 (1940).
Jones, C. M., and W. C. McC. Lewis, *Trans. Chem. Soc.*, **117**, 1120 (1920).
Kappana, *Z. physikal Chem.*, **158A**, 355 (1932).
Kendrew and E. A. Moelwyn-Hughes, *Proc. Roy. Soc.*, **176A**, 352 (1940). Important earlier work on the uncatalysed mutarotation of glucose in water was carried out by Trey, *Z. physikal Chem.*, **18**, 193 (1895); Osaka, *Z. physikal Chem.*, **35**, 661 (1900); Hudson and Dale, *J. Amer. Chem. Soc.*, **39**, 320 (1917).
Kistiakowsky and W. R. Smith, *J. Amer. Chem. Soc.*, **58**, 1042 (1936).
Kuhn and Albrecht, *Ann.*, **458**, 221 (1927).
Lapworth, *Trans. Chem. Soc.*, **85**, 30 (1904).
Lowry, T. M., and R. Traill, *Proc. Roy. Soc.*, **132A**, 416 (1931).
Li and Adams, *J. Amer. Chem. Soc.*, **57**, 1565 (1935).
Ling, C. C. K., and M. M. Harris, *Trans. Chem. Soc.*, 1825 (1964).
Lowry, T. M., *Trans. Chem. Soc.*, **75**, 212 (1899).
Lueck, *J. Amer. Chem. Soc.*, **44**, 757 (1922).
Mackay, H. A. C., *Nature*, **142**, 997 (1938).
Mellor, J. W., *Chemical Statics and Dynamics*, p. 100, Longmans (1904).
Moelwyn-Hughes, E. A., *Physical Chemistry*, 2nd edn., p. 1146 (1964).
Moelwyn-Hughes, E. A. *Proc. Roy. Soc.*, **196A**, 540 (1949).
Moelwyn-Hughes, E. A., *Trans. Faraday Soc.*, **34**, 91 (1938).
Moelwyn-Hughes, E. A., *Trans. Chem. Soc.*, 4301 (1962).
Moelwyn-Hughes, E. A., *Trans. Chem. Soc.*, 204 (1938).
Moelwyn-Hughes, E. A., and Hinshelwood, *Proc. Roy. Soc.*, **131A**, 186 (1931).
Moelwyn-Hughes, E. A., and P. Johnson, *Trans. Faraday Soc.*, **36**, 948 (1940).
Moran and Lewis, *Trans. Chem. Soc.*, **121**, 1613 (1922).
Natanson, *Ann. Physik*, **27**, 606 (1886).
Peters, R., *Physical Chemistry of War Gases*, H.M. Stationery Office, 1950.
Reicher, *Annalen*, **228**, 257 (1885); **232**, 211 (1886).

Richards, W. T., and J. A. Reid, *J. Chem. Physics*, **1**, 114 (1933).
Salomon, *Helv. Chim. Acta*, **19**, 743 (1936).
Salomon, *Trans. Faraday Soc.*, **32**, 153 (1936).
Scatchard, G., *J. Amer. Chem. Soc.*, **43**, 2387 (1921).
Selle, H., *Z. physikal Chem.*, **104**, 1 (1923).
Sickman, *J. Chem. Physics*, **4**, 297 (1936).
Smith, G. F., and T. M. Lowry, *Trans. Chem. Soc.*, 666 (1928).
Stern, *Z. physikal Chem.*, **50**, 513 (1905).
Swart and Le Roux, *Trans. Chem. Soc.*, 2110 (1956); *Trans. Chem. Soc.*, 406 (1957).
Thomson, J. J. and Thomson, G. P., *The Conduction of Electricity through Gases*, 3rd edn., Cambridge (1928).
Tommila, M. Tiilikainer and A. Voipio, *Am. Acad. Sci. Fennicae*, **A**, Chem. II, 65 (1955).
Trautz and Dalal, *Z. anorg. Chem.*, **110**, 1 (1920).
von Halban, *Z. physikal Chem.*, **67**, 129 (1909).
von Halban and Kirsch, *Z. physikal Chem.*, **82**, 325 (1913).
Warder, *Berichte*, **14**, 1361 (1881).
White and Tolman, *J. Amer. Chem. Soc.*, **47**, 1240 (1925).
Wiig, *J. physical Chem.*, **32**, 961 (1928); **34**, 596 (1930).
Wilhelmy, *Pogg. Annalen*, **81**, 413 (1850).

7

IONIC REACTIONS

Ionic reactions in solution may be divided into two categories. There are those between elementary ions, such as $Ag^+ + Cl^- \rightarrow AgCl$, where, provided there is a pronounced decrease in standard free energy, the rate of reaction may be determined by the rate at which cations and anions come into contact or by the rate at which either type becomes partially desolvated. A chemically important and carefully investigated reaction of this type is the union of hydrogen and hydroxyl ions in aqueous solution: $H^+ + OH^- \rightarrow H_2O$. As stated in Chapter 5, the rate is controlled by diffusion, aided by consecutive proton jumps, and is reproduced over a range of temperatures by Debye's equation (5.28). There are also reactions such as $I^- + CH_2Br \cdot COO^- \rightarrow Br^- + CH_2I \cdot COO^-$, which, though ostensibly interionic are, in essence, reactions between ions such as I^- and polar groups such as $C-Br$, the carbon atom of which is covalently attached to an electrically charged group. This interionic reaction resembles its parent change $I^- + CH_3Br \rightarrow Br^- + CH_3I$, and differs from it in that it takes place in the presence of the charged group $(-COO^-)$ which witnesses, but does not directly participate in, the chemical change. It is with this type of reaction that we shall mainly deal in this chapter. We shall first briefly trace the steps which have led to the present state of knowledge concerning ionic reactions in solution (G. N. Lewis, 1907).

After the introduction of a corrected concentration, called activity (G. N. Lewis, 1907), physical chemists experienced no little difficulty in assessing its kinetic relevance. Their verdict could be reached only by comparing the instantaneous rate of reaction with the product of the activities of the reactants, determined extrakinetically, as, for example, from electrometric or solubility data. H. S. Harned (1918) and W. C. McC. Lewis (C. M. Jones and W. C. McC. Lewis, 1920; W. C. McC. Lewis and Moran, 1922) established a direct proportionality between them in numerous reactions. J. N. Brönsted (1922, 1925) gave a satisfactory treatment of the kinetic significance of activity in his treatment of ionic reactions.

The Brönsted–Bjerrum Equation

According to Brönsted, the first step in a reaction between ions A and B in solution is the formation of a complex C with which they establish thermodynamic equilibrium:

$$A + B \rightleftarrows C \tag{1}$$

N. Bjerrum (1924) treats C as a collision complex. The relationship between the two derivations is summarized in equation (4.4). In both the equilibrium constant at constant temperature and pressure is given in terms of the activities, a:

$$K = \frac{a_C}{a_A a_B}. \tag{2}$$

By definition, each activity is the product of the concentration, c, and an activity coefficient, γ; hence

$$c_C = K c_A c_B \left(\frac{\gamma_A \gamma_B}{\gamma_C} \right) \tag{3}$$

The bracketed term is known as the kinetic activity factor, and, like each of the γ's, becomes unity at infinite dilution. The crucial postulates in the Harned–Lewis theory are that the rate of breakdown of the complex into its products, P, is proportional to its concentration, and is not so fast as to disturb the equilibrium. Then,

$$-\frac{dc_C}{dt} = +\frac{dc_P}{dt} = k c_C = k K c_A c_B \left(\frac{\gamma_A \gamma_B}{\gamma_C} \right), \tag{4}$$

and the observed bimolecular constant becomes

$$k_2 = \frac{1}{c_A c_B} \cdot \frac{dc_P}{dt} = k K \left(\frac{\gamma_A \gamma_B}{\gamma_C} \right), \tag{5}$$

or, since the kinetic activity factor is, by definition, unity at infinite dilution,

$$k_2 = k_2^0 \frac{\gamma_A \gamma_B}{\gamma_C}, \tag{6}$$

where k_2^0 is the experimental bimolecular constant extrapolated to zero concentration, or infinite dilution.

It has long been known, from the solubilities of electrolytes and the vapour pressures of electrolytic solutions, that the activity coefficient of an ion depends on the ionic strength, I, of the solution, defined as follows in terms of the ionic concentrations, c_i (gram-ions per litre of solution) and the electrovalencies, z_i, the sum being taken over all kinds of ions in the solution.

$$I = \tfrac{1}{2}\Sigma c_i z_i^2. \tag{7}$$

The theory of P. Debye and E. Hückel (1924) provides the following expression for the logarithm of the activity coefficiency, γ_i, of an ion with charge e_i in a medium of permittivity D, at a temperature, T:

$$\ln \gamma_i = -\frac{e_i^2 \kappa}{2DkT}, \tag{8}$$

where

$$\kappa = \left[\frac{4\pi\Sigma(n_i e_i^2)}{DkT}\right]^{1/2}. \tag{8a}$$

The summation is to be taken over all kinds of ions, and their concentrations, n_i, expressed in ions per cm^3. The charge e_i is the product of the protonic charge, ε, and the electrovalency, z_i. Hence, recalling that $n_i = (N_0/1000)c_i$, we may write

$$\ln \gamma_i = -\alpha z_i^2 \sqrt{I}, \tag{8b}$$

where

$$\alpha = \left(\frac{\varepsilon^2}{2DkT}\right)\left(\frac{8\pi N_0 \varepsilon^2}{1{,}000DkT}\right)^{1/2}. \tag{9}$$

By combining equations (6) and (8), we have

$$\ln k_2 = \ln k_2^0 - \alpha\sqrt{I}[z_A^2 + z_B^2 - (z_A + z_B)^2],$$

or

$$\ln (k_2/k_2^0) = 2\alpha a_A z_B \sqrt{I}. \tag{10}$$

Hence

$$\log_{10} (k_2/k_2^0) = \left(\frac{\varepsilon^2}{2\cdot303DkT}\right)\left(\frac{8\pi N_0 \varepsilon^2}{1{,}000DkT}\right)^{1/2} . z_A z_B \sqrt{I}. \tag{11}$$

According to the Brönsted–Bjerrum theory, therefore, the logarithm of the bimolecular constant for a particular reaction should vary linearly with respect to the square root of the ionic strength, and the gradients found for various ionic reactions in the same solvent at the same temperature should bear integral relationships, one with another.

Experimental Tests of the Brönsted–Bjerrum Equation

Most of the early practical tests of equation (11) were made on ionic reactions in aqueous solution at 25°C, for which the numerical value of the term preceding $z_A z_B \sqrt{I}$ in this equation is 1·0124. In this solvent at this temperature, therefore,

$$\log_{10} (k_2/k_2^0) = 1·0124 \, z_A z_B \sqrt{I}. \tag{12}$$

The validity of equation (8) is confined, under these conditions, to ionic concentrations not exceeding 1 millimole per litre. Data on bimolecular reactions of various ionic types were first summarized by R. Livingston (1930)* in graphical form (Fig. 1). The sloping lines on the graph are the nearly integral gradients required by equation (12). Despite the scatter in some of the data, they conform in a general way with the requirements of the Brönsted–Bjerrum equation, and provide a convincing proof of its soundness.

Apparent Exceptions to the Brönsted-Bjerrum Relationship

Equation (10) applies strictly only to the chemical change occurring between ions A and B, and ignores the possibility that the ions may simultaneously react with the solvent, and form ion-pairs with ions of opposite sign. The departure from linearity in the plot of $\ln k_2$ versus $\sqrt{I}$ for the thiosulphate–bromacetate reaction, which is evident from Livingston's diagram, exceeds the experimental error, and we have now to require whether, and, if so to what extent, either of these possibilities may explain the deviation. We shall deal with the solvolytic complication first.

The Role of Solvolysis in Ionic Reactions

In aqueous solution the rate of attack of the α-brompropionate ion by the thiosulphate ion resembles the rate of the corresponding bromacetate-thiosulphate reaction in its response to ionic environment. With the β-brompropionate ion, surprisingly enough, the salt effect appeared to be in the reverse direction (V. K. La Mer and M. E. Kamner, 1931). Moreover, the rate of variation of the apparent energy of activation with temperature, instead of being negative as required by equation (40), was found to be positive and of a higher order of magnitude (V. K. La Mer, 1933). A similar

* We have corrected an obvious slip of writing by replacing the formula of a disaccharide with that of ethyl acetate. We have also altered Livingston's μ to I, and his k to k_2/k_2^0.

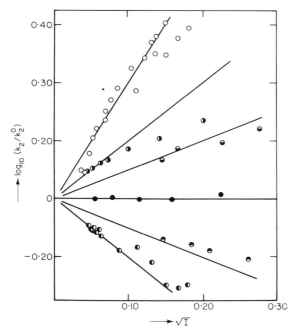

FIG. 1 (after Livingston). The effect of ionic strength on the velocity of ionic reactions.

KEY TO FIG. 1

	Reaction		Reference
O	$2[Co(NH_3)_5Br]^{++} + Hg^{++} + 2H_2O \rightarrow 2[CO(NH_3)_5H_2O]^{+++} + HgBr_2$	*	$z_A z_B = 4$
◗	$CH_2Br \cdot COO^- + S_2O_3^{--} \rightarrow [CH_2S_2O_3 \cdot COO]^{--} + Br^-$	†	$z_A z_B = 2$
⊖	$NO_2 : N \cdot COOC_2H_5^- + OH^- \rightarrow N_2O + CO_3^{--} + C_2H_5OH$	‡	$z_A z_B = 1$
●	$CH_3 \cdot COOC_2H_5 + OH^- \rightarrow CH_3 \cdot COO^- + C_2H_5OH$	§	$z_A z_B = 0$
⊕	$H_2O_2 + 2H^+ + 2Br^- \rightarrow 2H_2O + Br_2$	‖	$z_A z_B = -1$
◑	$[Co(NH_3)_5Br]^{++} + OH^- \rightarrow [Co(NH_3)_5OH]^{++} + Br^-$	¶	$z_A z_B = -2$

REFERENCES

* J. N. Brönsted and R. Livingston, *J. Amer. Chem. Soc.*, **49**, 435 (1927).
† V. K. La Mer, *ibid.*, **51**, 3341, 3678 (1929).
‡ J. N. Brönsted and Delbanco, *Z. anorg. Chem.*, **144**, 248 (1925).
§ S. Arrhenius, *Z. physikal. Chem.*, **1**, 110 (1887).
‖ R. Livingston, *J. Amer. Chem. Soc.*, **48**, 53 (1926).
¶ As for *.

anomaly in the salt effect at constant temperature has been reported for the reaction of the thiosulphate ion with the monobrom-malonate and monobrom-succinate ions (Bedford, Mason and Morrell, 1934; Bedford,

Austin and Webb, 1935).* These apparently anomalous results were interpreted by V. K. La Mer (1932) in terms of a postulated asymmetry of the
electrical field surrounding the charged organic reactant. It has been shown,
however, by a rather elaborate mathematical argument (Sturtevant, 1935)
that when only one of the ions is spherically symmetrical while the other is
prolate spheroidal, the effect of asymmetry will be experimentally undectable
under ordinary observational conditions. While this conclusion rules out
the orientation hypothesis as inadmissible, it gives no clue to the real cause
of the alleged deviation from the Brönsted–Bjerrum equation.

When alkyl halides react with the thiosulphate ion in dilute aqueous
solution, the kinetic scheme is as follows (E. A. Moelwyn-Hughes, 1933)

$$-d[RX]/dt = k_2[RX][S_2O_3^{--}] + k_1[RX] = k_2[RX]\{[S_2O_3^{--}] + (k_1/k_2)\},$$

$$(13)$$

where k_2 is a bimolecular, and k_1 a pseudo-unimolecular constant. Square
brackets denote concentrations. With $R = C_2H_5$ at 25°C, k_1/k_2 is 0·6
millimole per litre. Concentrations of the same order of magnitude are found
for numerous reactions of this type, and it is clear that, when ionic reactions
are investigated at the low concentrations to which the square-root law
applies, the omission of the hydrolytic correction may lead to serious error.
A re-examination of the β-brompropionate–thiosulphate reaction by Nielsen
(1936) showed, in fact, that when allowance is made for the concurrent
hydrolytic reaction, in accordance with equation (13), the gradient of the
plot of $(\ln k_2)/I^{1/2}$ has the right sign.

Qualitative Description of the Role of Ion-pairs in Ionic Reactions

In discussing the kinetics of the reaction between solutes A and B, let us
provisionally refer to the bimolecular constant obtaining when either solute
is uncharged as the normal value. Then, all other things being equal, we are
led on electrostatic grounds to expect a rate higher than the normal value
when the solutes are oppositely charged, and a rate lower than the normal
value when A and B have charges of the same sign. Let us apply these considerations to the reactions

$$\text{OH}^- + \begin{array}{c} \text{CHBr·COO}^- \\ | \\ \text{CHBr·COO}^- \end{array} \rightarrow \text{Br}^- + \begin{array}{c} \text{CH(OH)·COO}^- \\ | \\ \text{CHBr·COO}^- \end{array} \rightarrow \begin{array}{c} \text{CH·COO}^- \\ || \\ \text{CBr·COO}^- \end{array} + \text{H}_2\text{O}$$

---

* Bedford, Mason, and Morrell, *ibid.*, **56**, 280 (1934); Bedford, Austin, and Webb, *ibid.*, **57**,
1408 (1935). The claim by these authors that an investigation by Conant and his collaborators
on another type of reaction revealed a contradiction to the Brönsted relation is not well founded,
for Conant clearly stated that further work was necessary before his results could be interpreted
in terms of the theory.

and

$$OH^- + [Co(NH_3)_5Br]^{++} \rightarrow Br^- + [Co(NH_3)_5OH]^{++}.$$

The rate of reaction between the hydroxyl and dibromsuccinate ions is found to be enhanced by the presence of divalent cations, such as Ca^{++}, and the rate of reaction between the hydroxyl and brompentamminecobaltic ions is found to be reduced by the presence of divalent anions, such as SO_4^{--}. If, in the first reaction, ion pairs such as

$$\begin{bmatrix} CHBr\cdot COO^- \\ | \\ CHBr\cdot COO^- \end{bmatrix}, Ca^{++} \end{bmatrix}^0$$

are formed, and in the second reaction, ion pairs such as

$$[Co(NH_3)_5Br^{++}, SO_4^{--}]^0$$

are formed, we have an immediate explanation of the phenomena for, in the first reaction the ion-pair will react more readily with OH^- and in the second reaction less readily with OH^- than do the free ions.

The effect of divalent cations discussed here was initially referred to as cation catalysis (B. Holmberg, 1912).

Quantitative Formulation of the Role of Ion-pairs in Ionic Reactions

We shall deal with this subject by referring to the reaction between the thiosulphate ($S_2O_3^{--}$) and bromacetate ($BrAc^-$) ions in aqueous solution, on which there are extensive data (Slator, 1905; Krapiwin, 1913; Kapanna, 1929; Kapanna and Patwardhan, 1932; La Mer and Fessenden, 1932; von Kiss and Vass, 1934; La Mer and Kamner, 1935; Künnap and A. Parts, 1949; Wyatt and C. W. Davies, 1949; C. W. Davies and I. W. Williams, 1958). The bimolecular constants found by Slator at ionic strengths of 0·04 and 0·05 indicate a gradient of $d\log_{10} k_2/d\sqrt{I} = 1·84$, which is within 8 per cent of the theoretical value. Later workers showed that, particularly in the presence of divalent cations, M^{++}, we must, neglecting reactions between uncharged solutes, allow for the simultaneous occurrence of three reactions, namely

$$S_2O_3^{--} + CH_2Br\cdot COO^- \rightarrow [CH_2(S_2O_3)\cdot COO]^{--} + Br^- ; k_2$$

$$MS_2O_3 + CH_2Br\cdot COO^- \rightarrow [CH_2(S_2O_3)\cdot COO]^{--} + M^{++} + Br^- ; k_3$$

$$S_2O_3^{--} + CH_2Br\cdot COOM^+ \rightarrow [CH_2(S_2O_3)\cdot COO]^{--} + M^{++} + Br^- ; k_4.$$

The method of resolving the net rate of reaction into its three components is due to C. W. Davies (1962), to whom, and C. B. Monk (1961), we owe most of our knowledge of the dissociation constants of these ion pairs in aqueous solution. In this system, the ion pairs to be reckoned with are MS_2O_3 and $MBrAc^+$, for which the dissociation constants are, respectively,

$$K_1 = \frac{[M^{++}][S_2O_3^{--}]\gamma_2^2}{[MS_2O_3]} \tag{14}$$

and

$$K_2 = \frac{[M^{++}][BrAc^-]\gamma_2}{[MBrAc^+]}. \tag{15}$$

The subscript to γ, here as below, denotes the electrovalency. By applying the Brönsted–Bjerrum treatment to each of the three intermediates formed, the rate equation becomes:

$$-\frac{d[S_2O_3^{--}]}{dt} = k_2[S_2O_3^{--}][BrAc^-]\left(\frac{\gamma_1\gamma_2}{\gamma_3}\right) + k_3[MS_2O_3][BrAc^-]\left(\frac{\gamma_0\gamma_1}{\gamma_1}\right)$$

$$+ k_4[S_2O_3^{--}][MBrAc^+]\frac{\gamma_2\gamma_1}{\gamma_1}, \tag{16}$$

which can be re-written as

$$-\frac{d[S_2O_3]}{dt} = k_2[S_2O_3^{--}][BrAc^-]\left\{\frac{\gamma_1\gamma_2}{\gamma_3} + \frac{1}{k_2}\left(\frac{k_3}{K_1} + \frac{k_4}{K_2}\right)[M^{++}]\gamma_2^2\right\}. \tag{17}$$

During the course of a run, $[M^{++}]$ is maintained at a constant value, so that the reaction is throughout bimolecular. Values of the dissociation constants, K_1 and K_2, can be determined from precise conductimetric measurements in the absence of chemical reaction, as described in Chapter 3. For example, K_1 is 0·0159 gram-ions/litre for MgS_2O_3 at 25°C (Table III, 9), and K_2 is 0·28 for $MgBrAc^+$. Such data make it possible to evaluate the three bimolecular constants separately. The values of $\log k_2$ and $\log k_3$ are shown in Fig. 2.

Extensive data have led C. W. Davies (1928) to replace equation (8) by the improved equation

$$\ln \gamma_i = -\alpha z_i^2 [\sqrt{I}(1 + \sqrt{I})^{-1} - bI] = -\alpha z_i^2 f(I), \tag{18}$$

where b in water at 25°C is 0·30. With the details of the steps leading to the evaluation of k_4 we shall not here be concerned. We shall be content with summarizing Davies' findings in Table I, where the superscript o denotes that the bimolecular constants (in litres/mole-sec) have been extrapolated to zero concentration. It will be observed that the various k^o values are in the sequence anticipated on electrostatic grounds, and that the gradients given in the last column are the theoretical ones.

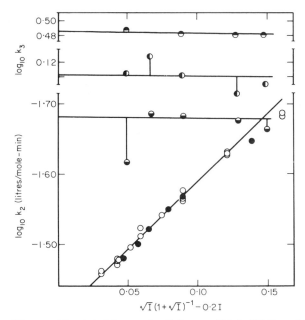

FIG. 2 (after Wyatt and Davies). The reaction $S_2O_3^{--} + CH_2Br\cdot COO^-$ at 25°C. Data of La Mer and Fessended (○) and of von Kiss and Vass (●). The reaction $BaS_2O_3 + CH_2Br\cdot COO^-$. Data of Wyatt and Davies at 15°C (⊕), 25°C (⊕) and 35°C (◓).

C. W. Davies and I. W. Williams have dealt in the same way with the kinetics of the reaction between the brompentammine–cobaltic and hydroxide ions in the presence of sulphates. From the abnormally high solubility of halogenopentammine-cobaltic halides in the presence of sodium and magnesium sulphates, the dissociation constant for the brompentammine–cobaltic–sulphate ion-pair is found to be 0·0028 at 0°C. The total rate of reaction can then be resolved into rates due to the concurrent attack of the complex ion by hydroxyl and sulphate ions. The logarithms of the

TABLE I

THE KINETIC COMPONENTS AND ELECTROLYTE EFFECTS OF THE THIO-SULPHATE–BROMACETATE REACTION IN WATER AT 25°C

Reactants	Bimolecular constants (litres/mole-minute)	$d \log_{10} k_2/df(I)$
$S_2O_3^{--} + CH_2Br\cdot COO^-$	$k_2^0 = 0\cdot247$	$+2$
$MS_2O_3 + CH_2Br\cdot COO^-$	$k_3^0 = 1\cdot23$	0
$S_2O_3^{--} + CH_2Br\cdot COOM^+$	$k_4^0 = 7\cdot63$	-2

separate rate constants are found to vary linearly with respect to $f(I)$ with gradients of -2 and -4 respectively in accordance with equation (11) as modified by equation (18). Davies' final conclusion is that the varied salt effects predicted by Brönsted's theory are strictly obeyed when quantitative allowance, based on extra-kinetic data, is made for the incidence of ion-association.

Let us, reverting to equation (16), and omitting the last term, denote by c_A and c_B the total concentration of $BrAc^-$ and MS_2O_3, respectively, and by α the degree of dissociation of the latter. Then

$$-\frac{dc_A}{dt} = k_2 c_B \alpha c_A \frac{\gamma_1 \gamma_2}{\gamma_3} + k_3 c_B (1 - \alpha) c_A,$$

and the observed bimolecular constant becomes

$$-\frac{1}{c_A c_B} \cdot \frac{dc_A}{dt} = k_3 + \left[k_2 \frac{\gamma_1 \gamma_2}{\gamma_3} - k_3 \right] \alpha, \qquad (19)$$

the form of which is reminiscent of the dual hypothesis of Bredig, Snethlage and Acree (Robertson and S. F. Acree, 1915), which is dealt with more fully in Chapter 9.

The Reaction between Potassium Persulphate and Potassium Ferrocyanide

The instantaneous rate of this reaction $K_2S_2O_8 + 2K_4Fe(CN)_6 \rightarrow 2K_2SO_4 + 2K_3Fe(CN)_6$ in aqueous solution is found to be proportional to the total concentration of each salt:

$$-\frac{d[K_2S_2O_8]}{dt} = -\frac{1}{2}\frac{d[K_4Fe(CN)_6]}{dt} = k_{obs}[K_2S_2O_8]_{total}[K_4Fe(CN)_6]_{total}.$$

The reaction is therefore bimolecular. The apparent bimolecular constant, k_{obs}, is defined by this equation, irrespective of whether the reactants are ions or ion-pairs. At a fixed ionic strength, k_{obs} is found to increase as the concentration of potassium ions is increased, so that the bimolecular constant is not determined solely by the ionic strength of the solution (J. Holluta and W. Herrmann, 1933). Chlebek and Lister (1966) have made a thorough investigation of the kinetics of this reaction at various concentrations and temperatures. In order to apply the theory of C. W. Davies, the association constants of the salts concerned were first obtained electrometrically, with the results given in Table II. The authors incidentally note that $d(\Delta H^0)/d(\Delta S^0)$ is approximately 220°. Accordingly, at 25°C, $T\Delta^0 S = 1{,}310 + 1{\cdot}35\,\Delta^0 H$ cal.

TABLE II
ASSOCIATION CONSTANTS AT ZERO IONIC STRENGTH AND AT 24·65°C
(CHLEBEK AND LISTER, 1966)

Salt	K^0(litre-mole^{-1})	ΔH^0(kcal-mole^{-1})	ΔS^0(cal-mole^{-1}-deg^{-1})
KNO_3	0·69	$-3·22$	$-11·5$
K_2SO_4	6·90	3·11	14·3
$K_2S_2O_8$	8·23	1·62	9·6
$K_3Fe(CN)_6$	25·1	2·57	15·0
$K_4Fe(CN)_6$	170	3·86	23·2

From a knowledge of these association constants and of the total concentrations, it is possible, by successive approximations, to evaluate the concentration of each kind of ion in solution, and to select from the following possibilities which pair of ions is mainly responsible:

$$Fe(CN)_6^{4-} + S_2O_8^{2-}\; ; k_1,$$

$$Fe(CN)_6^{3-} + S_2O_8^{2-}\; ; k_2,$$

$$Fe(CN)_6^{4-} + KS_2O_8^{-}\; ; k_3,$$

$$KFe(CN)_6^{3-} + KS_2O_8^{-}\; ; k_4.$$

The anticipated product $z_A z_B$ for these pairs is seen to be successively 8, 6, 4 and 3. Each of the four bimolecular constants can be related to k_{obs} and the equilibrium constants

$$K_1 = \frac{[KFe(CN)_6^{3-}]}{[K^+][Fe(CN)_6^{4-}]} \quad \text{and} \quad K_2 = \frac{[KS_2O_8^-]}{[K^+][S_2O_8^{2-}]}.$$

If we denote the concentration of free potassium ions by x, we have

$$K_1 = [KFe(CN)_6^{3-}]/[Fe(CN)_6^{4-}]x,$$

and

$$[K_4Fe(CN)_6]_{total} = [KFe(CN)_6^{3-}] + [Fe(CN)_6^{4-}].$$

Hence

$$[K_4Fe(CN)_6]_{total} = [KFe(CN)_6^{3-}]\left(\frac{1 + K_1 x}{K_1 x}\right).$$

Similarly

$$[K_2S_2O_8]_{total} = [KS_2O_8^-]\left(\frac{1 + K_2 x}{K_2 x}\right).$$

The total concentration of free iodide ion does not change perceptibly during a run, so that, if, for example, the rate is given by $k_4[KFe(CN)_6^{3-}] \times [KS_2O_8^-]$,

$$k_4 = k_{obs} \cdot \frac{(1 + K_1 x)(1 + K_2 x)}{K_1 K_2 x^2}. \tag{19a}$$

The bimolecular constants thus found are reproduced by the equations $\log_{10} k_4$ (litre-mole^{-1}-min^{-1}) $= 0.376 + 3.17 f(I)$ at 25°C and $0.693 + 3.33 f(I)$ at 40°C, which correspond to $z_A z_B$ values of 3.11 and 2.94 respectively. The same procedure applied to the logarithms of the constants k_1, k_2 and k_3 yield Brönsted slopes consistent with $z_A z_B$ values of 17.1, 11.5 and 8.9 respectively, which conflict with the mechanisms. It is therefore concluded that the rate of chemical change in this complicated system is determined by the concentration of the ions $KFe(CN)_6^{3-}$ and $KS_2O_8^-$ and that the three other pairs of ions do not play a significant kinetic role.

The bimolecular constant at zero concentration varies with respect to temperature according to the equation k_4^0 (litre-mole^{-1}-sec^{-1}) $= 1.63 \times 10^5 . \exp(-9{,}025/RT)$. The low value of the pre-exponential term in reactions between ions of the same charge sign is discussed later.

This reaction has been studied kinetically in the presence of various cations at 25°C by Kershaw and Prue (1967). In the absence of static data on the relevant association constants no detailed analysis of the net rates can be given.

The Reaction between the Iodide and Persulphate Ions

An early investigation of the kinetics of the reaction between the iodide and persulphate ions, $2I^- + S_2O_8^{2-} \rightarrow I_2 + 2SO_4^{2-}$, based on the titration of the liberated iodine by sodium thiosulphate, showed the order of reaction to be 2 by the method of isolation, and 2.4 by the method of initial rates. Price (1898) and King and Jacobs (1931) measured the rate of reaction by adding thiosulphate and measuring the time taken for the iodine colour to reappear. They found the Brönsted–Bjerrum equation to hold with $z_A z_B = +2$. When the decadic logarithm of the bimolecular constants obtained by R. G. Soper and E. Williams (1933) are plotted against $I^{1/2}$ at 25°C, the least-squares gradient over the whole range covered, i.e., up to an ionic strength of 0.022, is 2.42 ± 0.06. The data at ionic strengths below 0.006, however, yield a gradient of 1.97, confirming the work of King and Jacobs and suggesting that, in very dilute solution, the rate-determining step is that between the iodide and persulphate ions. Howells (1939, 1941, 1964), in a series of studies, finds that $k_2^0 = 1.09 \times 10^6 . \exp[-(12{,}425 \pm 125)/RT]$.

At a constant temperature and ionic strength, however, chlorides and nitrates of the alkali metals increase the rate constant in the descending order Cs^+, Rb^+, K^+, NH_4^+, Na^+, Li^+. This sequence has been confirmed by Indelli and Prue (1959), who establish the same order of the cationic effect in the persulphate-ferrocyanide reaction. As far as may be judged, the sequence is that of the association constants, K, for these univalent cations with a common anion (Table II, 9).

Let us suppose that complexes are formed between the iodide ion and the anions $S_2O_8^{2-}$ and $MS_2O_8^-$. By the Brönsted–Davies method, the reaction rate becomes

$$\frac{d[I_2]}{dt} = k_2[I^-][S_2O_8^{2-}]\frac{\gamma_1\gamma_2}{\gamma_3} + k_2'[I^-][MS_2O_8^-]\frac{\gamma_1^2}{\gamma_2}.$$

In terms of the observed rate constant, defined by the equation

$$\frac{d[I_2]}{dt} = k_{obs}[I^-]_{total}[S_2O_8^{2-}]_{total},$$

and the association constant

$$K = \frac{[MS_3O_8^-]}{[S_2O_8^{2-}][M^+]\gamma_2},$$

we see that

$$k_{obs} = \frac{(k_2\gamma_1\gamma_2/\gamma_3) + k_2'K[M^+]\gamma_1^2}{1 + K[M^+]\gamma_2}. \tag{19b}$$

The first term in the numerator accounts for the fact that the experimental value of the charge product $z_A z_B$ is nearly 2. The second term allows for the specific influence of the cation.

The approach of two similarly charged ions to each other is facilitated by the presence of an ion of opposite charge; and this is the qualitative description of the so-called cationic catalysis. These experiments, however, show that the mechanism whereby the cation is brought into the sphere of action is prior ion-pairing.

We have seen that bridges between anions may also be formed by a divalent metal, M or by two univalent metal atoms, as in the reactions of persulphates with the iodide and ferricyanide ions respectively, the structures of whose intermediate complexes may be written thus:

$$[I, MS_2O_8]^{2-} \quad \text{and} \quad \left[S_2O_8 \begin{smallmatrix} K \\ \diagup \\ \diagdown \\ K \end{smallmatrix} Fe(CN)_6 \right]^{4-}.$$

The rate determining step in the hydrolysis of metaphosphates in alkaline solution containing salts of the alkaline earth metals, M, has been shown to be $MP_3O_9^- + OH^-$, followed by rapid reactions, and preceded by the ion-pair equilibria $P_3O_9^{3-} + M^{2+} \rightleftarrows MP_3O_9^-$ (Indelli, 1956).

A Criticism of the Brönsted–Bjerrum Equation

A. R. Olson and T. R. Simonson (1949), discussing Livingston's diagram in detail, and adding supplementary data of their own on the two reactions involving the brompentammine cobaltic ion, conclude that the effect of the addition of inert salts on the rates of reactions between ions of the same charge sign "is caused almost exclusively by the concentration and character of salt ions of charge sign opposite to that of the reactants", and that "*the rate is not dependent upon the ionic strength of the solution*". "The [salt] effects are quantitatively interpretable in terms of an ion association constant and specific rate constants for the associated and non-associated reactants. *The further introduction of activity coefficients is not necessary.*"

The work of C. W. Davies provides a complete rebuttal of the two allegations which we have italicized. Olson and Simonson clearly failed to realize that the effect of ion-pairing is to supplement rather than to replace the effect of ionic activities.

Application of the Brönsted–Bjerrum Equation to the Kinetics of Nitration

Equation (10) may be used to detect the electrovalency z_A of an unknown ion when it reacts with another ion of known electrovalency, z_B. Thus, for example, when benzene sulphonate derivatives are nitrated in the presence of sulphuric acid, the electrolyte effect shows that the charge product $z_A z_B$ is -1. Since the charge on the sulphonate ion is known to be -1, it follows that the charge on the other reacting ion is $+1$. This ion is now known from extra-kinetic sources to be the nitronium ion, NO_2^+, formed by the basic ionization of nitric acid: $HNO_3 \rightarrow HO^- + NO_2^+$. It is this application (G. M. Bennett, J. C. D. Brand, and G. Williams, 1946) of equation (10) that led to the elucidation of the mechanism of aromatic nitration, a subject which is dealt with in Chapter 9, when we shall describe Chandra and Coulson's treatment.

Solvated Electrons

The alkali metals dissolve freely in liquid ammonia giving solutions which conduct electricity as well as do aqueous solutions of salts. The process of dissolution is attended by an abnormally large increase in the partial molar volume of the solute, and by the development of an inky blue colour (Onsager, 1964). More intense absorption appears in the infra-red region of the spectrum at a wave number of 7,000 cm^{-1}. These facts have been interpreted in terms of the partial ionization of the dissolved alkali metal atom, such as $Na \rightarrow Na^+ + e^-$. The increase in volume, due mainly to the dissolved electron, is about three times the molar volume of the solvent, indicating that the electron has dug a hole of considerable size for itself. As Onsager remarks, why it should do this is not clear. A possible electrostatic explanation is that the electron induces positive charges in the solvent molecules surrounding it, much as an elementary anion does in aqueous solution. Thereafter it moves in the field which it has created. The optical absorption represents its transfer to a new orbit in the field of the trapping charge. A similar model has been used by London to explain the trapping of electrons in ionic crystals. A wave-mechanical theory of the stability of solvated electrons has been given by Jorntner (1964). Attempts to measure their volume are described in a later section of this chapter.

When water is irradiated with hard X-rays, γ-rays or high-energy electrons, several chemical species are formed, e.g.

$$H_2O \rightsquigarrow H_2, H_2O_2, H, OH, H_3O^+, e_{aq}^-.$$

The existence of the hydrated electron, which is denoted by the symbol e_{aq}^-, was established spectroscopically by Hart and Boag (1962). The rate at which it reacts with the solvent and with numerous solutes has been determined, chiefly at 23°C, by very rapid spectrophotometry. Since the concentration of hydrated electrons is small in comparison with the concentration of either solute or solvent, the latter may be regarded as constant during a run. It is then found that the rate at which the hydrated electron disappears is proportional to its concentration. The pseudo-unimolecular constant varies linearly with respect to the concentration of solute, so that the reaction is bimolecular (Baxendale, Fielden and Keene, 1963; E. J. Hart and M. Anbar, 1970). Electron spin resonance, electrical conductivity and light absorption measurements give concordant rate constants. (Table III.) Omitting the relatively slow reaction with water ($k_2 = \sim 10$; $E_A = 4.5 \pm 1$ kcal) the rates have the order of magnitude to be expected for reactions not requiring an energy of activation, and vary in directions consistent with the sign of the charge product.

TABLE III
BIMOLECULAR CONSTANTS OF SOME REACTIONS OF THE HYDRATED ELECTRON*
(LITRE/MOLE-SEC) AT 22–23°C

Reaction	$k_2 \times 10^{-10}$	Reaction	$k_2 \times 10^{-10}$
$e_{aq}^- + H_3O_{aq}^+ \rightarrow H + H_2O$	2·36	$e_{aq}^- + Fe^{2+}$	0·016
$e_{aq}^- + Cu^{2+} \rightarrow Cu^+$	3·3	$e_{aq}^- + C_0^{2+}$	1·2
$e_{aq}^- + Fe(CN)_6^{3-} \rightarrow Fe(CN)_6^{4-}$	0·3	$e_{aq}^- + Ni^{2+}$	2·2
$e_{aq}^- + [Cr·en_3]^{3+} \rightarrow [Cr·en_3]^{2+}$	13	$e_{aq}^- + Ag^+$	3·2
$e_{aq}^- + e_{aq}^- \rightarrow H_2 + 20H_{aq}^-$	1	$e_{aq}^- + Pb^{2+}$	3·9
$e_{aq}^- + H_2O \rightarrow H + OH_{aq}^-$	$\sim 10^{-9}$	$e_{aq}^- + Co(NH_3)_6^{3+}$	9
		$e_{aq}^- + Co(NO_2)_6^{3-}$	5·8

REFERENCE

* M. S. Matheson, *Radiation Research Supplement*, **4**, 1 (1964); Boxendale *et al.*, *Proc. Roy. Soc.* A, 286, 322 (1965).

Collinson, Dainton, D. R. Smith and Tazuki (1962) found the velocity constant for the reduction of the ferricyanide ion by the hydrated electron $(e_{aq}^- + Fe(CN)_6^{3-} \rightarrow Fe(CN)_6^{4-})$ to obey equation (11), with the theoretical gradient of $+3$. Czapski and H. A. Schwarz (1962), cited by D. C. Walker (1967), studying the rate of reaction of the hydrated electron with ONO_2^-, O_2 and H^+ found equation (10) to be obeyed in each case, with $z_A z_B$ values of $+1$, 0 and -1 respectively, in agreement with the requirements of the Brönsted–Bjerrum theory.

Christiansen's Formulation of the Kinetics of Ionic Reactions in Solution

In a treatment of the kinetics of ionic reactions, Christiansen (1924) takes as his basis the gas collision theory expression for the instantaneous velocity of reaction:

$$\frac{dc_P}{dt} = Z^0 c_A c_B . \exp\left(-E_n/kT\right),$$

where E_n is the minimum energy which a pair of uncharged molecules, A and B, must acquire before being able to react, and Z^0 is the standard collision frequency (equation 4.38). To allow for the necessity of a special orientation, the numerical factor, o is introduced, which is probably nearly unity for small and symmetrical molecules. In his first adaptation of this equation to reactions in solution, Christiansen allows for the possibility of deactivations of the active complex by collisions with other types of molecules, including those of the solvent. Letting p stand for the probability per second that an activated complex will undergo chemical change, and $\Sigma k_j c_j$ the total probability per second that it will be deactivated by various species, the rate equation becomes

$$\frac{dc_P}{dt} = Z^0 c_A c_B \cdot o\left(\frac{p}{p + \Sigma k_j c_j}\right) \cdot \exp\left(-E_n/kT\right).$$

It was thought at one time that all bimolecular reactions in solution were collisionally inefficient due to frequent deactivations by solvent molecules, but the experiments described in Chapter 4 have shown that they are, on the whole, as normal in this respect as are bimolecular reactions in gases. To adapt this equation so as to allow for the effect of the electrical field on the frequency of activating collisions, we consider first the collisions made by all molecules of type A against one molecule of type B. To do so, we recognize that, if Boltzmann's distribution law is maintained, the concentration of ions of type A at a distance r from the central ion of type B is not the bulk concentration, c_A, but $c_A \exp\left(-z_A \varepsilon \psi_B/kT\right)$, where ψ_B is the electric potential at a distance r from the ion B, and is given, according to Debye and Hückel, as

$$\psi_B = \frac{z_B \varepsilon}{Dr} \cdot \exp\left(-\kappa r\right),$$

where z is the electrovalency, ε the charge on a proton, and D the permittivity of the medium. Finally

$$\kappa = \left(\frac{4\pi N_0 \varepsilon^2 \Sigma c_i z_i^2}{1,000 DkT}\right)^{1/2} = \varepsilon\left(\frac{8\pi N_0 I}{1,000 DkT}\right)^{1/2},$$

where I is the ionic strength. Omitting the orientational and deactivational factors, the instantaneous rate of reaction thus becomes

$$\frac{dc_P}{dt} = Z^0 c_A c_B \cdot \exp\left(-E_n/kT\right) \cdot \exp\left(-\frac{z_A z_B \varepsilon^2}{DrkT} \cdot e^{-\kappa r}\right)$$

G. Scatchard (1932) has argued that the electric potential should be expressed by the fuller form of the Debye–Hückel theory, which is

$$\psi_B = \frac{z_B \varepsilon}{Dr} \cdot \exp\left(-\kappa r\right)\left(\frac{\exp\left(\kappa a\right)}{1 + \kappa a}\right),$$

where a denotes the nearest distance to which the ions can approach each other. Christiansen, however, points out that, within the region of approximation made, the last term in this expression is unity, and its inclusion is therefore superfluous (J. A. Christiansen, 1951). Moreover, if κr is much less than unity, and this is the validity condition of our expression for ψ, we can write $1 - \kappa r$ for $\exp(-\kappa r)$. The theoretical expression for the bimolecular velocity constant thus becomes

$$k_2 = Z^0 . \exp(-E_n/kT) . \exp[-(z_A z_B \varepsilon^2/DrkT)(1 - \kappa r)]. \tag{20}$$

It need hardly be added that we would have obtained the same equation if we began by considering all the collisions made in one second by molecules of type B against one molecule of type A. We would then have to replace c_B of the gas collision equation by $c_B \exp(-z_B \varepsilon \psi_A/kT)$, where

$$\psi_A = \frac{z_A \varepsilon}{Dr} . \exp(-\kappa r).$$

The assumptions of Christiansen's theory may be summarized as follows: (i) chemical reaction occurs between ions when, while possessing a total energy E between them, their charge centres come to within a critical distance, r, of each other; (ii) the total energy, E, may be resolved into a non-electrostatic component, E_n, and an electrostatic component, E_e; (iii) the electrostatic energy of interaction of the ions at a distance r apart is given by Coulomb's expression, modified, in dilute solution, by the Debye–Hückel correction. Then

$$E = E_n + E_e = E_n + \frac{z_A z_B \varepsilon^2}{Dr}(1 - \kappa r). \tag{21}$$

Equation (20) gives for the rate constant at zero ionic strength

$$k_2^0 = Z^0 . \exp(-E_n/kT) . \exp(-z_A z_B \varepsilon^2/DrkT). \tag{22}$$

Hence

$$k_2 = k_2^0 . \exp(z_A z_B \varepsilon^2 \kappa/DkT), \tag{23}$$

which is the equation of Brönsted and Bjerrum, derived in another way and formulated in a different notation.

Variation of the Rate of Reaction with Respect to the Permittivity of the Solvent

If, therefore, for a given reaction Z^0 and E_n are independent of the solvent

$$\left[\frac{d \log_{10} k_2^0}{d(1/D)}\right]_T = -\frac{z_A z_B \varepsilon^2}{2 \cdot 303 rkT}. \tag{24}$$

This equation (G. Scatchard, 1932; N. Bjerrum, 1923) provides a method for evaluating the critical interionic separation, r. At 30°C, for example, the gradient for a reaction between univalent ions of opposite sign should be $+239\cdot2/r$ (Å). Amis and Price (1944), using the data of Svirbely and Schramm (1938) and of Lander and Svirbely (1938) on the reaction between the ammonium and cyanate ions in ethylene glycol–water mixtures at this temperature, find a gradient of $+93\cdot3$. Hence $r = 2\cdot56$ Å. Janelli (1953, 1958), following Warner and Stitt (1933), finds values of $2\cdot52$ and $2\cdot69$ Å in various water–alcohol mixtures. The sum of the Stokes radii is $1\cdot225$ Å (NH_4^+) + $1\cdot38$ Å (NCO^-) = $2\cdot61$ Å. Equation (24) when applied to the bromacetate–thiosulphate data, uncorrected for ion-pairing, yields a value (K. J. Laidler and H. Eyring, 1940) of $5\cdot1$ Å for r. These reactions, like most others, are not free from complications, and the various estimates of r may be regarded as entirely satisfactory. Further instances of the use of equation (24) are given by Amis (1949).

The Relationship between the True and Apparent Energies of Activation

Before equation (20) can be compared with experimental data expressed empirically as

$$k_2 = A \cdot \exp(-E_A/kT), \tag{25}$$

the true energy of activation, E, or $E_n + E_e$, must be related to the apparent energy of activation, generally defined as

$$E_A = kT^2\left(\frac{d \ln k_2}{dT}\right)_P. \tag{4.14}$$

Because the theoretical equation (20) contains the permittivity, D, allowance must be made for the fact that D decreases as the temperature is raised, according to the empirical equation

$$D = D_0 \cdot \exp(-LT), \tag{26}$$

where L is a constant, specific to each liquid (Abegg, 1897; Lowry and Jessop, 1930; Akerlof, 1932; Morgan and Smyth, 1928; Wyman, 1930). We then find the required relationship (E. A. Moelwyn-Hughes, 1933, 1936)[*]

$$E_A = E_n + \frac{z_A z_B \varepsilon^2}{Dr}(1 - LT)(1 - \tfrac{3}{2}\kappa r). \tag{27}$$

$$E_A = E_A^0 - \frac{3}{2}\frac{z_A z_B \varepsilon^2 \kappa}{D}(1 - LT). \tag{28}$$

---

[*] Independent derivations have been given by La Mer and Kamner, *J. Amer. Chem. Soc.*, **57**, 2622 (1935) and by Svirbely and Warner, *ibid*, **57**, 1883 (1935).

TABLE IV*
THE PERMITTIVITY OF LIQUIDS AT VARIOUS TEMPERATURES

Liquid	D_0	$D_{298 \cdot 16^0}$	$L \times 10^3$	$(1/L)(°K)$	$(1 - LT)$
$C_6H_5 \cdot CH_3$	2·86	2·340	0·673	1486	+0·799
nC_6H_{14}	2·33	1·885	0·714	1401	+0·788
CCl_4	2·88	2·238	0·843	1187	+0·748
C_6H_6	2·95	2·272	0·876	1142	+0·739
C_6H_5Cl	15·52	6·555	2·89	346	+0·138
$CHCl_3$	12·0	4·45	3·33	300	+0·007
$C_6H_5 \cdot CO \cdot CH_3$	60·2	17·7	4·1	244	−0·223
$(CH_3)_2CO$	120·4	30·3	4·63	216	−0·380
H_2O	311·17	78·53	4·63	216	−0·380
C_2H_5Br	39·45	9·128	4·91	204	−0·462
$C_6H_5 \cdot O \cdot CH_3$	19·97	4·24	5·20	192	−0·550
$C_6H_5 \cdot NO_2$	164·7	34·85	5·21	192	−0·553
CH_3OH	157·6	31·6	5·39	186	−0·607
C_2H_5OH	146·0	24·3	6·02	166	−0·795
HCl	34·67	4·97	6·52	153	−0·944
$(CH_3)_3OH$	127·9	9·268	8·80	114	−1·623
SO_2	270	12·03	10·43	98	−2·11

REFERENCE

* This is Table 20 from Moelwyn-Hughes, *Physical Chemistry*, 2nd Ed., p. 881.

From the numerical data summarized in Table IV, we see that, for a given reaction, E_A may increase or decrease with respect to the ionic strength, depending on the sign of the term $(1 - LT)$, and, for a given solvent, E_A may increase or decrease with respect to ionic strength, depending on the sign of the electrovalency product $z_A z_B$. In water at 25°C, the special form of equation (28), giving E_A in calories per mole, becomes

$$E_A = E_A^0 + 792 z_A z_B \sqrt{I}. \tag{29}$$

In other solvents and at other temperatures, we may write

$$E_A = E_A^0 + B z_A z_B \sqrt{I} \tag{30}$$

where

$$B = \frac{3}{2} \frac{z_A z_B \varepsilon^2}{D} (1 - LT) \left(\frac{8\pi N_0 \varepsilon^2}{1,000 D k T} \right)^{1/2}. \tag{31}$$

The application of this equation requires data of considerable accuracy on uncomplicated reactions. The instances cited in Table V show that, in all cases, the sign of the variation of E_A with respect to the ionic strength is

TABLE V

OBSERVED AND CALCULATED CHANGES IN E_A WITH RESPECT TO I IN AQUEOUS SOLUTION

Reaction	Reference	$z_A z_B$	T	B	I	E_A (observed)	Difference in E_A	
							Observed	Calculated
$NH_4^+ + NCO^-$	(a)	-1	313·2	1103	0	$(24{,}455 \pm 8)$		
					0·05	24,200	-230	-246
					0·20	23,970		
$NH_4^+ + NCO^-$	(b)	-1	323·1	1187	0	$(23{,}580)$		
					0·0376	23,240	-340	-230
$BrAc^- + S_2O_3^{--}$	(c)	$+2$	285·6	631	0	$(15{,}799 \pm 31)$		
					0·008	$15{,}881 \pm 30$	$+127$	$+66$
					0·020	$16{,}008 \pm 64$		
$BrY^{--} + OH^-$	(d)	$+2$	288·1	662	0	$(12{,}361 \pm 14)$		
					0·0023	12,439	$+31$	$+59$
					0·0086	12,470		

REFERENCES

(a) Doyle, *Thesis*, Liverpool (1922). Data cited in *Chem. Rev.*, **10**, 241 (1932).

(b) Svirbely and Warner, *J. Amer. Chem. Soc.*, **57**, 1883 (1935).

(c) La Mer and Kamner, *J. Amer. Chem. Soc.*, **57**, p. 1674.

(d) La Mer and Amis, *J. Franklin Inst.*, 225, 709 (1938); Panepinto and Kilpatrick, *J. Amer. Chem. Soc.*, **59**, 1871 (1937). *Y* is explained in the text following Equation (53).

correct, and that, in the ammonium cyanate reaction, the numerical agreement is not only good, but is better than could be reasonably expected at the
ionic strengths used. Of the two values of E_A^0 quoted under (a) and (b),
the later work of L. Janelli (1953) confirms Doyle's datum, being 24,430 cal.
Reaction (c), as explained above, is now known to be composite. The agreement in reaction (d) is wholly satisfactory, allowing an uncertainty of 14
calories per mole in the observed energy of activation.

The Pre-exponential Term A of the Arrhenius Equation for Ionic Reactions

By eliminating E_n from equations (20) and (27) the empirical term A of
the Arrhenius equation,

$$k_2 = A \cdot \exp\left(-E_A/kT\right) \tag{25}$$

can be related to the theoretical term Z^0 of Christiansen's equation,

$$k_2 = Z^0 \cdot \exp\left(-E_n/kT\right) \cdot \exp\left[-\frac{z_A z_B \varepsilon^2}{DrkT}(1 - \kappa r)\right], \tag{20}$$

yielding the result

$$A = Z^0 \cdot \exp\left(-\frac{z_A z_B \varepsilon^2 L}{kDr}\right) \cdot \exp\left[-\frac{z_A z_B \varepsilon^2}{2DkT}(1 - 3LT)\kappa\right]. \tag{32}$$

As stated in discussing the collision theory, the ratio of the pre-exponential
term of the empirical equation of Arrhenius for bimolecular reactions to
the standard collision frequency is denoted by the symbol P:

$$P = A/Z^0. \tag{33}$$

In particular, at zero ionic strength,

$$P^0 = A^0/Z^0.$$

From equation (32), we see that

$$P^0 = A^0/Z^0 = \exp\left[-\frac{z_A z_B \varepsilon^2 L}{kDr}\right]. \tag{34}$$

Before this equation can be used to evaluate r, accurate data are necessary
on systems where temperatures and ionic strengths are known over wide
ranges of both variables. Such are very seldom available, as revealed by a
detailed examination of 18 ionic reactions in water (E. A. Moelwyn-Hughes,
1947). A few of the results are shown in Table VI, from which we note that,
allowing for the inadequacy of most of the data, reasonable values of the

TABLE VI

Reaction	Ionic type z_A	z_B	P (observed)	r in Å	ΔS^*
$[Cr(H_2O)_6]^{+++} + CNS^-$	$+3$	-1	$1{\cdot}09 \times 10^{+8}$	$1{\cdot}56$	39
$[Co(NH_3)_5Br]^{++} + OH^-$	$+2$	-1	$1{\cdot}39 \times 10^{+6}$	$1{\cdot}36$	28
$NH_4^+ + CNO^-$	$+1$	-1	$3{\cdot}83 \times 10^{+1}$	$2{\cdot}63$	$7{\cdot}2$
$CH_2Br{\cdot}COO^- + S_2O_3^{--}$	-1	-2	$7{\cdot}64 \times 10^{-3}$	$3{\cdot}94$	$-9{\cdot}7$
$[Co(NH_3)_5Br]^{++} + Hg^{++}$	$+2$	$+2$	$4{\cdot}30 \times 10^{-4}$	$4{\cdot}96$	-15
$AsO_3^{--} + TeO_4^-$	-2	-2	$6{\cdot}77 \times 10^{-6}$	$3{\cdot}23$	-24

critical interionic separation, r, are sufficient to explain values of P ranging from 10^{+8} to 10^{-5}. Of greater interest than this numerically satisfactory result is the *a posteriori* vindication for resolving the energy of activation into electrostatic and non-electrostatic components.

From equations (25) and (32), as applied to reactions at zero ionic strength, we have

$$k_2 = Z^0 . \exp\left(-\frac{z_A z_B \varepsilon^2 L}{kDr}\right) . \exp\left(-\frac{E_A}{RT}\right).$$

Comparison with the equation

$$k_2 = Z^0 . \exp\,(1/2) . \exp\left(\frac{\Delta S^*}{R}\right) . \exp\left(-\frac{E_A}{RT}\right) \tag{4.44}$$

offers another means of summarizing the results. Ignoring the difference between $e^{1/2}$ and unity, we have

$$\Delta S^* = R \ln P^0. \tag{35}$$

It is in this form that these results have most frequently been reproduced in the literature. Typical data are given in the last column of Table VI, in the units of calories per mole-degree.

The variation of the term A with respect to ionic strength has been found to obey equation (32) (Amis and La Mer, 1939; Amis and Cook, 1941; Amis, 1966).

Methods for Estimating the Critical Interionic Separation

By introducing assumptions into the foregoing treatment, various estimates of r can be made which cohere despite the simplicity of the theory.

(1) If Z^0 and E_n of equation (22) are assumed to be the same for a pair of similar reactions such as

$$CH_2Br \cdot COO^- + S_2O_3^{--} \rightarrow CH_2(S_2O_3^-) \cdot COO^- + Br^-$$

and

$$CH_2Br \cdot COOCH_3 + S_2O_3^{--} \rightarrow CH_2(S_2O_3^-) \cdot COOCH_3 + Br^-,$$

we can write

$$\log_{10}\left[\frac{k_2^0 \text{ (ion–ester)}}{k_2^0 \text{ (ion–ion)}}\right] = \frac{z_A z_B \varepsilon^2}{2 \cdot 303 D r k T}. \tag{36}$$

In water at 25°C, we have, denoting by $\mathring{r}$ the value of r in Ångström units,

$$\log_{10}\left[\frac{k_2^0 \text{ (ion–ester)}}{k_2^0 \text{ (ion–ion)}}\right] = \frac{3 \cdot 10 z_A z_B}{\mathring{r}}.$$

The limiting value of the ratio of the velocity constants found by V. K. La Mer (1932) under these conditions is 58. Hence, $\mathring{r} = 3 \cdot 52$ Å. Generally, this method yields values of $\mathring{r}$ that are unreasonably large.

(2) Without making any assumption concerning Z_0, but by taking r and E_n to be the same for a similar pair of reactions, we have, from equation (27),

$$E_A^0 \text{ (ion–ion)} - E_A^0 \text{ (ion–molecule)} = (1 - LT)\frac{z_A z_B \varepsilon^2}{Dr}, \tag{37}$$

which, in calories per mole for reactions in water at 25°C, becomes

$$E_A^0 \text{ (ion–ion)} - E_A^0 \text{ (ion–molecule)} = -\frac{1,606 z_A z_B}{\mathring{r}}.$$

The data of numerous workers on the reactions $BrAc^- + S_2O_3^{--}$ and $BrAc^- + MS_2O_3$ provide the equations

$$k_2^0 \text{ (ion–ion)} = 1 \cdot 36 \times 10^9 . \exp\left(-15,710/RT\right), \tag{38}$$

and

$$k_2^0 \text{ (ion–molecule)} = 2 \cdot 95 \times 10^{10} . \exp\left[-(16,415 \pm 45)/RT\right]. \tag{39}$$

If we allow equal margins of error in the activation energies of the two reactions, we see that E_A^0 (ion–ion) $- E_A^0$ (ion–molecule) is -705 ± 90 cal. Hence, from equation (37), $\mathring{r} = 4 \cdot 56 \pm 0 \cdot 59$ Å.

(3) If E_n is independent of temperature, the variation of the apparent energy of activation for an ionic reaction with respect to temperature at constant ionic strength is given, according to equation (27), as

$$\left(\frac{dE_A}{dT}\right)_j = -\frac{z_A z_B \varepsilon^2 L^2 T}{Dr} + \frac{3}{2}\frac{z_A z_B \varepsilon^2 L^2 T}{D}\left[1 - \frac{1}{2}\left(\frac{1 - LT}{LT}\right)^2\right]\kappa. \tag{40}$$

On applying it to aqueous solutions at $291 \cdot 85°K$, and expressing E_A in cal, we obtain the result:

$$\left(\frac{dE_A}{dT}\right)_j = -25 \cdot 7\frac{z_A z_B}{\mathring{r}} + 12 \cdot 2 z_A z_B \sqrt{I}. \tag{41}$$

La Mer and Kamner (1935) find that dE_A/dT for the $BrAc^- - S_2O_3^{--}$ reaction at $I = 0 \cdot 02$ is $-8 \cdot 7 \pm 1 \cdot 5$; when I is $0 \cdot 20$, the gradient is $-19 \cdot 6 \pm 2 \cdot 2$. According to equation (41), the values of $\mathring{r}$ indicated are $4 \cdot 3$ and $1 \cdot 7$, respectively. The mean value is $3 \cdot 0 \pm 1 \cdot 3$ Å, with a bias in favour of the higher limit.

(4) We shall next apply equation (34) to a pair of reactions of different ionic types, for which we shall assume common values of r and Z^0. Then, from equation (34),

$$\frac{d \ln A^0}{d(z_A z_B)} = -\frac{\varepsilon^2 L}{kDr}. \tag{42}$$

Using the data previously cited, we have

$$(NH^+ + NCO^-; z_A z_B = -1); \quad k_2^0 = 1 \cdot 09 \times 10^{13} . \exp(-23{,}580/RT);$$

$$(BrAc^- + S_2O_3^{--}; z_A z_B = +2); \quad k_2^0 = 1 \cdot 36 \times 10^9 . \exp(-15{,}710/RT).$$

The energies of activation are not required in this method, and we find $\mathring{r} = 3 \cdot 29$ Å, and $Z^0 = 5 \cdot 45 \times 10^{11}$ litres per mole-sec. It follows that the values of P^0 are $2 \cdot 00 \times 10^1$ and $2 \cdot 50 \times 10^{-3}$ respectively.

The results summarized in Table VII indicate the degree of consistency afforded by the various methods. Because no correction for the effects of ion-pairs has been applied in using equations (24) and (12.18), the high values of r may be omitted, leaving $3 \cdot 41$ Å as the most probable value based on the five other estimates. The sum of the Stokes radii for the ions SO_4^{2-} and CH_3COO^- is $1 \cdot 15 + 2 \cdot 25 = 3 \cdot 40$ Å.

TABLE VII

THE CRITICAL IONIC SEPARATION IN THE REACTION BETWEEN THE THIOSULPHATE AND BROMACETATE IONS IN AQUEOUS SOLUTION

Method	Equation	$r \times 10^8$ (cm)
The absolute magnitude of P^0	(34)	$3 \cdot 27$
The comparison of k_2^0 for ion-ion and ion-molecule reactions	(36)	$3 \cdot 52$
The difference between E_A^0 values for a pair of reactions	(37)	$4 \cdot 56 \pm 0 \cdot 59$
The variation of E_A with respect to temperature	(40)	$3 \cdot 0 \pm 1 \cdot 3$
The comparison of A^0 for a pair of reactions	(42)	$3 \cdot 29$
The variation of $\ln k_2^0$ with respect to permittivity	(24)	$5 \cdot 1$
The variation of $\ln k_2^0$ with respect to pressure	(12, 18)	$4 \cdot 46$

An Alternative Derivation of the Brönsted–Bjerrum Law

V. K. La Mer and M. E. Kamner (1935) have adopted Christiansen's device of resolving the total energy of activation of ionic reactions into electrostatic and non-electrostatic terms, and have rightly pointed out that the kinetics of interionic reactions may be formulated without reference to the collision theory. It is sufficient to assume that the free energy of activation includes an electrostatic term, ΔG_e^*, and that the velocity constant is proportional to $\exp(-\Delta G_e^*/RT)$. Then

$$\ln k_2 = \text{constant} - \Delta G_e^*/RT.$$

In terms of the chemical potentials of the complex, c, and the reacting ions, A and B, we have

$$\Delta G_e^*/N_0 = \mu_C - (\mu_A + \mu_B).$$

For each ion of species i, we have

$$\mu_i = \mu_i^0 - z_i^2 \varepsilon^2 \kappa/2D,$$

where μ_i^0 is its chemical potential at unit concentration in a medium devoid of the ion atmosphere effect. It follows that

$$\Delta G_e^*/N_0 = \mu_C^0 - (\mu_A^0 + \mu_B^0) - (\kappa\varepsilon^2/2D)[z_C^2 - (z_A^2 + z_B^2)],$$

or, since $z_C = z_A + z_B$,

$$\Delta G_e^* = \Delta G_0^* - \frac{N_0\varepsilon^2 z_A z_B \kappa}{DRT},$$

and therefore

$$\ln k_2 = \ln k_2^0 + \varepsilon^2 z_A z_B \kappa/DkT. \tag{23}$$

By means of the Kelvin–Helmholtz relationship, the enthalpy of activation becomes

$$\Delta H_e^* = \Delta H_0^* + \frac{N_0\varepsilon^2 z_A z_B}{Dr}(1 - LT)(1 - \tfrac{3}{2}\kappa r).$$

This is the quantity loosely referred to as the Arrhenius, or apparent, energy of activation, as given in equation (27).

Determination of the Radius of the Solvated Electron

(i) IN METHANOL

A more satisfactory expression than equation (8) can be derived for the logarithm of the activity coefficient of an ion with charge e_i:

$$\ln \gamma_i = -\frac{e_i^2 \kappa}{2DkT} \left(\frac{1}{1 + \kappa \sigma_i} \right), \tag{8c}$$

where σ_i denotes the characteristic diameter of the ion (H. S. Harned and Owen, 1958). On combining with equation (6), we now have

$$\ln \frac{k_2}{k_2^0} = \frac{\varepsilon^2 \kappa}{2DkT} \left[\frac{z_C^2}{1 + \kappa \sigma_C} - \frac{z_A^2}{1 + \kappa \sigma_A} - \frac{z_B^2}{1 + \kappa \sigma_B} \right]. \tag{10b}$$

When we apply this equation to the kinetics of a reaction between univalent ions of opposite charge ($z_A = +1$; $z_B = -1$; $z_C = 0$), we see that

$$\log_{10} \frac{k_2}{k_2^0} = -\frac{\varepsilon^2 \kappa}{(2 \cdot 303) 2DkT} \left[\frac{1}{1 + \kappa \sigma_+} + \frac{1}{1 + \kappa \sigma_-} \right]. \tag{11b}$$

G. V. Buxton, F. S. Dainton and M. Hamerli (1967) have tested this equation in the light of their data on the reactions of H^+ and Ag^+ with the hydrated electron in methanol solution or 0°C. The plot of $\log_{10} k_2$ against the square root of the ionic strength, $\sqrt{I}$, is now not linear, since κ, which is proportional to $\sqrt{I}$, appears in the square-bracketed term. The radii of the cations, Ag^+ and H^+, determined from conductivity data in methanol solution are 8·3 Å and 2·73 Å respectively. The trial and error method is next used to find the value of the diameter, σ_-, of the solvated electron, e_s^-, which best reproduces the variation of $\log_{10} k_2$ with respect to κ up to ionic strengths of about 0·5 molar. It is found to be 5 Å. The radius of the solvated electron thus found in methanol is 2·5 Å, in agreement with Jorntner's (1964) calculated value. It may be doubted whether equation (8c) is reliable enough to stand the strain of the ingenious use made of it here. Falkenhagen has quoted instances where it leads to fictitious values of σ.

(ii) IN WATER

If the only attraction energy concerned in ionic hydration is the electrostatic energy due to the ion–dipole interactions, the theoretical expression for the gain in potential energy attending the removal of the ion from solution is

$$\Delta U^0 \text{ (kcal)} = 127 c |z| (1 - 2/n) (\mathring{a}_0)^2, \tag{43}$$

where c is the co-ordination number, n is the integer in the repulsion energy Aa^{-n}, and a_0 is the equilibrium value of a, the distance between the centre of the ion and the centres of the water dipoles (E. A. Moelwyn-Hughes, 1965). If the ions are treated as hard spheres, n may be taken as infinite. The experimental heats of escape of the halogen ions from water at 25°C are given empirically by the equation

$$\Delta H^0 \text{ (kcal)} = 774/(\mathring{r}_i + \mathring{r}_w)^2 \tag{44}$$

where the sum of the crystal radii, r_i, and the radius of the water molecule, $r_w (= 1 \cdot 343 \text{ Å})$, replaces a_0. The observed values of ΔH^0 and those reproduced by this equation are compared in Table VIII. It is to be noticed that the theoretical and empirical equations agree if c is taken as $6 \cdot 09$, which is not an unreasonable value for the co-ordination number. The crystal radii are those given by Waddington (1966). The heats of escape are based on the thermal equivalence of Na^+ and F^- ions.

TABLE VIII

Ion	r_i (Å)	ΔH^0 (obs)	ΔH^0 (calc)
F^-	1·322	109	109·2
Cl^-	1·822	77	77·2
Br^-	1·983	69	68·3
I^-	2·241	60	60·2
e^-	(3·056)	40	—

Baxendale (1964) has estimated ΔH^0 for the electron as 40 kcal. This value when substituted into the empirical equation requires the radius of the hydrated electron to be $3 \cdot 056$ Å. The basis of comparison between the electron, which is diffusely distributed, and the halogen ions, may, however, be unsound (J. H. Baxendale, private communication).

The Iso-dielectric Energy of Activation

The differentiation of equation (20) with respect to temperature at constant pressure leads to equation (27), which allows for the fact that a change in temperature alters the permittivity, D, of the medium. It is possible, by changing the composition of the solvent, to alter the temperature while keeping the permittivity at a fixed value. We then have

$$E_D = kT^2 \left(\frac{d \ln k_2}{dT} \right)_{P,D} = E_n + \frac{z_A z_B \varepsilon^2}{Dr} (1 - \tfrac{3}{2}\kappa r), \tag{27a}$$

which corresponds to equation (27), with L set as zero. This variable has been termed by Amis (1949, 1966) the iso-dielectric energy of activation. Comparison with equation (27) shows that

$$E_D = E_A + \frac{z_A z_B \varepsilon^2 LT}{Dr}(1 + \tfrac{3}{2}\kappa r). \tag{45}$$

Ionic Reactions in Non-aqueous Solvents

Relatively little information is available on the kinetics of ionic reactions in non-aqueous solvents. C. C. Miller (1934, 1935) has investigated kinetically the reaction between the cyanate and monomethylammonium ions in aqueous and in ethanolic solutions. In the alcoholic solution, the rate of reaction at $313°K$ is 540 times as fast as in aqueous solution and the effect of ionic strength at concentrations below one millimole per litre approximates to the theoretical value. The apparent energy of activation, however, increases with ionic strength, contrary to the requirements of equation (30). At zero ionic strength,

$$k_2^0 \text{ (litre/mole}^{-1}\text{-sec}^{-1}) = 6·73 \times 10^{10} . \exp(-16,370/RT),$$

the first term of which agrees with the gas collision estimate, using Stokes radii. The complete data for this reaction in the two solvents, when compared with the corresponding data for the ammonium cyanate reaction in water, indicate clearly the relative effects of changing the solvent and the basicity of one of the reactants.

The Conversion of Ammonium Cyanate into Urea: An Apparent Paradox

Despite the numerous investigations of the kinetics of this reaction, there has been no general agreement among chemists concerning its mechanism. That it is bimolecular is not in doubt, nor has it been denied that the bimolecular constant decreases as the initial concentration of salt increases, in agreement with equation (12). J. Walker and his collaborators (J. Walker and Hambly, 1895; J. Walker and S. A. Kay, 1897; Ross, 1914) accepted the ionic mechanism $NH_4^+ + NCO^- \rightarrow CO(NH_2)_2$. Chattaway and others (Chattaway and D. L. Chapman, 1912; Werner, 1913; E. E. Walker, 1912) favoured the non-ionic mechanism involving reaction between the ammonia and cyanic acid molecules $NH_3 + HNCO \rightarrow CO(NH_2)_2$. The fact that the instantaneous rate of reaction is known to be given in terms of the ionic concentrations and activities by the equation (Warner and Stitt, 1933;

Warner and Warwick, 1935)

$$\frac{dc}{dt} = k_2[NH_4^+][NCO^-](\gamma_+\gamma_-)$$

would seem at first sight to settle the issue, but such is not the case, because of the possible establishment of the equilibrium

$$K = \frac{[NH_3][HNCO]}{[NH_4^+][NCO^-]} \cdot \frac{1}{\gamma_+\gamma_-} = \frac{K_bK_a}{K_w}, \tag{46}$$

where K_b and K_a are, respectively, the ionization constants of the base and the acid, and K_w is the ionic product of water. If we now postulate that the rate of reaction is due to collisions between neutral molecules of acid and base, we have

$$\frac{dc}{dt} = k_2'[HNCO][NH_3]$$

which, on using equation (46), may be written as

$$\frac{dc}{dt} = k_2'K[NCO^-][NH_4^+]\left(\frac{\gamma_+\gamma_-}{\gamma_0}\right). \tag{47}$$

Thus, as Chattaway and Chapman have pointed out, the instantaneous rate, though due to collision between molecules, would still be proportional to the product of the activities of the ions. Herein lies the paradox.

We do not agree, however, with the conclusion (Frost and Pearson, 1961) that the two mechanisms are "equally supported by the available kinetic evidence" and "seem to be indistinguishable by any kinetic method".

It is true that isothermal data on the salt effect cannot solve the problem, but an answer can be given when the kinetic information available on this reaction is judged against the background of the more extensive knowledge of reactions in solution generally. In the first place, the pre-exponential term, A^0, of the Arrhenius equation for the reaction under discussion is $1 \cdot 09 \times 10^{13}$ litre/mole-sec, which is of the same order of magnitude as the A^0 values for other reactions of the same charge type, such as $6 \cdot 2 \times 10^{13}$ for the reaction (Sirs, 1958) $H^+ + HCO_3^- \rightarrow H_2O + CO_2$, but is higher by a factor of at least 100 than the value generally found for reactions between neutral molecules in solution. In the second place, the variation of E_A with respect to the permittivity is in quantitative agreement with equation (24), and far exceeds in magnitude what could reasonably be expected for reactions between uncharged molecules. The weight of the complete kinetic evidence is therefore in favour of the ionic mechanism.

It remains to be mentioned that the accurately determined velocity constants on which this conclusion is based have been corrected by Warner and his collaborators Warner and Stitt (1933); Warner and Warwick (1935) for the incidence of the reverse reaction, according to equation (6.54), and for the decrease in ionic strength which occurs during each run. The latter correction is made by applying, as follows, a convenient approximate method due to G. Scatchard (1930). The instantaneous rate of reaction is

$$dx/dt = k_2(a - x)^2\gamma_1^2 = k_2(a - x)^2 . \exp(-2\alpha\sqrt{I})$$

where α is defined by equation (9). In this reaction $I = a - x$, so that

$$dx/dt = k_2(a - x)^2 . \exp(-2\alpha(a - x)^{1/2}). \tag{48}$$

The exponential term may be successively approximated to $1 - 2\alpha(a - x)^{1/2}$, and $[1 + 2\alpha(a - x)^{1/2}]^{-1}$. Then

$$\frac{dx}{dt} = k_2 \frac{(a - x)^2}{1 + 2\alpha(a - x)^{1/2}},$$

which gives on integration

$$k_2 t = \frac{x}{a(a - x)} + 4\alpha\left[\frac{1}{(a - x)^{1/2}} - \frac{1}{a^{1/2}}\right],$$

or

$$k_2 t = \frac{1}{(a - x)}[1 + 4\alpha(a - x)^{1/2}] - \frac{1}{a}[1 + 4\alpha a^{1/2}]. \tag{49}$$

The simultaneous correction for reversibility and decrease in ionic strength can only be made graphically.

An Alternative Treatment of Ionic Reactions in Solution

Laidler and Eyring (1940; Laidler, 1965; Glasstane, Laidler and Eyring, 1941) have formulated the kinetics of ionic reactions in solution using the assumption that the reacting ions and the complex are all spherical, with symmetrically distributed charges. The chemical potential of each ion contains the two electrostatic terms $e_i^2/2Dr_i$ and $e_i^2\kappa/D$. The equilibrium constant for complex formation can thus be derived (equation 3.68), and the bimolecular velocity constant obtained by the method of Brönsted and Bjerrum. We have, at zero ionic strength,

$$\ln k_2^0 = \text{constant} + \frac{\varepsilon^2}{2DkT}\left[\frac{z_A^2}{r_A} + \frac{z_B^2}{r_B} - \frac{(z_A + z_B)^2}{r_C}\right]$$

$$= \text{constant} - \frac{z_A z_B \varepsilon^2}{Dr_c kT}\left[1 - \frac{1}{2}\frac{z_A}{z_B}\left(\frac{r_C}{r_A} - 1\right) - \frac{1}{2}\frac{z_B}{z_A}\left(\frac{r_C}{r_B} - 1\right)\right], \tag{50}$$

which is to be contrasted with equation (20), according to which

$$\ln k_2^0 = \text{constant} - \frac{E_n}{kT} - \frac{z_A z_B \varepsilon^2}{DrkT},$$ (51)

where r is the critical separation of point charges.

Equation (46) may be used to evaluate r_C from the experimental value of $d \ln k_2^0/d(1/D)$ and Stokes' values for r_A and r_B. For the ammonium cyanate reaction, r_C is found by this means to be 5·26 Å. We have seen that equation (47), in the hands of Svirbely and collaborators, of Janelli, and of Amis and collaborators yields the value of $r = 2·63 \pm 0·06$ Å, in closer agreement with that given in Table VII.

Let us consider the form of equation (46) when the complex is regarded as two charged spheres in contact, assuming that $r_C = r_A + r_B$. It is

$$\ln k_2^0 = \text{constant} + \frac{\varepsilon^2}{2D(r_A + r_B)kT}\left(z_A\sqrt{\frac{r_B}{r_A}} - z_B\sqrt{\frac{r_A}{r_B}}\right)^2,$$ (52)

from which we see that $d \ln k_2^0/d(1/D)$ should be positive for reactions of all ionic types. This, as we have seen, is contrary to the facts, for the experimental slope is positive for the NH_4^+–NCO^- reaction, and negative for the $BrAc^-$–$S_2O_3^{--}$ reaction, in agreement with Scatchard's equation (24).

It could be argued that equations (50) and (51), though demonstrably inapplicable to reactions between small ions, would hold when one of the ions is small compared with the other, e.g., when $r_A \ll r_B$, for then, approximately, we have

$$\ln k_2^0 = \text{constant} + \frac{z_A^2 \varepsilon^2}{2Dr_A kT}.$$ (53)

A reaction approximating to these conditions is that between the hydroxyl ion and the anion of tetrabromphenolsulphonephthalein (brom phenol blue) in aqueous solution, according to the scheme

where R_1 is

$-SO_3^-$ and R_2 is

$-O^-$. It will be

observed that the triple charge on the intermediate complex is widely spread.

Nevertheless, the effect of decreasing the permittivity by additions of methanol and ethanol does not give positive values for the gradients $d \ln k_2^0 / d(1/D)$, as required by equation (49), but negative values, as required by equation (24), with r lying between 2·81 and 3·43 Å (Amis and La Mer 1939). By means of equation (34), using the gas collision expression for Z^0, higher values of 5·5 and 7·1 Å have been derived by Amis and La Mer for r; and these are doubtless more acceptable.

It is thus clear that Eyring and Laidler's original version of the theory of the kinetics of ionic reactions is incapable of accounting for the influence of ionic strength and charge product on the rate. It has also been found (Buchanan and Hamann, 1953) inadequate to describe the effect of pressure on the velocities of ionic reactions (see Chapter 13).

Extensions to the Simple Electrostatic Theory

(i) An improved expression for $\ln \gamma_i$ is obtained by adding to equation (8) or (8b) a term bI (Hückel, 1925). While this extends the concentration range, it does not affect the limiting gradient of $\ln k_2$ versus $I^{1/2}$, which is what is generally required.

(ii) It is clear that the distribution of negative electricity in an asymmetric ion like $O_2N.C_6H_4.O^-$ is not uniform. Kirkwood (1934) has derived an expression for the chemical potential of a spherical solute with net charge $z\varepsilon$, made up of a number of charge elements arbitrarily distributed within it. In its simplest form, it reduces to

$$\mu_i(\kappa = 0) = \mu_i^0 + \frac{z_i^2 \varepsilon^2}{2Dr_i} - \frac{3}{8} \frac{(D-1)}{(2D+1)} \frac{\mu_i^2}{r_i^3}. \tag{54}$$

Here r_i is the radius of the dipolar ion, and μ is the dipole embedded in it, and arising from the uneven distribution of charges. When applied to reactions between dipolar ions by the method which led to equation (46), we recover that equation with the algebraic addition of three terms of the form $(3/8kT)f(D)(\mu_i^2/r_i^3)$ for the reactants A and B and the complex C (Laidler and Eyring, 1940; Hiromi, 1960).

(iii) No account has been taken, in the treatment so far given, of the structure of the solvent, which has been treated as a continuous medium with a uniform permittivity. There seems to be little doubt that the macroscopic value of D, is the appropriate one for most of the kinetic effects discussed here, as it is for the static effects deduced from the Debye–Hückel theory. Otherwise, it would be difficult to explain the limiting laws governing the variation, with respect to $\sqrt{I}$, of the activity coefficient, the diffusion coefficient, the heat of dilution, and the partial molecular volume of ions in aqueous solution. These square-root laws, however, tell us nothing of the absolute properties of ions at infinite dilution. For that purpose, we must enquire into the intimate interaction of the ions with the solvent molecules in contact with them. As an introduction to this wide and difficult subject, we shall consider, from an electrostatic point of view, the calcium hexahydrate ion $[Ca, 6H_2O]^{++}$.

Electrostatic Treatment of Complex Ions

The principal properties of complex ions can be adequately predicted by electrostatic theory, as Kossel (1924), van Arkel and de Boer (1931) and Garrick (1930) have shown. To illustrate their method, we shall consider a regular octahedron formed from a divalent cation, such as Ca^{2+} ($z_A = 2$) and six water molecules ($\mu_B = 1.83 \times 10^{-18}$ e.s.u.; $\alpha_B = 1.444 \times 10^{-24}$ cc), each at a distance a from the central ion (Fig. 3). As there are 6 ion–ligand interactions, the energy of the system is

$$u = 6\left[\frac{A}{a^9} + \frac{3\mu_B^2}{2\sqrt{2}a^3}\left(1 + \frac{\alpha_B z_A \varepsilon}{\mu_B a^2}\right)^2 - \frac{z_A \varepsilon \mu_B}{a^2} - \frac{\alpha_B(z_A \varepsilon)^2}{2a^4} - \frac{B}{a^6}\right]. \quad (55)$$

The first term is the intrinsic repulsion between the electronic shells, and has been computed from the constants for neon–neon and argon–argon pairs, using Lennard-Jones' equation (Fowler, 1936)

$$A_{12}^{1/9} = \tfrac{1}{2}(A_{11}^{1/9} + A_{22}^{1/9}). \quad (56)$$

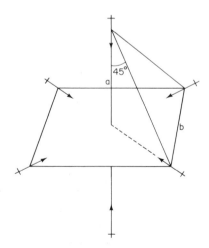

FIG. 3. Structure of the ion $[Ca, 6H_2O]^{++}$.

With $A_{11} = 0.459 \times 10^{-81}$ and $A_{22} = 6.18 \times 10^{-81}$, A_{12} becomes 1.85×10^{-81} erg-cm^9. The last term is London's dispersion energy for the neon–argon pair, obtained in terms of the polarizabilities of the atoms and the B terms for similar pairs (Moelwyn-Hughes, 1961):

$$B_{12} = \frac{2B_{11}B_{22}}{B_{11}(\alpha_2/\alpha_1) + B_{22}(\alpha_1/\alpha_2)}. \tag{57}$$

We find $B_{12} = 5.00 \times 10^{-59}$ erg/cm^6. The second positive term in equation (55) represents the repulsion between the dipoles of the ligands, and is obtained as follows. The energy of repulsion between two dipoles of moment μ each, at a distance b apart, when lying in the same plane, and inclined as follows

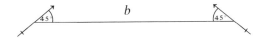

FIG. 4. Dipolar interaction.

is $(3/2)\mu^2/b^3$. Hence the ion–ligand distance, a, is related to the ligand–ligand distance, b, as follows: $a = b/\sqrt{2}$. Hence the dipole–dipole interaction per pair is $(3/4\sqrt{2})(\mu^2/a^3)$. There are 12 such interactions, one for each edge of the octahedron, so that, since 6 appears outside the square bracket in equation (55), the numerical factor inside the bracket is $3/2\sqrt{2}$. The moment, μ, is the sum of the permanent dipole moment, μ_B, and the moment, $\alpha_B z_A \varepsilon/a^2$, induced in the ligand by the field, $z_A \varepsilon/a^2$, due to the central ion. The effective moment is thus $\mu_B + \alpha_B z_A \varepsilon/a^2 = \mu_B(1 + \alpha_b z_A \varepsilon/\mu_B a^2)$. The third term in equation (55) gives the energy of attraction of the dipole of the ligand to the ion; the following term is the net effect of the interaction of the central ion with the moment it has induced in the ligand. The stable structure is that which corresponds to the minimum value of u, obtained by equating du/da to zero. With the numerical values given here, this minimum is found to occur when the ion–ligand distance is 2.32 Å, and the energy of the system is -284 kcal/gram-ion. It can be shown that the calcium ion vibrates with respect to its first co-ordination shell with a frequency of 2.12×10^{13} sec^{-1} ($\omega = 706$ cm^{-1}). The components of the energy of the stable configuration are as follows:

$$u_0 = f(a^{-9}) + f(a^{-3}) - f(a^{-2}) - f(a^{-4}) - f(a^{-6})$$

$$= 82 \quad\quad + 143 \quad\quad - 282 \quad\quad - 199 \quad\quad - 28 \quad\quad = -284 \,(\text{kcal}) \tag{58}$$

$-u_0$ is the energy released when one gram-mole of complex ion is formed from the elementary ion and gaseous water molecules. It comprises the major part of the heat of hydration. By adding 63 kcal, which is the heat of vaporization of 6 moles of water, the total energy loss is 347, which is close to the observed heat of dissolution (358 kcal) of the ion in water, as given by Born's cycle of operations.

If, in the same structure, and at the same internuclear distances, the ligand were ammonia ($\mu_B = 1.48 \times 10^{-18}$ e.s.u.; $\alpha_B = 2.145 \times 10^{-24}$ cc), $-u_0$ would be 361 kcal. The effect of a higher polarizability has thus outweighed that of a lower dipole moment: the true equilibrium separation in the ammine complex is smaller than in the aquo complex, and the actual value of u_0 is less than -361. Garrick (1930) has computed the values of u_0 for certain ammine complexes, and has compared them with those obtained in a thermodynamic cycle. His results (Table IX) show that a simple electrostatic

TABLE IX

OBSERVED AND COMPUTED VALUES OF $-u_0$ FOR COMPLEX AMMINES (GARRICK)

Complex ion	$-u_0$(kcal/gram-ion)	
	Observed	Computed
$[Mn(NH_3)_6]^{2+}$	374	391
$[Fe(NH_3)_6]^{2+}$	395	423
$[Zn(NH_3)_6]^{2+}$	438	439

treatment accounts reasonably well for the absolute magnitudes of u_0 and for their trend as the central ion is changed.

The largest term in equation (58) is the third term, which gives the interaction energy between the central ion and the permanent dipole of the ligand. It can therefore be expected that the stability of these complex ions towards a common elementary cation should vary roughly in proportion to the dipole moment of the ligand. This is frequently true. When ammonia ($\mu = 1.48 \times 10^{-18}$ e.s.u.) is changed for triethylamine ($\mu = 0.90 \times 10^{-18}$), the complex ion becomes less stable; when phosphine ($\mu = 0.55 \times 10^{-18}$) is replaced by triethylphosphine ($\mu = 1.45 \times 10^{-18}$), the complex becomes more stable (R. W. Parry and R. N. Keller, 1956; Mann, Wells and Purdie, 1937; A. E. Martell and M. Calvin, 1952).

It will be appreciated that, in deriving equation (55), no term was included to account for the additional (attractive) energy arising from the induction of dipoles in the central cation by the field due to the ligands. The omission can be justified only if the combined field due to the various ligands vanishes at the centre. We shall see later that the quantum theory imposes certain limitations on the effect of the ligand field on the d electrons in the uncompleted shells of transition metal ions.

An Approximate Electrostatic Estimate of the Energy of Activation

The energy required to remove one ligand from the 6-co-ordinated structure, while the internuclear distances remain the same, is

$$u_5 - u_6 = -\left[\frac{A}{a^9} + \frac{3\mu_B^2}{\sqrt{2}a^3}\left(1 + \frac{\alpha_B z_A \varepsilon}{\mu_B a^2}\right)^2 - \frac{z_A \varepsilon \mu_B}{a^2} - \frac{\alpha_B(z_A \varepsilon)^2}{2a^4} - \frac{B}{a^6}\right]. \quad (59)$$

It is to be noted that the expression for μ_5 is not found simply by replacing the 6 of equation (55) by 5: allowance must be made for the fact that the removal of one ligand reduces the number of ligand–ligand interactions from 12 to 8. The numerical value of $u_5 - u_6$ in the instance under consideration is 23·3 kilocalories per gramme-ion, a quantity which, as we shall see, lies near to the energy of activation of many reactions undergone by complex ions in solution. The 5-co-ordinated complex is unstable with respect to the 6-co-ordinated complex and with respect to the planar, 4-co-ordinated complex, whose potential energy is

$$u_4 = 4\left[\frac{A}{a^9} + \frac{3\mu^2}{4\sqrt{2}a^3}\left(1 + \frac{\alpha_B z_A \varepsilon}{\mu_B a^2}\right)^2 - \frac{z_A \varepsilon \mu_B}{a^2} - \frac{\alpha_B(z_A \varepsilon)^2}{a^4} - \frac{B}{a^6}\right]. \quad (60)$$

In this complex, there is only one dipole–dipole interaction for each ligand.

Eigen (1957), using a relaxation method, has studied the kinetics of the reactions of the Ca^{2+} ion with various anions, like SO_4^{2-}, in water. He concludes that the rate-determining step is the partial de-hydration $[Ca(H_2O)_6]^{2+} \rightarrow [Ca(H_2O)_5]^{2+} + H_2O$. It has been proposed that in certain other reactions of ions in solution, the energy of activation consists of the energy required to remove a water molecule from the first co-ordination shell (Glew and E. A. Moelwyn-Hughes, 1952).

The Kinetics of Ionic Substitution at Octahedral Complexes

Two reactions of this category have already been discussed (Key to Fig. VII, 1). We shall here examine some kinetic data on reactions of the type

$$[ML_5X]^{2+} + OH^- \rightarrow [ML_5OH]^+ + X^- ; k_2$$

where M is Cr, Co or Rh; L is H_2O or NH_3 and X is F, Cl, Br, I, N_3 or NO_2. These substitutions are frequently attended, in aqueous or water-methanol solutions, by spontaneous reactions of the complex ions with

TABLE X

Pseudo-unimolecular Constants for the Aquations $[ML_5X]^{2+} + H_2O \rightarrow [ML_5H_2O]^{3+} + X^-$ where $M = Cr$ or Co; $L = H_2O$ or NH_3; $X = $ Halogen or NCS

Reacting ion	Reference	$k_1(sec^{-1})$ at 25°C	$\log_{10} A_1$	E_A(kcal/mole)	s
$[Cr(H_2O)_5F]^{2+}$	*	6.22×10^{-10}	12.27	29.3 ± 0.6	1
$[Cr(H_2O)_5Cl]^{2+}$	**	2.77×10^{-7}	11.69	24.9 ± 0.2	1
$[Cr(NH_3)_5Cl]^{2+}$	†	9.4×10^{-6}	10.85	21.7	1
$[Cr(H_2O)_5Br]^{2+}$	‡	4.37×10^{-6}	12.52	24.4 ± 0.3	1
$[Cr(H_2O)_5I]^{2+}$	**	8.41×10^{-5}	9.67	23.6 ± 0.3	0
$[Co(NH_3)_5N_3]^{2+}$	§	2.1×10^{-9}	16.06	33.7 ± 1.0	4
$[Co(NH_3)_5Cl]^{2+}$	‖	1.67×10^{-6}	11.73	23.9 ± 0.15	1
$[Co(NH_3)_5Br]^{2+}$	¶	6.5×10^{-6}	12.40	23.9	1
$[Co(NH_3)_5NCS]^{2+}$	**	3.5×10^{-10}	13.06	30.7 ± 1.0	
$[Cr(NH_3)_5NCS]^{2+}$	**	8.5×10^{-8}	11.49	25.3 ± 0.6	

REFERENCES

* T. W. Swaddle and E. L. King, Inorg. Chem., 4, 532 (1965).
† Freundlich and Pape, Z. physikal Chem., 86, 458 (1924).
‡ F. A. Guthrie and E. L. King, Inorg. Chem., 3, 916 (1964).
§ G. C. Lalor and E. A. Moelwyn-Hughes, Trans. Chem. Soc., 1560 (1963).
‖ Garrick, Trans. Faraday Soc., 33, 486 (1937); Lamb and Marden, J. Amer. Chem. Soc., 33, 1873 (1911).
¶ Brönsted and Livingston, J. Amer. Chem. Soc., 49, 435 (1927).
** Gay and Lalor, Trans. Chem. Soc., 1179 (1966).

water, according to the equation

$$[ML_5X]^{2+} + H_2O \rightarrow [ML_5H_2O]^{3+} + X^- ; k_1$$

so that the instantaneous rate of reaction is

$$dx/dt = k_1(a - x) + k_2(a - x)(b - x), \tag{6.44}$$

where x is the extent of chemical change, and a and b are the initial concentrations of complex ion and hydroxyl ion respectively. Conditions are usually chosen so that $a \ll b$, and therefore $b - x$ may be taken as constant. b is often a few millimoles per litre, and a of the order of magnitude 10^{-6} to 10^{-5} moles per litre. Such low concentrations of the complex ion are readily measured spectrophotometrically. Under these conditions, the rate law approximates to

$$dx/dt = (k_1 + k_2b)(a - x).$$

The observed first-order constant is consequently $k_{obs} = k_1 + k_2b$. The aquation constant k_1 has long been known to be independent of the initial concentration of the complex ion (Freundlich and Pape, 1914). Some recent data have been brought together in Table XII. The values of s listed are those required to make the data fit an equation of the Herzfeld–Berthoud type:

$$k_1 = \frac{kT}{h} \cdot \exp\left(-E/RT\right)\frac{(E/RT)^{s-1}}{(s-1)!}. \tag{3.35}$$

Judged on this basis, the replacements of the elementary ions by water molecules appear to be slightly simpler reactions than their replacement by other ions. These estimates give the minimum values of s. The true values are probably higher. S. C. Chan (1967) finds $dE_A/AT = -50$ cal/mole-deg for the aquation of $[Co(NH_3)_5Cl]^{2+}$ and related ions.

The mechanism of aquation has been abundantly discussed (M. A. Levine, J. P. Jones, W. E. Harries and W. J. Wallace, 1961; S. A. Johnson, F. Basolo and R. G. Pearson, 1963; Basolo and Pearson, 1960; M. L. Tobe, 1966; A. G. Sykes, 1966). We may, for example, think of the activated state as the distorted structure $[ML_5XH_2O]^{2+}$, in which the co-ordination number is temporarily seven, and the X atom has been drawn partly away from M, while the water molecule has thrust itself partly into the inner co-ordination shell. The energy of activation then would be the sum of the energies expended in the processes of partial desolvation and partial intrusion.

* M. A. Levine, J. P. Jones, W. E. Harries and W. J. Wallace, *J. Amer. Chem. Soc.*, **83**, 2453 (1961); S. A. Johnson, F. Basolo and R. G. Pearson, *ibid.*, **85**, 1741 (1963); Basolo and Pearson, *Mechanisms of Inorganic Reactions*, Wiley, New York (1960); M. L. Tobe, *Mechanisms of Inorganic Reactions*, Amer. Chem. Soc. Publ. (1966); A. G. Sykes, *Kinetics of Inorganic Reactions*, Pergamon, Oxford (1966).

The bimolecular constant is found to vary with respect to ionic strength at all temperatures according to the following equation, based on equation (18):

$$\log_{10} k_2 = \log_{10} k_2^0 + 2\alpha z_A z_B [\sqrt{I}(1 + \sqrt{I})^{-1} - bI],$$

where α is given by equation (9), and the charge product is -2. Values of k_2^0 obtained for various reactions at 25°C are given in Table XI, along with the corresponding Arrhenius parameters. The rates of reaction of the hydroxyl ion with the halogenopentammine cobaltic ions form a sequence which is the reverse of that found for the halogenopentammine rhodic ions. Neither sequence tallies with that for the halogenomethanes (Table VIII, 10). The most notable feature of Table XI is the high value of the A_2^0 term in the integrated form of the Arrhenius equation,

$$k_2^0 = A_2^0 . \exp(-E_A/RT). \tag{25}$$

TABLE XI

BIMOLECULAR CONSTANTS GOVERNING THE SUBSTITUTION OF OH⁻ FOR OTHER ANIONS IN CERTAIN OCTAHEDRAL COMPLEX IONS IN AQUEOUS SOLUTION

Ref.	Reaction	k_2^0 (litre/ mole^{-1}-sec^{-1}) at 25°C	$\log_{10} A_2^0$	E_A (kcal/mole^{-1})	s
*	$[Co(NH_3)_5NO_2]^{2+} + OH^-$	4.2×10^{-6}	22.57	38.0	13
†	$[Co(NH_3)_5NCS]^{2+} + OH^-$	5.0×10^{-4}	22.10	34.6 ± 0.6	12
†	$[Co(NH_3)_5N_3]^{2+} + OH^-$	3.0×10^{-4}	20.80	33.2 ± 0.8	10
‡	$[Co(NH_3)_5F]^{2+} + OH^-$	1.3×10^{-2}	17.70	26.7	6
§	$[Co(NH_3)_5Cl]^{2+} + OH^-$	1.66	20.18	27.24	14
‡	$[Co(NH_3)_5Br]^{2+} + OH^-$	6.20	21.40	28.2	16
‡	$[Co(NH_3)_5I]^{2+} + OH^-$	2.29×10^1	22.60	28.9	19
‖	$[Rh(NH_3)_5Cl]^{2+} + OH^-$	4.06×10^{-4}	17.50	28.49 ± 0.33	6
‖	$[Rh(NH_3)_5Br]^{2+} + OH^-$	3.37×10^{-4}	18.94	30.55 ± 0.20	8
‖	$[Rh(NH_3)_5I]^{2+} + OH^-$	7.26×10^{-5}	19.95	32.85 ± 0.50	10
†	$[Cr(NH_3)_5NCS]^{2+} + OH^-$	7.61×10^{-6}	21.02	35.65 ± 0.9	10

REFERENCES

* Lalor and J. Lang, *Trans. Chem. Soc.*, 5620 (1963).
† Gay and Lalor, *Trans. Chem. Soc.*, 1179 (1966); Lalor and Moelwyn-Hughes, *Trans. Chem. Soc.*, 1560 (1963).
‡ S. C. Chan, K. Y. Hui, I. Miller, and W. S. Tsang, *Trans. Chem. Soc.*, 3207 (1965).
‡ G. W. Bushnell and G. C. Lalor, *J. Inorg. Nucl. Chem.*, **30**, 219 (1968). Earlier work by Adamson and Basolo (*Acta. Chem. Scand.*, **9**, 1261 (1955)) gave a higher value for E_A, and revealed no primary electrolyte effect.
‖ G. W. Bushnell, G. C. Lalor and E. A. Moelwyn-Hughes, *Trans. Chem. Soc.*, 719 (1966).

Values of the same high order of magnitude are found for the reaction of the hydroxyl ion with the trisethylenediaminecobalt(III) ion (Friend and Nunn, 1958) and for the reaction of the hydroxyl ion with numerous complex ions of the type $[CoX(H_2NCH_2CH_2NH_2)_2Y]^{+1}$, where X and Y are identical or dissimilar unidentate ligands, and ethylene diamine acts as a bidentate ligand (S. C. Chan and M. L. Tobe, 1962). It is impossible to account for these high values on purely electrostatic grounds, by means of the equation

$$A_2^0 = Z^0 . \exp\left(-z_A z_B \varepsilon^2 L / kDr\right), \tag{34}$$

without accepting unreasonably small values of r, as in Table VI. If, however, allowance is made for the distribution of an energy E between s classical oscillators, we may accept the following theoretical expression for the bimolecular constant at zero concentration:

$$k_2^0 = Z^0 . \exp\left(-E/RT\right) \frac{(E/RT)^{s-1}}{(s-1)!} . \exp\left(-z_A z_B \varepsilon^2 / DrkT\right), \tag{61}$$

according to which

$$E_A = E - (s-1)kT + z_A z_B \varepsilon^2 (1 - LT)/Dr, \tag{62}$$

and

$$A_2^0 = Z^0 . \frac{\exp\left(-z_A z_B \varepsilon^2 L / kDr\right)}{\exp(s-1).(s-1)!}\left[\frac{E_A}{RT} + (s-1) - \frac{z_A z_B \varepsilon^2}{DrkT}(1 - LT)\right]^{s-1}. \tag{63}$$

Integral values of s required to account for the rates are given in the last column of Table XIII. They are seen to be greater than those required to account for the aquation constants (Table XII) and, in each series, to increase among the halogens from fluorine to iodine. The first result suggests that the activation process in the bimolecular substitutions is more complicated than in the aquations, as would be true if the water of hydration of the hydroxyl ion also played a part. The second result merely indicates that the number of oscillators increases with the size of the ion that is to be replaced.

The temperature coefficient of the apparent energy of activation according to equation (62), is

$$dE_A/dT = -k[(s-1) + (z_A z_B \varepsilon^2 L^2 T)/Drk]. \tag{64}$$

In the present set of reactions, $z_A z_B$ is negative, and the effect of energy distribution and the electrostatic effect act in opposite directions. In fact, with some of the estimated values of r and s, dE_A/dT can be zero. The two-constant equation of Arrhenius would then hold strictly but fortuitously.

In attempting to calculate the electrostatic contribution to the energy of the ion-pair systems here considered, terms other than those due to the interaction of the ions are to be included. The electric field due to the dipoles of the polar molecules or anions situated at the apices of the octahedron surrounding the central ion directs the approach of the attacking ion towards the centre of a facet and is responsible for an energy which, as we shall see, is comparable in magnitude with the Coulombic contribution.

In Fig. 5 an ion (charge $z_A\varepsilon$) is shown at a perpendicular distance r_2 from one of the facets of a regular octahedron. If the length of the dipoles is small

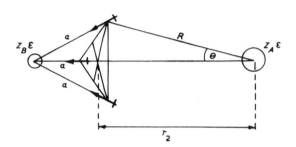

FIG. 5. Electrostatic interaction between an ion and one face of a 6-co-ordinated octahedral complex ion.

compared with the internuclear distance, the energy of interaction between the approaching ion and the three dipoles at the corners of the facet is approximately $u = 3z_A\varepsilon\mu \sin\phi \cos\theta/DR^2$, where μ is the dipole moment of the ligand, ϕ is the angle subtended between the polar axis and the plane of the facet, R is the distance between the centre of the ion (charge $z_A\varepsilon$) and each dipole, and θ is the angle of inclination of the line of centres to the normal. From the properties of the regular octahedron, we have $\sin\phi = (2/3)^{1/2}$, and $R^2 = r_2^2 + (2/3)a^2$, a being the distance between the centre of the complex ion and the corners of the octahedron. It follows that

$$u = \sqrt{3} \cdot z_A\varepsilon\mu/\{Dr_2^2[1 + \tfrac{2}{3}(a/r_2)^2]^{3/2}\}.$$

When the two ions are at the same distance $(a/\sqrt{3})$ from the facet, $u = \sqrt{2} \cdot z_A\varepsilon\mu/Da^2$. The ratio of the ion–dipole interaction to the Coulombic interaction in this configuration is

$$\frac{u\,\text{(ion–dipole)}}{u\,\text{(ion–ion)}} = \frac{2(2/3)^{1/2}\mu}{z_B\varepsilon a},$$

which, even with the gas values of μ for such ligands as ammonia and water, is of the order of magnitude unity. It is, however, the higher dipole moment possessed by the ligand in the presence of the central field that should be employed, and a value of $\sin\phi$ nearer unity than $1/\sqrt{3}$.

The more general expression for the energy of activation of reactions of the kind now considered is the extended form of equation (21):

$$E = E_n + \frac{z_A z_B \varepsilon^2}{D r_1} + \frac{\sqrt{6} \cdot z_A \varepsilon \mu}{D r_2^2} \left[1 + \frac{2}{3} \left(\frac{a}{r_2} \right)^{-2} \right]^{-3/2} \tag{65}$$

where r_1 is the critical distance between the central ion and the attacking ion, and r_2 the critical distance of the latter from the facet. They are related as follows:

$$r_2 = r_1 - a/\sqrt{3}. \tag{66}$$

The theoretical expression for A_2^0 must now be adjusted. With precise numerical values, and a knowledge of the distance a, it becomes possible to evaluate the electrostatic contribution to the energy of intrusion.

The Kinetic Determination of an Association Constant

We have seen that the complex ion $[Co(NH_3)_5Cl]^{2+}$ in aqueous solution reacts pseudo-unimolecularly with water, and in alkaline solution reacts bimolecularly with hydroxyl ions. When the initial concentration, b, of hydroxyl ion greatly exceeds the initial concentration, a, of the complex ion, we have the composite rate constant $k_{obs} = k_1 + k_2 b$, where k_2 is a function of the ionic strength. From the numerical values given in Tables XII and XIII ($k_1 = 1.67 \times 10^{-6}$; $k_2^0 = 1.49$ litre-mole^{-1} sec^{-1} at 25°C) it is clear that in moderately alkaline solution at this temperature the term k_1 can be ignored, leaving $k_{obs} = k_2 b$. If, therefore, k_{obs} is measured for several solutions of varying hydroxyl ion concentration but at a common ionic strength, it should vary linearly with respect to b. Such is often found to be the case. In the present instance, S. C. Chan (1966) finds a slight curvature when k_{obs} is plotted against b. The ionic strength of the solutions was kept constant by addition of $NaClO_4$, and b ranged from 0.02 to 0.08M. His explanation is as follows. The reacting ions first rapidly form intimate ion pairs:

$[Co(NH_3)_5Cl]^{2+}$
${\scriptstyle a-x-y}$

$+ \ OH^- \ \underset{k_1}{\overset{k_2}{\rightleftharpoons}} \ \{[Co(NH_3)_5Cl]^{2+}, OH^-\} \overset{k_3}{\rightarrow} [Co(NH_3)_5OH]^{2+} + Cl^-.$
${\scriptstyle b-x-y} {\scriptstyle x} {\scriptstyle y}$

These may revert to their progenitors or be converted by the rate-determining step into products. Since $b \gg (x + y)$, the stationary concentration of ion pairs is

$$x = \frac{k_2(a - y)b}{k_2 b + k_1 + k_3}.$$

By hypothesis, $k_1 \gg k_3$. Hence, effectively

$$x = \frac{k_2(a - y)b}{k_2b + k_1}.$$

The rate of reaction is k_3x

$$\frac{dy}{dt} = \frac{k_3k_2b(a - y)}{k_2b + k_1} \tag{67}$$

and

$$k_{\text{obs}} = \frac{k_3Kb}{1 + Kb},$$

where $K(=k_2/k_1)$ is the association constant of the ion pair. This equation is to be compared with the Michaelis equation (1.5). The reciprocal of k_{obs} should now vary linearly with respect to $1/b$, since

$$1/k_{\text{obs}} = 1/k_3 + 1/k_3Kb. \tag{68}$$

Chan has established the linear relationship at $0°$ and $25°C$, obtaining k_3 from the intercept, and thence K from the gradient. His results can be summarized as follows:

$$k_3 \, (\text{sec}^{-1}) = 9\cdot9 \times 10^{19} \cdot \exp(-28,450/RT),$$

$$K \, (\text{litre-mole}^{-1}) = 0\cdot12 \cdot \exp(+670/RT)$$

$$k_3K \, (\text{litre-mole}^{-1}\text{-sec}^{-1}) = 1\cdot19 \times 10^{19} \cdot \exp(-27,780/RT).$$

Clearly, k_3K is the limiting value of k_{obs}/b which holds at zero ionic strength, and is what we have denoted throughout as k_2^0.

An alternative mechanism (Basolo and Pearson, 1958) postulates the pre-equilibrium as $[Co(NH_3)_5Cl]^{2+} + OH^- \rightleftarrows [Co(NH_3)_4NH_2, Cl]^+ + H_2O$, which involves the abstraction of hydrogen ion from the complex. Chan considers the kinetic and spectroscopic evidence to conflict with this scheme. His data, however, have been called in question (Buckingham, 1968).

Electronic Structure of Complex Ions of the First Transition Period

The chemical, magnetic and optical properties of elements in the first transition period derive chiefly from the number and state of electrons in the $3d$ and $4s$ shells. Starting with manganese, which has five d electrons, the occupancy of the $4s$ shell remains at two electrons, except for copper, where there is only one electron in the $4s$ shell. The maxima which occur as many of the properties of these elements are plotted as functions of the atomic number occur at this point.

Atomic number	Atom	Electronic structure, excluding argon core	Number of unpaired electrons in a shell of ion M^{2+}	Energy of escape of ion M^{2+} from water (kcal/gram-ion)		
				Observed	Ligand field correction	Difference
20	Ca	— $4s^2$	0	358	0	358
25	Mn	$3d^5$ $4s^2$	5	407	0	407
26	Fe	$3d^6$ $4s^2$	4	447	10	437
27	Co	$3d^7$ $4s^2$	3	469	20	449
28	Ni	$3d^8$ $4s^2$	2	496	25	471
29	Cu	$3d^{10}$ $4s^1$	1	496	20	476
30	Zn	$3d^{10}$ $4s^2$	0	482	0	482

The electronic states of the atoms, as revealed by their spectra, are shown in Table XIV. The number, n, of unpaired electrons in the d shell of the ions M^{2+} is obtained from Langevin's equation relating the total magnetic moment, M, with the magnetic susceptibility, κ:

$$\kappa = \kappa_0 + \frac{1}{3}\frac{M^2}{kT} \qquad (69)$$

where M is a multiple of the Bohr magneton, $he/4\pi m_e c$

$$M = \left(\frac{he}{4\pi m_e c}\right)[n(n+2)]^{1/2}. \qquad (70)$$

It is found that this number, for a given ion, is not always the same for the free ion as for the complex ion, and the change gives an indication of the nature of the bond uniting the ion and the ligands. The magnetic moment of the cobaltic ion, for example, indicates that four out of the six d electrons are unpaired. This fact, and the emptiness of the 4s and 4p shells can be represented as follows:

Co^{3+} $\underline{}$ $\overset{\uparrow\downarrow}{\underline{}}$ $\overset{3d}{\overset{\uparrow}{\underline{}}}$ $\overset{\uparrow}{\underline{}}$ $\overset{\uparrow}{\underline{}}$ $\overset{\uparrow}{\underline{}}$ $\overset{4s}{\underline{}}$ $\overset{4p}{\underline{}}$.

The magnetic moment of the complex ion $[CoF_6]^{3-}$ is the same as that of the elementary ion. None of the d electrons has therefore changed its state or entered into combination with ligand electrons to form covalent bonds. That is the reason that the properties of this complex ion like those of Ca^{++} can be almost wholly understood in terms of electrostatic interactions between the central cobaltic ion Co^{3+} and the six F^- fluoride ions surrounding it. The magnetic moments of the ferric ion Fe^{3+} and the ferricyanide ion $[Fe(CN)_6]^{3-}$, on the other hand, are different, indicating that in the elementary ion, the five d electrons are unpaired, whereas in the complex ion only one d electron remains unpaired. Their states may be represented schematically as follows:

Each of the CN^- ions has two paired electrons, so that the six ligands bring twelve electrons to the system. Optical evidence shows that the highest level occupied by them in the complex ion is the $4p$ level, i.e. no electrons appear in $4d$ levels or above. This can be explained by supposing that four of the previously unpaired d electrons of the elementary ion have coupled up, leaving one unpaired d electron and two empty d orbitals. In terms of Pauling's theory of atomic orbitals, these two d orbitals can hybridize with one s and three p orbitals to give six equivalent orbitals, denoted by the symbol d^2sp^3, whose directions are those of the six lines joining the centre of a regular octahedron to its corners. Such a complex is termed a six-covalent spin paired complex. When the energy difference between a d electron and an s electron is greater than the small value assumed in this example, planar and tetrahedral structures can, according to this theory, be more stable than the octahedral structures.

Ligand Field Theory

When hydrogen atoms are placed in a strong electric field, their energy levels are split, i.e. new energy levels, above and below the original ones, become available (Stark effect). The difference between any two successive levels is proportional to the field strength. Penny and Schlapp (1932) and van Vleck (1932) have shown that the electric field exerted by the ligands around a central ion of the transition metals produces a similar effect. In particular, the cubic field due to six water molecules at the corners of a regular octahedron surrounding the Co^{2+} ion splits its original single energy level into three levels, the separation between successive levels being about $10^4 cm^{-1}$ or 28 kilocalories. The effect of a slight distortion in the regular octahedron can be allowed for by superimposing a rhombic field on the perfect cubic field. There results a further splitting on the two lower levels into triplets, with a much smaller separation, as shown schematically below:

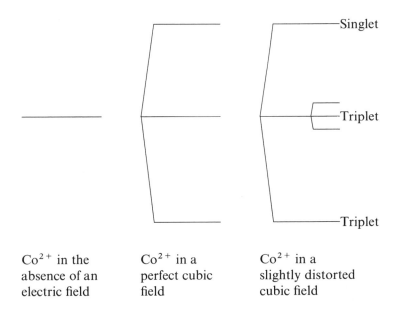

| Co²⁺ in the absence of an electric field | Co²⁺ in a perfect cubic field | Co²⁺ in a slightly distorted cubic field |

As the occupancy of the seven levels is determined by Boltzmann's law, most of the ions in a system at equilibrium are found in the lowest levels. The net effect of the ligand field, therefore, is to lower the total energy of the complex ion and thus to confer on it a greater stability than it would have in the absence of the field. Similar results are found in fields of different symmetry.

Originally advanced to account for certain anomalous magnetic properties of transition metal ions in crystals, the ligand field theory has found numerous other applications. The absorption of light by salts of the transition metals is due to the excitation of a d electron from a low energy level E_2 to a higher energy level, E_1. For example, the first two maxima in the absorption spectrum of $[Co(H_2O)_6]^{2+}$ are at 8,100 and 11,300 cm^{-1}. With the ion $[Co(NH_3)_6]^{2+}$ the positions are 9,000 and 18,500 cm^{-1}. Nyholm (1956) on analysing the complete spectra, finds that $E_1 - E_2$ is 9,700 cm^{-1} for the aquo complex, and 10,500 cm^{-1} for the ammine complex, both of which lie near to Penny and Schlapp's theoretical estimate of the split.

Orgel (1952) has examined certain properties of transition metal ion complexes in terms of both the molecular orbital and the ligand field theories. We shall here be concerned chiefly with his application of ligand field theory to the determination of the energy of escape of the bivalent ions from aqueous solution. The decrease in energy of the complex ion due to the ligand field is determined by the symmetry and strength of the field (i.e. the arrangement and kind of ligand) and by the number and state of the d electrons. The theory predicts that there is no ligand field effect when the d shell is full, as in Zn^{2+}, or when it is exactly half full, as in Mn^{2+}. For the intervening ions, the decrease in energy due to the ligand field is, to a first approximation, proportional to $(v - 5)$, where v is the number of unpaired d electrons. After making allowance for certain complicating factors, especially in the case of the Cu^{2+} ion, Orgel has estimated the energy decreases shown, to the nearest 5 kilocalories, in Table XIV. When these are deducted from the observed energies of escape, corrected energies are found which increase with the atomic number. When, from the sum of the first two ionization potentials of the cupric ion Cu^{2+} the energy required to raise an electron from the d to the s state is deducted, the maximum in the curve found by plotting ionizations potentials against atomic numbers also vanishes. It is thus concluded that the erratic variation of the heats of escape with respect to atomic number is due to the ligand field effect. Its magnitude, it may be noted, may amount to as much as 5 per cent of the total heat of escape.

A Possible Kinetic Role of the Ligand Field Effect in Aqueous Solution

Although the contribution of the ligand field effect to the stability (i.e. to the reduction in the potential energy) of complex ions in aqueous solution is a small fraction of their total stability as measured by the heats of formation of the complex ions (R. J. P. Williams, 1960) it may be a considerable fraction of the energy of activation. There appears to be no prospect at present of predicting the kinetic consequences of ligand field effects in bimolecular reactions of the type $[ML_5X]^{2+} + Y^- \rightarrow [ML_5Y]^{2+} + X^-$. It nevertheless appears worthwhile to examine an approximate correlation which has been detected (G. C. Lalor and J. Lang, 1963) between the apparent energy of activation, E_A, and the wave number, ω, of the absorption peak. In octahedral complexes, $\Delta = h\nu = hc\omega$ is a rough measure of the ligand field effect. In three reactions of the type $[Co(NH_3)_5X]^{2+} + OH^- \rightarrow [Co(NH_3)_5OH]^{2+} + X^-$, where the Co—X links are all Co—N links there appears a linear relationship between E_A and ω, given empirically by the equation:

$$E_A(\text{cal-mole}^{-1}) = -1,270 + 1.79\omega\ \text{cm}^{-1}),\qquad(71)$$

as shown in Table XV:

TABLE XIII

Apparent Energies of Activation (kcal/mole^{-1}) for the Reactions

$[Co(NH_3)_5X]^{2+} + OH^- \to [Co(NH_3)_5OH]^{2+} + X^-$

X	(cm^{-1})	E_A (obs)	E_A (calc)
N$_3$	19,305‡	33·2 ± 0·8*	33·28
NCS	20,000	34·6 ± 0·6†	34·53
NO$_2$	21,934	38·0‡	37·99

REFERENCES

* Lalor and Moelwyn-Hughes, *Trans. Chem. Soc.*, 2560 (1963).
† D. L. Gay and G. C. Lalor, *Trans. Chem. Soc.*, 1180 (1966).
‡ C. C. Lalor and J. Lang, *Trans. Chem. Soc.*, 5620 (1963).

The first term, $-1,270$ cal, is the apparent energy of activation for the hypothetical reaction in this series where the ligand field effect has been eliminated. The true energy of activation, E, has been given by the equation

$$E = E_A + (s - 1)RT - \frac{z_A z_B \varepsilon^2}{Dr}(1 - LT).\qquad(62)$$

Here s is the number of classical harmonic oscillators among which the non-electrostatic component of the energy of activation is thought to be distributed; z_A and z_B are the electrovalencies of the reacting ions; ε is the protonic charge; D is the permittivity; L is $(\partial \ln D/\partial T)_p$; r is the distance apart of the point charges in the critical complex. Using numerical values appropriate for these reactions ($z_A z_B = -2$) in water at 25°C, and expressing r in Ångström units, we have

$$E = E_A + (s - 1)RT - \frac{4,227}{\mathring{r}}.$$

The minimum values of s required to account for the absolute rates of these reactions according to the collision theory has been estimated as 13, 10 and 19 respectively. The actual values may well be greater. Adopting a mean value of $s = 14$, and assuming that $\mathring{r} = 3\,\text{Å}$,

$$E = E_A + 7{,}702 - 1{,}409$$

$$= E_A + 6{,}293$$

According to equation (71) $E_A(\omega = 0)$ is $-1{,}270$. Hence

$$E(\omega = 0) = 5{,}020. \tag{72}$$

This is, probably fortuitously, about the energy of activation for a diffusion-controlled reaction in water. The higher values of E found for reactions with real values of ω could thus be interpreted as energies expended in undoing the effect of the ligand field and in temporarily restoring in the critical complex the symmetry which would have prevailed in the ground state ion in the absence of the ligand field.

Oxidation–Reduction and Electron-transfer Reactions

Electrons have been described by their discoverer as bricks out of which matter is built. The conversion of one kind of matter to another, that is, chemical change must therefore resemble the shifting of bricks. Thus, for example, in the homolytic change represented by $DD + H \rightarrow D + DH$, it is obvious that one of the electrons in the D_2 molecule has moved to form the DH molecule. Though electron transfer has taken place, however, the rate of chemical change is not determined by the movement of the electron but by the relative motions of the nuclei. Electrons, being much more mobile than nuclei, accompany their movements. By an electron-transfer reaction it meant a reaction the rate of which is determined by the rate of electronic transmission.

Oxidation consists of the removal of an electron: reduction consists of the addition of an electron. Oxidation of the ferrous ion, by the ceric ion, for example, involves the transfer of an electron from the former to the latter:

$$Fe^{2+} + Ce^{4+} \rightarrow Fe^{3+} + Ce^{3+}.$$

There is, however, no certainty that the rate of this reaction is governed by the rate of transfer of the electron. In many similar reactions, the rate-determining step has been shown to involve charged intermediates, such as $[Fe(OH)]^{2+}$, which serve to hand on an electron on encountering another reactant. The rate of reaction in such cases may be simply the rate at which appropriate ions come into contact, the actual transfer of the electron being virtually instantaneous.

Investigations of the kinetics of oxidation-reduction reactions and of reactions controlled by, or thought to be controlled by, electron jumps have been greatly facilitated by the use of isotopes and the many techniques now available for the study of very rapid reactions. We shall consider the kinetics of some typical reactions in these two related fields.*

(i) Oxygen Transfers

We have seen that the reaction between the arsenite and tellurate ions

$$AsO_3^{3-} + TeO_4^{2-} \rightarrow AsO_4^{3-} + TeO_3^{2-}$$

is bimolecular (Stroup and Meloche, 1931) with an extremely low value of A_2 which can be explained on a collisional basis if the interionic distance in the critical complex is 3·23 Å. As there is no evidence that the reacting ions first form complexes with the solvent, direct exchange of oxygen atoms appears to occur. In the comparable reaction $ClO_3^- + SO_3^{2-} \rightarrow ClO_2^- + SO_4^{2-}$, Halperin and Taube (1952) have proved, by using a heavy isotope of oxygen, that direct oxygen transfer does in fact take place

$$*OClO_2^- + SO_3^{2-} \rightarrow ClO_2^- + *OSO_3^{2-}.$$

(ii) Oxidation by Ferric Ions; Retardation by Hydrogen Ions

The rate of oxidation of hydrogen peroxide by ferric ions in aqueous solution

$$2Fe^{3+} + H_2O_2 \rightarrow 2Fe^{2+} + 2H^+ + O_2$$

is governed principally by the empirical equation

$$\frac{d[O_2]}{dt} = k_0 \frac{[Fe^{3+}][H_2O_2]}{[H^+]},$$

which can be interpreted on the assumption that the rate-determining step is a bimolecular reaction between the ferric ion and the anion HO_2^- formed by the ionization of hydrogen peroxide (Barb, Baxendale, George and Hargrave, 1951);

$$H_2O_2 \rightleftarrows HO_2^- + H^+; \quad K = [HO_2^-][H^+]/[H_2O_2].$$

Then

$$\frac{d[O^2]}{dt} = k_2[Fe^{3+}][HO_2^-] = k_2 K \frac{[Fe^{3+}][H_2O_2]}{[H^+]},$$

---

* Review articles by Taube, *Chem. Rev.*, **50**, 69 (1952); Amphlett, *Quart. Rev.*, **8**, 219 (1954); Baxendale, Dainton and Irvine, *Special Publ. Chem. Soc.*, **1** (1954); Halpern, *Quart. Rev.*, **10**, 463 (1956); **15**, 207 (1961). A. G. Sykes, *Kinetics of Inorganic Reactions*, Pergamon, Oxford (1966).

which agrees with the experimental equation provided $k_2 = k_0/K$. At 20°C $k_2 = 2 \times 10^3$ and $E_A = 28$ kcal. The rate constant for the corresponding oxidation with the cobaltic ion Co^{3+} is greater by a factor of 10^9, and $E_A = 18$. Thus the principal effect on the rate is reflected in the change in activation energy. There are, as the authors emphasize, subsidiary and even alternative mechanisms.

Irvine (see Laidler (1965)) has conveniently summarized the rate law for the exchange reaction

$$*Fe^{2+} + Fe^{3+} \rightarrow *Fe^{3+} + Fe^{2+}$$

in the general form

$$\frac{d[Fe^{2+}]}{dt} = [*Fe^{2+}][Fe^{3+}]\left(k_1 + \frac{k_2}{[H^+]} + \Sigma k_{X^-}[X^-]\right).$$

The first term represents the bimolecular constant for the direct ion–ion reaction (rate constant k_1), for which $E_A = 16.8$ kcal (Silverman and Dobson, 1952). The second term indicates a mechanism similar to that discussed above, involving the participation of the intermediate $FeOH^{2+}$. The last term suggests anionic catalysis (X = halogen), which is always a possibility in reactions between two cations, just as cationic catalysis is prevalent in reactions between two anions.

Bridged Complexes

The kinetics of the exchange of electrovalencies between the cobaltic and cobaltous hexammines has been studied by Biradar, Stranks and Vaidya (1962) using ^{60}Co as the tagged atom (here denoted by Co*):

$$Co*(NH_3)_6^{3+} + Co(NH_3)_6^{2+} \rightarrow Co*(NH_3)_6^{2+} + Co(NH_3)_6^{3+}.$$

The net change is seen to consist of the transfer of an electron from Co* to Co. The observed rate of reaction is reproduced by the empirical equation

$$dx/dt = k_0[*Co(NH_3)_6^{3+}][Co(NH_3)_6^{2+}]/[H^+],$$

which can be interpreted by assuming that the rate-determining step is a bimolecular reaction between the hexamminecobaltous ion and an ion-pair formed between the hexamminecobaltic and the hydroxyl ions for which the association constant is known (Caton and Prue, 1956):

$$K_a = \frac{[Co(NH_3)_6^{3+}, OH^-]}{[Co(NH_3)_6^{3+}][OH^-]}.$$

According to this mechanism,

$$dx/dt = k_2[Co(NH_3)_6^{3+}, OH^-][Co(NH_3)_6^{2+}]$$
$$= k_2 K_a[Co(NH_3)_6^{3+}][Co(NH_3)_6^{2+}][OH^-]$$
$$= k_2 K_a K_w[Co(NH_3)_6^{3+}][Co(NH_3)_6^{2+}]/[H^+],$$

where K_w is the ionic product of water. It follows that $k_2 = k_0/K_a K_w$. From the temperature coefficients of these constants, the energy of activation for the bimolecular rate-determining step is $E_2 = E_0 - \Delta H_a - \Delta H_w = 30.1 - 11.7 - 5.8 = 12.6 \pm 1.6$ kcal. On combining with the observed value of $k_2 = 5.7 \times 10^{-3}$ litre-mole^{-1}-sec^{-1} at 64.5°C, we have, for solutions of unit ionic strength $k_2 = 1.27 \times 10^6 . \exp(-12,600/RT)$. The pre-exponential term, which would be lower at zero ionic strength, has the order of magnitude as found for other ionic reactions with the same charge product of $z_A z_B = 4$ (Table 6).

Tracer studies have shown (Taube, H. Myers and R. L. Rich, 1953) that, in the following reaction $[(NH_3)_5CoCl]^{2+} + Cr^{2+} \rightarrow [(NH_3)_5CoClCr]^{4+} \rightarrow 5NH_3 + Co^{2+} + CrCl^{2+}$ the direct CrCl link exists in the complex ion because the product $CrCl^{2+}$ cannot be formed directly from Cr^{3+} and Cl^-. The chlorine atom in the complex thus forms a bridge between the two metal atoms.

Direct Electron Transfer

The data cited in Table III show that, with but minor deviations, the rate at which the hydrated electron reacts with cations and anions in solution is determined by diffusion. In the reactions $e_{aq}^- + Co(NH_3)_6^{3+} \rightarrow Co(NH_3)_6^{2+}$ and $e_{aq}^- + Co(NO_2)_6^{3-} \rightarrow Co(NO_2)_6^{4-}$, the electron must break through the shell of water molecules which form its cage and penetrate the inner shells formed by the ligands attached to the cobalt ion. Irrespective of whether the complex ion is positively or negatively charged, the rate of reaction is nearly the same. It is thus clear that the rate of reaction is simply the rate at which the hydrated electron and the complex ions come into contact by the process of diffusion: the subsequent electron jump from the centre of one cell to the centre of the other is relatively rapid. Neither the charge on the complex ion nor the presence of the ligands offers any hindrance to the transmission. Obstruction doubtless exists in the form of a potential energy barrier which, according to classical mechanics the electron must surmount, but which according to quantum mechanics the electron can penetrate. The transmission coefficient, κ_t, depends on the height, V, of the barrier, its width, l, at the point of transmission, the kinetic energy, E, and mass, m_e, of the electron (Mott and Sneddon, 1948). For a parabolic barrier, more fully discussed in Chapter 11 (equations 33 and 34),

$$\kappa_t = \exp - \{(\pi^2 l/h)[2m_e(V - E)]^{1/2}\}. \tag{72}$$

When E is equal to or greater than V, there is no tunnelling, and the transmission coefficient is unity. This appears to be the case in these reactions. We can reasonably expect small values of κ_t when large ligands surround the two cores. Such, however, is not generally found. Halpern, Legare and Lumry (1963) have measured the rate of the following reaction in aqueous solution:

$$[Fe(DMP)_3]^{2+} + [IrCl_6]^{2-} \rightarrow [Fe(DMP)_3]^{3+} + [IrCl_6]^{3-}.$$

DMP stands for the very voluminous ligand 4,7 dimethyl-1,10 phenantholine. At 25°C, the bimolecular constant is $1\cdot1 \times 10^9$ litre-mole^{-1}-sec^{-1}, and $E_A = 0\cdot5 \pm 0\cdot5$ kcal. These results are consistent with a diffusion-controlled reaction, and the absence of a barrier to electronic transmission. In the somewhat similar system

$$[IrBr_6]^{3-} + [IrCl_6]^{2-} \underset{k_4}{\overset{k_2}{\rightleftharpoons}} [IrBr_6]^{2-} + [IrCl_6]^{3-},$$

the bimolecular constants are $k_2 = 2\cdot2 \times 10^{11} \cdot \exp(-5,300/RT)$ and $k_4 = 7\cdot1 \times 10^{11} \cdot \exp(-7,500/RT)$ in aqueous solutions of ionic strength 0·1 molar (Hurwitz and Kustin, 1966). At 25°C the bimolecular constants are lower than the limiting value for diffusion-controlled processes by factors of 10^2 and 10^3 respectively. Despite the relatively high energies of activation, it

Outer–sphere pair; both ions retain their
solvation shells intact

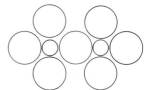

Bridged pair; the two ions have at least
one solvent molecule in common

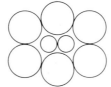

Intimate pair; the two ions are in contact, surrounded
by a common shell of solvent molecules

FIG. 6. Various types of ion pairs.

is thought that diffusion is the principal factor controlling the reaction rate (see equation 3.129).

In both these reactions, the ligands remain attached to the metal ions, and the Brönsted complex is referred to as an outer sphere complex, in contrast to the inner sphere complex discussed above, which consisted of a pair of solvated or ligandated ions sharing between them a single solvent molecule or a ligand (see Figure 6). In certain systems, e.g. $[Co(NH_3)_5H_2O]^{3+}$ + SO_4^{2-}, the rate at which an outer sphere complex changes into an inner sphere complex has been measured (Taube and Posey, 1953; M. Eigen and L. de Maeyer, 1955). This subject is further discussed in Chapter 14.

Energetics of Electron Transfers

Numerous energy changes may accompany the transfer of an electron from membership of one solute system to another in aqueous solution, but it is by no means clear which of these changes, or which combination of them, constitutes the process of activation. In the one-electron transfers denoted by the equation

$$A^{z_A} + B^{z_B} \rightarrow A^{z_A+1} + B^{z_B-1},$$

we have the following possible contributions:

(i) Electron transition energies. The total reaction may be thought of as consisting of two steps,

$$A^{z_A} \rightarrow A^{z_A+1} + e^-,$$

and

$$B^{z_B} + e^- \rightarrow B^{z_B-1},$$

the first of which necessitates the absorption of energy necessary, for example, to convert a $3d$ electron into a $4s$ electron. The step written second is attended by the release of energy during the capture of the electron. In the symmetrical case, when $A = B$, the net change in free energy is zero, but the system may have to pass through a state of high energy in the transitory activated complex.

(ii) Intrinsic repulsion energy. As the reaction proceeds in either direction, the electronic shells repel each other on close approach.

(iii) Coulombic energy. The Coulombic energy of the reacting ions increases or decreases as they get near together, depending on whether their charges are like or unlike.

(iv) Shell reorganization energy. The shells surrounding the ions (not indicated in the equations written above) consist of solvent molecules or other ligands. Their number and distance from the ion change as the ion changes its electrovalency, involving the expansion of one solvated reacting ion and the contraction of another. If, for example, the energy of interaction of A with a single ligand of electric moment μ, were to change from $\phi_1 = -z_A \varepsilon\mu/r_1^2$ to $\phi_2 = -(z_A + 1)\varepsilon\mu/r_2^2$, the gain in potential energy would be

$$\phi_2 - \phi_1 = -\frac{\varepsilon\mu}{r_2^2}\left\{1 + z_A\left[1 - \left(\frac{r_2}{r_1}\right)^2\right]\right\}.$$

Provided the co-ordination number, c, remained unchanged, the total gain in potential energy would be $c(\phi_2 - \phi_1)$. A more complicated expression results if, during the electronic transfer, c changes from c_1 to c_2. Although one, all or none of these energy changes may be relevant to the kinetics of electron transfer reactions in solution, let us consider some of the possibilities.

Release of the electron from the ion of charge z_A and its acceptance by the ion of charge z_B may be hindered by the ligands surrounding them. The magnitude of the hindrance may be lowered by ligand re-orientation around both ions, and, by analogy with certain established aspects of gas kinetics, the transfer may be facilitated when both ions approach a similar structure. The electron will not then make its jump until, as Laidler (1965) puts it, the stage has been set for its release from A and its acceptance by B. Solvent or ligand re-orientation is a complicated process which is relatively slow; and, under these conditions, the rate of transfer of the electron will be controlled by the rate of solvent or ligand re-orientation. Secondly, the transfer of an electron from one core to another will be facilitated by the proximity of the cores, and the consequent narrowing of the barrier over which it has to jump or through which it has to tunnel. From the diversity and complexity of the mechanisms possible, it is not surprising that many theories have been developed (R. J. Marcus, Zwolinski and H. Eyring, 1954; Libby, 1952; Laidler, 1959). Hush (1961) has extended his treatment of adiabatic electron transfers at metal electrodes to transfers in solution. In accordance with wave-mechanical principles, he deals less with the electron as a discrete particle than with the most probable density of negative electricity, which varies continuously during the reaction along the stages represented by $A^{z_A} \rightarrow A^{z_A + x} + xe^-$, where x increases from zero to unity.

REFERENCES

Abegg, *Ann. Physik*, **60**, 54 (1897).
Adamson and Basolo, *Acta Chem. Scand.*, **9**, 1261 (1955).
Akerlof, *J. Amer. Chem. Soc.*, **54**, 4125 (1932).
Amis, *Kinetics of Chemical Change in Solution*, Macmillan, New York (1949).
Amis, *Solvent Effects on Reaction Rates and Mechanisms*, Academic Press, New York (1966).
Amis and Cook, *J. Amer. Chem. Soc.*, **63**, 2621 (1941).

Amis and Price, *J. Phys. Chem.*, **47**, 338 (1944).
Amis and V. K. La Mer, *J. Amer. Chem. Soc.*, **61**, 905 (1939).
Barb, Baxendale, George and Hargrave, *Trans. Faraday Soc.*, **47**, 591 (1951).
Basolo, F., and R. G. Pearson, *Mechanisms of Inorganic Reactions*, Wiley, New York (1960).
Basolo, F., and R. G. Pearson, *Mechanisms of Inorganic Reactions*, Wiley, New York, (1958).
Baxendale, J. H., private communication.
Baxendale, J. H., Fielden and Keene, *Trans. Chem. Soc.*, 242 (1963).
Baxendale, J. H., *Rad. Res. Suppl.*, **4**, 139 (1964).
Bedford, Austin and Webb, *J. Amer. Chem. Soc.*, **57**, 1408 (1935).
Bedford, Mason and Morrell, *J. Amer. Chem. Soc.*, **56**, 280 (1934).
Bennett, G. M., J. C. D. Brand and G. Williams, *Trans. Chem. Soc.*, 869 (1946).
Biradar, Stranks and Vaidya, *Trans. Faraday Soc.*, **58**, 2421 (1962).
Bjerrum, N., *Z. physikal Chem.*, **108**, 82 (1924).
Bjerrum, N., *Z. physikal Chem.*, **106**, 219 (1923).
Brönsted, J. N., *Z. physikal Chem.*, **102**, 69 (1922); **115**, 337 (1925).
Buchanan and Hamann, *Trans. Faraday Soc.*, **49**, 1425 (1953).
Buckingham, *J. Inorg. Nucl. Chem.*, **7**, 174 (1968).
Bushnell, G. W., and G. C. Lalor, *J. Inorg. Nucl. Chem.*, **30**, 219 (1968).
Buxton, G. V., F. S. Dainton and M. Hamerli, *Trans. Faraday Soc.*, **63**, 1191 (1967).
Caton and Prue, *Trans. Chem. Soc.*, 671 (1956).
Chan, S. C., *Trans. Chem. Soc.* A, 291 (1967).
Chan, S. C., *Trans. Chem. Soc.*, A, 1124 (1966).
Chan, S. C., H. Y. Hui, I. Miller, and W. S. Tsang, *Trans. Chem. Soc.*, 3207 (1965).
Chan, S. C., and M. L. Tobe, *Trans. Chem. Soc.*, 4531 (1962).
Chattaway and D. L. Chapman, *Trans. Chem. Soc.*, **101**, 170 (1912).
Chlebek and Lister, *Can. J. Chem.*, **44**, 437 (1966).
Christiansen, J. A., *J. Colloid Science*, **6**, 213 (1951).
Christiansen, J. A., *Z. physikal Chem.*, **113**, 35 (1924).
Collinson, Dainton, D. R. Smith and Tazuki, *Trans. Chem. Soc.*, 140 (1962).
Czapski and H. A. Schwarz, *J. Phys. Chem.*, **66**, 471 (1962).
Davies, C. W., *Ion Association*, Butterworth, Washington (1962).
Davies, C. W., *Trans. Chem. Soc.*, 2093 (1928).
Davies, C. W., and I. W. Williams, *Trans. Faraday Soc.*, **54**, 1547 (1958).
Day, P., and C. K. Jørgensen, *Trans. Chem. Soc.*, 6226 (1965).
Debye, P., and E. Hückel, *Physikal Z.*, **25**, 97 (1924).
Eigen, *Faraday Soc. Disc.*, **24**, 25 (1957).
Eigen, M., and L. de Maeyer, *Z. Electrochem.*, **59**, 986 (1955).
Fowler, R. H., *Statistical Mechanics*, 2nd edn., Cambridge (1936).
Freundlich and Pape, *Z. physikal Chem.*, **86**, 458 (1914).
Friend and Nunn, *Trans. Chem. Soc.*, 1567 (1958).
Frost and Pearson, *Kinetics and Mechanism*, 2nd edn., Wiley, New York (1961).
Gay and Lalor, *Trans. Chem. Soc.*, 1179 (1966).
Garrick, *Phil. Mag.*, **14**, 912 (1932); **9**, 131 (1930).
Glasstane, Laidler and Eyring, *The Theory of Rate Processes*, p. 427, McGraw-Hill, New York (1941).
Glew and E. A. Moelwyn-Hughes, *Proc. Roy. Soc.* **A211**, 254 (1952).
Halperin and Taube, *J. Amer. Chem. Soc.*, **74**, 375 (1952).
Halperin, Legare and Lumry, *J. Amer. Chem. Soc.*, **85**, 680 (1963).

Harned, H. S., *J. Amer. Chem. Soc.*, **40**, 1461 (1918).
Harned, H. S., and B. B. Owen, *The Physical Chemistry of Electrolytic Solutions*, 3rd edn., Reinhold, New York, 1958.
Hart and Boag, *J. Amer. Chem. Soc.*, **84**, 4090 (1962).
Hart, E. J., and M. Anbar, *The Hydrated Electron*, Wiley, London (1970).
Hiromi, *Bull. Chem. Soc. Japan*, **33**, 1251 (1960).
Holluta, J., and W. Herrmann, *Z. physikal Chem.*, A, **166**, 433 (1933).
Holmberg, B., *Z. physikal Chem.*, **79**, 147 (1912).
Howells, *Trans. Chem. Soc.*, 463 (1939); 641 (1941): 5844 (1964).
Hückel, *Physikal Z.*, **26**, 93 (1925).
Hurwitz and Kustin, *Trans. Faraday Soc.*, **62**, 427 (1966).
Hush, *Trans. Faraday Soc.*, **57**, 557 (1961).
Indelli, *Annali di Chimica*, **46**, 717 (1965).
Indelli and Prue, *Trans. Chem. Soc.*, 107 (1959).
Irvine; See Laidler, *Chemical Kinetics*, McGraw-Hill, New York (1965).
Janelli, *Gazz. Chim. Ital.*, **83**, 983 (1953); **88**, 443 (1958).
Johnson, S. A., F. Basolo and R. G. Pearson, *J. Amer. Chem. Soc.*, **85**, 1741 (1963).
Jones, C. M., and W. C. McC. Lewis, *Trans. Chem. Soc.*, **117**, 1120 (1920).
Jorntner, *Rad. Res. Suppl.*, **4**, 24 (1964).
Kappanna, *J. Indian Chem. Soc.*, **6**, 45 (1929).
Kappanna and Patwardhan, *J. Indian Chem. Soc.*, **9**, 379 (1932).
Kershaw and Prue, *Trans. Faraday Soc.*, **63**, 1198 (1967).
King and Jacobs, *J. Amer. Chem. Soc.*, **53**, 1704 (1931).
Kirkwood, *J. Chem. Physics*, **2**, 351 (1934).
Kossel, *Z. Physik*, **23**, 403 (1924).
Krapiwin, *Z. physikal Chem.*, **83**, 439 (1913).
Künnap and A. Parts, *Apophereta Tartuensis*, p. 377 (1949).
Laidler, *Chemical Kinetics*, p. 217, 2nd edn., McGraw-Hill, New York (1965).
Laidler, *Chemical Kinetics*, p. 532, McGraw-Hill, New York (1965).
Laidler, *Can. J. Chem.*, **37**, 138 (1959).
Laidler and Eyring, *Ann. N.Y. Acad. Sci.*, **39**, 303 (1940).
Lalor, G. C., and J. Lang, *Trans. Chem. Soc.*, 5620 (1963).
La Mer, V. K., *Chem. Rev.*, **10**, 179 (1932).
La Mer, V. K., *J. Amer. Chem. Soc.*, **55**, 1739 (1933).
La Mer, V. K., and Fessenden, *J. Amer. Chem. Soc.*, **54**, 2351 (1932).
La Mer, V. K., and M. E. Kamner, *J. Amer. Chem. Soc.*, **53**, 2832 (1931).
La Mer, V. K., and M. E. Kamner, *J. Amer. Chem. Soc.*, **57**, 2662 (1935).
Lander and Svirbely, *J. Amer. Chem. Soc.*, 60, 1623 (1938).
Levine, M. A., J. P. Jones, W. E. Harries and W. J. Wallace, *J. Amer. Chem. Soc.*, **83**, 2453 (1961).
Lewis, G. N., *Z. physikal Chem.*, **61**, 129 (1907).
Libby, *J. Phys. Chem.*, **56**, 863 (1952).
Livingston, R., *J. Chem. Education* (*Canada*), **7**, 2887 (1930).
Lowry and Jessop, *Trans. Chem. Soc.*, 782 (1930).
Mann, Wells and Purdie, *Trans. Chem. Soc.*, 1828 (1937).
Marcus, R. J., Zwolinski and H. Eyring, *Chem. Rev.*, **55**, 157 (1954).
Martell, A. E., and M. Calvin, *Chemistry of Metal Chelate Compounds*, Prentice-Hall, New York, 1952.
McC. Lewis and Moran, *Trans. Chem. Soc.*, **121**, 1613 (1922).
Miller, C. C., *Proc. Roy. Soc.* A, **145**, 288 (1934); **151**, 189 (1935).

Moelwyn-Hughes, E. A., *Kinetics of Reactions in Solution*, 1st edn., p. 199 (1933); *Proc. Roy. Soc.* A, **155**, 308 (1936).
Moelwyn-Hughes, E. A., *Kinetics of Reactions in Solution*, 2nd edn., p. 94 (1947).
Moelwyn-Hughes, E. A., *Physical Chemistry*, 2nd edn., Pergamon (1961).
Moelwyn-Hughes, E. A., *Physical Chemistry*, 2nd edn., p. 918, Pergamon (1965).
Moelwyn-Hughes, E. A., *Trans. Chem. Soc.*, 1576 (1933).
Monk, G. B., *Electrolytic Dissociation*, Academic Press (1961).
Morgan and Smyth, *J. Amer. Chem. Soc.*, **50**, 1547 (1928).
Mott and Sneddon, *Wave Mechanics and its Applications*, Oxford (1948).
Nielsen, *J. Amer. Chem. Soc.*, **58**, 206 (1936).
Nyholm, *Dixième Conseil de Chimie* (Solvay), Stoops, Bruxelles (1956).
Olsen, A. R., and T. R. Simonson, *J. Chem. Physics*, **17**, 1167 (1949).
Onsager, *Radiation Research Supplement*, **4**, 14 (1964).
Orgel, *Trans. Chem. Soc.*, 4756 (1952); see also P. Day and C. K. Jørgensen, *Dixième Conseil de Chemie* (Solvay), Stoops, Brussels (1956).
Parry, R. W., and R. N. Keller, *Chemistry of the Co-ordination Compounds*, edited by Bailar and Busch, Reinhold, New York (1956).
Price, *Z. physikal Chem.*, **27**, 474 (1898).
Robertson, A. C. and S. F. Acree, *J. Amer. Chem. Soc.*, **37**, 1902 (1915).
Ross, *Trans. Chem. Soc.*, **105**, 690 (1914).
Scatchard, G., *Chem. Rev.*, **10**, 229 (1932).
Scatchard, G., *J. Amer. Chem. Soc.*, **52**, 52 (1930).
Schlapp, *Phys. Rev.*, **41**, 194 (1932).
Silverman and Dobson, *J. Phys. Chem.*, **56**, 846 (1952), cited by Laidler, *Chemical Kinetics*, p. 532, McGraw-Hill (1965).
Sirs, *Trans. Faraday Soc.*, **54**, 207 (1958).
Slator, *Trans. Chem. Soc.*, **87**, 485 (1905).
Soper, R. G., and E. Williams, *Proc. Roy. Soc.*, A, **140**, 59 (1933).
Straup and Meloche, *J. Amer. Chem. Soc.*, **53**, 3331 (1931).
Sturtevant, *J. Chem. Phys.*, **3**, 295 (1935).
Svirbely and Schramm, *J. Amer. Chem. Soc.*, **60**, 330 (1938).
Sykes, A. G., *Kinetics of Inorganic Reactions*, Pergamon, Oxford (1966).
Taube, H. Myers and R. L. Rich, *J. Amer. Chem. Soc.*, **75**, 4118 (1953).
Tobe, M. L., "Mechanisms of Inorganic Reactions", *Amer. Chem. Soc. Publ.* (1966).
Traube and Posey, *J. Amer. Chem. Soc.*, **75**, 1463 (1953).
van Arkel and de Boer, *Chemische Bildung als elecktrostatische Ersheinung*, Hirzel, Leipzig (1931).
van Vleck, *Phys. Rev.*, **41**, 208 (1932).
von Kiss and Vass, *Z. anorg. Chem.*, **217**, 305 (1934).
Waddington, *Trans. Faraday Soc.*, **62**, 1483 (1966).
Walker, D. C., *Quart. Rev.*, **21**, 79 (1967).
Walker, E. E., *Proc. Roy. Soc.*, A, **87**, 539 (1912).
Walker, J., and Hambly, *Trans. Chem. Soc.*, **67**, 746 (1895).
Walker, J., and S. A. Kay, *Trans. Chem. Soc.*, **71**, 489 (1897).
Warner and Stitt, *J. Amer. Chem. Soc.*, **55**, 4807 (1933).
Warner and Warwick, *J. Amer. Chem. Soc.*, **57**, 1491 (1935).
Werner, *Trans. Chem. Soc.*, **103**, 1010 (1913).
Williams, R. J. P., *Ann. Rep. Chem. Soc.*, **56**, 87 (1960).
Wyatt and C. W. Davies, *Trans. Faraday Soc.*, **45**, 774 (1949).
Wyman, *Phys. Rev.*, **35**, 623 (1930).

8

SUBSTITUTION AT THE SATURATED
CARBON ATOM

The mechanism of substitution at the saturated carbon atom is of fundamental interest to chemistry. To avoid complications which may arise when the carbon atom is attached to another carbon atom, we shall restrict attention in this chapter to reactions of the methyl and methylene halides, beginning with the work of Hecht, Conrad and Brückner (1890) who studied the kinetics of the following reaction, and of many similar etherifications in ethanol solutions containing initially 0.5 mole/litre of each reactant: $CH_3I + C_2H_5ONa \rightarrow CH_3OC_2H_5 + NaI$. The change goes to completion, is free from side reactions and is bimolecular. Its rate was measured by determining the rate of increase in the concentration of inorganic iodide and the rate of decrease in the concentration of sodium ethoxide. Equation (6.35) is obeyed at all temperatures, and the bimolecular constants are given by the equation $k_2 = 2.46 \times 10^{11} . \exp(-19,490/RT)$. The pre-exponential term is seen to have the order of magnitude of the standard collision frequency, (4.40). This observation (Moelwyn-Hughes, 1932) dispelled the illusion, at one time prevalent, that bimolecular reactions were necessarily complicated in solution but simple in gases. The sole complication in this instance is that sodium ethoxide in ethanol is incompletely ionized. We shall show later that its degree of ionization at $24°C$ and at the concentration employed is 0.55, and that the un-ionized molecule does not react with methyl iodide. The real reaction is, therefore, $CH_3I + C_2H_5O^- \rightarrow CH_3OC_2H_5 + I^-$, and both constants of the Arrhenius equation given above are consequently in need of slight adjustments. Still simpler reactions are those where the halogen in the methyl halide is replaced by an elementary anion, such as that of another halogen: $CH_3X + Y^- \rightleftarrows CH_3Y + X^-$. The first reaction of this type to be studied experimentally was $CH_3Br + I^- \rightleftarrows CH_3I + Br^-$ in aqueous solution (Moelwyn-Hughes, 1938). During the reversible substitution, both methyl halides slowly hydrolyse, and the differential equation for the rate of reaction is not integrable. The reaction of these methyl halides with methanol is much slower than with water, so that the solvolysis complication is not so important in this solvent, and the rate of reaction is approximately reproduced by the standard equation (6.77) for reversible bimolecular reactions (Moelwyn-Hughes, 1939). In a nonhydroxylic medium like acetone, the solvolysis complication is absent, and work in such a solvent should, in principle, provide data of sufficient simplicity to justify theoretical treatments. Unfortunately, in this solvent, most inorganic salts are but slightly soluble and only partly ionized, so that, although the difficulty incidental to solvolysis is eliminated, two other difficulties arise. For both these, however, quantitative allowance can be made.

We shall in this chapter review such information as is now available on the kinetics and mechanism of the reactions of the methyl and methylene halides with the anions F^-, Cl^-, Br^-, I^-, CN^-, SCN^-, OH^-, OCH_3^-, $OC_2H_5^-$, $S_2O_3^{2-}$, and $CH_3SO_4^-$ in acetone, methanol, ethanol, ethylene glycol and water. One instance is also cited of the mechanism of the attack of methyl iodide by a cation, namely Ag^+. In order to describe the true role of the ions and the effect of dilution, occasional reference is made to organic solutes of greater complexity than the methyl halides, but the systematic discussion of their reactions with various anions is postponed to Chapter 9.

Reactions of the Methyl Halides with the Iodine Ion in Acetone Solution

The first reaction to be investigated in this category is $CH_3Br + KI \rightleftarrows CH_3I + KBr$ (Moelwyn-Hughes, 1949), the rate of which was measured by an analytical technique. During the early stages of reaction, crystalline KBr is precipitated, and the concentration, s, of bromide ion in its presence appeared to remain constant, so that the kinetic equation was derived as follows:

$$CH_3Br + I^- \underset{k_4}{\overset{k_2}{\rightleftarrows}} CH_3I + Br^-$$
$$_{(a-x)} _{(b-x)} _{(x)} _{(s)}$$

$$\frac{dx}{dt} = k_2(a - x)(b - x) - k_4 sx. \tag{1}$$

Integration gives, for the velocity coefficient

$$k_2 = \frac{1}{2\beta bt} \ln \frac{a - (\alpha - \beta)x}{a - (\alpha + \beta)x}, \tag{2}$$

and for the fractional change

$$\frac{x}{a} = [\alpha + \beta \coth (k_2 \beta b t)]^{-1},$$

where

$$\alpha = \frac{1}{2}\left(1 + \frac{a}{b} + \frac{sk_4}{bk_2}\right),$$

and

$$\beta^2 = \alpha^2 - (a/b).$$

These equations which, with the appropriate change of notation, are identical with equations (6.49–51), reproduce the observed rates of reaction when the initial concentrations of both reactants are changed by a factor of 10. After making allowance for the ionization of KI, the apparent energy of activation is found to be 15,140 cal. A. G. Evans and S. D. Hamann (1951) measuring the rate of reaction by a conductivity method, find $E_A = 16,700$ cal.

Using radioactive iodine of atomic weight 131, Swart and Le Roux (1957) measured the rate of the following symmetrical and reversible substitution reaction in acetone solution

$$NaI^* + CH_3I \underset{k_2}{\overset{k_2}{\rightleftarrows}} NaI + CH_3I^*$$
$$\downarrow^{\lambda} \qquad\qquad\qquad\qquad \downarrow^{\lambda}$$
$$NaI \qquad\qquad\qquad\qquad CH_3I$$

by the Geiger counter method, after separating the organic and inorganic reactants. Both substitution reactions were shown to be bimolecular, and were assumed to have a common rate constant, k_2, which was evaluated, with allowance for the spontaneous unimolecular decay, by means of equation 6.117. The effect of temperature on the rate constants is summarized by the equation $k_2 = 5.00 \times 10^{10} \times \exp(-13,500/RT)$. At a sodium iodide concentration of 10 millimoles/litre, the degree of ionization of sodium iodide is 0.623 (McBain and Coleman, 1919); hence the pre-exponential term for the reaction between the molecule and the ion becomes 8.03×10^{10}.

Farhat-Aziz and Moelwyn-Hughes (1959) studied the system $CH_3Cl + LiI \rightleftarrows CH_3I + LiCl$ in vapour-free reactors by an analytical technique. Lithium iodide was chosen as the inorganic reactant because of its relatively high solubility and degree of ionization, and, following the observations of S. Sugden on similar systems, the rate-determining steps were assumed to be the ionic replacements

$$CH_3Cl + I^- \underset{k_4}{\overset{k_2}{\rightleftarrows}} CH_3I + Cl^-$$

and the instantaneous rate of reaction to be

$$-\frac{d[CH_3Cl]_t}{dt} = k_2[CH_3Cl]_t[I^-]_t - k_4[CH_3I]_t[Cl^-]_t. \tag{3}$$

Because most of the chloride ions formed combine to give the molecules LiCl, the simple equation (6.77) for reversible bimolecular reactions must be amended as follows before it can be applied to the results.

Denoting by a the initial concentration of methyl chloride, by b the initial concentration of lithium iodide (ionized and un-ionized), by x the concentration of methyl iodide at time t and by α the degree of ionization of lithium iodide at time t, we see that

$$[CH_3Cl]_t = a - x$$

$$[I^-]_t = (b - x)\alpha$$

and

$$[CH_3I]_t = x = [LiCl]_t + [Cl^-]_t. \tag{4}$$

The rate law now becomes

$$dx/dt = k_2(a - x)(b - x)\alpha - k_4x[Cl^-]_t. \tag{5}$$

Because lithium chloride in acetone is much less ionized than lithium iodide, we can express its ionization constant approximately as follows:

$$K_1 = [Li^+]_t[Cl^-]_t/[LiCl]_t = [I^-]_t[Cl^-]_t/[LiCl]_t.$$

For the same reason, $[LiCl]_t$, according to equation (4), is very nearly x. Then $[Cl^-]_t = K_1x(b - x)\alpha$, and equation (5) reduces to

$$\frac{dx}{dt} = k_2(a - x)(b - x)\alpha - \frac{k_4K_1}{\alpha} \cdot \frac{x^2}{b - x}. \tag{6}$$

The equilibrium conditions are governed by the equation

$$K = \frac{k_2}{k_4} = \frac{K_1}{\alpha_\infty^2} \frac{x_\infty^2}{(a - x_\infty)(b - x_\infty)^2} \tag{7}$$

where x_∞ is the final concentration of methyl iodide. On combining equations (6) and (7), we have

$$\frac{dx}{dt} = k_2\alpha\left[(a - x)(b - x) - \frac{K_1}{K\alpha^2} \frac{x^2}{(b - x)}\right]. \tag{8}$$

Because α is a function of the concentration, this equation is not integrable. When α is treated as a constant, there results a highly complicated integrated form (Farhat-Aziz, 1959) which is not here reproduced. During the course of a run, α varies little from the arithmetic mean of its initial and final values, and the term $K_1/K\alpha^2(b - x)$ has the approximately constant value of $2K_1/K\alpha^2(2b - x_\infty) = W$. These facts allow equation (8) to be integrated, giving

$$k_2 = \frac{1}{2\alpha\beta(1 - W)t} \ln \left[\frac{1 - x/(\gamma + \beta)}{1 - x/(\gamma - \beta)} \right]. \tag{9}$$

where

$$\gamma = \frac{a + b}{2(1 - W)} \quad \text{and} \quad \beta = [(a - b)^2 + 4Wab]^{1/2}/2(1 - W).$$

From the accurate conductivity data of Serkov (1908), Ross Kane (1930) and Blokker (1935) the ionization constant of lithium chloride in acetone at 25°C is found to be 1.97×10^{-5} gramme-ion/litre. More generally, $\ln K_1 = -(31.28/R) + (2.908/RT)$. K is 2.43×10^{-4} at 25°C and from

TABLE I

EQUILIBRIUM CONSTANTS FOR THE REACTION $CH_3Cl + I^- \rightleftarrows CH_3I + Cl^-$
IN ACETONE SOLUTION:

$$K = [CH_3I]_\infty[Cl^-]_\infty/[CH_3Cl]_\infty[I^-]_\infty$$

$t°$ C	$10^5 K_1$ (moles/litre)	$[CH_3Cl]_0$	$[LiI]_0 + [I^-]_0$ (mmoles/litre)	$[LiI]_\infty + [I^-]_\infty$	α_∞	$10^4 K$
20·00	2·15	44·25	52·26	44·71	0·284	2·08
20·00		39·41*	43·12	34·26	0·305	2·63
20·00		16·18	35·39	31·66	0·312	2·47
					Average =	2·37
30·00	1·83	44·25	52·26	43·99	0·280	2·29
30·00		39·41*	43·12	33·62	0·299	2·96
30·00		16·67	33·35	29·46	0·309	2·62
30·00		44·31	10·64	8·23	0·417	2·16
					Average =	2·47
40·00	1·57	16·43	62·17	55·86	0·255	3·04
40·00		44·25	52·26	43·01	0·276	2·72
40·00		16·74	30·04	26·04	0·316	2·91
40·00		17·75	14·50	12·12	0·381	2·71
40·00		44·31	10·64	7·92	0·422	2·50
					Average =	2·76
50·00	1·36	44·25	52·26	42·14	0·273	3·08
50·00		17·61	30·96	26·68	0·308	2·76
50·00		15·82	31·65	27·57	0·305	2·72
50·00		44·31	10·64	7·58	0·408	3·22
					Average =	2·93

values at other temperatures (Table I), we find $\ln K = -(11.71/R) - (1440/RT)$. Kinetic results on a specimen run are shown in Table II. The bimolecular constant k_2 was found to be, at a constant temperature, independent of the initial concentrations. The values of k_2 at various temperatures led to an apparent energy of activation of 18,610 cal. A repetition of the work using potassium iodide as reactant gave a value of 20,250 cals. With either salt at 50°C. $k_2 \times 10^2 = 1.50 \pm 0.07$. Giving equal weight to the two values of E_A, we accept 19,430 $\pm$ 810 as the most reasonable value.

TABLE II

BIMOLECULAR VELOCITY CONSTANTS FOR THE REACTION
$CH_3Cl + I^- \rightarrow CH_3I + Cl^-$ IN ACETONE AT 40·0°C;
$[CH_3Cl]_0 = 17.75$, $[CH_3Cl]_\infty = 15.37$ MMOLES/LITRE

| | | $10^3 k_2$ (litre/mole^{-1}-sec^{-1}) | |
| | | Completely integrated form of | |
t (sec)	$[LiI]_t + [I^-]_t$ (mmoles/litre)	equation (8)	equation (9)
0	14·50	—	—
552	14·16	6·58	6·60
1152	13·95	5·21	5·22
1752	13·74	4·86	4·87
2652	13·37	5·08	5·10
4152	12·89	5·26	5·26
7152	12·43	5·07	4·93
1284($t_{1/4}$)	13·91	5·08	5·09
2832($t_{1/2}$)	13·31	5·08	5·09
5136($t_{3/4}$)	12·72	5·07	5·04
∞	12·12	—	—

The reaction between methyl fluoride and lithium iodide is free from complications, and has been investigated by means of gas chromatography, yielding results which are included in Table III.

The reaction between the iodide ion and methyl iodide proceeds 9.44×10^5 times as fast as its reaction with methyl fluoride. To account for this ratio at 25°C in terms of a change in activation energy only, requires $E_A(CH_3F - I^-) - E_A(CH_3I - I^-)$ to be 8,150 cal, which tallies with the observed difference of 8,690 $\pm$ 700 cal. There is no doubt, therefore, that the factor determining the rate of reaction in this series is the energy of activation, a conclusion which, as we shall see, has been confirmed by data relating to other solvents. The trend in the A values may prove to be real, but data of greater accuracy are required before a decision can be reached. The E_A

TABLE III

REACTIONS OF THE METHYL HALIDES WITH IODINE IONS IN ACETONE SOLUTION

Ref.	Reaction	$k_2 \times 10^5$ at 25°C (litre/ mole-sec)	$A \times 10^{-11}$ (litre/ mole-sec)	$\mu \times 10^{18}$ (e.s.u.)	E_A (cal/ mole)	$D(CH_3X)$ kcals/mole
*	$CH_3F + I^-$	1·08	2·01	1·81	$22,190 \pm 200$	116·3
†	$CH_3Cl + I^-$	119	2·07	1·87	$19,430 \pm 800$	76·0
‡	$CH_3Br + I^-$	82,700	1·04	1·80	$15,140 \pm 400$	63·3
§	$CH_3I + I^-$	1,020,000	0·80	1·64	$13,500 \pm 500$	117·2

REFERENCES

* Farhat-Aziz and Moelwyn-Hughes, *Trans. Chem. Soc.*, 1523 (1961).
† *Idem., ibid.*, 2635 (1959); Evans and Hamann, *Trans. Faraday Soc.*, **47**, 30 (1951).
‡ Moelwyn-Hughes, *Trans. Faraday Soc.*, **45**, 167 (1949).
§ Swart and le Roux, *Trans. Chem. Soc.*, 2110 (1956); 406 (1957).

values change in the same direction as the dissociation energies (D) of the methyl halides into radicals or ions in the gas phase, but are not proportional to either. The variation in E_A is much greater than that in the ionization energies of the methyl halides in aqueous solution (Chapter 10).

The Kinetics and Mechanism of the Reaction $CH_3I + Cl^- \rightarrow CH_3Cl + I^-$ in Acetone Solution

The rate of this reaction, and of other homologous changes has been measured at various temperatures by E. D. Hughes, C. K. Ingold and J. D. H. Mackie (1955) using lithium chloride, and applying the familiar equation (6.35) for an irreversible bimolecular reaction to obtain the term k'_4, defined by the equation

$$\frac{d[I^-]}{dt} = k'_4[CH_3I][LiCl]_{total}. \qquad (10)$$

At 24·88°C, k'_4 falls from $1·97 \times 10^{-2}$ at 20 per cent reaction to $1·48 \times 10^{-2}$ at 48 per cent reaction, but is said to be constant at $4·60 \times 10^{-3}$ in the presence of 0·110 molar $LiClO_4$. Their repetition of the work at 25·20°C, with initial concentrations about one-half of the previous values shows k'_4 to fall from $3·78 \times 10^{-2}$ at 27 per cent reaction to $2·76 \times 10^{-2}$ at 71 per cent reaction, and to reach $5·70 \times 10^{-3}$ in the presence of 0·09475 molar

LiClO₄ (E. D. Hughes, C. K. Ingold and A. J. Parker, 1960). The fall in k_4' during a given run, and its dependence on the initial concentration of electrolyte recall the data on Arrhenius who, before the discovery of ionization in solution, represented the kinetics of the reaction between ethyl acetate and ammonia by a similar equation, namely,

$$d[C_2H_5OH]/dt = k_4'[CH_3COOC_2H_5][NH_4OH]_{total}.$$

Farhat-Aziz and Moelwyn-Hughes define their rate constants by the equation

$$d[I^-]/dt = k_4[CH_3I][Cl^-] - k_2[CH_3Cl][I^-] \tag{3}$$

Since $k_2 = 1.19 \times 10^{-3}$ (Table III) and $K = 2.43 \times 10^{-4}$ (Table I) at 25°C, k_4 becomes 4.90 litre/mole-sec, which exceeds Ingold's "stoichiometric second order rate constant", k_4', by a factor of more than 1,000.

To formulate a kinetic scheme consistent with this mechanism, while allowing for the reversibility of the reaction and the ionization of the lithium salts, we may let

$$[CH_3I]_0 = a,$$

$$[LiCl]_0 + [Cl^-]_0 = b,$$

$$[CH_3I]_t = a - x$$

$$[CH_3Cl]_t = x = [I^-]_t + [LiI]_t,$$

$$[I^-]_t = x\alpha,$$

where α is the degree of ionization of lithium iodide. Then

$$dx/dt = k_4(a - x)[Cl^-] - k_2\alpha x^2.$$

Because lithium chloride in acetone is much less ionized than lithium iodide, its ionization constant may be expressed, as in the preceding section, as follows:

$$K_1 = [Li^+]_t[Cl^-]_t/[LiCl]_t$$

$$= [I^-]_t[Cl^-]_t/[LiCl]_t.$$

For the same reason, $[LiCl]_t$ is approximately $(b - x)$, so that

$$[Cl^-]_t = K_1(b - x)/\alpha x,$$

and

$$dx/dt = k_4 K_1(a - x)(b - x)/\alpha x - k_2\alpha x^2. \tag{11}$$

This equation, applied to the data of Hughes, Ingold and Parker's run yields the values of k_4 given in column 3 of Table IV. Over the greater part of the run, they differ but little from the value of 4.90 given by Farhat-Aziz and

TABLE IV

KINETIC ANALYSIS OF THE REACTION $CH_3I + Cl^- \rightarrow CH_3Cl + I^-$
IN ACETONE AT $25.2°C$

[$a = 28.62$ mmoles/litre. $b = 14.80$. $\alpha = 0.44$. $K_1 = 1.97 \times 10^{-5}$ gram-ions/litre.
$k_2/k_4 = 2.43 \times 10^{-4}$]

| | | k_4 (litre/mole-sec) | |
| | | --- | --- |
t (min)	x (mmoles/litre)	Corrected for ionization and reversal (equation 11)	Corrected for ionization only (equation 14)
0	3.05	—	—
1	3.66	3.62	3.94
2	4.24	3.72	3.63
4	5.04	4.09	3.41
7	6.15	4.63	3.47
9	6.59	5.23	3.75
13	7.81	5.10	3.80
21	9.10	5.10	3.73
27	9.77	—	3.94

Moelwyn-Hughes at $25.00°C$. As the correction necessary for the reverse reaction in this instance is relatively slight, we may neglect the last term in equation (11), and write

$$dx/dt = k_4 K_1(a - x)(b - x)/\alpha x, \tag{12}$$

which gives, on integration

$$k_4 = \frac{\alpha}{2K_1 t}\left\{\ln\left[1 - \left(\frac{a + b}{ab}\right)x + \frac{x^2}{ab}\right] + \frac{1}{F}\left[\frac{1 + \frac{1}{2}\left(\frac{a + b}{ab}\right)(F - 1)x}{1 - \frac{1}{2}\left(\frac{a + b}{ab}\right)(F + 1)x}\right]\right\}, \tag{13}$$

where

$$F = \pm(a - b)/(a + b).$$

In the present case, F is positive, and

$$k_4 = \frac{\alpha}{2K_1 t}\left\{\ln\left[1 - \left(\frac{a + b}{ab}\right)x + \frac{x^2}{ab}\right] + \left(\frac{a + b}{a - b}\right)\ln\frac{b(a - x)}{a(b - x)}\right\}. \tag{14}$$

The sum of the logarithmic terms varies linearly with respect to t, indicating a zero time error of 1 min, which has therefore been added to the observed times. The mean value of k_4 found in this way is 3·71, which is naturally less than the fully corrected value. Moreover, k_4 does not drift.

This formulation (Farhat-Aziz and Moelwyn-Hughes, 1962) therefore satisfactorily accounts for the principal kinetic features of the reaction, whether it is studied from the direct or the reverse direction. It is consistent with the constancy of k_4 during any given run and with its independence of the initial concentrations; and it explains why the k_4' term of Ingold falls during the course of each run, and has different initial values for solutions with . different initial concentrations. That the mechanism adopted here is the correct one has been established beyond reasonable doubt by Winstein, Savedoff, S. Smith, I. D. R. Stephens and Gall (1960), who have measured the rates of reaction of *n*-butyl *p*-brombenzene sulphonate with lithium chloride, bromide and iodide and with tetra-*n*-butyl-ammonium chloride, bromide and iodide. The ionization constants of these six salts range from $2·7 \times 10^{-6}$ to $6·9 \times 10^{-3}$ gram-ions/litre. Column 2 of Table V gives the 'stoichiometric second order rate constant' as calculated according to Ingold's method. Column 3 gives the bimolecular constant for the reaction between the organic bromide and the ions, Cl^-, Br^- and I^-, as the term is understood here. The relative rates at which *n*-butyl *p*-brombenzene sulphonate reacts with the ions are seen to bear the following sequence $Cl^- : Br^- : I^- : : 1·00 : 0·23 : 0·05$. It is instructive, as Winstein points out, to contrast this sequence with that found from the relative rates of reaction of methyl bromide with these ions in the same solvent and at the same temperature according to Ingold's interpretation of their kinetics and mechanism. His sequence, $Cl^- : Br^- : I^- : : 1·00 : 21·7 : 41·7$, is based on stoichiometric second order rate constants obtained using lithium salts and published (E. D. Hughes, C. K. Ingold and J. D. H. Mackie, 1955; de la Mare, 1955; Fowden, E. D. Hughes and C. K. Ingold, 1955) under the general title "Kinetics of the interaction of chloride, bromide and iodide *ions* [italics are ours] with simple alkyl bromides in acetone".

TABLE V

VELOCITY CONSTANT (IN LITRES/MOLE-SEC) FOR THE REACTION OF *n*-BUTYL *p*-BROMBENZENE SULPHONATE WITH 0·04M SALTS IN ACETONE AT 25°C

Salt	$k_4' \times 10^3$	$k_4 \times 10^3$	Average $k_4 \times 10^3$
$(n\text{-Bu})_4NCl$	33·5	58·0	
LiCl	0·493	51·0	54·5
$(n\text{-Bu})_4NBr$	9·09	13·0	
LiBr	2·81	12·0	12·5
$(n\text{-Bu})_4NI$	1·68	2·0	
LiI	2·97	3·6	2·8

The Effect of Dilution: The Dual Hypothesis

The bimolecular constants found by Hecht, Conrad and Brückner (1890) for the reaction between methyl iodide and sodium ethoxide at equimolar concentrations in ethanol solution at 24°C are shown in Fig. 1. k_2 is the instantaneous rate of reaction, dx/dt, divided by the product of the total concentration of each reactant, and is thus what has latterly been referred to as the stoichiometric second-order rate constant. The values given have been converted into standard units from the time unit of one minute and the rather unusual concentration unit adopted by the investigators. The velocity constant is seen to decrease as the concentration increases, an effect found also for many similar systems where one of the reactants is an electrolyte. The constants at any temperature have been well reproduced by empirical equations in a form which, although applied by later workers (Cox, 1918, 1922; Gibson, Fawcett and Williams, 1935) has no discernible physical content. Acree (H. C. Robertson and Acree, 1915), suggested that the dilution effect was due to incomplete ionization of the electrolyte, and to a difference between the rate constant, k_m, with which the undissociated base reacts, and the rate constant, k_i, with which the ion reacts. In this instance, we would then write for the total rate

$$+ d[I^-]/dt = k_m[CH_3I][C_2H_5ONa] + k_i[CH_3I][C_2H_5O^-],$$

so that, in the general case where α is the degree of ionization of the electrolyte, we have

$$+ dx/dt = k_m(a - x)(b - x)(1 - \alpha) + k_i(a - x)(b - x)\alpha.$$

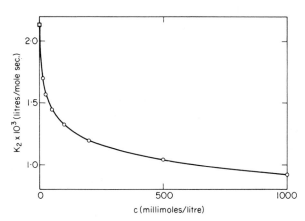

FIG. 1. The influence of the initial concentration of electrolyte on the stoichiometric second-order constant for the reaction $CH_3I + C_2H_5ONa \rightarrow CH_3OC_2H_5 + NaI$ in ethanol solution at 24°C.

Since the observed second order constant, k_2, is $(dx/dt)/(a - x)(b - x)$ we have

$$k_2 = k_m(1 - \alpha) + k_1\alpha$$
$$= k_m + (k_i - k_m)\alpha. \tag{15}$$

k_2 should thus vary linearly with respect to α. Acree's idea is often referred to as the dual hypothesis. As applied by him (1919), k_i/k_m appeared to be about 2 or 3 for certain reactions, but generally no linear relationship was established between k_2 and α. Moreover, the true value of k_i/k_m is known from other kinetic sources to have a higher order of magnitude, $10^{3 \pm 1}$, for reactions of the kind here considered. The reason is now known to be due, not to any flaw in the argument, but to wrong estimates of the degree of ionization. Arrhenius' expression for α in terms of the equivalent conductance, Λ, and the sum of the ionic mobilities, U and V,

$$\alpha = \frac{\Lambda}{U + V}$$

is still valid, but values of U and V at the concentrations concerned must be used and not Λ_∞, which is their sum at infinite dilution. Later work on the reactions of alkyl iodides, in ethanolic solution, with sodium m-4-xylyloxide (Hardwick, 1935) sodium eugenoxide (Woolf, 1937) and sodium guaiacoxide (Mitchell, 1937) indicate that, when the degree of ionization is properly computed, k_m for these reactions is negligibly small. A specimen of Mitchell's data is shown in Table VII. The solution of several simultaneous equations

TABLE VII

THE INFLUENCE OF DILUTION ON THE REACTION BETWEEN ETHYL IODIDE (0·24M) AND SODIUM GUAIACOXIDE IN ETHYLALCOHOLIC SOLUTION AT 55°C (MITCHELL)

Concentration of base (mole/litre)	α	$k_2 \times 10^4$ (litre/gram mol-sec)	$(k_2/\alpha) \times 10^2$
0·24	0·0600	7·65	1·275
0·22	0·0622	7·82	1·257
0·20	0·0648	8·03	1·240
0·17	0·0697	8·42	1·208
0·14	0·0753	9·07	1·204
0·12	0·0801	9·68	1·209
0·10	0·0855	10·5	1·232
0·08	0·0938	11·6	1·233
0·06	0·1065	13·2	1·241
0·04	0·1266	15·59	1·260
			Average 1·239

relating to propyl iodide at 75° yielded Woolf the values $k_i = 1.9$ and $k_m = -0.001$. By this is meant, of course, that the reaction is, within the limits of error, solely between propyl iodide and the eugenoxide ion.

A still more extensive test of the dual hypothesis has been carried out by C. C. Evans and S. Sugden (1949) on the isotopic reaction

$$nC_4H_9Br + LiBr* \rightleftarrows nC_4H_9Br* + LiBr$$

in anhydrous acetone solution, in which the salt is a weak electrolyte, with an ionization constant of

$$K = c\alpha^2/(1 - \alpha) = 5.22 \times 10^{-4} \text{ gram–ions/litre}$$

at 25°C. By applying the method of least squares to their data at 26.2°C (see Table VIII) they find, in standard units, $10^4 k_2 = 0.12 \pm 0.05 + (4.41 \pm 0.17)\alpha$. Hence k_m is small, and possibly insignificant. If, as seems likely, only the bromide ion reacts with the butyl bromide molecule, k_m is zero, and the

TABLE VIII

THE REACTION $nC_4H_9Br + LiBr* \rightleftarrows nC_4H_9Br* + LiBr$
IN ACETONE AT 26.2°C

$[LiBr]_0 \times 10^4$ (moles/litre)	α	$k_2 \times 10^3$ (litres/mole-sec)	$(k_1/\alpha) \times 10^3$
0	1.000	—	4.61
0.504	0.925	3.93	4.3
0.595	0.916	3.69	4.0
1.28	0.848	3.99	4.7
1.53	0.831	4.13	5.0
2.49	0.774	3.13	4.0
3.10	0.741	3.57	4.8
4.38	0.696	3.48	5.0
17.1	0.497	2.56	5.2
22.7	0.457	2.22	4.9
43.4	0.378	1.84	4.9
79.4	0.314	1.44	4.6
153.4	0.257	1.08	4.2
191.0	0.241	1.07	4.4
			Average 4.6 $\pm$ 0.3

average value of k_i is 4.61×10^{-3}. It is to be observed that the salt concentration has been altered by a factor of 400. The same behaviour is found in the isotopic displacement of bromine from secondary octyl bromide $(CH_3)(C_6H_{13})HCBr$ in the same solvent at 65.5°C, where $K = 2.4 \times 10^{-4}$, and the concentration of salt has been varied by a factor of 533.

There remains little doubt that in the reactions of normal alkyl halides with bases in alcoholic solution and with halogen ions in acetone solution, the rate is exclusively that of the reaction between the alkyl halide and the anion.

When k_m is zero, $k_2 = k_i\alpha$ from equation (15), and since the bimolecular constant extrapolated to infinite dilution is $k_2^0 = k_i$, we have a simple kinetic method for the determination of the degree of ionization, $\alpha = k_2/k_2^0$, and thence the ionization constant, K. From Fig. 1, for example, k_2 is seen to reach one half of its limiting value, k_2^0, at a concentration of 0·40 mole per litre. The ionization constant of sodium chloride in ethanol at 24°C is thus 0·20 gramme–ion per litre.

The Dilution Effect in Aqueous Solution

Kinetic measurements have been made on aqueous solutions of methyl iodide in the range of 10^{-2} to 10^{-8} molar (R. A. Hasty, 1969). k_2 for the reaction between methyl iodide and the thiosulphate ion is found to be strictly independent of the ionic strength, I. For the reaction of methyl iodide with the sulphite ion, however, there is a negative electrolyte effect:

$$k_2 = k_2^0(1 - 0·3I).$$

Reactions of the Methyl Halides in Methanol Solution

In studying exchange reactions of the methyl halides with halogen ions in alcoholic solutions, allowance must be made for the simultaneous solvolyses of the two halides. As in aqueous solution, the bimolecular substitutions and the pseudo-unimolecular solvolyses proceed side by side, but, on account of the reversibility of the reaction, equation (6.45) is no longer applicable. With methyl bromide and methyl iodide in methanol solution, for example, we have the following set of reactions:

$$CH_3Br + I^- \xrightleftharpoons[k_4]{k_2} CH_3I + Br^-$$

$$k_1 \downarrow CH_3OH \qquad\qquad k_3 \downarrow CH_3OH$$

$$CH_3OCH_3 + H^+ + Br^- \qquad CH_3OCH_3 + H^+ + I^-,$$

according to which, if we start from the top left side, the concentration of iodide ion reaches a minimum when

$$\frac{[Br^-][CH_3I]}{[I^-][CH_3Br]} = \frac{k_2}{k_4} - \frac{k_3}{k_4}\frac{[CH_3I]}{[CH_3Br][I^-]}.$$

Starting from the top right side, the concentration of bromide ion reaches a minimum when

$$\frac{[\text{Br}^-][\text{CH}_3\text{I}]}{[\text{I}^-][\text{CH}_3\text{Br}]} = \frac{k_2}{k_4} + \frac{k_1}{k_4[\text{I}^-]}.$$

By analysing samples for the halogen ions and hydrogen ion, the four rate constants can be evaluated. The bimolecular constants are given in Table IX. No difference could be detected due to the presence of the cations (Li^+, Na^+ and K^+), and the data have consequently not been adjusted to allow for the role of undissociated molecules of the electrolytes or for ion-pairs. For the exchange reaction $\text{CH}_3\text{I} + {}^*\text{I}^- \rightleftarrows \text{CH}_3^*\text{I} + \text{I}^-$, slightly different results have been found by Swart and Le Roux, using the radioactive technique, and by Beronius and Lamm, using an electrolytic precipitation procedure developed by them for the study of fast reactions. From the summary of the scant data

TABLE IX

REACTIONS OF THE METHYL HALIDES IN METHANOL SOLUTION

Reaction	k_2 at 25°C (litres/mole-sec)	$\log_{10} A_2$	E_A (kcal/mole)
$\text{CH}_3\text{Br} + \text{I}^-$	$9 \cdot 42 \times 10^{-4}$	$10 \cdot 354$	$18 \cdot 25 \pm 0 \cdot 25$*
$\text{CH}_3\text{I} + \text{Br}^-$	$7 \cdot 35 \times 10^{-5}$	$10 \cdot 806$	$20 \cdot 37 \pm 0 \cdot 50$*
$\text{CH}_3\text{I} + \text{I}^-$	$7 \cdot 73 \times 10^{-3}$	$9 \cdot 623$	$16 \cdot 00 \pm 0 \cdot 50$†
$\text{CH}_3\text{I} + \text{I}^-$	$3 \cdot 42 \times 10^{-3}$	$9 \cdot 114$	$16 \cdot 18$‡
$\text{CH}_3\text{Cl} + \text{CH}_3\text{O}^-$	$7 \cdot 64 \times 10^{-6}$	$12 \cdot 246$	$23 \cdot 69 \pm 0 \cdot 29$§
$\text{CH}_3\text{Br} + \text{CH}_3\text{O}^-$	$2 \cdot 89 \times 10^{-4}$	$12 \cdot 241$	$21 \cdot 53 \pm 0 \cdot 35$§
$\text{CH}_3\text{I} + \text{CH}_3\text{O}^-$	$2 \cdot 78 \times 10^{-4}$	$12 \cdot 303$	$21 \cdot 64 \pm 0 \cdot 20$‖
$(\text{CH}_3)_2\text{SO}_4 + \text{CH}_3\text{O}^-$	$2 \cdot 32 \times 10^{-2}$	$13 \cdot 404$	$20 \cdot 52 \pm 0 \cdot 22$§

REFERENCES

* Moelwyn-Hughes, *Trans. Faraday Soc.*, **35**, 368 (1939).
† Swart and le Roux, *Trans. Chem. Soc.*, 406 (1957).
‡ Beronius and Lamm, *Trans. Faraday Soc.*, **56**, 1793 (1960).
§ R. Hurst and Moelwyn-Hughes, *Academia Nazionale dei Lincei*, 109 (1957).
‖ *Cf.* J. Hine, C. H. Thomas and S. J. Ehrenson (*J. Amer. Chem. Soc.*, **77**, 3886 (1955)) who give $E_A = 21 \cdot 4 \pm 0 \cdot 4$ and J. Murto (*Annales Academiae Scientarum Fennicae*, **AII**, 117 (1962)) who gives $E_A = 21 \cdot 75$.

of the reactions of the methyl halides in methanol solution given in Table IX it is clear that, with one exception, changes in the rates of comparable reactions are due mainly to changes in the energy of activation.

Reactions of the Methyl Halides with the Cyanide Ion in Aqueous Solution

Attending the substitution of CN for a halogen atom, X,

$$k_2: \quad CH_3X + CN^- \rightarrow CH_3CN + X^- \tag{1}$$

the following subsidiary reactions can, in principle, occur in water:

$$CH_3CN + X^- \rightarrow CH_3X + CN^- \tag{2}$$

$$CH_3X + CN^- \rightarrow CH_3NC + X^- \tag{3}$$

$$k_1: CH_3X + H_2O \rightarrow CH_3OH + X^- + H^+ \tag{4}$$

$$K: CN^- + H_2O \rightleftharpoons HCN + OH^- \tag{5}$$

$$CH_3X + OH^- \rightarrow CH_3OH + X^- \tag{6}$$

$$CH_3CN + 2H_2O \rightarrow CH_3COONH_4 \tag{7}$$

Thermodynamic calculations show that the equilibrium constants $[CH_3CN][X^-]/[CH_3X][CN^-]$ at 25°C lie in the range 10^{24}–10^{27}. The substitution is thus effectively complete, and reaction 2 can be ignored. Chemical analysis of the product solution showed a complete absence of isocyanide: reaction (3), therefore, does not take place. Quantitative allowance can be made for reactions (4) and (6), of which only the former need be applied. Constants for the equilibrium represented by equation (5) are known at the relevant temperatures (Madsen, 1901; H. F. Brown and Cranston, 1940). Separate experiments with aqueous solutions of acetonitrile alone showed the rate of reaction (7) to be extremely slow. The extent of hydrolysis found at 85°C was less than one per cent. The rate of production of halogen ions can thus be formulated as follows:

$$d[X^-]/dt = k_2[CH_3X][CN^-] + k_1[CH_3X] \tag{16}$$

$$[CH_3X]_0 = a, \quad [CH_3X]_t = a - x, \quad [X^-]_t = x$$

$$[KCN]_0 = b, \quad [KCN]_t \simeq b - x$$

$$[HCN]_t \simeq [OH^-]_t \simeq K^{1/2} . [CN^-]^{1/2} \simeq K^{1/2} . (b - x)^{1/2}$$

$$[CN^-]_t \simeq b - x - K^{1/2} . (b - x)^{1/2}$$

Then

$$dx/dt \simeq k_2 . (a - x)[(b - x) + (k_1/k_2) - K^{1/2} . (b - x)^{1/2}] \tag{17}$$

This equation may be simplified by substituting $B - k_1/k_2$ for b, and by ignoring $B - b$ in the correction term for the hydrolysis of the cyanide ion. This expedient is justified by the smallness of the correction term and of k_1/k_2. Then, $dx/dt = k_2(a - x)[B - x - K^{1/2}(B - x)^{1/2}]$, which may be integrated to give,

$$k_2 t = [1/(B - a - K^{1/2})]\{\log_e [a/(a - x)] + 2 . \log_e [(z - K^{1/2})/(z_0 - K^{1/2})]$$
$$+ [K/(B - a)]^{1/2} . \log_e [(z_0 - z_\infty).(z + z_\infty)/(z - z_\infty).(z_0 + z_\infty)]\}   (18)$$

TABLE X

REACTIONS OF THE METHYL HALIDES IN WATER

Reaction	k_2 at 25°C (litres/mole-sec)	$\log_{10} A_2$	E_A (kcal/mole)
$CH_3Cl + F^-$	1.40×10^{-8}	11·867	26·90*
$CH_3Br + F^-$	3.02×10^{-7}	11·960	25·20
$CH_3I + F^-$	6.92×10^{-8}	11·320	25·20
$CH_3F + I^-$	2.08×10^{-8}	9·218	23·06
$CH_3Cl + I^-$	2.00×10^{-5}	9·980	20·02
$CH_3Br + I^-$	7.02×10^{-4}	10·225	18·26†
$CH_3I + I^-$	4.85×10^{-4}	9·954	18·10‡
$CH_3CN + I^-$	1.90×10^{-30}	10·6	55·00¶
$CH_3F + OH^-$	5.86×10^{-7}	9·608	21·60§
$CH_3Cl + OH^-$	6.67×10^{-6}	12·614	24·28‖
$CH_3Br + OH^-$	1.435×10^{-4}	13·021	23·00
$CH_3I + OH^-$	6.36×10^{-5}	12·093	22·22
$CH_3F + CN^-$	1.60×10^{-6}	11·05	23¶
$CH_3Cl + CN^-$	4.90×10^{-5}	11·50	21·56
$CH_3Br + CN^-$	1.26×10^{-3}	12·27	20·70
$CH_3I + CN^-$	5.90×10^{-4}	11·78	20·47
$CH_3Cl + S_2O_3^{--}$	9.14×10^{-4}	12·004	20·52**‖
$CH_3Br + S_2O_3^{--}$	3.24×10^{-2}	12·834	19·54
$CH_3I + S_2O_3^{--}$	2.84×10^{-2}	12·303	18·88

REFERENCES

* R. H. Bathgate and E. A. Moelwyn-Hughes, *Trans. Chem. Soc.*, 2642 (1959).
† Moelwyn-Hughes, *Trans. Chem. Soc.*, 779 (1938).
‡ Swart and le Roux, *Trans. Chem. Soc.*, 406 (1957).
§ Glew and Moelwyn-Hughes, *Proc. Roy. Soc.* **A, 211**, 254 (1952).
‖ Moelwyn-Hughes, *Proc. Roy. Soc.*, **A, 196**, 540 (1949).
¶ Marshall and Moelwyn-Hughes, *Trans. Chem. Soc.*, 7119 (1965).
** Slator, *Trans. Chem. Soc.*, **85**, 1286 (1904).

where $z_0 = B^{1/2}$, $z = (B - x)^{1/2}$, and $z_\infty = (B - a)^{1/2}$. This equation adequately reproduces the results over the complete runs at all temperatures. The results (Marshall and Moelwyn-Hughes, 1959) are included in Table X.

Reactions of the Methyl Halides with Various Ions in Aqueous Solution

The data summarized in Table X have been corrected for the rates of hydrolysis of the methyl halides, which are known with precision at all temperatures, and, where necessary, for the rates of the reverse reactions. The reactions of the methyl halides with the hydroxyl ion are effectively irreversible, and their kinetic course is accurately reproduced by equation (6.35). On examining the data of Table X, we again note the predominant influence of E_A, which is particularly evident in the reactions of the methyl halides with the iodide ion, where a rate ratio of about 2.5×10^4 is accounted for almost entirely by an energy difference of 5 kcal per mole. The high values of A_2 for the reactions of the methyl halides with the hydroxyl ion resemble those for their reactions with methoxide ion in methanol solution (Table IX). A_2 for the reactions $CH_3X + OH^-$ are about 400 times as great as for the reactions $CH_3X + I^-$, which is far greater than can be explained in terms of the ionic mobilities, the ratio of which is only 2·63. The following approximate energy differences (in kcal) suggest that the total energy of activation may consist of contributions made separately by the two reactants:

$$E_A(CH_3F + X^-) - E_A(CH_3Cl + X^-) = 2.1 \pm 0.9$$

$$E_A(CH_3Cl + X^-) - E_A(CH_3I + X^-) = 1.7 \pm 0.6$$

$$E_A(CH_3Cl + X^-) - E_A(CH_3Br + X^-) = 1.3 \pm 0.4$$

$$E_A(CH_3Br + X^-) - E_A(CH_3I + X^-) = 0.4 \pm 0.3$$

$$E_A(CH_3X + OH^-) - E_A(CH_3X + S_2O_3^{--}) = 3.5 \pm 0.3$$

The reaction of methyl bromide with any given ion is always faster than the reactions of the other methyl halides with it, and the constant A_2 is invariably higher than other values in the series. From the five sets of reactions given in the table, we have the approximate relative rates at 25°C:

CH_3F	CH_3Cl	CH_3Br	CH_3I
0·04	1	28 ± 7	16,

which is the familiar sequence of the rates of hydrolysis.

Reactions of Methyl Iodide with Various Ions in Aqueous Solution

In this section we shall consider the rate of reaction of a given molecule, namely methyl iodide, with a variety of ions in aqueous solution. The first four entries in Table XI suggest a general trend in both the constants of the

TABLE XI

REACTIONS OF METHYL IODIDE WITH VARIOUS IONS IN AQUEOUS SOLUTION

Reaction	k_2 at 25°C	Relative k_2	$\log_{10} A_2$	E_A
$CH_3I + F^-$	6.92×10^{-8}	0.021	11.320	25.20*
$CH_3I + Cl^-$	3.24×10^{-6}	1	10.620	21.97
$CH_3I + Br^-$	8.93×10^{-5}	28	10.111	19.31†
$CH_3I + I^-$	4.85×10^{-4}	150	9.954	18.10‡
$CH_3I + OH^-$	6.36×10^{-5}	20	12.093	22.22§
$CH_3I + SCN^-$	3.58×10^{-4}	110	11.173	19.95‖
$CH_3I + CN^-$	5.90×10^{-4}	170	11.78	20.47¶
$CH_3I + Ag^+$	2.64×10^{-3}	815	11.642	19.40**
$CH_3I + S_2O_3^{--}$	2.84×10^{-2}	8,770	12.303	18.88††
$CH_3I + SO_3^{--}$	4.40×10^{-2}	—	12.3	18.9‡‡

REFERENCES

* R. H. Bathgate and Moelwyn-Hughes, *Trans. Chem. Soc.*, 2642 (1959).
† Moelwyn-Hughes, *Trans. Chem. Soc.* (1938).
‡ Swart and le Roux, *Trans. Chem. Soc.* (1957).
§ Moelwyn-Hughes, *Trans. Chem. Soc.* (1949); Puenté, *Anales de la Asociación Química, Argentina*, **38**, 355 (1950).
‖ Lalor and Moelwyn-Hughes, *Trans. Chem. Soc.*, 2201 (1965).
¶ B. W. Marshall and Moelwyn-Hughes, *Trans. Chem. Soc.*, 2640 (1959); 7119 (1965).
** D. N. Colcleugh and Moelwyn-Hughes, *Trans. Chem. Soc.*, 2542 (1964).
†† Slator, *Trans. Chem. Soc.*, (1904).
‡‡ R. A. Hasty and S. L. Sutter, *Can. J. Chem.*, in Press, 1969.

Arrhenius equation, with $dE_A/d \log A_2 = 5.30$ kcal but it is difficult to decide whether the trend is real, because, within the limits of experimental errors in E_A, the data for the reactions of the four halogen ions with methyl iodide can be reproduced with the common value of 10.33 for $\log_{10} A_2$.

TABLE XII

APPARENT ENERGIES OF ACTIVATION OF THE REACTIONS
$CH_3I + X^- \rightarrow CH_3X + I^-$ IN WATER, COMPARED WITH
THE HEATS OF ESCAPE OF THE IONS INTO VACUO (KCAL)

Ion	E_A	E_S	E_A/E_S
F^-	25·20	109	0·231
Cl^-	21·97	77	0·285
Br^-	19·31	69	0·280
I^-	18·10	60	0·302
CN^-	20·47	68	0·301
OH^-	22·22	84·5	0·263

Average 0·277

The Role of Solvation

There appears to be a roughly constant ratio between the apparent energy of activation and the heat of escape of the ions from water (Table XII), suggesting that partial desolvation of the ion may contribute significantly to the energy of activation. This hypothesis can be applied without much conjecture or computation.

Each ion in solution, as in crystalline hydrates, is surrounded by solvent molecules, of number c, to which it adheres with a known tenacity. If the chemical reaction of such a solvated ion with a molecule in solution necessitates its partial exposure, it must rid itself of at least one of the solvent molecules forming its sheath, and present itself to the molecule it seeks to attack with only $c - 1$ solvent molecules around it. The ratio of the concentration, n^*, of such partly de-solvated ions to the concentration, n, of completely solvated ions is proportional to the Boltzmann factor $\exp(-\psi^*/kT)/\exp(-\psi/kT) = \exp[-(\psi^* - \psi)/kT]$, where ψ and ψ^* respectively are the energies of interaction of the normal and the activated ion with its solvent neighbours. If c and c^* denote the corresponding co-ordination numbers, and D the energy required to remove one solvent molecule to infinity from its equilibrium position around the ion, then n^*/n is proportional to $\exp[-(c - c^*)D/kT]$. A similar expression holds for the ratio of the concentration of partly de-solvated solute molecules to the total concentration. The rate of reaction is $dn/dt = Zn_1^*n_2^* = Zn_1n_2 \exp\{-[(c_1 - c_1^*)D_1 + (c_2 - c_2^*)D_2]/kT\}$, where Z is proportional to the collision frequency. The bimolecular velocity coefficient is then $k_2 = (dn/dt)/n_1n_2 = Z \exp\{-[(c_1 - $

$c_1^*)D_1 + (c_2 - c_2^*)D_2]/kT\}$ and, details being omitted, the true energy of activation, E, becomes

$$E = (c_1 - c_1^*)D_1 + (c_2 - c_2^*)D_2. \tag{19}$$

The energy required to deprive an ion of its sheath molecules is E_s/c_1, where E_s is the gain in energy when the ion is removed from solution. The energy required partly to desolvate the ion is

$$E_1 = E_s(c_1 - c_1^*)/c_1. \tag{20}$$

From molecular models of diffusion in solids, c_1^* may be taken as 4 when c_1 is 6, and as 3 when c_1 is 4. The energy term D_2 can be estimated from the dipole moments of the solute and solvent molecules and the permittivity D, of the solvent

$$D_2 = 2\left[\mu_1\mu_s\left(\frac{D + 2}{9D}\right)\right]^{3/2} . A^{-1/2} \tag{21}$$

where $A = 1.66 \times 10^{-82}$ erg-cm^9. Table XIII gives a comparison of observed energies of activation (E_A), in kcal/mole, with those calculated in this way.

TABLE XIII

OBSERVED AND CALCULATED VALUES OF THE APPARENT
ENERGY OF ACTIVATION (KCAL/MOLE) OF CERTAIN
SUBSTITUTION REACTIONS OF THE METHYL HALIDES IN WATER

Reactants	E_A (obs)	E (calc)
$CH_3Cl + F^-$	26.9 ± 0.5	25.9
$CH_3Br + F^-$	25.2 ± 0.5	25.8
$CH_3I + F^-$	25.2 ± 0.5	25.6
$CH_3I + Cl^-$	22.0 ± 0.3	23.5
$CH_3I + Br^-$	19.3 ± 0.5	20.9
$CH_3F + I^-$	22.9 ± 0.3	17.7
$CH_3Cl + I^-$	20.0 ± 0.3	17.8
$CH_3Br + I^-$	18.3 ± 0.1	17.7
$CH_3I + I^-$	18.1 ± 0.2	17.5
$CH_3F + OH^-$	21.6 ± 0.1	22.5
$CH_3Cl + OH^-$	24.3 ± 0.1	22.6
$CH_3Br + OH^-$	23.0 ± 0.15	22.6
$CH_3I + OH^-$	22.2 ± 0.23	22.3

A little reflection will show that such a simple hypothesis, attributing the whole of the activation energy to partial desolvation, can be logically applied only to reactions which are athermic or exothermic. Let us consider the reactions $CH_3X + CN^- \rightleftarrows CH_3CN + X^-$ in aqueous solution, X being a halogen atom. The increase in heat content attending the reaction in the gas phase is given by the expression

$$\Delta H_g^0 = D_{C-X} - D_{C-CN} + E_{CN} - E_X, \tag{22}$$

where D denotes bond energy, and E electron affinity (Table XIV). The corresponding increase for the reaction in solution is obtained from the equation:

$$\Delta H_s^0 = \Delta H_g^0 + \Sigma L_i - \Sigma L_j, \tag{3.22}$$

where ΣL_i is the sum of the heats of escape of the reactants from solution to the gas phase, and ΣL_j is the corresponding sum for the resultants. It is seen

TABLE XIV

THERMODYNAMIC EVALUATION OF ΔH_S^0 (KCAL) FOR THE REACTIONS
$CH_3X + CN^- \rightleftarrows CH_3CN + X^-$ IN AQUEOUS SOLUTION

	D_{CH_3X}	E_X	ΔH_g^0	L (escape of molecule)	L (escape of ion)	$\Sigma L_i - \Sigma L_j$	ΔH_s^0
CH_3F	116·3	82·5	14·9	4·4	109	−43·0	−28·1
CH_3Cl	76·0	87·8	−30·7	5·7	77	−9·7	−40·4
CH_3Br	63·3	82·1	−37·7	6·3	69	−1·1	−38·8
CH_3I	52·6	75·9	−42·2	6·4	60	+8·0	−34·2
CH_3CN	104·2	85·3	0	6·4	68	0	0

that these substitution reactions are strongly exothermic. The energy of activation, E_2, of the reverse reaction is consequently very large. For the reaction $CH_3CN + I^- \rightarrow CH_3I + CN^-$, for example, $E_2 = E_1 - \Delta H_s^0 = 20 \cdot 5 + 34 \cdot 2 = 55$ kcal. From the standard entropies of the four solutes, we thus derived the expression $k_2 = 4 \times 10^{10} . \exp(-55,000/RT)$. At 25°C, $k_2 = 1 \cdot 9 \times 10^{-30}$ litre/mole-sec. The normal value of the pre-exponential term for this extremely slow reaction is to be noted.

On applying the same considerations to the reactions of the iodide ion with CH_3F, CH_3Cl, CH_3Br and CH_3I in aqueous solution, Marshall (1965), using slightly different thermodynamic data, obtains the following values of ΔH_s^0 at 25°C: 4·4, 2·5, 2·2 and 0. These reactions are consequently slightly endothermic or athermic, and the simple desolvation hypothesis seems to be justified. In the schematic representation of the various heat effects shown in Fig. 2, the asterisks denote partially desolvated solutes and the quadrupole

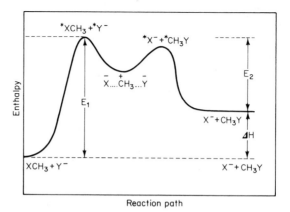

FIG. 2. Schematic representation of the role of partial desolvation (*).

is regarded as stable with respect to both pairs of partially desolvated ion plus molecule.

The Reaction Between Methyl Iodide and the Silver Ion in Aqueous Solution

As far as the magnitude of the Arrhenius parameters is concerned, there is nothing to distinguish between the reactions of anions and cations with the methyl halides (see Table XI). The mechanism, however, is radically different. With anions, we have the well known substitution, with inversion of the configuration, such as is shown in Fig. 3.

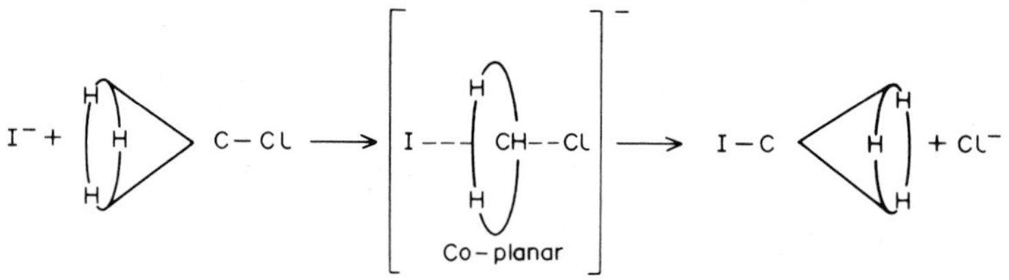

FIG. 3. From "The Kinetics of Reactions in Solution," by permission of The Clarendon Press, Oxford.

The chemical change between methyl iodide and the silver ion is represented quantitatively by the equation $CH_3I + Ag^+ + H_2O \rightarrow CH_3OH + AgI + H^+$, and there is no inversion of configuration:

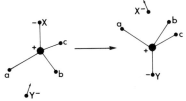

Positive mechanism : no steric hindrance:no optical inversion

Negative mechanism: steric hindrance : optical inversion

FIG. 4. The two types of substitution (after M. Polony; *Atomic Reactions*, Williams and Norgate, 1932).

The reaction has long presented some puzzling kinetic features. Burke and Donnan (1904, 1909) found it to be bimolecular during the course of each run, but the value of the bimolecular constant increased with an increase in the initial concentration of salt. Hammond, Hawthorne, Waters and Graybill (1960) concluded that the kinetic facts cannot be rationalized without postulating a role of the anions in the rate-determining step. Colclengh and Moelwyn-Hughes (1964) measured the rate of reaction analytically and conductimetrically, at various concentrations of silver nitrate and silver perchlorate, and, by addition of potassium nitrate, at various concentrations of silver ion in solutions of constant ionic strength. The fact that silver ions disappear at the same rate as hydrogen ions are formed proves that no methyl nitrate is formed—a conclusion consistent with the known rate at which this ester is hydrolysed (McKinley-McKee and E. A. Moelwyn-Hughes, 1952). The bimolecular constant extrapolated to zero ionic strength has the same value with silver nitrate and silver perchlorate (Fig. 5). To interpret the results, it has been postulated that binary and ternary complexes are formed, as follows:

$$CH_3I + Ag^+ \rightleftarrows (CH_3I,Ag)^+ ; K_\alpha$$

$$NO_3^- + CH_3I + Ag^+ \rightleftarrows (NO_3^-,CH_3I,Ag^+)^0 ; K_\beta,$$

where the equilibrium constants are given in terms of activities. The rate of reaction is the sum of two terms, each of which is assumed to be proportional to the concentration of complex, as in Brönsted and Davies' theories. Then in terms of concentrations, [ ], and activity coefficients, γ, we have

$$+\frac{d[H^+]}{dt} = k_\alpha K_\alpha [CH_3I][Ag^+]\left(\frac{\gamma_0\gamma_+}{\gamma_+}\right) + k_\beta K_\beta [CH_3I][Ag^+][NO_3^-]\frac{\gamma_0\gamma_+ + \gamma_-}{\gamma_0}.$$

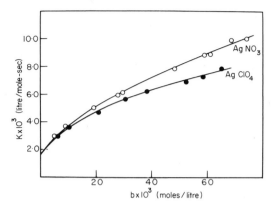

FIG. 5. Bimolecular constants for the reaction between methyl iodide and silver salts, at initial concentrations, b, in aqueous solution.

As usual, γ_0 is assumed to be unity. The observed bimolecular constant is thus

$$k_2 = k_2^0 + k_3[NO_3^-]\gamma^2, \tag{23}$$

where γ is the mean ionic activity coefficient. Because the reaction proceeds without change in ionic strength, γ remains constant during the course of any given run. If this formulation is correct, k_2 should vary linearly with respect to the product of $[NO_3^-]$ and the square of the mean activity coefficient. Values of the latter at 25°C are known (McInnes, 1939) and are shown, at rounded concentrations, in Table XV. The calculated values of k_2 are those reproduced by the equation

$$k_2 = 2.61 \times 10^{-3} + 1.72 \times 10^{-1}[NO_3^-]\gamma^2.$$

TABLE XV

THE INFLUENCE OF IONIC STRENGTH ON THE RATE OF REACTION
BETWEEN $AgNO_3$ AND CH_3I AT 25°C

[$AgNO_3$] (moles/litre)	γ	$k_2 \times 10^3$ (litres/mole-sec) observed	calculated
0	1·000	(2·61)	2·61
20	0·858	5·15	5·15
40	0·813	7·17	7·16
60	0·778	8·85	8·86
80	0·752	10·40	10·40

It follows that, at concentrations exceeding 0·021 molar, the number of molecules reacting per second by the termolecular mechanism exceeds the number reacting by the bimolecular mechanism. This can come about only if the energy of activation of the termolecular process is much lower than that of the bimolecular process. Such a lowering is to be expected if the anion, approaching the methyl end of the molecule, without reacting chemically with it, gives coulombic assistance in drawing the silver ion, approaching the halide end of the molecule, to within a critical distance of the iodine atom.

The Influence of the Solvent

Methyl bromide reacts with the iodide ion about 1,200 times more rapidly in acetone solution than in aqueous solution at 25°C. The difference in the E_A values accounts for a factor of 200 and that in the A_2 values for about 6.

TABLE XVI

THE INFLUENCE OF THE SOLVENT ON THE REACTION
$CH_3Br + I^- \rightarrow CH_3I + Br^-$

Solvent	$k_2 \times 10^4$ at 25°C (litres/mole-sec)	$\log_{10}A_2$	E_A (kcal/mole)
* H_2O	7·02	10·225	18·26 ± 0·13
† CH_3OH	9·42	10·354	18·25 ± 0·20
‡ $C_2H_4(OH)_2$	32·6	10·659	17·93 ± 0·26
§ $CO(CH_3)_2$	8,270	11·017	15·14 ± 0·40

REFERENCES

* Moelwyn-Hughes, *Trans. Chem. Soc.*, 779 (1938).
† *Idem.*, *Trans. Faraday Soc.*, **35**, 368 (1939).
‡ McKinley-McKee and Moelwyn-Hughes, *Trans. Chem. Soc.*, 838 (1952).
§ Moelwyn-Hughes, *Trans. Faraday Soc.*, **45**, 167 (1949); A. G. Evans and Hamann., *ibid.*, **47**, 30 (1951).

Hence the principal effect of a change in solvent is reflected in the energy of activation (Table XVI). This conclusion has been confirmed in the more extensive work of Swart and Le Roux (1957) on the reaction $CH_3I + *I^- \rightleftarrows CH_3I* + I^-$. They find that the bimolecular velocity coefficients in different solvents (Table XVII) can be reproduced, with a mean deviation of 3 per cent, by the equation $k_2 = A_2[\exp(-E_A/RT)]$ with a common value of A_2.

TABLE XVII

THE INFLUENCE OF THE SOLVENT ON THE REACTION $CH_3I + *I^- \rightarrow CH_3*I + I^-$

Solvent	$k_2 \times 10^4$ at 298·2°K (litres/mole-sec)	$A \times 10^{-10}$	E_A (kcal/mole)	$r \times 10^8$ (cm)
H_2O	4·85	0·90	18·1	3·16
$C_2H_4(OH)_2$	82·8	1·73	16·8	3·55
CH_3OH	77·3	0·42	16·0	2·46
C_2H_5OH	211	0·58	15·6	2·76
$(CH_3)_2CO$	64,000	5·00	13·5	4·73

Substitution by Monatomic and Diatomic Anions

It is noteworthy that the terms A_2 for reactions of the methyl halides with the methoxide ion in methanol solution (Table XVIII) and for reactions of

TABLE XVIII

Reaction	A_2 (obs)	$E = E_A + (1/2)RT$	$r \times 10^8$ (cm)
$OCH_3^- + CH_3Cl$	$1·76 \times 10^{12}$	23,990 cal	2·53
$OCH_3^- + CH_3Br$	$1·74 \times 10^{12}$	21,830 cal	2·76
$OCH_3^- + CH_3I$	$2·01 \times 10^{12}$	21,940 cal	3·03
$OCH_3^- + (CH_3)_2SO_4$	$2·53 \times 10^{12}$	20,820 cal	3·47

the methyl halides with the hydroxyl ion in water (Table X) should be from 200 to 340 times as great as for their reactions with the iodide ion (Table IX). A simple explanation is available by assuming that, with the hydroxide and methoxide ions, one internal degree of freedom more contributes to the activation energy than in the case of the I^-. We may therefore use equation (4.46) with $s = 2$:

$$k_2 = Z^0\left(1 + \frac{E}{RT}\right) . \exp(-E/RT).$$

If Z^0 is proportional to $T^{1/2}$, it follows that, approximately, $E_A = E - (1/2)RT$, and that

$$A_2 = Z^0(\tfrac{3}{2} + E_A/RT) . \exp(-1/2). \tag{24}$$

Using the observed values of A_2, we may solve for r (in Z^0). The critical separations thus found in methanol are given in Table XVIII (Hurst, 1957). Similar results are found for the reactions in aqueous solution when the attacking ion is OH^- or $S_2O_3^{--}$.

Electrostatic Considerations on the Solvent Effect

The theory of electrostatic effects on the kinetics of reactions between ions (charge $z_A\varepsilon$) and polar molecules (dipole moment of μ_B) in a medium of uniform permittivity, D, is treated systematically in Chapter 9, using the method suggested by Christiansen in his theory of interionic reactions (Chapter 7). The total energy of activation is regarded, at infinite dilution, as

$$E = E_n + E_e$$

$$= E_n + z_A\varepsilon\mu_B \cos\theta/Dr^2 = E_n + b\cos\theta,$$

where r is the distance between the centres of the ion and the dipole at the instant of reaction, and θ is the angle subtended between the directions of the polar axis and the line of centres (Fig. 9.1). By plotting $\ln k_2^0$ against $1/D$, the data of Table XVII show a linear dependence, with $r = 1\cdot15$ Å, which is unreasonably small. The corresponding plot against $(D + 2)/3D$ yields a more reasonable value of r. A real difficulty exists in deciding on the right value of D to use. Kacser (1952) has derived the following expression for the average value of the bimolecular constant by allowing for all angles of approach from zero to π:

$$k_2 = Z^0 . \exp\left(-E_n/RT\right)\frac{\sinh b}{b}$$

$$= Z^0 . \exp\left(-E_n/RT\right)\left(\frac{Dr^2kT}{2z_A\varepsilon\mu_B}\right)\left[\exp\left(\frac{z_A\varepsilon\mu_B}{Dr^2kT}\right) - \exp\left(-\frac{z_A\varepsilon\mu_B}{Dr^2kT}\right)\right]. \quad (23a)$$

In order to obtain the corresponding theoretical expression for the pre-exponential parameter of the Arrhenius equation, it is necessary to know the values of D and of their temperature coefficients. The necessary steps have been taken by Kacser, using values of D ranging from unity to the value for the pure solvent. Because Z^0 is proportional to r^2, the pre-exponential term for the reactions of univalent anions in media of low values of D varies as r^4. The most reasonable estimates of r found in this way are given in Table XVII.

Bimolecular Substitution Reactions of Methylene Halides in Solution

It is a matter of interest to find what effect the presence of a second halogen atom Y has on the rate at which the other halogen atom X is replaced by ion Z in reactions of the general type $YCH_2X + Z^- \rightarrow YCH_2Z + X^-$. As in our study of the methyl halides, we shall refer first to reactions carried out in acetone solution. The data of Table XIX on substitution reactions of the methylene compounds in acetone and in methanol are due to Hine, C. W. Thomas and Ehrenson (1955). The inorganic reactant used was sodium iodide, and the results were fitted to the equation of an uncomplicated, irreversible bimolecular reaction. The investigators found this to be satisfactory except with bromiodomethane, where k_2 fell as the reaction progressed. Equation (6.77) fitted the results more closely but the values calculated are not cited. The results for the reaction of methylene dichloride with sodium iodide, being based on two temperatues only, may be liable to considerable error. When it is recalled that no allowance has been made for the extent of ionization of sodium iodide or for its temperature variation, the data indicate clearly that here, as in other instances, the pre-exponential term, A_2, is reasonably constant for each series, and that the main effect of substitution is reflected in E_A.

Hine's results on the rate of reaction of the methyl halides with the methoxide ion in methanol solution agree with those of Hurst, cited in Table IX. In this set of reactions also, which are effectively irreversible and free from complications, change of rate is due, as previously noted, to changes in the activation energy.

TABLE XIX

BIMOLECULAR SUBSTITUTION REACTIONS OF METHYLENE HALIDES IN SOLUTION

Solvent	Reaction	$k_2 \times 10^7$ at 25°C (litre/mole-sec)	$\log_{10} A_2$	E_A (kcal/mole)
$(CH_3)_2CO$	$CH_3Br + I^-$	8,270,000	11·02	15·14 ± 0·8
	$FCH_2Br + I^-$	10,700	11·25	19·4 ± 1·2
	$ClCH_2Br + I^-$	1,290	11·88	21·5 ± 0·5
	$BrCH_2Br + I^-$	380	11·93	22·3 ± 1·2
	$CH_3Cl + I^-$	11,900	11·30	19·4 ± 0·8
	$ClCH_2Cl + I^-$	0·44	14·35	29·6
CH_3OH	$CH_3I + CH_3O^-$	2,570	12·11	21·4 ± 0·4
	$ClCH_2I + CH_3O^-$	17·0	12·56	25·0 ± 0·7
	$BrCH_2I + CH_3O^-$	5·50	10·03	22·2 ± 1·0
	$ICH_2I + CH_3O^-$	1·95	11·85	25·3 ± 2·0
	$CH_3Br + CH_3O^-$	2,950	11·94	21·1 ± 0·7
	$FCH_2Br + CH_3O^-$	1,290	12·03	21·7 ± 0·7
	$ClCH_2Br + CH_3O^-$	8·13	12·68	25·6 ± 0·4
	$BrCH_2Br + CH_3O^-$	1·23	12·15	26·0 ± 0·7
H_2O	$CH_3Cl + OH^-$	66·7	12·61	24·28 ± 0·12
	$ClCH_2Cl + OH^-$	0·22	11·57	26·22 ± 0·45

In aqueous solution, the chemical reaction between methylene dichloride and the hydroxyl ion gives a quantitative yield of formaldehyde (Fells and E. A. Moelwyn-Hughes, 1958) according to the equation $CH_2Cl_2 + 2OH^- \to 2Cl^- + CH_2(OH)_2 \rightleftarrows CH_2O + H_2O$, but allowance must be made for the concurrent reaction with the solvent, using equation (6.45), and for the Cannizzaro reaction $2CH_2O + OH^- \to H \cdot COO^- + CH_3OH$.

Steric Hindrance in Terms of Intermolecular Force Theory

Although substitution of any atom Y for a hydrogen atom in a methyl halide can result in a number of kinetic consequences, the most obvious one is that it makes the approach of an attacking ion more difficult by attempting to block the ion's passage to the seat of attack, as may be seen from Fig. 6.

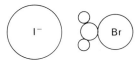

Reaction between the iodide ion and methyl bromide (one hydrogen atom not shown) in acetone solution. $E_A = 15 \cdot 1$ kcal/mole.

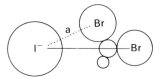

Reaction between the iodide ion and methylene dibromide (one hydrogen atom not shown) in acetone solution. $E_A = 22 \cdot 3$ k/mole.

FIG. 6. Schematic representation of steric hindrance.

The phenomenon is known as steric hindrance, and is not confined to substitution reactions at the saturated carbon atom. (See, for example, racemizations of substituted biphenyls, Chapter 10.) Hindrance is a household name for a strong repulsive force, and we naturally expect it to increase with the size of the hindering obstacle. Hine's data on the effect of substituents on the rate of reaction of the iodide ion with substituted methyl bromides show that the successive replacement of a hydrogen atom by atoms of fluorine, chlorine and bromine successively reduce the rate by factors of 773, 6,410 and 21,760 respectively, and increase the apparent energy of activation by 4·3, 6·4 and 7·2 kcal-mole^{-1}.

If the distance between the centres of the attacking ion, Z^-, and the substituted atom, Y, in the activated complex were known, it would be possible to estimate quantitatively the energy of repulsion. Such information is generally not available, but intermolecular force constants can be combined with experimental values of the repulsion energy to determine the critical distance, a, between Z^- and Y. We shall assume the energy of repulsion per pair to be given by the equation

$$\phi = A_{12}a^{-n},\tag{25}$$

where A_{12} for unlike pairs is given by Lennard-Jones' (1937) equation in terms of the A values of like pairs:

$$A_{12}^{1/n} = \tfrac{1}{2}(A_{11}^{1/n} + A_{22}^{1/n}).\tag{26}$$

Because of the similarity in the electronic structure of xenon and the iodide ion, and the similarity of the covalently bound fluorine, chlorine and bromine atoms with the electronic structures of neon, argon and krypton, respectively, we shall use the force constants of the inert elements, assuming that $n = 9$ (E. A. Moelwyn-Hughes, 1965). These yield the repulsion constants given in column 3 of Table XX. From Hine's data on the apparent energies of activation of the reactions between the iodide ion and three substituted methyl bromides in acetone solution, we obtain the steric energies given in column 2. The computed values of the critical separation, a (column 4), are seen to exceed the crystal radius of the iodide ion (2·23 Å) by distances which lie very near to the covalent radii of the F, Cl and Br atoms. These results,

TABLE XX

NUMERICAL COMPUTATION OF STERIC HINDRANCE IN CERTAIN REACTIONS
IN ACETONE SOLUTION

Reaction	E (repulsion) (kcal/mole^{-1})	$A_{12} \times 10^{81}$ (erg-cm^9/pair)	$a \times 10^8$ (cm)	$(a - r_{1-}) \times 10^8$ (cm)	Covalent radius (cm)
$I^- + FCH_2Br$	4·3	6·08 (Xe–Ne)	3·01	0·78	0·71 (F)
$I^- + ClCH_2Br$	6·4	17·68 (Xe–A)	3·24	1·01	1·01 (Cl)
$I^- + BrCH_2Br$	7·2	27·83 (Xe–Kr)	3·37	1·14	1·14 (Br)

necessarily obtained after making simplifying assumptions, warrant two conclusions: (1) intermolecular force constants obtained from the virial coefficients of gases suffice to explain the magnitude of the repulsion energies responsible for steric hindrance, (2) the identity of the critical separation with the sum of the ionic and covalent radii leaves no doubt that the attacking ion and the hindering atom are in intimate contact in the reactive complex, and are not separated by solvent molecules.

Similar conclusions may be drawn by comparing the apparent energies of activation for the reactions of methylene dichloride and methyl chloride with the hydroxyl ion in aqueous solution. We have $\Delta E = 2 \cdot 04 \pm 0 \cdot 57$ kcal-mole^{-1}. With $A = 6 \cdot 18 \times 10^{-81}$ (argon–argon), a becomes $3 \cdot 30 \pm 0 \cdot 10$ Å. This exceeds the sum of the ionic and covalent radii ($1 \cdot 81$ and $1 \cdot 01$ Å) of chlorine by $0 \cdot 48$ Å, which is too small to accommodate a water molecule.

In the reaction between the hydroxyl ion and chloroform, we anticipate the presence of the three chlorine atoms to provide a repulsive energy of about 6 kcal, which can reduce the rate of reaction by a factor of 25,000. The rate of reaction of methyl chloride with the hydroxyl ion at 25°C is already very slow ($k_2 = 6 \cdot 67 \times 10^{-6}$), and the estimated rate for the chloroform reaction becomes $2 \cdot 67 \times 10^{-10}$. Steric hindrance in this case is so great as to prohibit the reaction from proceeding by the usual mechanism, as illustrated in Fig. 7. The alternative is for the hydroxyl ion to approach the

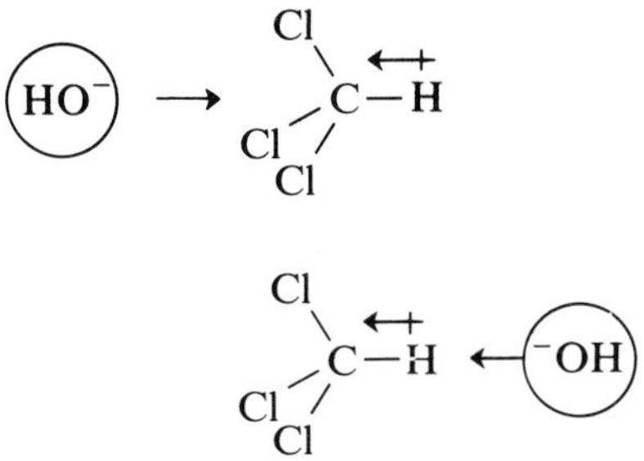

FIG. 7. The reaction between chloroform and the hydroxyl ion.
(7a) Steric hindrance of the three chlorine atoms, and the electrostatic field exerted by the C—H bond inhibit attack in the direction indicated.
(7b) There is no steric hindrance, and the field of the dipole facilitates ionic attack from this direction.

hydrogen atom directly, leading to its abstraction:

$$Cl_3CH + {}^-OH \rightleftarrows Cl_3C^- + H_2O; \text{ rapid,}$$

$$Cl_3C^- \rightarrow Cl_2C + Cl^-; \text{ rate-determining.}$$

The radical produced in the rate-determining step subsequently reacts with the hydroxyl ion to produce carbon monoxide and the formate ion (Hine, 1950; Hine, Burske, Hine and Langford, 1957; Horiuti and Tanabe, 1952; Fells and E. A. Moelwyn-Hughes, 1959):

$$CCl_2 + 2OH^- \rightarrow CO + 2Cl^- + H_2O,$$

$$CCl_2 + 3OH^- \rightarrow H \cdot COO^- + 2Cl^- + H_2O,$$

Energies of Deformation

The three hydrogen atoms in unsubstituted methyl halides exert, on an approaching anion, a force of repulsion similar in magnitude to the forces discussed above. The inversion depicted in Fig. 3 can take place only through the formation of an atomic arrangement of higher symmetry (Lowry, 1923) which implies that the hydrogen atoms in the activated complex lie in the same plane as the carbon atom (Penney, 1935). The energy required to fan out the three hydrogen atoms from their position in the stable methyl halide to the coplanar configuration of the critical complex may be estimated from bending force constants (Herzberg, 1945) and internuclear distances. It is seen from Table XXI that this energy is a considerable, and in these instances an approximately constant, fraction of the observed energy of activation.

TABLE XXI

DEFORMATION ENERGIES AND ACTIVATION ENERGIES FOR REACTIONS OF THE METHYL HALIDES WITH THE IODIDE ION IN ACETONE SOLUTION (KCAL/MOLE)

Molecule	$V - V^0 = (3/2)Kr^2_{C-H}(\theta - \theta_0)^2$	E_A	Ratio
CH_3F	9.95	22.19	0.449
CH_3Cl	7.75	19.43	0.399
CH_3Br	6.85	15.14	0.452
CH_3I	6.12	13.50	0.453
		Average	0.438

On confining attention to the same set of reactions, we may subtract from the observed energy of activation the contribution due to the fanning out, and divide the difference by the quantum of energy, $h\omega c$, associated with the fundamental vibration, ωc, of the carbon–halogen link. The resulting number is seen from Table XXII to be approximately 5. The mechanism of the excitation of vibrational energy is discussed in Chapter 13. In the present instances the variation in the activation energies is more likely to be due to classically computed changes in the polarity of the reactants and the complexes, which in turn affect the magnitude of the solute–solvent interactions.

TABLE XXII

ESTIMATED NUMBER OF VIBRATIONAL QUANTA CONTRIBUTING TO THE ENERGY OF ACTIVATION IN THE REACTIONS $CH_3X + I^-$ IN ACETONE SOLUTION (KCAL/MOLE)

Molecule	$E_A - (V - V^0)$	$N_0 h\omega c$	$[E_A - (V - V^0)]/N_0 h\omega c$
CH_3F	12·24	3·00	4·1
CH_3Cl	11·68	2·09	5·6
CH_3Br	8·29	1·75	4·7
CH_3I	7·38	1·53	4·8

Calculation of the Energy of Activation

An interesting and relatively early attempt has been made by R. Ogg and M. Polanyi (1935) to calculate the energy of activation, in acetone solution, of reactions of the type $CH_3Cl + I^- \rightarrow CH_3I + Cl^-$. The interaction energy of the iodide–methyl chloride system (Fig. 8) in terms of the internuclear

FIG. 8. From "Kinetics of Reactions in Solution," 2nd ed., by permission of The Clarendon Press, Oxford.

distances, r_1 and r_2, the dipole moment, μ, and dissociation energy, D, of the molecule, and of the polarizability, α, of the carbon atom, is given by the expression

$$E_1 = S_{I^-} + B_1 \cdot \exp\left(-r_1/b_1\right) - \frac{\varepsilon\mu_{CH_3Cl}}{(r_1 + \Delta)^2} - \frac{1}{2}\frac{\alpha\varepsilon^2}{r_1^4}$$

$$+ D_{CH_3Cl}\{1 - \exp\left[-a_2(r_2 - r_2^0)\right]\}^2, \tag{27}$$

where s_{I^-} is the work necessary to free the iodide ion from one of its solvent neighbours. The next three terms represent the total interaction energy of the ion–dipole system; and the last term is Morse's expression for the energy required to stretch the carbon–chlorine distance from its equilibrium value of r_2^0 to r_2. The electrical dipole is taken to be situated at the rim of the carbon atom, at a distance Δ from its nucleus, where Δ is the atomic radius. The interaction energy of the chloride ion–methyl iodide system (Fig. 9) is given by the complementary expression

$$E_2 = S_{Cl} + B_2 \cdot \exp\left(-r_2/b_2\right) - \frac{\varepsilon\mu_{CH_3I}}{(r_2 + \Delta)^2} - \frac{1}{2}\frac{\alpha\varepsilon^2}{r_2^4}$$

$$+ D_{CH_3I}\{1 - \exp\left[-a_1(r_1 - r_1^0)\right]\}^2. \tag{28}$$

The problem is to determine values of r_1, r_2, E_1 and E_2 such that r_1 and r_2 may simultaneous and independently have the same values in the two equations, while E_1 and E_2 are both equal and minimal. The solution is found graphically in the following manner. Selecting an arbitrary value of r_1, we represent E_1 as a Morse function of r_2, displaced vertically by an amount

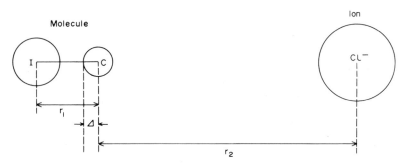

Fig. 9. From "Kinetics of Reactions in Solutions," 2nd ed., by permission of The Clarendon Press, Oxford.

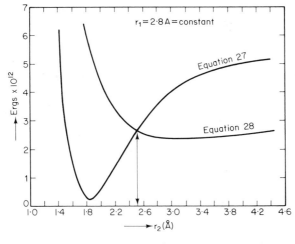

Fig. 10.

dependent on the constant value of r_1 chosen. Similarly, still with the same fixed value of r_1, we express E_2 as a Born–Mayer (Born and J. E. Mayer, 1932) function in r_2, displaced vertically by an amount dependent on the value of r_1 chosen. The type of curve found is shown in Fig. 10, where r_1 was kept at a constant value of 2·80 Å. The point of intersection of the two curves gives a "pass height" of $2·29 \times 10^{-12}$ ergs per molecule. Proceeding with a new value of r_1, we construct another pair of curves, and find another value for the energy corresponding to the intersection; the minimum value of this quantity is assumed to be the energy of activation.

Values thus calculated for the activation energies of certain reactions in acetone solution are compared in Table XXIII with the experimental values since obtained (Table III). Their order of magnitude and trend are correct.

TABLE XXIII

Calculated and Observed Energies of Activation (kcal/mole) in Acetone Solution

Reaction	E (calc)	E (obs.)
$CH_3I + F^- \rightarrow CH_3F + I^-$	31·6	22·2
$CH_3I + Cl^- \rightarrow CH_3Cl + I^-$	28·5	19·4
$CH_3I + Br^- \rightarrow CH_3Br + I^-$	26·5	15·1
$CH_3I + I^- \rightarrow CH_3I + I^-$	25·0	13·5

Two factors for which it has not been possible to make quantitative allowance are the decrease in the energy of the reactive complex due to resonance at the crossing point, and the increase in energy of the complex due to the spreading out of the hydrogen atoms into the carbon plane. Both energy contributions are regarded as small and the net effect as virtually nil.

The principal deficiency of the Ogg–Polanyi theory arises from their assumption that it is sufficient to consider the crossing of curves in the neighbourhood of the critical activation region only, without attempting to formulate the energy of the system as a continuous function of the variables r_1 and r_2. The assumption has led to some thermodynamically inconsistent conclusions. According to their equations, the initial and final energies are

$$E_1 \, (r_1 = \infty ; r_2 = r_2^0) = S_{1^-}$$

and

$$E_2 \, (r_2 = \infty ; r_1 = r_1^0) = S_{Cl^-} \, .$$

The standard gain in energy attending the reaction in solution should thus be

$$\Delta E_s = E_2 - E_1 = S_{Cl^-} - S_{1^-}, \qquad (29)$$

and the corresponding gain for the reaction in the gaseous phase should be $\Delta E_g = 0$. The treatment is thus confined to athermal systems. According to the principles illustrated in equation (22),

$$\Delta E_g = D_{C-Cl} - D_{C-I} - (E_{Cl} - E_I)$$

$$= 76 - 52 \cdot 6 - 87 \cdot 8 + 75 \cdot 9 = 11 \cdot 5 \, \text{kcal,}$$

and

$$\Delta E_s = \Delta E_g + (L_{CH_3Cl} - L_{CH_3I}) - (L_{Cl^-} - L_{1^-}). \qquad (30)$$

Only approximate estimates are available for the heats of escape of these four solutes from acetone. The term in the first brackets may be taken as the difference between the heats of vaporization, which is $-0 \cdot 7$. From the difference between the heats of dissolution of the potassium salts in acetone ($+1$ and -2 kcal), and from the constants quoted in Table XIV, the term in the second brackets is approximately -14. Hence $\Delta E_s = -3 \cdot 2$. The observed value is $+1 \cdot 44$ (Table I). In view of the large and speculative quantities used, the difference is not unreasonable.

A later attempt (Hurst and Moelwyn-Hughes, 1957) differs from the earlier one by referring the energy of the reactants and resultants to a common level, which is that of the isolated ions at infinite separation. The model is that of a linear combination of ions $X^- - CH_3^+ - Y^-$ which retains an ionic character throughout the change from $XCH_3 + Y^-$ to $YCH_3 + X^-$, *via* the critical state of the quadrupole $\bar{X} - \overset{+}{C}H_3 - \bar{Y}$. Following Born and Heisenberg (1924), the energy of a methyl halide molecule, considered as a pair of mutually polarizable univalent ions, may be expressed as follows in terms of the internuclear separation, a:

$$\phi = Aa^{-n} - Ba^{-4} - Ca^{-1}, \tag{31}$$

where A is a constant determining the repulsion between the electronic shells, C is ε^2, and B is given in terms of the polarizabilities of the ions as $(1/2)(\alpha_1 + \alpha_2)\varepsilon^2$. The dipole moment of such a molecule is given in terms of the inter-nuclear separation, a_e, by the equation

$$\mu = a_e\varepsilon[1 - (\alpha_1 + \alpha_2)/a_e^3], \tag{32}$$

and its energy of ionization in the gas phase is

$$D_e = [(n-4)/n]Ba_e^{n-4} + [(n-1)/n]Ca_e^{n-1}. \tag{33}$$

With methyl chloride, for example, for which $\mu = 1.86 \times 10^{-18}$ e.s.u., $a_e = 1.71 \times 10^{-8}$ cm, $\alpha_1(Cl^-) = 3.54 \times 10^{-24}$ cc, and $D_e = 227$ kcal/mole^{-1}, we find $\alpha_2(CH_3^+) = 0.33 \times 10^{-24}$ cc, and $n = 11.7$, which may be rounded to $n = 12$. Then $A = 1.79 \times 10^{-105}$ erg-cm^{12}, which is about one half of A for a Ne–Ne pair and about $\frac{1}{10}$ of A for the A–A pair, and is probably too low. For the reacting system shown in Fig. 11.

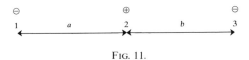

Fig. 11.

we have

$$\phi = A_{12}\, a^{-n_{12}} + A_{23}\, b^{-n_{23}} + A_{13}\, (a+b)^{-n_{13}}$$

$$-\left(\frac{1}{2}\right)\alpha_1\varepsilon^2\left[\frac{1}{a^2} - \frac{1}{(a+b)^2}\right]^2 - \left(\frac{1}{2}\right)\alpha_2\varepsilon^2\left[\frac{1}{a^2} - \frac{1}{b^2}\right]^2$$

$$-\left(\frac{1}{2}\right)\alpha_3\varepsilon^2\left[\frac{1}{b^2} - \frac{1}{(a+b)^2}\right]^2 - \varepsilon^2 a^{-1} - \varepsilon^2 b^{-1} + \varepsilon^2(a+b)^{-1}. \tag{34}$$

When the two anions are identical and the repulsion between them can be ignored, we have a simpler equation, which, when $a = b$, becomes

$$\phi = 2A_{12}\, a^{-n} - \tfrac{9}{16}\alpha\varepsilon^2 a^{-4} - \tfrac{3}{2}\varepsilon^2 a^{-1}. \tag{35}$$

This equation represents the energy of the symmetrical complex, and its value when $(d\phi/da)$ is zero gives the energy of the most stable configuration of the quadrupole. In the $\overset{-}{Cl}$—$\overset{+}{CH_3}$—$\overset{-}{Cl}$ system, this is found to be -291 kcal, compared with -227 for the initial system $CH_3Cl + Cl^-$. The energy of the complex, compared with that of the reactants, is accordingly -64, denoting considerable stability, as in the complex $F\overset{-}{H}F$. Similar values are found for other complexes $[X—CH_3—X]^-$. It thus appears that racemizations of this kind in the gas phase should occur without any energy of activation, except in endothermic reactions, when the energy of activation is simply the endothermicity. The role of the solvent appears to be that of hindering a reaction which in its absence would take place readily.

The energy of activation in the present instance is -64 plus the fanning energy of 8, i.e. -55 cal/mole. To obtain the energy of activation of the reaction in solution, allowance must be made for the heats of escape of CH_3Cl, Cl^- and $[Cl\cdots C\cdots Cl]^-$ from acetone solution, not all of which are known. Without a knowledge of the heat of escape of the complex, it is found that the trends in the calculated values of E_A are in the right order. The computed value for the CH_3Cl—Cl^- system is 14 kcal greater than for the CH_3Br—Br^- system, and the latter value lies near to E_A for the CH_3I—I^- system. It also becomes clear that the heat of escape of the symmetrical complex must be relatively small. This method is thought to be on the right lines, and capable of refinement. The replacement of an exponential for the inverse power expression for the repulsion energy is advocated by Kirkwood. A correction may also be needed for a change in the electronic contribution to the total energy.

Concurrent Substitution Reactions in Mixed Solvents

Many investigations have been carried out on substitution reactions in mixed solvents, the compositions of which have been expressed in terms of the volumes of the two pure solvents used to make the mixture. Ethanol is frequently added to water to provide a solvent in which organic solutes are usually more soluble. Seldom, however, have static measurements preceded the kinetic measurements. A notable exception is provided by Murto (1962) in his work, now to be considered, on the reactions of methyl iodide in alkaline water–methanol solutions.

The bimolecular velocity constants at 25°C for the reactions

$$CH_3I + HO^- \rightarrow CH_3OH + I^- \qquad \text{(in water)}$$

and

$$CH_3I + CH_3O^- \rightarrow CH_3OCH_3 + I^- \qquad \text{(in methanol)}$$

are $k_h^0 = 6\cdot35 \times 10^{-5}$ and $k_m^0 = 27\cdot75 \times 10^{-5}$ litre/mole-sec respectively. Murto finds that when methyl iodide reacts concurrently with both these anions in methanol–water mixtures, the observed bimolecular constant, k_2, reaches a maximum value of about 40×10^{-5} when the molar fraction of methanol is approximately 0·55 (Fig. 12). To find to what extent k_h and k_m

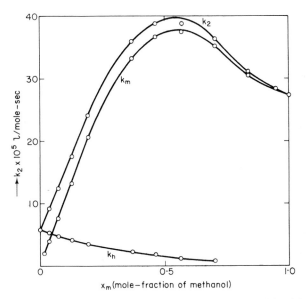

FIG. 12. Bimolecular constants, at 25°C, for the reaction of methyl iodide with the hydroxide and methoxide ions in methanol–water mixtures (data of Murto).

may vary with respect to the solvent composition he first considers the proton-transfer equilibrium $HO^- + CH_3OH \rightleftarrows H_2O + CH_3O^-$. In terms of the activities, a, of the solvent molecules, and the fraction, s, of the base which is present as methoxide ion, we have the equilibrium constant

$$K = \frac{a_w}{a_m} \cdot \frac{s}{1-s}, \qquad (36)$$

provided the activity coefficients of the anions are equal. Denoting by a the initial concentration of methyl iodide, by b the total initial concentration of base, and by x the concentration of iodide ion present at time t, we have the following kinetic scheme:

$$CH_3I + HO^- \rightarrow CH_3OH + I^-; k_h$$
$$\quad (a-x) \quad (b-x)(1-s) \quad x(1-s) \quad x$$

$$CH_3I + CH_3O^- \rightarrow CH_3OCH_3 + I^-; k_m.$$
$$\quad (a-x) \quad (b-x)s \quad xs \quad x$$

The instantaneous rates of reaction are

$$+d[CH_3OH]/dt = k_h(a-x)(b-x)(1-s),$$
$$+d[CH_3OCH_3]/dt = k_m(a-x)(b-x)s.$$

Hence, if the proton transfers are rapid, dimethylether and methanol are produced in a constant ratio, r, which, being independent of time, must also be the ratio of their concentrations at the end of the reaction:

$$\frac{[CH_3OCH_3]_\infty}{[CH_3OH]_\infty} = \frac{k_m}{k_h} \cdot \frac{s}{1-s} = r. \tag{37}$$

On combining equations (36) and (37), we have

$$\frac{k_m}{k_h} K = \frac{a_w}{a_m} \cdot r. \tag{38}$$

In terms of the molar fraction of methanol, the ratio of the activities may be expressed as follows:

$$\frac{a_w}{a_m} = \frac{1-x_m}{x_m} \cdot \exp(2x_m - 1)\alpha, \tag{39}$$

where α can be derived from the data of J. A. V. Butler, D. W. Thomson and W. H. Maclennan (1933) as 0·378. Murto has measured r by gas chromatography, over the complete range of solvent composition at various temperatures. By means of equations (38) and (39), the product $(k_m/k_h)K$ may thus be found: it increases rapidly as x_m is increased. If K, estimated from various sources as 4·5, is independent of the solvent composition, the values of the ratio k_m/k_h may be determined. From the kinetic expression

$$dx/dt = k_m(a-x)(b-x)s + k_h(a-x)(b-x)(1-s),$$

the observed bimolecular constant is

$$k_2 = k_h + (k_m - k_h)s. \tag{40}$$

On eliminating s from equations (37) and (40), we have

$$k_2 = \frac{k_h k_m(1+r)}{k_m + k_h r},$$

which may be rearranged to yield k_m from the known values of k_2, r and k_m/k_h:

$$k_m = \frac{k_2[(k_m/k_h) + r]}{(1 + r)}. \tag{41}$$

It is found that k_m also passes through a maximum (Fig. 12), and that k_h decreases as x_m is increased. The problem still remains of explaining why k_h and k_m should vary with the composition of the solvent.

REFERENCES

Acree, *Trans. Faraday Soc.*, **15**, 18 (1919).

Blokker, *Rec. Trav. Chim.*, **54**, 957 (1935).

Born and J. E. Mayer, *Z. Physik*, **75**, 1 (1932).

Born and Heisenberg, *Z. Physik*, **23**, 403 (1924).

Brown, H. F., and Cranston, *Trans. Chem. Soc.*, 578 (1940).

Burke and Donnan, *Trans. Chem. Soc.*, **85**, 555 (1904); *Z. physikal Chem.*, **69**, 148 (1909).

Butler, J. A. V., D. W. Thomson and W. H. Maclennan, *Trans. Chem. Soc.*, 674 (1933).

Colcleugh and A. E. Moelwyn-Hughes, *Trans. Chem. Soc.*, 2542 (1964).

Cox, *Trans. Chem. Soc.*, **113**, 666 (1918); **121**, 1904 (1922).

de la Mare, *Trans. Chem. Soc.*, 3180 (1955).

Evans, A. G., and S. D. Hamann, *Trans. Faraday Soc.*, **47**, 30 (1951).

Evans, C. C., and S. Sugden, *Trans. Chem. Soc.*, 270 (1949).

Farhat-Aziz, *Dissertation*, Cambridge (1959).

Farhat-Aziz and E. A. Moelwyn-Hughes, *Trans. Chem. Soc.*, 2635 (1959).

Farhat-Aziz and E. A. Moelwyn-Hughes, *Pakistan J. Sci.*, **14**, 16 (1962).

Fells and E. A. Moelwyn-Hughes, *Trans. Chem. Soc.*, 398 (1959).

Fells and E. A. Moelwyn-Hughes, *Trans. Chem. Soc.*, 1326 (1958).

Fowden, E. D. Hughes and C. K. Ingold, *Trans. Chem. Soc.*, 3187 (1955).

Gibson, Fawcett and Williams, *Proc. Roy. Soc.*, **A**, **150**, 223 (1935).

Hammond, Hawthorne, Waters and Graybill, *J. Amer. Chem. Soc.*, **82**, 704 (1960).

Hardwick, *Trans. Chem. Soc.*, 141 (1935).

Hasty, R. A., *J. Chem. Physics*, **73**, 319 (1969).

Hecht, Conrad and Brückner, *Z. physikal Chem.*, **5**, 289 (1890).

Herzberg, *Infra-Red and Raman Spectra of Polyatomic Molecules*, Van Nostrand, New York, 1945.

Hine, *J. Amer. Chem. Soc.*, **72**, 2438 (1950).

Hine, Burske, Hine and Langford, *J. Amer. Chem. Soc.*, **79**, 1406 (1957).

Hine, C. W. Thomas and Ehrenson, *J. Amer. Chem. Soc.*, **77**, 3886 (1955).

Horiuti and Tanabe, *Proc. Japan. Acad. Sci.*, **28**, 127 (1952).

Hughes, E. D., C. K. Ingold and J. D. H. Mackie, *Trans. Chem. Soc.*, 3173 (1955).

Hughes, E. D., C. K. Ingold and J. D. H. Mackie, *Trans. Chem. Soc.*, 3177 (1955).

Hughes, E. D., C. K. Ingold and A. J. Parker, *Trans. Chem. Soc.*, 4400 (1960).

Hurst and E. A. Moelwyn-Hughes, *Academia Nazionale dei Leici*, Varenna (1957).

Kacser, *J. Phys. Chem.*, **56**, 1101 (1952).

Lennard-Jones, *Physica*, **IV**, **10**, 947 (1937).

Lowry, *Trans. Chem. Soc.*, 828 (1923).
Madsen, *Z. physikal Chem.*, **36**, 290 (1901).
Marshall and E. A. Moelwyn-Hughes, *Trans. Chem. Soc.*, 2640 (1959); 7119 (1965).
McBain and Coleman, *Trans. Faraday Soc.*, **15**, 27 (1919).
McInnes, *The Principles of Electrochemistry*, Reinhold, New York (1939).
McKinley-McKee and E. A. Moelwyn-Hughes, *Trans. Faraday Soc.*, **48**, 247 (1952).
Mitchell, *Trans. Chem. Soc.*, 1792 (1937).
Moelwyn-Hughes, E. A., *Physical Chemistry*, 2nd end., Pergamon Press, 1965.
Moelwyn-Hughes, E. A., *Trans. Chem. Soc.*, 779 (1938).
Moelwyn-Hughes, E. A., *Chem. Rev.*, **10**, 241 (1932).
Moelwyn-Hughes, E. A., *Trans. Faraday Soc.*, **35**, 368 (1939).
Moelwyn-Hughes, E. A., *Trans. Faraday Soc.*, **45**, 167 (1949).
Murto, *Ann. Acad. Scient. Fennicae*, **A, II**, 117 (1962).
Ogg, R., and M. Polanyi, *Trans. Faraday Soc.*, **31**, 604 (1935).
Penney, *Trans. Faraday Soc.*, **31**, 734 (1935).
Robertson, H. C., and Acree, *J. Amer. Chem. Soc.*, **37**, 1902 (1915)—who refer to twenty
    papers of the same series.
Ross Kane, *Ann. Reports*, **27**, 351 (1930).
Serkov, *J. Russ. Phys. Chem.*, **40**, 413 (1908).
Swart and Le Roux, *Trans. Chem. Soc.*, 406 (1957).
Winstein, Savedoff, S. Smith, J. D. R. Stephens and Gall, *Tetrahedron Letters*, **9**, 24
    (1960).
Woolf, *Trans. Chem. Soc.*, 1172 (1937).

9

IONS AND POLAR MOLECULES; AROMATIC SUBSTITUTION

The reactions between ions and the methyl and methylene halides, studied in Chapter 8, constitute but a small fraction of reactions which have been examined between ions and other solutes—chiefly polar molecules—in solution. With the methyl halides the complications were relatively few, namely reaction with the solvent, reversibility, the partial ionization of the electrolyte and its limited solubility. These are present in reactions of ions with aryl halides and with higher alkyl halides which, moreover, can undergo eliminations, such as $RCH_2CH_2X \rightarrow RCH:CH_2 + H^+ + X^-$. Finally, substitution reactions can, as we have seen, take place by two distinct mechanisms. To limit the scope of our inquiry, we shall, as far as possible, deal with the kinetics of some representative ion–dipole reactions in pure solvents.

The Reproducibility and Constancy of the Apparent Energy of Activation

The values of E_A originally given for the reaction between ethyl iodide and sodium ethoxide in ethanol solution, and for the enolization of acetone in aqueous solution were 20,880 and 20,580 cal respectively.* Accurate repetitions gave 20,980 (Gibson, Fawcett and Perrin, 1935) and 20,680 respectively (G. F. Smith, 1934). Different workers can thus reproduce activation energies to within ± 50 cal of the mean. Over a temperature

---

* *The Kinetics of Reactions in Solution*, pp. 34 and 47, 1st Ed., 1933.

range of 40°C, a value of ± 2.5 cal/mole-deg for dE_A/dT thus lies within the limits of reproducibility. Most reactions between ions and polar molecules have values of dE_A/dT within these limits. There have appeared some exceptions which, after logical resolution of the bimolecular constants, have been shown to be apparent only. We shall consider one example.

The bimolecular character of the reaction between hydriodic acid and hydrogen peroxide, $2H^+ + I^- + H_2O_2 \rightarrow 2H_2O + I_3^-$, and the accelerating influence of acids have long been established (Harcourt and Esson, 1867).

TABLE I

[NaI] $\times 10^3$ (gramme mole/litre)	$k_{25°C} \times 10^3$ (min^{-1}; log$_e$)	$\dfrac{k_{obs}}{[NaI]}$
6·16	8·13	1·32
9·21	11·99	1·30
12·26	16·21	1·32
18·40	24·19	1·31
24·52	31·95	1·30
30·60	39·47	1·29
36·80	48·10	1·31

In neutral solutions, the unimolecular velocity constant governing the decomposition of hydrogen peroxide is proportional to the concentration of iodide ion, as shown in Table I (Walton, 1904). The average bimolecular constant is 1.31 ± 0.02 in the units adopted. With potassium iodide, it is 1.33 ± 0.01. There is thus no marked specific cation effect. When Walton's data at other temperatures are examined, it is found that the apparent energy of activation decreases as the temperature is raised: $dE_A/dT = -88 \pm 22$ cal/mole-deg. While such an effect has since proved to be common in solvolytic reactions, it is rarely found in bimolecular reactions. The effect is here due to the failure to allow for concurrent processes, characterized by measurable velocity coefficients, k_j:

$$-d[H_2O_2]/dt = k_2[H_2O_2][I^-] + \Sigma k_j[H_2O_2][J],$$

where J denotes a catalyst other than the iodide ion (Magnanini, 1891; Noyes, 1896; Brode, 1904; Abel, 1908, 1920, 1928; Harned, 1918). When, for example, J is OI^-, k_j at 25°C is 2×10^{11} litre/mole^{-1}-sec^{-1} (Liebhafsky, 1932), indicating, as the investigator realized, that each collision between the hydrogen peroxide molecule and the hypoiodous ion is chemically effective. More important, from the present point of view, is the fact that, when correction has been made for the incidence of concomitant reactions, the energy of activation, E_2, is constant. Christiansen's data (1925) give $E_2 = 13{,}430$ cal. Between 0 and 50°C, $E_2 = 13{,}400 \pm 105$ cal (Liebhafsky and Mohammad, 1933; Hender and R. A. Robinson, 1933). Thus, from the catalytic coefficient k_2, we find $A_2 = 1{\cdot}5 \times 10^8$ in the standard units.

Homologous Reactions

When a given ion in a given solvent reacts with monohalogeno-substituted straight-chained hydrocarbons, the highest rate constant and the lowest energy of activation are those of the first member, CH_3X. Reaction with the second member, C_2H_5X, is usually slower by a factor of about 10, and the energy of activation is higher by about 1 kcal. Lengthening the chain further has only a slight effect. Thus, for example, normal butyl and hexadecyl halides react at nearly the same rate. Table II provides some instances (Segallar, 1913; Haywood, 1922). Lengthening the chain by substituting an oxygen atom or a sulphur atom for a methylene group at a point removed from the seat of attack has likewise only a minor effect (Conant and Kirner, 1924; Conant and Hussey, 1925). The explanation is obvious; substitution of any atom or group for a hydrogen atom attached to the carbon atom which is being attacked has a greater effect, positive or negative, than a remote substitution. It follows that, in straight-chained homologous reactions, a roughly constant energy of activation is found for all the members

TABLE II

HOMOLOGOUS REACTIONS IN ETHANOLIC SOLUTION

R	RI + C_6H_5ONa			RI + $C_6H_5CH_2ONa$		
	$k_{42\cdot5°C} \times 10^5$	E_A	$A \times 10^{-11}$	$k_{50°C} \times 10^5$	E_A	$A \times 10^{-11}$
CH_3	101	22,120	22·1	266	20,570	20·65
C_2H_5	22·5	22,000	4·00	23·70	21,860	15·67
C_3H_7	8·67	22,450	3·45	9·24	21,730	4·86
C_4H_9	8·08	22,090	1·65	6·69	21,560	2·92
C_7H_{15}	7·52	22,230	0·96	6·68	21,510	2·48
C_8H_{17}	7·23	22,500	1·37	6·68	21,450	2·21
$C_{16}H_{33}$	7·15	22,430	1·93	6·90	21,090	1·26
iso C_3H_7	7·58	22,100	2·88	5·52	21,410	1·75
iso C_4H_9	3·22	21,790	2·31	8·68	21,350	2·45
iso C_5H_{11}	4·75	22,250	1·56	3·39	21,410	1·07
sec. C_6H_{13}	7·22	22,190	0·41	9·61	21,580	4·01
sec. C_8H_{17}	6·85	22,110	1·28	7·25	21,450	2·40

TABLE III

GROUP ENERGIES OF ACTIVATION FOR STRAIGHT-CHAINED HOMOLOGOUS REACTIONS

Reaction	Solvent	Temperature Range (°C)	E_A (cal/gramme-mole)
$ROOC \cdot CH_3 + OH^-$	H_2O	40	$12{,}300 \pm 300$
$ROOC \cdot CH:CH \cdot CH_3 + OH^-$	H_2O	40	$12{,}720 \pm 150$
$ROOC \cdot CH_3 + H_3O^+$	H_2O	30	$16{,}880 \pm 50$
$RCl + I^-$	$(CH_3)_2CO$	10	$18{,}900 \pm 300$
$RCONH_2 + H_3O^+$	H_2O	40	$19{,}200 \pm 300$
$RI + C_2H_5O^-$	C_2H_5OH	30	$19{,}950 \pm 700$
$RI + C_6H_5CH_2O^-$	C_2H_5OH	30	$21{,}500 \pm 400$
$RI + C_6H_5O^-$	C_2H_5OH	50	$22{,}200 \pm 350$

after the first two (Table III). Bifurcation of the chain leads to a lowering of the rate of reaction in the instances cited in Table II. Replacement of the three hydrogen atoms in CH_3X by three methyl groups, to yield $C(CH_3)_3X$, usually has a pronounced effect, positive or negative, on the bimolecular constant, and may lead to a change of mechanism.

Some General Features

The general characteristics of reactions between ions and polar molecules in hydroxylic solvents appear from an inspection of the data given in Table IV. Except for the saponification of esters in water, the constant A_2 is nearly the same for all the reactions. Comparison of the first entry with the data of Table 8.X shows that the substitution of the carboxyl group for a hydrogen atom in the reaction $CH_3Cl + I^- \rightarrow CH_3I + Cl^-$ has doubled the rate constant and lowered the energy of activation by 0·25 kcal. The data cited on the saponification of ethyl acetate confirm those of Reicher (1885, 1887) and Warder (1881), according to whom $\log_{10} A_2 = 7 \cdot 22 \pm 0 \cdot 07$, and $E_A = 11{,}205 \pm 5$ cal. For reactions in this category, $\log_{10} A_2$ is seen to be lower by about 3 than the value found for the other reactions. It has been suggested that at least part of the explanation lies in the difference between the types of collision involved. If the solvent acts as a third partner, an expression of the form of equation (5.52) appears to be applicable. The fourth entry in Table IV shows that the replacement of the ethyl group, C_2H_5, by the tertiary butyl group, $(CH_3)_3$, brings about a marked decrease in the rate of reaction, in contrast to its effect in other reactions to be discussed later. When Steger's data in Table IV are compared with the results discussed in Chapter 8, we see that, as far as the magnitude of the two Arrhenius parameters are concerned, there is no significant difference between the kinetics of substitution in

TABLE IV

ARRHENIUS CONSTANTS FOR SOME REPRESENTATIVE REACTIONS BETWEEN IONS AND POLAR MOLECULES. k_2 IN LITRE/MOLE-SEC. E_A IN CAL/MOLE

Reaction	Solvent	k_2 (25°C)	$\log_{10} A_2$	E_A	Reference
$CH_2Cl \cdot COOH + I^-$	H_2O	$4 \cdot 07 \times 10^{-5}$	10·11	19,770	*
$CH_3COOC_2H_5 + OH^-$	H_2O	$1 \cdot 13 \times 10^{-1}$	7·27	11,210	†
$CH_3COOC_2H_5 + OH^-$	H_2O	$1 \cdot 11 \times 10^{-1}$	7·38	11,370	‡
$CH_3COOC(CH_3)_3 + OH^-$	H_2O	$2 \cdot 14 \times 10^{-3}$	6·89	13,040	†
$(CH_3)_2SO_4 + NCS^-$	CH_3OH	$2 \cdot 95 \times 10^{-3}$	10·58	17,880	§
$oC_6H_4(NO_2)_2 + CH_3O^-$	CH_3OH	$4 \cdot 90 \times 10^{-4}$	11·79	20,590	‖
$C_6H_5CH_2Cl + C_2H_5O^-$	C_2H_5OH	$3 \cdot 72 \times 10^{-5}$	10·16	19,900	¶
$C_3H_5Cl + C_2H_5O^-$	C_2H_5OH	$1 \cdot 91 \times 10^{-5}$	10·11	20,220	¶
$CH_3I + oCH_3C_6H_4O^-$	C_2H_5OH	$1 \cdot 86 \times 10^{-4}$	11·11	20,240	**
$CH_3I + mCH_3C_6H_4O^-$	C_2H_5OH	$2 \cdot 09 \times 10^{-4}$	11·36	20,510	**
$CH_3I + pCH_3C_6H_4O^-$	C_2H_5OH	$2 \cdot 40 \times 10^{-4}$	11·93	21,220	**

REFERENCES

* Wagner, *Z. physikal. Chem.*, **115**, 121 (1925).
† L. Smith and Olsson, *ibid.*, **118**, 99 (1925).
‡ Tommila, Koivisto, Lyyra, Antell and Heimo, *Ann. Acad. Sci., Fennicae*, II, **47**, 1 (1952).
§ Walden and Centnerszwer, *Z. Elektrochem.*, **15**, 310 (1909).
‖ Steger, *Z. physikal. Chem.*, **49**, 329 (1904).
¶ Hecht, Conrad and Brückner, *ibid.*, **4**, 273 (1889).
** Conrad and Brückner, *ibid.*, 7, 274 (1891).

aliphatic and aromatic reactions, such as $CH_3X + Y^-$ and $XC_6H_4X + Y^-$. Finally, we note that the relative rates at which methyl iodide reacts with the three cresolate ions stand in the ratios

$$o : m : p : : 1 \cdot 00 : 1 \cdot 15 : 1 \cdot 29.$$

To the intriguing and much discussed problem of the *ortho, meta* and *para* sequence we shall return later. In the meantime, it should be emphasized that the rate constants cited here are the ever-doubtful "stoichiometric second order" constants. No allowance has been made for the difference in the ionization constants of the bases, on which ultimately a good deal of the differences may depend.

Substitution and Inversion of Configuration

Emil Fischer's (1911) conclusion that "the Walden inversion seems to be a general phenomenon which stands in intimate relationship to the nature of substitution processes" has, from time to time, been re-stated in various forms (Werner, 1912; Meisenheimer, 1927; Meer and Polanyi, 1932; Olson, 1933). Its application to certain simple systems has been illustrated in Figs. 3 and 4 of Chapter 8, where it is seen that substitution by an anion leads to an inversion of configuration and a consequent change in optical rotatory power, while substitution by a cation leaves the configuration and optical rotatory power unchanged. There have been several experimental confirmations of the identity of the rate at which a negative ion reacts with a molecule possessing an asymmetrically saturated carbon atom and the rate of optical inversion.

Szabo (1933) dissolved ordinary lithium iodide in an acetone solution of levorotatory methylpropyliodomethane, and measured the rate of change of optical rotatory power. The first-order constant given by equation (6.19) was found to be, in dilute solutions, proportional to the concentration of iodide ion, so that the reaction is essentially a bimolecular one, represented schematically as follows:

$$
\underset{\underset{\text{C}_3\text{H}_7}{|}}{\overset{\overset{\text{H}}{|}}{\text{I}-\text{C}-\text{CH}_3}} + \text{I}^- \rightleftarrows \underset{\underset{\text{C}_3\text{H}_7}{|}}{\overset{\overset{\text{H}}{|}}{\text{CH}_3-\text{C}-\text{I}}} + \text{I}^-
$$

His data may be compared with those later found for a simpler system by Swart and Le Roux (Table 8.III) in the same solvent:

Reaction	k_2 (25°C)	$\log_{10} A_2$	E_A
$(\text{CH}_3)(\text{C}_3\text{H}_7)\text{CHI} + \text{I}^-$	1.33×10^{-3}	9.81	17,300
$\text{CH}_3\text{I} + \text{I}^-$	1.02×10^1	10.90	13,500

The ratio of the rate constants is 7,700, of which the difference in the A_2 values contributes about 12. The difference in rates is thus due mainly to the difference in the E_A terms.

Olson and Young (1936) have measured polarimetrically, in aqueous solution, the rates of reactions of the type

$$
\underset{\underset{\underset{\text{COOH}}{|}}{\overset{\text{CH}_2}{|}}}{\overset{\overset{\text{H}}{|}}{\text{Br}-\text{C}-\text{COOH}}} + \text{Br}^- \rightleftarrows \underset{\underset{\underset{\text{COOH}}{|}}{\overset{\text{CH}_2}{|}}}{\overset{\overset{\text{H}}{|}}{\text{HOOC}-\text{C}-\text{Br}}} + \text{Br}^-
$$

TABLE V

OPTICAL INVERSIONS AND RACEMIZATIONS IN AQUEOUS SOLUTION

Reaction	$A \times 10^{-10}$	E_A
l-Bromsuccinic acid $+ Br^-$	1·49	21,800 $\pm$ 50
l-Chlorsuccinic acid $+ Cl^-$	0·842	24,770 $\pm$ 150
l-Chlorsuccinic acid $+ Br^-$	0·624	23,540 $\pm$ 150
l-Bromsuccinic acid $+ Cl^-$	3·95	23,650 $\pm$ 40

with the results summarized in Table V. Similar kinetic effects have been found with certain α-halogeno-phenylacetic acids (Olson and Long, 1936). The rates of reactions of the type

$$\underset{\underset{C_2H_5}{|}}{\overset{\overset{H}{|}}{I-C-CH_3}} + *I^- \rightleftarrows \underset{\underset{C_2H_5}{|}}{\overset{\overset{H}{|}}{CH_3-C-*I}} + I^-$$

have been measured by following the rate of inversion polarimetrically and the rate of substitution by the removal of samples from time to time, and, after separating the reactants estimating the radioactivity by counts. Within the limits of experimental errors, the two rates at a common temperature coincide (Hughes, Juliusberger, Scott, Topley and Weiss, 1936). The same conclusion has been reached by Olson, Frashier and Spieth (1951) from their measurements of the rates of racemization and substitution of the di-methyl ester of l-bromsuccinic acid in anhydrous acetone. The effect of dilution is accurately given by equation (8.15), with k_i/k_m having an unusually low value of 21 at 24·9°C.

Electrostatic Considerations

From our discussion of the kinetics of the reactions between ions and the methyl halides, it is clear that the energy of activation consists of a number of contributions attributable to a variety of effects—bond stretching and bending, induction, resonance, intrinsic repulsion, solvent re-organization around all the solutes, and various electrostatic terms. Experimentalists persist in the hope that conditions can be so chosen as to enable them to isolate the various effects. While acknowledging the complexity of the problem, we shall here seek to see how far simple electrostatic considerations can lead us. We shall therefore assume, as in our treatment of reactions between ions, that the total energy of activation for a given reaction can be resolved into two components,

$$E = E_n + E_e,\qquad(1)$$

where E_e is the electrostatic contribution, the principal part of which can be ascribed to the direct interaction between the charge, $z_A \varepsilon$, on the ion, and the dipole moment, μ_B, of the bond which it attacks. We shall adopt the following approximate expression for E_e:

$$E_e = \frac{z_A \varepsilon \mu_B \cos\theta}{Dr^2}(1 - \kappa^2 r^2),\qquad(2)$$

where D is the permittivity, r is the distance between the charge on the ion and the centre of the dipole in the critical complex, and θ is the angle subtended between the line of centres and the polar axis, reckoned positive from the positive end of the dipole, as in Fig. 1 (Spong, 1934; Waters, 1929;

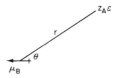

Fig. 1.

E. A. Moelwyn-Hughes, 1936). The Debye-Hückel term κ for uni-univalent electrolytes, is defined as follows in terms of the concentration, c, in moles per litre, the protonic charge, ε, and the Avogadro number, N_0:

$$\kappa^2 = \left(\frac{8\pi N_0 \varepsilon^2}{1{,}000 DkT}\right)c.\qquad(3)$$

The expression for the bimolecular constant thus becomes

$$k_2 = Z^0 \exp(-E/kT)$$

$$= Z^0 \exp(-E_n/kT) \cdot \exp\left[\frac{-z_A \varepsilon \mu_B \cos\theta}{Dr^2 kT}(1 - \kappa^2 r^2)\right],\qquad(4)$$

where Z^0 has the dimensions of a standard collision frequency. The effect of dilution on the rate constant should then be an exponential one, since

$$k_2 = k_2^0 \cdot \exp\left(\frac{+z_A \varepsilon \mu_B \cos\theta}{DkT}\right)\kappa^2,\qquad(5)$$

which reduces to a linear relationship at low ionic strengths:

$$k_2 = k_2^0 \left[1 + \left(\frac{z_A \varepsilon \mu_B \cos \theta}{DkT} \right) \kappa^2 \right]. \tag{6}$$

While these conclusions agree with the empirical equation of Grube and Schmidt (1926), it is seldom, after the correction for ionization has been made, that there remains a genuine electrolyte effect. On the whole it appears that, in conformity with the theory of Brönsted and Bjerrum, the bimolecular constant for most reactions between ions and polar molecules in dilute solution is independent of the electrolyte concentration, as found by Arrhenius (1887) and Burgarsky (1891) in the reaction between ethyl acetate and alkali hydroxides. More general treatments have been advanced for the effect of ionic strength on the rates of reactions between ions and polar molecules (Cross, 1935; Scatchard, 1932).

The relationship between the observed energy of activation and the non-electrostatic energy, E_n, is obtained as in Chapter 7:

$$E_A = E_n + \frac{z_A \mu_B \cos \theta}{Dr^2} (1 - LT). \tag{7}$$

In attacks by negative ions in hydroxylic solvents both z_A and $(1 - LT)$ are negative, so that plots of E_A against $(1/D)$ should be positive; and this, as a rule, is found to be the case, although the value of r estimated from the gradient is generally too small, e.g. 0.73 Å for the reaction $CH_3Br + I^- \rightarrow CH_3I + Br^-$ in various solvents.

By eliminating E_n from equations (4) and (7), we find that

$$k_2^0 = Z^0 . \exp \left[-(z_A \mu_B \cos \theta / Dr^2 k)L \right] . \exp \left(-E_A/kT \right). \tag{8}$$

The central expression on the right hand side may be identified with the efficiency of collision term $P \ (= A/Z^0)$. The calculated values of P with $r = 3$ Å, θ varying from 0 to π, z_A ranging from $+2$ to -2, and D as for water and methanol, range from 0.16 to 6.42, in agreement with most of the experimental values given in Tables II and IV.

For reacting systems within the same range of conditions, the theoretical values of

$$\frac{dE_A}{dT} = -\frac{L^2 T z_A \mu_B \cos \theta}{Dr^2} \tag{9}$$

range from zero to $+2.65$ cal/mole-deg, which, as we have seen, again lie within the experimental error.

It remains to show to what extent the electrostatic treatment of the reactions between ions and polar molecules in its present simple form can explain the effects of changes in the permittivity of the solvent. Writing equation (4) (as applied to solutions of zero ionic strength) as follows

$$\ln k_2^0 = \ln Z^0 - \frac{E_n}{kT} - \frac{z_A \mu_B \cos \theta}{Dr^2 kT},$$ (10)

we see that

$$\left[\frac{d \ln k_2^0}{d(1/D)} \right]_T = \frac{-z_A \mu_B \cos \theta}{r^2 kT}$$ (11)

according to which, for the most favourable approaches (i.e., z_A negative and $\theta = 0$), the plot of $\ln k_2$ against $1/D$ should be linear and positive. Experiments show that the slope is usually positive, but that linearity is limited to solvents with relatively high values of D. An example is shown in Fig. 2

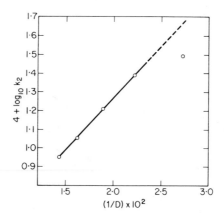

FIG. 2. The influence of the permittivity of the solvent on the rate of reaction between sodium hydroxide and methyl iodide in water-dioxane solution at 323·41°K.

(Moelwyn-Hughes and Hurst, 1951), from which $(r^2/\cos \theta)$ is found to be $1·31 \times 10^{-16} \, \text{cm}^2$. As with the corresponding evaluation of the critical separation between reacting ions, values of r found in this way are frequently too small. More reasonable results are found from data for a given reaction in a variety of pure solvents. The results of Swart and le Roux, for example, on the reaction $CH_3I + *I^- \rightleftarrows CH_3I* + I^-$ in five solvents (Table 8.XVII) yield the reasonable value of $r = 2·98 \, \text{Å}$.

An alternative expression for the electrostatic component of the activation energy has been obtained by Laidler and Eyring (1940), Laidler (1965), using a method suggested by Scatchard (1932). At infinite dilution the electrostatic energy of a sphere of radius r_A, with charge $z_A \varepsilon$, is $z^2 \varepsilon^2 / 2Dr_A$, and that of an uncharged solute which is spherical and has a point dipole of moment μ_B at its centre is

$$-\frac{\mu_B^2}{r_B^3}\frac{D-1}{2D+1}$$

(Bell, 1931; Kirkwood, 1934). Laidler and Eyring regard the ion-dipole complex also as a sphere, on which resides the same charge $z_A\varepsilon$ as on the free ion, and at the centre of which the point dipole is situated. The electrostatic component of the energy of activation is the difference between the electrostatic energy of the charged complex and the sum of the electrostatic energies of the ion A and the molecule B, i.e.

$$E_e = \frac{z_A^2\varepsilon^2}{2Dr_x} - \frac{D-1}{2D+1}\frac{\mu_x^2}{r_x^3} - \frac{z_A^2\varepsilon^2}{2Dr_A} + \frac{D-1}{2D+1}\frac{\mu_B^2}{r_B^3}, \tag{12}$$

where r_x is the radius of the complex. Recalling that $\ln k_2$ is a constant minus E_e/kT, we obtain, after slight rearrangement

$$\ln k_2^0 = \text{constant}\,(T) + \frac{z^2\varepsilon^2}{2DkT}\left(\frac{1}{r_A} - \frac{1}{r_x}\right) - \frac{D-1}{2D+1}\frac{1}{kT}\left(\frac{\mu_B^2}{r_B^3} - \frac{\mu_x^2}{r_x^3}\right). \tag{13}$$

Because r_A and r_B are naturally less than r_x, both bracketed terms are positive. Of the two energy contributions in this equation, the first is the more important. Quantitative applications are not possible, in the absence of information relating to r_x and μ_x. Numerically, there is probably little difference between this theory and the earlier one, as may be seen when the principal term of equation (13) is written as $z_A\varepsilon(r_x - r_A)\varepsilon/2DkTr_Ar_x$, which is to be compared with the term $z_A\varepsilon\mu_B/DkTr^2$ of equation (10).

E. S. Amis (1953) has improved the earlier treatment by adopting Onsager's expression for the effective dipole moment of a solute in terms of the value, μ_g, ordinarily obtained, and the permittivity and refractive index of the solvent:

$$\mu_s = \mu_g\frac{D(n^2+2)}{2D+n^2}. \tag{14}$$

By this means, he estimates, for the hydrogen-ion catalysed hydrolysis of sucrose a critical distance of approach of 5·5 Å.

There are examples, some of which we shall here discuss, where no simple electrostatic theory can explain the effect of changes in the permittivity of the solvent. Cavell and Speed (1961) have measured the rates of the reactions $nC_4H_9I + {}^*I^- \rightleftarrows nC_4H_9I^* + I^-$ at various temperatures in mixtures of methanol and methyl cyanide of all compositions. While the permittivity changes by a factor of only 1·2, the bimolecular velocity constant changes by a factor of 50, and not in the direction anticipated. Similar results have been found for the reaction of the iodide ion with nC_4H_9Br in acetone-water mixtures. The hypothesis advanced by the investigators is that an equilibrium $I^- + H_2O \rightleftarrows (I,H_2O)^-$ is established, providing two types of ions with different specific reactivities. Reasonable values of the equilibrium constants can, with some approximations, be derived, giving a consistent interpretation of the effect of changing the solvent. There is also some indirect evidence, from the absorption of infra-red radiation, that there is a small change in the restoring force constant of the O—H group with solvent composition. The hypothesis could be logically extended to include all the various solvated ions in solution, from $(I,mCH_3OH)^-$ to $(I,nCH_3CN)^-$, including ions partly solvated by both types of solvent molecules, such as $[I,(n - m)CH_3OH,mCH_3CN]^-$, where n is the common co-ordination number.

Doubt remains as to whether the macroscopic permittivity is the right one to use. Kacser (1952) has examined numerous experimental values in the light of his theory, taking the effective dielectric constant to be, in the first place, unity, and, in the second place $3D/(D + 2)$. His equation, (8.23a), reduces to the standard equation when E_e is small. When $D = 1$, E_e is large, and the equation for univalent ions becomes

$$k_2 = Z_0 \exp\left(-E_n/kT\right)\left(\frac{r^2kT}{2\varepsilon\mu}\right) . \exp\left(+\varepsilon\mu/r^2kT\right). \tag{15}$$

It then follows that the term A of the Arrhenius equation becomes $Z_0(r^2kT/2\varepsilon\mu)\exp(3/2)$, which, since Z_0 varies as r^2, is proportional to r^4. The corresponding expression which results using $3D/(D + 2)$ is more complicated, but tractable. Surprisingly enough, the calculated values of the critical separation are approximately equal whichever function of D is used. Thus, for the reaction $CH_3Br + I^- \rightarrow CH_3I + Br^-$ in the first three solvents considered, Kacser finds $r = 3.95 \pm 0.34$ Å when D is taken as unity, and 3.04 ± 0.25 Å on the other basis (Table 8.XVII). We shall discuss later the difficult problem of deciding on the permittivity appropriate to chemical kinetics.

Substitution by Two Mechanisms; Carbonium Ions

We have seen (Chapter 4) that the rate of reaction of the ion $S_2O_3^{2-}$ with C_2H_4BrI is bimolecular, and with C_2H_4ClI is unimolecular, and that Slator (1905), among his speculations on the mechanism, suggested that the intermediate formed in the latter case was an organic cation. The process could then be formulated as follows:

$$C_2H_4ClI \rightarrow C_2H_4Cl^+ + I^- \quad \text{(slow and rate-determining)}$$
$$\underline{S_2O_3^{--} + C_2H_4Cl^+ \rightarrow C_2H_4ClS_2O_3^-} \quad \text{(rapid)}$$
$$S_2O_3^{--} + C_2H_4ClI \rightarrow C_2H_4ClS_2O_3^- + I^- \quad \text{(net chemical change).}$$

In this schematic representation, it must not be supposed that the ionization is irreversible, or that the ions are unsolvated. The mechanism is seen to be identical with that postulated by Stewart and Bradley for the reactions of disubstituted aminomethyl sulphonic acids with various oxidizing agents (Chapter 4). The substitution of one group for another may thus take place by direct replacement in a bimolecular reaction between the anion and the polar molecule, or as the result of a unimolecular ionization followed by a relatively rapid addition. We have seen that, with the methyl halides in water, substitution of OH for X takes place by the two concurrent mechanisms, each of which is rate-determining (equation 6.45):

$$CH_3X + OH^- \rightarrow CH_3OH + X^- \quad \text{(bimolecular)}; k_2$$

$$CH_3X + HOH \rightarrow CH_3OH + X^- + H^+ \quad \text{(pseudo-unimolecular)}: k_1.$$

Schematically, the mechanism of the latter change could be written as follows:

$$CH_3X \rightarrow CH_3^+ + X^- \quad \text{(unimolecular and rate-determining; } k_1\text{)}$$

followed by $\quad CH_3^+ + HOH \rightarrow CH_3OH + H^+ \quad$ (rapid)

$$\overline{CH_3X + HOH \rightarrow CH_3OH + H^+ + X^-} \quad \text{(net chemical change).}$$

Since the ions produced in aqueous solution are known to be hydrated, it is necessary to postulate the subsequent rapid reaction of H^+ and X^- with water molecules. Because of the high energy required to ionize the methyl halides, the scheme as represented here—i.e. ionization into free ions followed by rapid solvation—is an improbable one. It is more likely that the critical complex in its transitory state consists of a closely-knit system of one highly polar solute molecule surrounded by $m + n + 1$ water molecules, which eventually breaks up into the organic product CH_3OH, and the solvated ions, $[H,mH_2O]^+$ and $[X,nH_2O]^-$. We shall revert to this hypothesis in dealing with hydrolysis.

With certain organic solutes, the unimolecular mechanism alone plays a part. This is true, for example, for the hydrolysis of benzhydryl chloride (Ward, 1927) and of tertiary butyl chloride (Hughes, 1935) which hydrolyse in aqueous ethanol and aqueous acetone at rates unaffected by the concentration of the ions OH^- and $C_2H_5O^-$.

The mechanisms described here as bimolecular reactions between anions and polar molecules and as pseudo-unimolecular reactions with the solvent have been labelled (Ingold, 1953) with the symbols S_N2 and S_N1, respectively, where S denotes substitution (and not elimination), the numerals give the kinetic order, and N stands for "nucleophilic" or negative. It is seldom possible to predict which mechanism, under given circumstances, is likely to be followed by any reaction, though there are a few guiding rules. For example, since the point of anionic attack on a polar molecule is its positive end, anions of high basicity, like OH^-, can be expected to favour the bimolecular mechanism. The structure of the polar molecule may also have a determining effect. The failure of the hydroxyl ion to attack $(CH_3)_3CX$ bimolecularly under circumstances when it readily attacks CH_3X bimolecularly can be explained in terms of the stronger hindrance offered by the three methyl groups to the approach of the ion towards the positive end of the dipole $C-X$ (see Fig. 8.6). With iso-propyl bromide there is a similar hindrance, but not sufficient to suppress the bimolecular attack altogether. The reaction of this halide with anions in aqueous ethanol solutions can be resolved into (1) the elimination of iso-propylene, (2) the pseudo-unimolecular reactions of the organic halide with the two types of solvent molecules, (3) the substitution of Br by OH and by C_2H_5O according to the two mechanisms.

The first three entries in Table VI are examples of uncomplicated bimolecular substitutions of *iso*-propyl and *n*-butyl halides in acetone and acetonitrile solutions. The substitution reactions of tertiary butyl bromide in acetone occur by two mechanisms, like the substitution reactions of the methyl halides in aqueous alkaline solution. Tertiary butyl fluoride in alkaline ethanolic solution suffers no substitution but only elimination (*e*). The last entry in Table VI gives Arrhenius parameters for three of the processes taking place when *iso*-propyl bromide reacts in alkaline water–ethanol mixtures.

Extra-kinetic evidence for the existence of stable carbonium ions in hydroxylic media is not abundant. Their properties are generally inferred from the composition of the products of reaction, combined with kinetic expressions derived on the supposition that such ions are significant intermediaries. Conductimetric evidence for the existence of the triphenylmethyl ion in sulphur dioxide solution has been discussed in Chapter 3 (equation 95).

Let us attempt to formulate the course and kinetics of a simple reaction thought of as taking place *via* the formation of a carbonium ion. Let RX be an ionizable organic halide, aliphatic or aromatic, in an aqueous solution containing an inorganic anion Y^- derived from a completely dissociated salt MY. If the rates of substitution and solvolysis can be attributed solely

TABLE VI

SIMULTANEOUS FIRST ORDER AND SECOND ORDER SUBSTITUTIONS AND
ELIMINATIONS

($E_{A,2}$ and $E_{A,1}$ in kcal/mole. A_2 in litres/mole-sec. A_1 in sec^{-1})

Reference	Reaction	Solvent	$\log_{10} A_2$	$E_{A,2}$	$\log_{10} A_1$	$E_{A,1}$
*	isoC$_3$H$_7$Br + Br$^-$	(CH$_3$)$_2$CO	10·6	20·0	—	—
†	nC$_4$H$_9$Br + I$^-$	(CH$_3$)$_2$CO	11·5	19·4	—	—
‡	nC$_4$H$_9$I + I$^-$	CH$_3$CN	11·1	18·5	—	—
*	$tert$C$_4$H$_9$Br + Br$^-$	(CH$_3$)$_2$CO	11·0	21·8	7·2	19·7
§	$tert$C$_4$H$_9$Cl + OH$^-$	H$_2$O	—	—	15·6	23·4
‖	CH$_3$Cl + OH$^-$	H$_2$O	12·6	24·3	11·8	26·5
¶	$tert$C$_4$H$_9$F + C$_2$H$_5$O$^-$	C$_2$H$_5$OH	8·0(e)	24·8	—	—
**	isoC$_3$H$_7$Br + OH$^-$	H$_2$O + C$_2$H$_5$OH	10·1	21·7	9·8	23·2
	+ CH$_2$H$_5$O$^-$					
			10·9(e)	22·6		

REFERENCES

* Le Roux and Swart, *Trans. Chem. Soc.*, 1475 (1955).
† Cavell, *ibid.*, 4217 (1958).
‡ Cavell and Speed, *ibid.*, 226 (1961).
§ Winstein and Fainberg, *J. Amer. Chem. Soc.* **79**, 5937 (1957); Moelwyn-Hughes, *Trans. Chem. Soc.*, 1517 (1961); Moelwyn-Hughes, Robertson and Sugamori, *ibid.*, 1965 (1965).
‖ Table VI, 10.
¶ N. B. Chapman and J. L. Levy, *Trans. Chem. Soc.*, 1673 (1952): >95% elimination (e).
** E. D. Hughes, C. K. Ingold and Shapiro, *ibid.*, 225 (1936).

to the carbonium ion R$^+$, we have the following scheme:

$$RX \underset{k_2}{\overset{k_1}{\rightleftarrows}} R^+ + X^-$$
$$_{a-x} _{x}$$

$$R^+ + Y^- \underset{k_3}{\overset{k_4}{\rightleftarrows}} RY$$
$$_{b-y} _{y}$$

$$R^+ + H_2O \overset{k_5}{\rightarrow} ROH + H^+,$$
$$_{z}$$

where a and b are the initial concentrations of RX and Y$^-$ respectively. The last step is regarded as irreversible. At any instant, x, the decrease in the concentration of RX is clearly $y + z$. The stationary concentration of the carbonium ion is

$$[R^+] = \frac{k_1(a - x) + k_3 y}{k_2 x + k_4(b - y) + k_5}. \tag{16}$$

We have, assuming all rate constants to be independent of the composition,

$$dy/dt = k_4[R^+](b - y) - k_3y$$

and

$$dz/dt = k_5[R^+].$$

Let us further assume that the rate of ionization of the product RY is much less than that of RX, so that the term k_3y may be dropped. Then

$$dy/dz = (k_4/k_5)(b - y),$$

and

$$z = (k_5/k_4) \ln [b/(b - y)]. \tag{17}$$

The ratio of the constants k_5/k_4 can thus be found from an analysis, at any stage of the reaction, of the concentrations of RY and ROH. An approximation of this kind is often made, and many published rate constant ratios have been based on final analyses only. We next have

$$dx/dt = k_1(a - x) - k_2[R^+]x$$

$$= \frac{k_1(a - x)[k_5 + k_4(b - y)] - k_2k_3xy}{k_2x + k_4(b - y) + k_5} \tag{18}$$

from which we see that, at all stages of reaction, the rate is retarded by the anion generated in the primary ionization. The first order constant with respect to RX, when extrapolated to zero time, is

$$\frac{1}{a}\left(\frac{dx}{dt}\right)_0 = k_1 + \left(\frac{k_1k_5}{k_4}\right)\left(\frac{1}{b}\right), \tag{19}$$

from which the rate of the assumed ionization may be found. Exact values are not known for the systems considered here, but an approximate estimate may be made as follows.

We refer to the exact equation governing the rate of formation of ions in small concentrations, x:

$$x = (Ka)^{1/2} \tanh [k_2(Ka)^{1/2} . t], \tag{6.65}$$

where K is the ionization constant, k_1/k_2. We may confidently take K in the present problem to be extremely small, so that, approximately,

$$-dx/dt = k_1a.$$

Let us now suppose that in the system considered a stationary rate of ionization

$$+d[R^+]/dt = -d[RX]/dt = k_1[RX]$$

is maintained, despite the presence of the ions Y^- and the solvent molecules H_2O, both of which subsequently react rapidly with the carbonium ion. The rates of the subsequent reactions do not then enter into the problem, and the rate at which the products are produced are

$$d[RY]/dt = k_1[RX] \times \text{(probability that } R^+ \text{ reacts with } Y^-),$$

and

$$d[ROH]/dt = k_1[RX] \times \text{(probability that } R^+ \text{ reacts with } H_2O).$$

(20)

The probabilities may be taken as the volume fraction, θ. The volume fraction of the ion is $[Y^-]V_i$, where $[Y^-]$ is the ionic concentration in mole/litre, and V_i the partial ionic volume, in cc/gramme–ion. In dilute solutions, the volume fraction of the solvent is effectively unity. Hence the bimolecular constant governing the reaction between RX and the ion Y^- should be greater than the pseudo-unimolecular constant for reaction of RX with water by V_i. The partial ionic volumes of elementary ions are often about 20 cc, which is less by a factor of about ten than the experimental values of k_2/k_1 given in Table 6.III.

This formulation, it will be realized, is an over-simplified one. A mathematically exact treatment is unobtainable even for simple systems. If the present conclusion is qualitatively correct, it means that the rate of ionization of the solute RX is nearly equal to the observed bimolecular constant for the reaction RX + Y^-, divided by the partial molar volume, V_i. From the data given in Table 12.III, k_1 for methyl iodide in water at 25°C is seen to have the order of magnitude of $10^{-7} \sec^{-1}$. Since the rate of combination of the ions has the order of magnitude of 10^{+10}, the ionization constant of methyl iodide in water is thus about 10^{-17} gramme-ions per litre. An earlier estimate for its ionization constant in nitrobenzene at 25°C is 10^{-20} (Moelwyn-Hughes, 1936).

According to this theory, the key rate constant, k_1, is simply the absolute probability per second that the solute molecule RX should ionize in the pure solvent. This probability is different when the solute molecule ionizes in the electrostatic field of the ion Y^-. Such an effect on the reaction rate is clearly a specific one. A different interpretation is possible: the key rate constant may be that of the ionization of the water molecule. Specificity now enters in the effect which the presence of the added ions have on the rate of ionization of the solvent. To this idea we shall revert when dealing with hydrolysis (Chapter 12).

We shall next apply simple electrostatic theory to the kinetics of reactions between an ion and a molecule possessing at least two polar groups.

The Effect of a Resident Dipole

Let us consider the set of reactions represented by the equation

$$p\text{-}X\text{·}C_6H_4\text{·}B + A^{z_A} \rightarrow p\text{-}X\text{·}C_6H_4\text{·}A + B^{z_A}.$$

The energy of activation may reasonably be supposed to contain two electrostatic terms, so that the total activation energy may be expressed as follows:

$$E = E_n + E_e(r_1) + E_e(r_2),$$

where r_1 and r_2 are the distances between the centre of the ion A and the centres of the respective dipoles μ_B and μ_X in the reactive complex (Fig. 3).

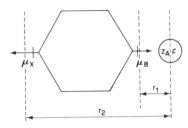

FIG. 3.

We shall first consider the simplest case, i.e. when the ion approaches the molecule in the plane of the benzene ring and along the direction of the polar axes. The energy of the reactive complex in an infinitely dilute solution is thus

$$E = E_n - \frac{z_A \varepsilon \mu_B}{D_1 r_1^2} + \frac{z_A \varepsilon \mu_X}{D_2 r_2^2}. \tag{22}$$

For a set of reactions in which μ_B is a constant and μ_X is a variable, we may relate the activation energy of the substituted complex to that of the unsubstituted complex as follows:

$$E_s = E_u + \frac{z_A \varepsilon \mu_X}{D_2 r_2^2}. \tag{23}$$

The corresponding expression for the Arrhenius, or observed, energies of activation is

$$E_{A,s} = E_{A,u} + \frac{z_A \varepsilon \mu_X}{D_2 r_2^2}(1 - L_2 T). \tag{24}$$

When, therefore, the apparent energy of activation of reactions in a given series is plotted against the dipole moment of the *para* substituent, the plot should be linear, and should have gradients of different signs for reactions of the molecules with ions of different charge signs.

Two sets of reactions which have been studied in aqueous ethanolic solution are the acid-catalysed enolization of *p*-substituted acetophenone (Nathan and Watson, 1933):

$$X\!\!-\!\!\langle\ \rangle\!\!-\!\!COCH_3 + H_3O^+ \rightarrow X\!\!-\!\!\langle\ \rangle\!\!-\!\!C(OH)\!:\!CH_2 + H_3O^+,$$

and the saponification of *p*-substituted ethyl benzoates (Ingold and Nathan, 1936; D. P. Evans, Gordon and Watson, 1937):

$$X\!\!-\!\!\langle\ \rangle\!\!-\!\!COOC_2H_5 + OH^- \rightarrow X\!\!-\!\!\langle\ \rangle\!\!-\!\!COO^- + C_2H_5OH.$$

The data plotted in Fig. 4 confirm both conclusions: the gradients are linear, within the limits of experimental error, and of opposite signs.

An analysis of the two sets of data for the saponification reactions, using the method of least squares, yields the following relationship between the apparent energy of activation and the dipole moments given by McAlpine

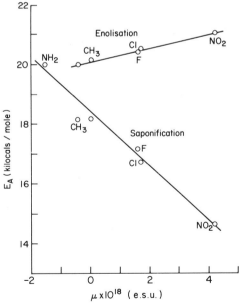

FIG. 4. The enolization of para-substituted acetophenones and the saponification of para-substituted ethyl benzoates.

and Smyth (1935):

$$E_{A,s} \text{(cal/mole}^{-1}) = 18,179 - 844 \cdot 132\mu \text{ (Debye units)}. \qquad (25)$$

The gradient in these units is therefore approximately

$$dE_{A,s}/d\mu = -844.$$

The theoretical gradient, obtained from equation (24), with $z_A = -1$, is

$$\frac{dE_{A,s}}{d\mu} = -\frac{69,130}{(\mathring{r}_2)^2}\left(\frac{1 - L_2 T}{D_2}\right), \qquad (26)$$

On combining the two equations,

$$(\mathring{r}_2)^2 = 81 \cdot 96\left(\frac{1 - L_2 T}{D_2}\right). \qquad (27)$$

We therefore conclude that $(1 - L_2 T)$ must be positive, and that D_2 cannot refer to the solvent but to the benzene ring. Let us provisionally work with the experimental value of $(1 - L_2 T)/D_2$ for toluene, which is 0·341. Then $\mathring{r}_2$ becomes 5·29 Å, which, according to any reasonable model, is too small. We shall, instead, estimate $\mathring{r}_2$ from a tentative model (Fig. 5), and solve for

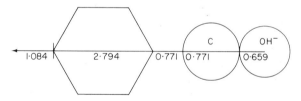

FIG. 5. Linear model of the hydroxyl ion-*para*-substituted ethyl benzoate complex. Distances in Ångström units.

$(1 - L_2 T)/D_2$. In the linear ion-molecule complex, the negative charge on the hydroxyl ion is assumed to be at a distance of 1·43 Å from the centre of the carbonyl carbon atom, and the point dipole in the polar substituent X is placed at a distance of 0·542 Å from the carbon atom in the ring. The distance $\mathring{r}_2$ between the point dipole and the charge on the hydroxyl ion is seen to be $0·542 + 2·794 + 1·542 + 1·430 = 6·31$ Å. From equation (27), it follows that $(1 - L_2 T)/D_2$ is 0·486, which is slightly higher than the experimental value (0·341) for pure toluene. The adoption of this value is equivalent to assuming a temperature-invariant permittivity of 2·058—a value which has been theoretically justified by D. M. Bishop and D. P. Craig (1963).

An illustration of the present conclusions which does not require mathematics is given in Fig. 6.

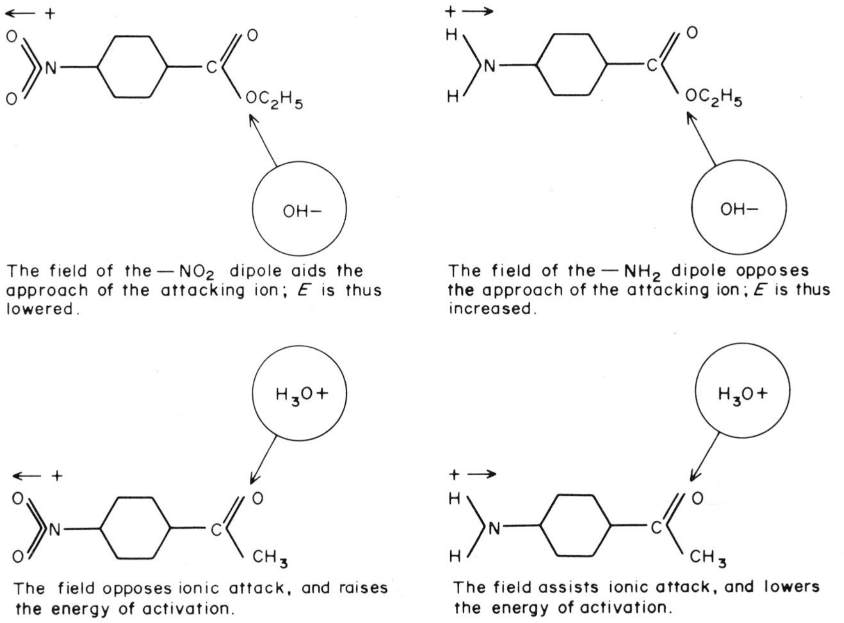

The field of the $-NO_2$ dipole aids the approach of the attacking ion; E is thus lowered.

The field of the $-NH_2$ dipole opposes the approach of the attacking ion; E is thus increased.

The field opposes ionic attack, and raises the energy of activation.

The field assists ionic attack, and lowers the energy of activation.

FIG. 6.

The Variation of (k_s/k_u) with Respect to Permittivity

Numerous isothermal ratios, k_s/k_u, of the rate constants for substituted and unsubstituted reactions have been measured by Tommila (1967) in mixed solvents, consisting of water with one of the following miscible solvents: methanol, ethanol, acetone, dioxane and sulpholane. The results obtained for the reactions of the hydroxyl ion with certain *para*-substituted ethyl benzoates in dioxane–water mixtures are shown in Fig. 7. The gradients of $\log (k_s/k_u)$ against $1/D$ are seen to be initially linear, and to decrease in solvents of low permittivity, as in Fig. 2.

From equation (23) we have, assuming a common collision frequency,

$$\ln (k_s^0/k_u^0) = -\frac{z_A \varepsilon \mu_x}{Dr_2^2 kT}. \tag{23a}$$

Since $z_A = -1$,

$$\ln (k_s^0/k_u^0) = \frac{\varepsilon \mu_x}{Dr_2^2 kT},$$

and

$$\frac{d \log_{10} (k_s^0/k_u^0)}{d(1/D)} = \frac{\varepsilon \mu_x}{2 \cdot 303 r_2^2 kT}. \tag{23b}$$

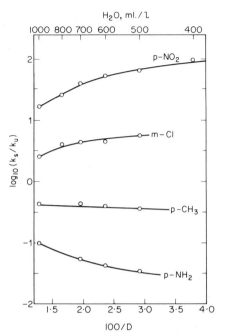

FIG. 7. The alkaline hydrolysis of some substituted ethyl benzoates in dioxane–water mixtures at 25°C (after E. Tommila, *Ann. Acad. Sci. Fennicae*, **A, II**, *Chem.*, 139 (1967).

The initial positive gradient found with *para* nitroethyl benzoate is 53. The theoretical value, with $\mu = 4 \cdot 19 \times 10^{-18}$ and $r_2 = 6 \cdot 31 \times 10^{-8}$ is $5 \cdot 33$. Hence the effective permittivity is, on an average, about one tenth of the macroscopic value. The initial negative gradient found with *para* aminoethyl benzoate is consistent with equation (23b) and again indicates the transmission of the electrostatic force through the benzene ring rather than through the solvent. The ratio D (macroscopic)/D (effective) naturally varies with the composition of the solvent. From equation (23a), written in the form

$$\log_{10} k_s^0 = \log_{10} k_u^0 - z_A \varepsilon \mu_x / 2 \cdot 303 D r_2^2 kT,$$

it is seen that the difference between the decadic logarithm of the rate constants of the substituted and unsubstituted reactions at a given temperature can be resolved into the product of two terms, one of which depends on the reactants only, and the other of which depends on the effective permittivity only. Later reference is made to equations of this type.

For the set of reactions considered here, we have, at 25°C,

$$\log_{10} k_s^0 = \log_{10} k_u^0 - 1 \cdot 273 z_A \mu \text{ (Debyes) } / D$$

where the effective value of D is about 2.

The Kinetic Effect of a Resident Dipole in the *ortho, meta* and *para* Positions

A tentative model for a planar complex formed between the hydroxyl ion and a *meta*-substituted ethyl benzoate molecule is shown in Fig. 8, from

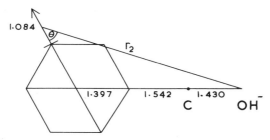

FIG. 8. Planar model of the hydroxyl ion-*meta*-substituted ethyl benzoate complex. The calculated energies of activation are in the order *ortho* > *meta* > *para*.

which the positions of the oxygen atoms attached to the ketonic carbon atom have been omitted. From the internuclear distances adopted, and the assumed position of the point dipole, the distance, r_2, between the ion and the centre of the dipole can be evaluated, as well as the angle θ subtended between its direction and that of the polar axis. Similar calculations can be made on the *ortho*-substituted system. The part of the observed energy of activation due to this interaction is given by the equation

$$E_{A,s} - E_{A,u} = \frac{z_A \varepsilon \mu_X}{r_2^2}\left(\frac{1 - L_2 T}{D_2}\right) \cos \theta, \qquad (28)$$

where $z_A = -1$ and $(1 - L_2 T)/D_2 = 0.486$. The results when the dipole moment is 1 Debye unit are given in Table VII. We note that the apparent

TABLE VII

CALCULATED EFFECTS OF *o*-, *m*- AND *p*-SUBSTITUTION ON THE APPARENT ENERGY OF ACTIVATION FOR THE REACTIONS
$X-C_6H_4COOC_2H_5 + OH^-$ IN ETHANOL–WATER MIXTURES

	$\mathring{r}_2$	θ	$E_{A,s} - E_{A,u}$ (cal; equation 28)
o	3·79	93° 24′	+139
m	5·60	42° 30′	−790
p	6·31	0	−844

energies of activation stand in the order

$$E_{A,s}; \quad o > m > p,$$

and that, on account of the compensating effects of the r_2 and θ values, the difference between the m and p energies is slight. When, as in the present series, the pre-exponential term of the Arrhenius equation is constant $[A_2 = (5\cdot36 \pm 0\cdot43) \times 10^9]$, the relative rate constants take the sequence

$$k_2; \quad p > m > o.$$

Other systems to which the present considerations are applicable include the reaction of the iodide ion with the four halogeno-substituted benzyl chlorides (Bennett and B. Jones, 1935; Baddeley and Bennett, 1935) and the reaction of methoxide and ethoxide ions with the dinitrobenzenes (Steger, 1904). The predicted sequences are sometimes, but by no means always, followed. When the charge on the attacking ion is positive, as in the hydrogen ion-catalysed enolizations, the theoretical orders given above take the reverse sequence, i.e. $k_2; o > m > p$.

The Equation of Nathan and Watson

Nathan and Watson (1933; see also Waters, 1933) have expressed as follows the effect of a substituted dipole, μ, on the apparent energy of activation for reactions between an ion and a series of molecules all possessing initially a fixed dipole, μ_B:

$$E_{A,s} = E_{A,u} + K_1\mu + K_2\mu^2. \tag{29}$$

K_1 and K_2 are constants for the series, the former related to the direct field effect of the ion, and the latter to induction. The full line in Fig. 9 represents the application of this equation to the hydrogen ion-catalysed enolization of substituted acetophenones, with $K_1 = 429$ and $K_2 = 53\cdot6$ in the units employed here. While the adoption of this equation leads to some interesting consequences, the data as they stand can, with one exception, be represented by two linear equations, each holding separately for the *meta-* and *para-*series. These are shown by the dotted lines which have been superimposed onto the original graph. It is to be noted that the signs of the ordinate and the polarity used by Evans, Morgan and Watson are inverted. The dotted lines are given by the equations $E_{A,s}$ (*meta*) $= 20,150 + 350\mu(\mathring{D})$ and $E_{A,s}$ (*para*) $= 20,150 + 240\mu(\mathring{D})$. According to equation (7), with $z_A = +1$, and $(1 - LT)/D = 0\cdot5$, the theoretical gradients are given by the equation

$$K_1 = \frac{dE_{A,s}}{d\mu} = \frac{34,600}{(\mathring{r}_2)^2} \cos\theta.$$

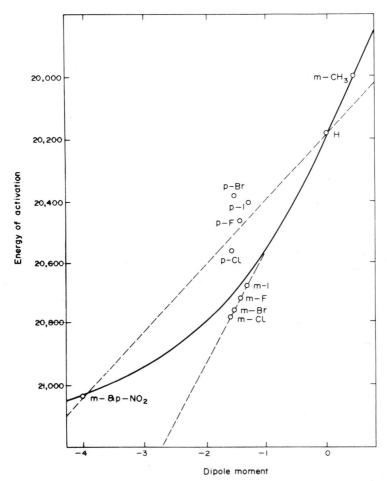

FIG. 9. The effect of *mesa-* and *para*-substituents on the apparent energy of activation for the enolization of acetophenone, catalysed by hydrogen ions (after Evans, Morgan and Watson, *Trans. Chem. Soc.*, 1174 (1935).)

The gradients K_m and K_p for the *meta* and *para* series should thus differ:

$$K_m = \frac{34{,}600}{(\mathring{r}_m)^2} \cos \theta_m \qquad K_p = \frac{34{,}600}{(\mathring{r}_p)^2} \cos \theta_p. \qquad (30)$$

These relationships have been used in an attempt (Moelwyn-Hughes, 1936) to gain some information concerning the geometry of the ion–molecule complex. The results were "consistent with the approach of the ion at an angle considerably inclined to the direction of the *para*-substituent". If we accept the approximate distances given in Table VII, we find that $\theta_p = 74°$ and that $\theta_m = 68°$. Later calculations by a much improved method have confirmed this conclusion.

The effects of substitution on the kinetics of aromatic reactions are further discussed in connection with correlation equations (Chapter 15; cf. J. Miller, 1968).

The Kinetic Effect of a Resident Charge in Simple Aliphatic Reactions

Let us consider pairs of reactions of the following type:

$$RH_2CX + Y^- \rightarrow RCH_2Y + X^-$$

and

$$RCHZ^-CX + Y^- \rightarrow RCHZ^-CY + X^-$$

in an attempt to discover the kinetic effect of the presence of the ion Z^- in the reacting molecule. Data referring to aqueous solutions are given in Table VIII. A theoretical equation for the effect of substituting the ion Z^-

TABLE VIII

THE KINETIC EFFECT OF A RESIDENT CHARGE IN SIMPLE ALIPHATIC REACTIONS

Reaction	k_2 at 25°C	A(litre/mole^{-1}-sec^{-1})	E_A(cal)	$E_{A,s} - E_{A,u}$
$ICH_2COOH + SCN^-$	1.58×10^{-3}	3.5×10^{10}	18,190	
$ICH_2COO^- + SCN^-$	4.45×10^{-4}	6.3×10^8	16,580	$-1,610$
$BrCH_3 + S_2O_3^{--}$	3.24×10^{-2}	6.82×10^{12}	19,540	
$BrCH_2COO^- + S_2O_3^{--}$	4.12×10^{-3}	1.36×10^9	15,710	$-3,830$

has already been derived:

$$E_{A,s} - E_{A,u} = \frac{(1 - LT)}{D} \cdot \frac{z_A z_B \varepsilon^2}{r}. \tag{7.37}$$

Here r is the distance between the centre of the attacking ion, Y^-, and the centre of the substituted ion, Z^-, in the reactive complex. In aqueous solutions, $(1 - LT)$ is negative, and since, in these instances, the charge product is positive, we anticipate that the substitution of the ion Z^- will cause reductions in the apparent energies of activation, roughly in the proportion of $1:2$. The data of Table VIII confirm this view. Using the values of D and L appropriate to water at 25°C, equation (7.37) becomes

$$(E_{A,s} - E_{A,u})\,(cal) = -\frac{1,606 z_A z_B}{r}.$$

The resulting values of r are about 1 Å which are unreasonably low. Such a result often appears, as we have seen in Chapter 7, when only the simple Coulombic expression is used. If we include the energy of induction, we have a more satisfactory equation, applicable when $z_A z_B = 1$:

$$E_{A,s} - E_{A,u} = -\frac{1,606}{\mathring{r}}\left(1 - \frac{\alpha_A + \alpha_B}{2r^3}\right). \tag{31}$$

Here α_A and α_B are the polarizabilities of the ions Y^- and Z^-. The inclusion of the induction energy has worsened an already poor result. Though the acceptance of the macroscopic D and L values for water is probably at fault, the simple treatment is capable of explaining a somewhat unusual phenomenon—a decrease in the apparent energy of activation accompanied by a decrease in the rate of reaction.

Substitution in the Saponification of Aliphatic Esters

We take as the unsubstituted hydrolysis that of ethyl acetate in aqueous alkaline solution

$$CH_3COOC_2H_5 + OH^- \rightarrow CH_3COO^- + C_2H_5OH,$$

and as the substituted reactions

$$XCH_2COOC_2H_5 + OH^- \rightarrow XCH_2COO^- + C_2H_5OH,$$

where X denotes a polar group, or an ion or the one superimposed on the other. From a dozen reactions of this type investigated electrometrically at 25°C by Bell and Coller (1965), some representative bimolecular constants, extrapolated to zero ionic strength, are given in Table IX. Since the charge

TABLE IX

BIMOLECULAR CONSTANTS FOR THE REACTIONS OF THE HYDROXYL ION
WITH SUBSTITUTED ETHYL ACETATES AT 25°C (LITRES/MOLE-SEC)

X	k_2^0	$RT\ln(k_s^0/k_u^0)$ (cals/mole)	$\mathring{r}$
$-H$	1.11×10^{-1}	0	—
$-S^-$	6.40×10^{-3}	$-1,690$	3.41 (3.92)
$-S(CH_3)_2^+$	$2.04 \times 10^{+2}$	$+4,450$	3.41 (3.09)

on the attacking ion is negative, we see that the electrostatic interaction between it and the substituted ion accounts in large measure for the effect of substitution as in the examples given in Table VIII. Equation (7.37) must now be amended to include the two electrostatic effects:

$$E_s - E_u = \frac{z_A z_B \varepsilon^2}{D_1 r_1} - \frac{z_A \mu_B \cos \theta}{D_2 r_2^2}. \tag{32}$$

As the temperature coefficients of these reactions have not been measured, we must assume that the pre-exponential term of the Arrhenius equation is the same for them all. Moreover, in the present instance, when the charge on the ion, B, and the dipole moment, μ_B, of its group are superimposed, the two r's are equal. Then, if the effect of the substitution is wholly electrostatic,

$$kT \ln \left(\frac{k_s}{k_u}\right) = -(E_s - E_u) = -\left(\frac{z_A z_B \varepsilon^2}{D_1 r} - \frac{z_A \mu_B \cos \theta}{D_2 r^2}\right). \tag{33}$$

A model of the substituted ester–hydroxyl-ion complex has been constructed by Bell and Coller (Fig. 10), partly on the basis of a conclusion

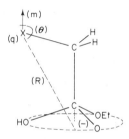

Fig. 10. A proposed model for the complex formed between the hydroxyl ion and an ion-substituted ethyl acetate molecule (after Bell and Coller, *Trans. Faraday Soc.*, **61**, 1445 (1965).

reached by M. L. Bender (1963) on the mechanism of hydrolysis. The negative charge, thought of as being distributed between the three oxygens shown at the base of the diagram, is represented by a point charge situated on the circle passing through the centres of the three oxygen atoms, at a distance R from the substituent, X. The distance is assumed to be the root-mean-square distance from X. The two permittivities necessary before equation (33) can be applied, are estimated partly from the kinetic data and partly by means of a theorem relating them to each other. D_1 is found to lie in the range 26–40, and D_2 in the range 6–9. The investigators conclude that "the effects of simple substituents on the rate of alkaline hydrolysis may be interpreted in terms of direct electrostatic interaction between charge distributions in the substituted groups and the negative charge associated with the reaction site in the hydrolysis complex". Their analysis is more detailed than any previously attempted, and raises in an acute form the problem of deciding on the right value to use for the permittivities and their temperature dependence.

The insistence on placing the effective negative charge on a fixed point on the basic circle, though reasonable, may not be realistic. If we assume free rotation around the C—C axis, the negative charge due to the attacking ion, OH^-, will find a position in the complex consistent with a minimum electrostatic energy. That is, when the substituent X is negatively charged, the two charges will be as far apart as possible; and when X is positively charged, the two charges will come as near together as possible. Let us illustrate the point by assuming that the interionic term alone need be considered. The theoretical equation then becomes

$$E_s - E_u \,(\text{kcal}) = \frac{z_A z_B \cdot 332}{D_1 \mathring{r}}. \tag{34}$$

The conventional model, using ordinary bond lengths, and taking the larger of the C—X distances as 1.85 Å, yields the distances $\mathring{r}_{--} = 3.92$ and $\mathring{r}_{+-} = 3.09$, rather than the common value of 3.41 given in Table IX. From the energy terms given in the same table, we have $D_1 \mathring{r}_{--} = 205$ and $D_1 \mathring{r}_{+-} = 78$, which require D_1 to be 54 and 25 respectively.

Wave-mechanical Treatment of the *ortho, meta* and *para* Problem

In wave mechanics, we need no longer confine attention to point electrical charges having magnitudes which are integral multiples, positive or negative, of the electronic charge. On the other hand, we deal with charge distributions at various points within any molecule, the effective charges being fractions of the fundamental unit charge. In the unsubstituted benzene ring, the effective amount of negative electricity surrounding each of the six carbon atoms is the same. When, however, one of the carbon atoms in the ring is directly attached to an ion or a polar group, the distribution of electrons around all six carbon atoms in the ring is altered, some of them becoming formally

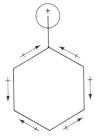

Fig. 11.

positive and some formally negative, depending on the direction and strength of the field exerted by the substituent ion or polar group and, on their relative position (*ortho, meta* or *para*) in the benzene ring. Consequently, not all the five remaining C—H groups are equally attractive to any attacking ion. To decide which position in the benzene ring is most likely to be attacked is an old and much-studied problem. If the influence of the substituted ion or dipole is due entirely to induction, a qualitative answer can be given, somewhat as follows. Let the substituent be a unit positive charge, as in Fig. 11. It attracts negative electricity to the carbon atom in the ring to which it is attached, making that atom effectively negative. As the total amount of negative electricity in the ring is constant, the carbon atoms in the *ortho* positions become effectively positive. The electrostatic induction is transmitted, with diminishing intensity, through the ring to the remaining carbon atoms (J. J. Thomson, 1923; L. E. Sutton, 1931; C. K. Ingold, 1953). In reactions of the mono-substituted molecule with cations (termed, in this connection, electrophils) substitution at the *meta* positions are seen to be the most favourable, and the rate of substitution at these positions should exceed the rate of substitution at the *para* position, which is seen to have acquired a formal positive charge. Experiments on the rates of nitration of aniline in acidic media show that the reverse is true (M. Brinkman, S. Johnson and P. Ridd, 1962). The induction hypothesis alone is therefore incapable of coping with the facts.

We turn next to two distinct and mutually satisfactory estimates of the π-electron charges at the *ortho, meta* and *para* carbon atoms of the organic ion shown in Fig. 11, where a unit positive charge (specifically the ammonio group $\overset{+}{N}H_3$) is attached to a ring carbon atom at a distance of 1·42 Å away from it in a radial direction. The wave-mechanical computations, due to Bishop and Craig (1963) and to Chandra and Coulson (1965) both indicate that the highest formal positive charge resides on the *para* carbon atom. In reactions with a cation, the *para* position would again be the least favourable —an unsatisfactory conclusion.

Chandra and Coulson next apply Pople's (1955) molecular-orbital perturbation theory to the complicated problem of determining the π-electron electrons have been perturbed from their normal state (i.e. their state in the unsubstituted benzene ring) under the joint electrostatic fields exerted by the ammonio group $\overset{+}{N}H_3$ (at a distance of 1·42 Å from the C atom) and by a nitronium ion $\overset{+}{N}O_2$ placed in the plane of the benzene ring at a distance of 4·2 Å from its centre in the *ortho, meta* and *para* directions. The effective permittivity adopted is 2. Their results are reproduced in Fig. 12, which shows that, once again, "the *meta* position carries the greatest net negative charge (i.e., −0·048), and the *para* position the least". Further calculations show

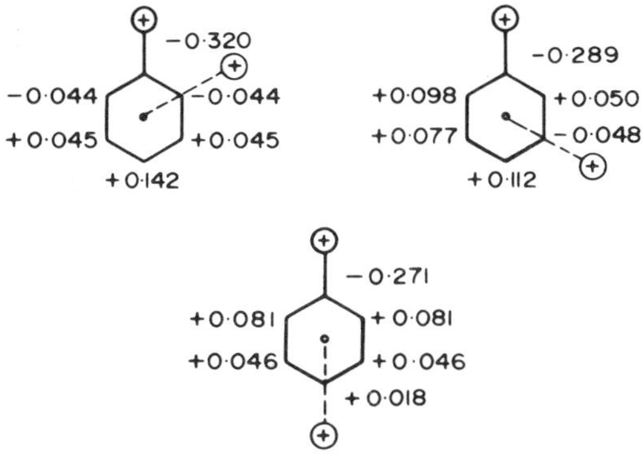

Fig. 12.

that the same result is found when the attacking nitronium ion $\overset{+}{N}O_2$ approaches the anilinium ion in the same three radial directions but inclined at various angles θ to the plane of the ring. Chandra and Coulson conclude "that the charge distribution under the joint perturbation of $\overset{+}{N}H_3$ and $\overset{+}{N}O_2$ does not, by itself, allow us to predict the position of nitration". This important deduction disposes of the hypothesis which has been used more than any other in attempts to interpret the experimental facts relating to the position at which a second substituent enters a mono-substituted benzene derivative.

The supposition that an attacking ion bearing a given charge sign should attack that particular position in the ring which has the highest charge of opposite sign or the smallest charge of the same sign seems at first to be so reasonable as to be incontestable. The fallacy of adopting it lies in the fact that in doing so we ignore the effect on the approaching ion of all the remaining charge elements in the system. The logical approach to the problem is to find the total electrostatic energies of the three ion pairs formed, in this instance, between the inorganic cation $\overset{+}{N}O_2$ and the three organic cations, *ortho*, *meta* and *para* $C_6H_5\overset{+}{N}H_3$. The most likely reaction is that between the pair for which the energy is the least. Expressed in another way, the attacking cation will most probably approach that corner of the benzene ring in the organic cation which exerts the lowest force of repulsion. This is the argument which leads Chandra and Coulson to their next move. Instead of concentrating on the formal charges, they calculate the total electrostatic energies (E_o, E_m and E_p) by summing the Coulombic terms for all pairs of charges in the three systems: $E = \Sigma(q_i q_j / r_{ij})$. The formal charges, q, are those which have been evaluated by the wave-mechanical methods, as given in Fig. 12. Similarly, the forces (F_o, F_m and F_p) acting on the inorganic cation when it approaches the organic cation in the three relevant directions are

TABLE X

TOTAL INTERACTION ENERGIES OF THE $\overset{+}{N}O_2$ AND $C_6H_5\overset{+}{N}H_3$ ION PAIRS AND
THE NET FORCE EXERTED BETWEEN THEM, IN ATOMIC UNITS* $\times 10^3$
(CHANDRA AND COULSON)

	E_0	E_m	E_p	F_0	F_m	F_p
0°	18·05	6·02	8·33	8·24	3·84	3·77
30°	18·27	6·25	8·26	7·75	3·96	4·05
45°	16·81	8·15	7·28	7·31	4·53	4·23
60°	10·97	9·49	8·55	6·55	4·81	4·48
90°	10·69	10·69	10·69	5·49	5·49	5·49

REFERENCE

* 1 atomic unit of energy $= 27\cdot21$ electron-volts $= 6\cdot274 \times 10^5$ cal/mole. Thus, when $\theta = 45°$, $E_m - E_p = 0\cdot87 \times 10^{-3}$ electron-volt $= 546$ cal/mole.

obtained by the vectorial addition of all terms of the form q_iq_j/r_{ij}^2. Some of their results are included in Table X. The entries in the last column show that, when the attacking ion is placed at a distance of $4\cdot2$ Å vertically above the centre of the benzene ring, the energy of interaction of the ions and the force exerted between them in this vertical direction are independent of the position occupied by the ammonio group in the ring. At all other angles of approach, E_o is so much greater than E_m or E_p as to render *ortho*-substitution almost prohibitive, and to confirm the greater rates of substitution in the *meta* and *para* positions. Attacks in the plane of the ring ($\theta = 0$) and at relatively small angles of approach ($\theta < 45°$) are more likely to be made in the *para* than in the *meta* position, as is found experimentally. Since E_p passes through a minimum when θ is about 45°, this is the most probable angle of approach of the inorganic ion to the plane of the benzene ring during *para* substitution. The substitution of any atom, ion or group for one of the hydrogen atoms in the benzene ring naturally entails a breaking of the C—H bond. In the nitration here considered it appears that this "chopping off" of the hydrogen atom from the C—H bond is most conveniently executed when the "axe" falls at an angle of less than 45° to the bond direction. Chandra and Coulson conclude that "Even if we do not accept the details of Table X, their general character is likely to be more accurate. We may therefore conclude with rather more confidence that since the minimum of E_o occurs at $\theta = 90°$, and of E_m and E_p at 0° and 45°, and the rates of change of these quantities are much bigger for E_o than for either E_m or E_p, therefore the geometry of the activated complex for *ortho*-substitution will differ from that for either *meta*- or *para*-substitution. Our conclusions . . . although they have been obtained for the particular case of attack by NO_2^+ on the anilinium ion, are likely, if correct, to apply much more generally. The position of electrophilic substitution in the anilinium cation is not adequately predicted either by the charge distribution of an isolated anilinium ion, nor by the charges on the *ortho, meta* and *para* positions under the joint influence of the original ammonio-group and the approaching NO_2^+; but it is correctly predicted by comparisons of the repulsive force acting on the approaching electrophil, or of the total electrostatic energy of the charges induced by the NH_3^+ and NO_2^+ ions."

It is to be emphasized that the calculation of the energies and forces in this treatment, as in most passages in this book, have been made according to classical electrostatics. The novelty of the modern treatment, its evident superiority over and its fundamental difference from the elementary treatment lie in the acceptance of wave-mechanically computed magnitudes of formal charges, and in the realization that satisfactory models of activated complexes must be constructed in three dimensions.

The earlier treatment, based on the supposition that the electrostatic interaction energy could be adequately given in terms of the ionic charges and the dipole moments, succeeded, despite its shortcomings, in deducing that the ionic approach leading to reaction was along a line inclined to the plane of the benzene ring.

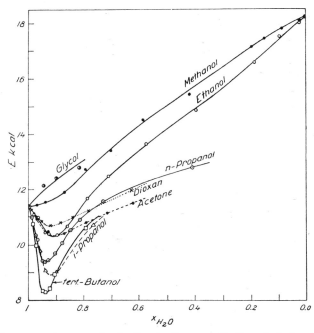

FIG. 13. The apparent energy of activation for the saponification of ethyl acetate in mixed solvents (after Tommila *et alia*).

Reactions between Ions and Polar Molecules in Mixed Solvents

The transfer of solutes capable of chemical interaction from one solvent to another is attended by a large number of changes. In the first place, bulk properties of the media, such as density, compressibility, viscosity and permittivity are altered. The number and nature of the solvent molecules surrounding the solute molecules are certainly changed, and, particularly in hydroxylic solvents, proton-transfer equilibria, such as $OH^- + ROH \rightleftarrows H_2O + RO^-$ are established. It is therefore not surprising to find that, when one solvent is gradually replaced by another, the apparent energy of activation varies in a complicated manner with respect to the composition of the medium.

When the saponification of ethyl acetate is carried out in methanol–water mixtures, E_A increases steadily, but not linearly, with respect to the molar fraction of methanol. In mixtures of water with other solvents, E_A passes through sharp minima when the molar fraction of water is about 0·95, as shown in Fig. 13 (Tommila, Koivisto, Lyyra, Antell and Heimo, 1952). Similar results have been found in the alkaline hydrolysis of ethyl benzoate, benzyl acetate (Tommila, 1952) and valerolactone (Tommila, Pentti and Ilomäki, 1952). We shall see later that the phenomenon of a minimum in E_A is by no means confined to reactions between ions and polar molecules, but extends to unimolecular reactions for which Arnett (see Chapter 12) finds curves of nearly the inverse shape for the gain in enthalpy attending dissolution of the solutes.

Reactions between Ions and Molecules Containing Three Polar Groups

Although there may be little prospect at present of interpreting kinetic information obtained in this category of chemical reactions, reliable experi-

TABLE XI

REACTIONS OF CH_3O^-, IN METHANOL SOLUTION, WITH $1,2,4X(NO_2)_2C_6H_3$

Reference	X	$k_2(0°C)$	$\log_{10} A_2$	E_A (kcals)
*	F	$1·74 \times 10^0$	11·44	14·0
†	Cl	$1·83 \times 10^{-3}$	10·63	$16·7 \pm 1·0$
‡	I	$3·08 \times 10^{-4}$	12·09	19·5

REFERENCES

* K. C. Ho, J. Miller and (Miss) K. W. Wong, *Trans. Chem. Soc.*, **B**, 310 (1966).
† Talen, *Thesis*, Leiden (1927). See also van der Kam, *Thesis*, Leiden (1925).
‡ K. H. Kendall and J. Miller, *Trans. Chem. Soc.*, **B**, 119 (1967); D. L. Hill, K. C. Ho and J. Miller, *ibid.*, 299 (1966).

mental information is available, and must be recorded. Let us, as an example, examine reactions of the general type $1X,2,4\text{-}(NO_2)_2C_6H_3 + CH_3O^-$, where the methoxide ion is substituted for the halogen, X, in methanol solution. The data of Table XI show that, once again, the changes in rate are due preponderantly to changes in the apparent energy of activation. From the rate ratio $k(F)/k(I)$, which is 5,650, only a ratio of 4·6 can be attributed to the A_2 terms.

A Comparison between Aliphatic and Aromatic Substitution

It is abundantly clear that the electrostatic interaction between ion and polar molecules is only one of the many components of the total interaction energy of the complex formed by their union. While bearing this in mind, we shall nevertheless seek to discover to what extent the magnitude and sign of this energy component can account for some of the differences between the behaviour of aliphatic and aromatic molecules towards the approach of a common univalent anion in a common solvent. Schematically, the mechanism of formation of the critical complex in the two cases may be represented crudely as follows, where the reacting ion is the methoxide ion, and the relevant polar group is $C{-}F$:

Since we are dealing with the substitution of a negative ion, in infinitely dilute solution, equation (2) becomes

$$E_e = -\varepsilon\mu\cos\theta/Da^2, \tag{35}$$

where θ, by definition, is the angle subtended between the line of centres and the polar axis, reckoned positive from the positive end of the dipole (Fig. 1). In the aliphatic substitution, θ is zero, and therefore

$$E_e = -\varepsilon\mu/Da^2.$$

In the aromatic substitution, $\theta = \pi$, and consequently

$$E_e = +\varepsilon\mu/Da^2.$$

It is assumed that, in the aliphatic complex, the three hydrogen atoms in the central methyl group lie in the same plane, which is perpendicular to the figure axis containing the two partial bonds. In the aromatic complex, it has been argued (Miller, 1969) that the directions of the two partial bonds are outside the plane of the benzene ring, one of them above and the other below it. Despite the superficial similarity in the modes of formation of the critical complexes, it is clear that the approach of the anion is aided by the electrostatic field of the aliphatic molecule, and opposed by the field of the aromatic molecule. The contrast is well illustrated by reference to the apparent energies of activation (given in kcal/mole) of the following reactions in methanol solution ($\phi = 2,4(NO_2)_2C_6H_3$):

	E_A
$CH_3O^- + CH_3F$	26
$CH_3O^- + CH_3I$	22
$CH_3O^- + \phi F$	14
$CH_3O^- + \phi I$	20

REFERENCES

Abel, *Z. Electrochem.*, **14**, 598 (1908); *Z. physikal Chem.*, **96**, pp. 1–180 (1920); **136**, 161 (1928).

Amis, E. S., *J. Chem. Ed.*, **30**, 351 (1953); *Kinetics of Chemical Change in Solution*, Macmillan, New York, 1949: *Solvent Effects on Reaction Rates and Mechanisms*, Academic Press, 1966.

Arrhenius, *Z. physikal Chem.*, **1**, 110 (1887).

Baddeley and Bennett, *Trans. Chem. Soc.*, 1819 (1935).

Bell, *Trans. Faraday Soc.*, **27**, 797 (1931).

Bell and Coller, *Trans. Faraday Soc.*, **61**, 1445 (1965).

Bender, M. L., *Technique of Organic Chemistry*, 2nd edn., VIII (2), 1499, Wiley, New York, 1963.

Bennett and B. Jones, *Trans. Chem. Soc.*, 1815 (1935).

Bishop, D. M., and D. P. Craig, *Mol. Phys.*, **6**, 139 (1963).

Brinkman, M., S. Johnson and P. Ridd, *Trans. Chem. Soc.*, 228 (1962).

Brode, *Z. physikal Chem.*, **49**, 208 (1904).

Burgarsky, *Z. physikal Chem.*, **8**, 398 (1891).

Cavell and Speed, *Trans. Chem. Soc.*, 226 (1961).

Chandra and Coulson, *Trans. Chem. Soc.*, 2210 (1965).

Christiansen, *Z. physikal Chem.*, **117**, 433 (1925).

Conant and Hussey, *J. Amer. Chem. Soc.*, **47**, 476 (1925).

Conant and Kirner, *J. Amer. Chem. Soc.*, **46**, 232 (1924).

Cross, *Acta Physiochem.*, *U.S.S.R.*, **5**, 583 (1935).

Evans, D. P., Gordon and Watson, *Trans. Chem. Soc.*, 1430 (1937).

Fischer, E., *Annalen*, **381**, 123 (1911).

Gibson, Fawsett and Perrin, *Proc. Roy. Soc.*, **A**, **150**, 323 (1935).

Grube and Schmidt, *Z. physikal Chem.*, **119**, 19 (1926).

Harcourt and Esson, *Phil. Trans.*, **157**, 117 (1867).

Harned, *J. Amer. Chem. Soc.*, **40**, 1461 (1918).

Haywood, *Trans. Chem. Soc.*, **121**, 1904 (1922).
Hender and R. A. Robinson, *Trans. Faraday Soc.*, **29**, 1300 (1933).
Hughes, *Trans. Chem. Soc.*, 255 (1935).
Hughes, Juliusberger, Scott, Topley and Weiss, *Trans. Chem. Soc.*, 1173 (1936).
Ingold, *Structure and Mechanism in Organic Chemistry*, Bell, London, 1953.
Ingold and Nathan, *Trans. Chem. Soc.*, 222 (1936).
Kacser, *J. Phys. Chem.*, **56**, 1101 (1952).
Kirkwood, *J. Chem. Physics*, **2**, 351 (1934).
Laidler, *Chemical Kinetics*, 2nd edn., McGraw-Hill, 1965.
Laidler and Eyring, *Ann. N.Y. Acad. Sci.*, **39**, 303 (1940).
Liebhafsky, *J. Amer. Chem. Soc.*, **54**, 3499 (1932).
Liebhafsky and Mohammad, *J. Amer. Chem. Soc.*, **55**, 3977 (1933).
Magnanini, *Gazz. chim. Ital.*, **21**, 476 (1891).
McAlpine and Smyth, *J. Chem. Physics*, **3**, 55 (1935).
Meer and Polanyi, *Z. physikal Chem.*, **B, 19**, 164 (1932).
Miller, J., *Aromatic Nucleophilic Substitution*, Elsevier (1969).
Moelwyn-Hughes, E. A., *Acta Physicochim. U.S.S.R.*, **IV**, 201 (1936).
Moelwyn-Hughes, E. A., *Proc. Roy. Soc.*, **A, 157**, 667 (1936).
Moelwyn-Hughes, E. A., and Hurst, *Academia Nazionale dei Lincei*, Varenna, 127 (1957).
Nathan and Watson, *Trans. Chem. Soc.*, 890 (1933).
Noyes, *Z. physikal Chem.*, **19**, 601 (1896).
Olson, *J. Chem. Physics*, **1**, 418 (1933).
Olson, Frashier and Spieth, *J. Phys. Chem.*, **55**, 860 (1951).
Olson and Long, *J. Amer. Chem. Soc.*, **58**, 383 (1936).
Olson and Young, *J. Amer. Chem. Soc.*, **58**, 1157 (1936).
Pople, *Proc. Roy. Soc.*, **A, 233**, 233 (1955).
Reicher, *Annalen*, **228**, 257 (1885); **238**, 276 (1887).
Scatchard, *Chem. Rev.*, **10**, 229 (1932).
Segallar, *Trans. Chem. Soc.*, **103**, 1421 (1913).
Slator, *Trans. Chem. Soc.*, **87**, 485 (1905).
Smith, G. F., *Trans. Chem. Soc.*, 1744 (1934).
Spong, *Trans. Chem. Soc.*, 1283 (1934).
Steger, *Z. physikal Chem.*, **49**, 329 (1904).
Sutton, L. E., *Proc. Roy. Soc.*, **A, 133**, 668 (1931).
Szabo, *Über den Mechanismus einfacher Substitutionenand die Waldensche Umkehrung*, Thomas and Hubert, Berlin, 1933.
Thomson, J. J., *Phil. Mag.*, **46**, 497 (1923).
Tommila, *Suomen Kemistilehti*, **B, 25**, 37 (1952).
Tommila, Koivisto, Lyyra, Antell and Heimo, *Ann. Acad. Sci. Fennicae*, **II**, 47 (1952).
Tommila, Pentti and Ilomäki, *Acta Chim. Scand.*, **6**, 1249 (1952).
Walton, *Z. physikal Chem.*, **47**, 185 (1904).
Ward, *Trans. Chem. Soc.*, 2285 (1927).
Warder, *Berichte*, **14**, 1361 (1881); *J. Amer. Chem. Soc.*, **3**, 203 (1881).
Waters, *Phil. Mag.*, **8**, 436 (1929).
Waters, *Trans. Chem. Soc.*, 1551 (1933).
Werner, *Annalen*, **386**, 70 (1912).

10

UNIMOLECULAR REACTIONS

The rate-determining step in most unimolecular reactions is governed by the rearrangement of bonds and the redistribution of energies within the molecule, leading finally to its disruption. There is no *a priori* reason for supposing that the rate of such adjustments should be in general affected by the presence of surrounding molecules. Numerous unimolecular reactions are in fact known to proceed with the same velocity and apparent energy of activation in the gas phase as in a variety of solvents. Such solvents are loosely referred to as inert, although it would be more accurate to say that they influence the solute molecule to the same extent in the ground state as in the activated state. Many solvents, on the other hand, modify the rate of reaction without altering its mechanism or changing the rate to a greater extent than can be accounted for in terms of changes in simple properties of the solvent. Finally, there are solvents which exert a profound influence on both rates and mechanisms, and even allow to proceed in solution chemical changes which do not take place in the homogeneous gas phase.

An attempt is made in this chapter to distinguish between the true unimolecular reaction of a solute in a solvent and the pseudo-unimolecular reaction of a solute with a solvent. Hydrolysis and other solvolyses, which belong to the latter category, are dealt with in the next chapter, where it will be seen that the distinction is not always a clear-cut one. We shall first give brief accounts of the kinetics of certain reactions which have been investigated in the homogeneous gas phase and in solution.

The Decomposition of Ozone: $2O_3 \rightarrow 3O_2$

In the gaseous state, ozone decomposes by a complicated mechanism (Jahn, 1906; D. L. Chapman and H. E. Clarke, 1908; Benson and Axworthy, 1957; McKenney and Laidler, 1962), the heterogeneous component of the rate being unimolecular and the heterogeneous component being predominantly bimolecular but partly inhibited by the product of reaction. Some of the apparent energies of activation found for the homogeneous bimolecular reaction include $26,700$ cal/mole^{-1} (Warburg, 1902), $26,100$ (Clement, 1904), $26,000$ (Perman and Greaves, 1908), and $27,700$ (Belton, Griffith and McKeown, 1926).

In carbon tetrachloride solution, the reaction is unimolecular (E. J. Bowen, E. A. Moelwyn-Hughes and C. N. Hinshelwood, 1931). At a given temperature, the unimolecular constant is independent of the initial concentration of ozone, of its method of preparation and of the presence of water. In the neighbourhood of $336°K$,

$$k_1 \text{ (sec}^{-1}) = 3 \cdot 64 \times 10^{12} \cdot \exp\left[-(25,960 \pm 1,440)/RT\right]. \tag{1}$$

The apparent energy of activation is clearly the same in the two media, despite an undisputed difference in mechanism. If the rate in solution is proportional to the viscosity of the solvent, E_A is to be increased by $2,375$ cal, and if 12 quadratic terms are involved, an additional $6RT$ is to be added, making a total energy of activation of $32,400 \pm 1,500$, which lies near to the oxygen–oxygen bond energy in hydrogen peroxide ($33,300$). It is thus possible that the rate-determining step in solution is simply the break-down $O_3 \rightarrow O_2 + O$, followed by rapid reactions of the oxygen atoms. There is, however, some oxidation of the solvent ($O_3 + CCl_4 \rightarrow COCl_2 + Cl_2 + O_2$), one molecule of which is attacked for every 17 molecules of ozone destroyed. This oxidation does not account for more than a small fraction of the total change, and is either a relatively unimportant side-reaction, or the first stage of a chain reaction in which 17 molecules of ozone are destroyed for each initial act. A chain mechanism has been suggested for the gas-phase reaction, but the weight of evidence is against it (see Kassel, 1932). According to Riesenfeld and Wassmuth (1929), there is a homogeneous unimolecular component of the reaction rate in the gas phase, with $k_1 = 1 \cdot 38 \times 10^{-4}$ sec^{-1}, which is within a factor of 6 of the unimolecular constant for the reaction in carbon tetrachloride at the same temperature ($90°C$). It would thus appear that the complexity of the decomposition in the gas phase is linked with its bimolecular component, which seems to be absent in solution, leaving the unimolecular velocity constant and the apparent energy of activation approximately the same in the two phases.

The Decomposition of Dinitrogen Pentoxide in the Gaseous Phase and in Solution

This classical example of a first-order homogeneous reaction (F. Daniels and E. H. Johnson, 1921) has been carefully investigated under a wide variety of conditions, using chemical analysis as well as the increase in pressure and colour which attend the decomposition. The results were first interpreted on a molecular basis and later in terms of free radicals.

The scheme, according to the molecular interpretation, consists of three steps:

$$N_2O_5 + N_2O_5 \underset{\substack{k_2 \\ \overline{k_4}}}{\rightleftharpoons} N_2O_5 + N_2O_5 \xrightarrow{k_3} \text{Products}.$$
$$\qquad\quad {}_n \qquad\quad {}_n \qquad\qquad {}_n \qquad\quad {}_a$$

Normal molecules are activated in binary collisions, attaining a stationary concentration, a. The fate of the activated molecules is either to lose their energy in binary collisions or to break down spontaneously into products. Under stationary conditions

$$da/dt = k_2 n^2 - k_4 na - k_3 a = 0, \tag{4.9}$$

and the rate of reaction is

$$-\frac{dn}{dt} = k_3 a = \frac{k_3 k_2 n^2}{k_4 n + k_3}. \tag{4.10}$$

This simple hypothesis can explain C. N. Hinshelwood and H. W. Thompson's (1926) observation that the apparent order of a gas reaction can be 1 at high pressures and 2 at low pressures. The equation for the observed velocity constant,

$$k_1 = -\frac{1}{n} \cdot \frac{dn}{dt} = \frac{k_3 k_2 n}{k_4 n + k_3} \tag{2}$$

reduces at high pressures, to

$$k_1 = k_3 \frac{k_2}{k_4} = k_3 K.$$

The Boltzmann distribution law is here maintained, and the rate of reaction is governed by the spontaneous decomposition of the activated molecules. At low pressures, however, we have

$$-\frac{dn}{dt} = k_2 n^2$$

showing that the rate of reaction is now the rate at which molecules are activated (Trautz, 1916; McC. Lewis, 1918; Lindemann, 1922; Christiansen, 1923). Equation (2) can be written in the form

$$\frac{1}{k_1} = \frac{k_4}{k_3 k_2} + \frac{1}{k_2} \cdot \frac{1}{n} \tag{2a}$$

from which k_2 and k_4/k_3 may be determined. The data of Schumacher and

Sprenger (1930) at 35°C, analysed in this way, yield the values $k_2 = 2.11 \times 10^1$ litres/mole-sec, and $k_4/k_2 = 1.37 \times 10^5$ litres/mole. The first of these constants can be examined in the light of equations (4.46) and (4.47), accepting the experimental value of $E_A = 24,710$ cal at 305.6°K. In order to account for the rate constant, it is necessary to take s as 19, so that $E = E_A + (35/2)RT = 36,420$. The unimolecular constants observed by various workers can be summarized empirically as follows:

$$\log_{10} k_1 \text{ (sec}^{-1}) = 66.7661 - 18.0533 \log_{10} T - 7,920.87/T, \tag{3}$$

according to which $E = E_A + 18.0533RT = 36,238$, in close agreement with the value found by the first method. If deactivations occur at every collision between normal and active molecules, k_4 may be identified with the standard collision frequency, Z^0, and k_3 becomes 7.74×10^5. The average life time of the activated molecules is $1/k_3 = 1.29 \times 10^{-6}$ sec.

The free radical mechanism, formulated for gas reactions in general by Rice and Herzfeld (1934), has been applied by Ogg (1947) to this reaction. With but a slight modification, it can be represented by the sequence:

$$\underset{n}{N_2O_5} + \underset{n}{N_2O_5} \rightarrow \underset{n}{N_2O_5} + \underset{a}{NO_3} + \underset{b}{NO_2} \quad k_1$$

$$\underset{a}{NO_3} + \underset{b}{NO_2} + \underset{n}{N_2O_5} \rightarrow \underset{n}{N_2O_5} + \underset{n}{N_2O_5} \quad k_2$$

$$\underset{a}{NO_3} + \underset{b}{NO_2} \rightarrow \underset{b}{NO_2} + O_2 + \underset{c}{NO} \quad k_3$$

$$\underset{c}{NO} + \underset{n}{N_2O_5} \rightarrow \underset{b}{3NO_2} \quad k_4.$$

The first step is now taken to be the dissociation of one molecule of reactant into NO_3 and NO_2 in a binary collision. The second step is the reverse of this reaction. Nitric oxide is assumed to be formed in the third step and destroyed in the last step. The stationary concentrations are obtained in the usual way:

$$da/dt = k_1 n^2 - k_2 nab - k_3 ab = 0 \quad \therefore \quad ab = k_1 n^2/(k_2 n + k_3)$$

$$dc/dt = k_3 ab - k_4 cn = 0 \quad \therefore \quad c = k_3 ab/k_4 n.$$

The rate of reaction is $db/dt = k_1 n^2 - k_2 nab + 3k_4 cn$. After eliminating ab and c, and noting that the rate of production of NO_2 is twice as great as the rate of disappearance of N_2O_5, we obtain the rate equation

$$-\frac{dn}{dt} = \frac{2k_1 k_3 n^2}{k_2 n + k_3} \tag{4}$$

which, like that derived on the molecular basis (equation 4.10) is in a form to account for kinetics of the first, second and intermediate orders.

In deciding which of the two interpretations is the right one, we note in the first place that the molecular mechanism is the simpler one and is self-consistent. Without supplementary evidence, it would, indeed, be difficult to justify its rejection. Such supplementary evidence, nevertheless, is provided by absorption spectra, which prove that the radical NO_3 does in fact exist in the reacting system. Moreover, reaction 4 has been separately examined, and the relative slowness of reaction 3 established by using the isotopes N^{13} and N^{15} (J. H. Smith and Daniels, 1933). The free radical mechanism thus wins the day, not only in this instance but in most other apparently unimolecular reactions in gaseous systems. Many methods are available for the detection of transient intermediaries. Their identification, by means of the mass spectrometer, for example, guides and limits the choice of reaction mechanism which, without collateral evidence, could be misleading.

Solvents in general have only a slight effect on the rate of decomposition (Table I). Nitromethane has no influence whatsoever. Most solvents cause a slight increase in k_1 over the gas value. The contrast between the effects of ethylene dichloride and propylene dichloride is striking. Since the rate of

TABLE I

THE DECOMPOSITION OF NITROGEN PENTOXIDE IN VARIOUS MEDIA

Reference	Medium	$k_{20°C} \times 10^5$ (sec^{-1})	E_A (kcal/mole)	ln A
*	Gas	1·65	24·7	31·48
†	N_2O_4	3·44	25·0	32·82
†	CH_3CHCl_2	3·22	24·9	32·56
†	$CHCl_3$	2·74	24·6	31·90
‡	$CHCl_3$	2·14	24·5	31·22
†	$C_2H_4Cl_2$	2·38	24·4	31·42
†	CCl_4	2·35	24·2	31·05
‡	CCl_4	2·34	24·5	33·09
†	$CHCl_2CCl_3$	2·20	25·0	32·35
†	Br_2	2·15	24·0	30·61
†	CH_3NO_2	1·67	24·5	31·13
†	$C_3H_6Cl_2$	0·24	28·3	35·72
†	HNO_3	0·05	28·3	34·11

REFERENCES

* Daniels and Johnston, *J. Amer. Chem. Soc.*, **43**, 53 (1921); White and Tolman, *ibid.*, **47**, 1240 (1925).

† Eyring and Daniels, *ibid.*, **52**, 1473 (1930).

‡ Lueck, *ibid.*, **44**, 757 (1922).

reaction generally is not greatly affected by the presence of a solvent, it may be concluded that the various bimolecular steps postulated in the mechanism are also unaffected, as would be expected if the standard collision frequency were independent of the medium. This conclusion, as we have seen, has been reached by all those experimentalists who have measured the velocity of bimolecular reactions in the gas phase and in solution.

The Simplest Expression for a Unimolecular Velocity Constant

When activation is confined to a single bond which is loosened to the breaking point, an approximate expression for the unimolecular constant is that derived by Herzfeld:

$$k_1 = \frac{kT}{h} \cdot \exp\left(-\Delta E_0/RT\right). \tag{4.31}$$

Since $E_A = \Delta E_0 + RT$, we may also write

$$k_1 = \frac{kTe}{h} \cdot \exp\left(-E_A/RT\right). \tag{4.32}$$

The first parameter of the Arrhenius equation in such a case becomes

$$A_1 = \frac{kTe}{h}. \tag{5}$$

Numerically at 25°C, $A_1 = 1\cdot689 \times 10^{13}$ sec^{-1}, and $\ln_e A_1 = 30\cdot47$. Such values are frequently found for unimolecular reactions in solution and in the gaseous phase, as may be seen from Table I, even when the mechanism of activation is known to be not so simple as that implied in the derivation.

Further Comparisons between Unimolecular Reactions in the Gas Phase and in Solution

The conversion of *d*-pinene into *dl*-limonene (Conant and Carlson, 1929) in the gasous phase is a unimolecular reaction, the rate constants of which, determined polarimetrically (D. F. Smith, 1927) are summarized by the equation

$$\ln k_1 = 33{\cdot}77 - \frac{43,710}{RT}. \tag{6}$$

The energy is probably reliable to within ± 2 kcal. In the pure liquid phase at $184{\cdot}5°$C, the reaction proceeds at a rate which is 36 per cent faster; and in petrolatum solution over a range of temperatures, k is 48 per cent greater than in the gas phase. E_A for the reaction in solution is $41{\cdot}2$ kcal. No real change has thus been detected between the Arrhenius parameters for the two systems.

The racemization of $2:2'$-diamino-$6:6'$-dimethylbiphenyl is a homogeneous unimolecular reaction in the gaseous phase and in diphenylether solution, with the same E_A value in both systems (Kistiakowsky and W. R. Smith, 1936). In the gas phase,

$$k_1 = 2{\cdot}9 \times 10^{11} \,.\, \exp\,(-45,100/RT). \tag{7}$$

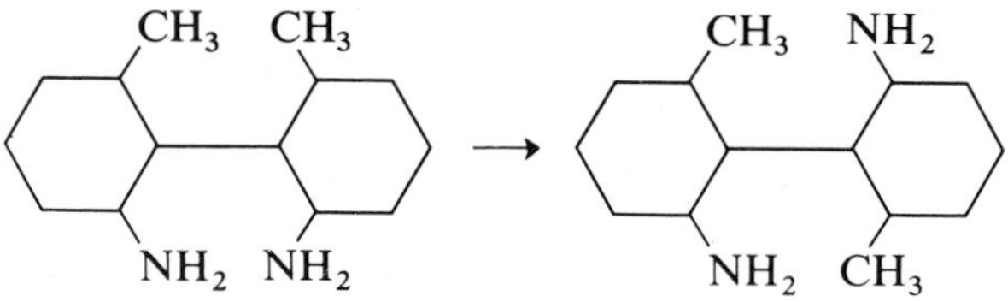

Bicyclo [3,2,0]-hept-6-ene is converted completely, unimolecularly and homogeneously into cyclohepta-1,3-diene:

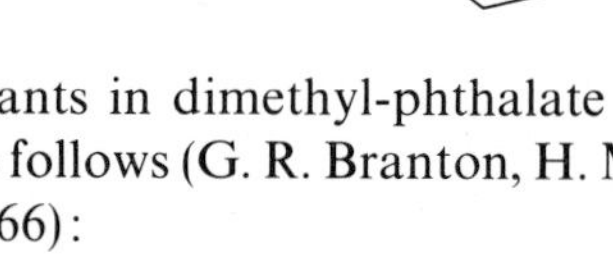

The unimolecular constants in dimethyl-phthalate solution and in the gas phase are summarized as follows (G. R. Branton, H. M. Frey, D. C. Montague and I. D. R. Stevens, 1966):

$$k_s = 4{\cdot}47 \times 10^{14} \,.\, \exp\,(-45,860/RT), \tag{8}$$

$$k_g = 2{\cdot}04 \times 10^{14} \,.\, \exp\,(-45,510/RT), \tag{9}$$

$$k_s/k_g = 2{\cdot}19 \,.\, \exp\,(-350/RT). \tag{10}$$

Over the range of temperature examined (529 to 600°K), this ratio increases slightly from $1{\cdot}57$ to $1{\cdot}63$. The virtual identity of the activation energies makes it clear that, if there is any kind of interaction between solute and solvent, it is the same for the normal and activated solute. The pre-exponential parameters in both equations exceed the term kTe/h of equation (5) by a factor of about 10. More satisfactory agreement is found by combining Herzfeld's equation (4.31) with the probability that the energy of activation is shared among 4 quadratic terms, i.e. by taking $s = 2$ in equation (10.55). The unimolecular constant then becomes

$$k_1 = \frac{kT}{h}\left(\frac{\Delta E_0}{RT} + 1\right) . \exp\left(-\Delta E_0/RT\right),$$

or, approximately,

$$k_1 = \frac{\Delta \varepsilon_0}{h} \cdot \exp\left(-\Delta \varepsilon_0/kT\right),$$

as first derived by Dushman (Chapter 4). The value of $\Delta\varepsilon_0/h$ for the reaction in solution is $4 \cdot 8 \times 10^{14}$ sec^{-1}.

The Kinetics of Decarboxylations (RCOOH → RH + CO₂) in Solution

Monocarboxylic and dicarboxylic acids decompose in solution uni-molecularly and irreversibly with rates that have been measured with concordant results from the decrease in the concentration of acid, $c_t = c_0 . \exp\left(-k_1 t\right)$, and the increase in the pressure of the system, $p_t = p_\infty - (p_\infty - p_0) . \exp\left(-k_1 t\right)$. (See Fig. 6.1.) The unimolecular constant has been found to be independent of a ten-fold change in the initial concentration in the case of $2:3:5$-trinitrobenzoic acid in water and to decrease with an increase in the initial concentration in the case of trichloracetic acid in the same solvent (Kappanna, 1932). The most noteworthy feature of decarboxylations is the complete failure of the two-constant equation of Arrhenius to account for the temperature coefficient of the velocity constant: at least three constants are required to summarize empirically the variation of $\ln k_1$ with respect to T. Those found by Johnson and Moelwyn-Hughes in aqueous solution are as follows:

for trichloracetic acid: $\ln k_1 = 110{,}824 - 42{,}910/RT - (20/R)\ln T$,

for tribomacetic acid: $\ln k_1 = 109{,}862 - 39{,}610/RT - (20/R)\ln T$,

for trinitrobenzoic acid: $\ln k_1 = 197{,}728 - 52{,}300/RT - (45/R)\ln T$.

$$(11)$$

The fast rate of the bromine compound compared with the chlorine analogue is clearly due to the difference of some 3,000 calories in the energy of activation. The tri-iodoacid proved to be too unstable to work with. It follows that, in these instances, the apparent energy of activation decreases linearly with respect to temperature:

$$CCl_3COOH : E_A = 42,910 - 20T,$$

$$CBr_3COOH : E_A = 39,610 - 20T,$$

$$C_6H_2(NO_2)_3COOH : E_A = 52,300 - 45T. \tag{12}$$

It is not possible to give more than rounded integral values to the coefficients of T. For the decarboxylation of oxalic acid in dioxane solution, the approximate equation obtained by Dinglinger and Schröer (1937) is

$$(COOH)_2 : E_A = 52,850 - 57T. \tag{13}$$

A still higher value of E_0 is afforded by the work of Bernouilli and Jakubowicz (1921) on the decarboxylation of malonic acid in water:

$$HOOC·CH_2·COOH : E_A = 58,750 - 76·5T. \tag{14}$$

These values of E_0 lie near to the bond energy of the carbon–carbon link.

The decomposition of acetonedicarboxylic acid is catalysed in water by weak bases (Wiig, 1928), such as aniline, the catalytic coefficient for which has been evaluated (equation 4.89). Trinitrobenzoic acid, however, is affected only slightly, if at all, by this base or by pyridine.

There has been considerable discussion as to whether the anion or the undissociated acid is responsible for the rate of elimination of carbon dioxide from aqueous solutions of carboxylic acids. There is little doubt that both contribute to varying extents, depending on the degree of ionization of the acid. Thus, for example, sodium tribromacetate decomposes in water 1·66 times as fast as the undissociated acid within the narrow temperature range where it can be measured without undue interference from the hydrolysis of the anion. The apparent energy of activation for the ionic decomposition at 345°K is 34 ± 2 kcal, which, within the limits of error, agrees with the value 32·95 found for the acid solution at the same temperature. In the case of trichloracetic acid, there is much closer agreement. For the anion at 345°K, Fairclough (1938) finds $E_A = 36·6$, which is only slightly greater than Johnson's value $E_A = 36·0$ for the acid. These results are consistent with the hypothesis that it is the ion which mainly decomposes. If that is true, the rate of decomposition of the acid in non-polar solvents may be expected to be much slower than in water. Difficulties, however, attend the experimental investigation of the decomposition of tribromacetic acid in toluene. Rough data suggest a rate slower than that in water by a factor of about 10^7, and an E_A value some 13·0 kcal higher. Unfortunately, this solute reacts with the solvent at high temperatures, yielding benzyl bromide and the three bromtoluenes (Johnson and Moelwyn-Hughes, 1940).

The rate at which trinitrobenzoic acid decomposes in aromatic solvents is extremely sensitive to traces of impurity. The bracketed data for acetophenone given in Table II refer to a solvent purified by fractional crystallization only, while the unbracketed figures refer to the solvent after further purification by storage over phosphorus pentoxide and distillation at 6 mm Hg. The final purification has halved the rate. With nitrobenzene, stringent purification to conform with conductivity standards and the complete exclusion of moisture have a much greater effect (Hinshelwood and Moelwyn-Hughes, 1931). There appears to be a very rough proportionality between E_A and ln A for the different solvents, but the results as a whole have proved too difficult to interpret. Ancillary static data are lacking.

TABLE II

THE DECOMPOSITION OF $1:3:5$-TRINITROBENZOIC ACID IN VARIOUS SOLVENTS

Solvent	k_1 (sec^{-1}) at 60°C	E_A (cal/mole^{-1}) at 80°C
Toluene	$1\cdot62 \times 10^{-9}$	31,600
Acetophenone	$5\cdot79 \times 10^{-7}$	25,450
	$(1\cdot35 \times 10^{-6})$	(24,130)
Anisole	$1\cdot97 \times 10^{-7}$	30,730
Nitrobenzene	$4\cdot07 \times 10^{-9}$	34,990
	$(6\cdot35 \times 10^{-6})$	(26,320)
Water	$2\cdot09 \times 10^{-6}$	36,400

G. A. Hall and F. H. Verhoek (1947) have suggested that the rate-determining step in the decomposition of trichloracetic acid is the break down of the ion CCl_3COO^- into CO_2 and CCl_3^-, and have supported their hypothesis by kinetic investigations of the decomposition of various salts of the acid in ethanol–water mixtures. Reasonable values of the association constants have been obtained from their kinetic measurements. When trinitrobenzoic acid is studied in dioxane–water mixtures, the order of reaction with respect to the solute is found to decrease from unity in water to one-half in dioxane, which is again consistent with the ionic mechanism (Trivich and Verhoek, 1943). With phenylmalonic acid in water, the singly charged ion decomposes three times as rapidly as the undissociated acid (Gelles, 1953).

The kinetics of the decarboxylation of two trihalogenocarboxylate ions in ethylene glycol solution (I. Auerbach, F. H. Verhoek and A. L. Henne,

1950) have been compared with the following results:

$$CCl_3COO^- \rightarrow CCl_3^- + CO_2;$$

$$k_1 \, (\sec^{-1}) = 2 \cdot 54 \times 10^{15} \cdot \exp(-31{,}600/RT), \quad (15)$$

$$CF_3COO^- \rightarrow CF_3^- + CO_2$$

$$k_1 \, (\sec^{-1}) = 1 \cdot 53 \times 10^{16} \cdot \exp(-42{,}000/RT). \quad (16)$$

At the mean experimental temperature of 525°K, $k_1(chloro)/k_1(fluoro)$ is $1 \cdot 3 \times 10^5$, and is due mainly to the difference in the apparent energies of activation, as the investigators point out. It is postulated that unstable ions formed subsequently react rapidly with the proton to yield chloroform and fluoroform. Attempts to establish further relationships between the energy of activation and the carbon–carbon bond strength by studying the difluoro-chloroacetate have been frustrated by hydrolyses.

The apparent energy of activation for the decomposition of trichloracetic acid in the fused state at 385°K is $E_A = 41 \cdot 7$ kcal (L. W. Clark, 1955, 1960), which is about 7,700 cal higher than its value in aqueous solution.

Kappanna (1932) has interpreted his results (Table III) on the concentration effect in aqueous trichloracetic acid according to equation (6.27). The

TABLE III

THE INFLUENCE OF CONCENTRATION ON THE RATE OF DECOMPOSITION
OF TRICHLORACETIC ACID IN WATER AT 80°C

[CCl$_3$·COOH] (moles/litre)	$k_1 \times 10^5$ (min^{-1}; log$_{10}$)	Density (gm/cc)	[H$_2$O] (moles/litre)	x_S (mol fraction of solvent)
0	183·5	0·972	54·00	1·000
0·25	156·6	0·988	52·61	0·995
0·50	136·9	1·007	51·40	0·990
1·00	98·5	1·043	48·87	0·980
2·00	51·0	1·1105	43·49	0·954
3·00	24·8	1·1800	38·33	0·927
4·00	10·1	1·2475	33·00	0·892
5·00	3·45	1·3150	27·68	0·847

order of reaction with respect to the water is 6, as found in certain other reactions discussed in Chapter 6.

Because most of the known instances of a decrease in E_A with a rise in temperature are reactions in aqueous solution, it can be argued that the effect is specific to water as a medium. That it occurs for decarboxylations in dioxane and ethylene glycol proves that the effect is more general.

Cyclizations

Saturated aliphatic hydrocarbons, though loosely referred to as straight-chained, are not in fact straight but flexible chains. When the two substituents attached to the terminal carbon atoms are capable of chemical reaction, the string of intervening methylene groups can be joined to form a closed ring. Such processes are generally unimolecular. Arrhenius parameters for the ring closures of certain amino-alkyl halides in aqueous solution

$$X-[CH_2]_n-N^+H_3 \rightarrow (CH_2)_n \quad N^+H_2 + H^+ + X^-$$

are given in Table IV, where X stands for a halogen atom and n is an integer. Thus, for example, the conversion of ε-chloramylamine into piperidine hydrochloride corresponds to X = Cl and $n = 5$. Latterly, chains containing as

TABLE IV

Arrhenius Parameters for Certain Cyclization
Reactions in Aqueous Alkaline Solution

X	n	E_A (kcal/mole)	$\ln_\varepsilon A$ (sec^{-1})
Br	2	24·9	34·7*
Br	3	22·3	26·1†
Cl	4	19·8	30·8‡
Cl	5	20·7	26·0§

REFERENCES

* Freundlich and Neumann, *Z. physikal Chem.*, **87**, 69 (1914).
† Freundlich and Kroepelin, *ibid.*, **122**, 39 (1926).
‡ Freundlich and Salomon, *Berichte*, **66**, 355 (1933).
§ Salomon, *Helv. Chim. Acta*, **19**, 743 (1936).

many as seventeen methylene groups have been investigated (Salomon, 1936a), some of them in various solvents. Salomon (1936b) has examined the energies of activation to discover what fraction may be represented by the energy of buckling, which it is not difficult to estimate by adopting the hypothesis of Langmuir and Butler, i.e. by summing the solute–solvent interaction energies for the various methylene groups. The total interaction energy is generally less for the coiled than for the drawn-out configuration.

Another set of cyclizations is afforded by the decomposition of ω-substituted halogenocarboxylate ions, such as

$$
\begin{array}{c}
\mathrm{O} \\
\diagup\!\!\!\diagdown \\
\mathrm{C} \\
\diagup \quad \diagdown \\
(\mathrm{CH_2})_n \quad \mathrm{O}^- \\
\diagdown \\
\mathrm{Br}
\end{array}
\;\rightarrow\;
\begin{array}{c}
(\mathrm{CH_2})_n
\begin{array}{|c}
\hline
\mathrm{C}\!=\!\mathrm{O} \\
\mid \\
\mathrm{O} \\
\hline
\end{array}
\end{array}
+\;\mathrm{Br}^-
$$

which have been extensively investigated (L. Ruzicka, W. Brugger, M. Pfeiffer, H. Schinz and M. Stoll, 1926; G. M. Bennett, 1941; A. G. Davies, Mansel Davies and M. Stoll, 1954). From solutions of the potassium salts of ω-bromundecanoic acid ($n = 10$) and ω-brompentadecanoic acid ($n = 14$) in methylethylketone, at least 80 per cent of the simple product lactones could be isolated. The remaining products are polyesters. The rate of reaction is probably that of the ionization

$$
\mathrm{BrCH_2(CH_2)_nCOO^-} \rightarrow \mathrm{Br^-} + \overset{+}{\mathrm{C}}\mathrm{H_2(CH_2)_nCOO^-},
$$

followed by rapid cyclization or by the formation of polyesters. The unimolecular constant for the brompentadecanoate ion is given by the equation $k_1\,(\mathrm{sec^{-1}}) = 6\cdot3 \times 10^9 \,.\, \exp\left[-(22{,}900 \pm 800/RT)\right]$. Not all cyclizations follow this course. The instantaneous rate of conversion of γ-hydroxyvaleric acid in aqueous solution is proportional to its concentration and to the concentration of hydrogen ion (Garrett and W. C. McC. Lewis, 1923):

$$
\begin{array}{c}
\mathrm{O} \\
\parallel \\
\mathrm{CH_2\!-\!C\!-\!OH} \\
\mid \\
\mathrm{CH_2\!-\!CH\!-\!OH} \\
\mid \\
\mathrm{CH_3}
\end{array}
\;\xrightarrow{+\mathrm{H}^{\pm}}\;
\begin{array}{c}
\mathrm{O} \\
\parallel \\
\mathrm{CH_2\!-\!C} \\
\mid \qquad\quad \diagdown \\
\qquad\qquad \mathrm{O} \;+\; \mathrm{H_2O}. \\
\mid \qquad\quad \diagup \\
\mathrm{CH_2\!-\!CH} \\
\mid \\
\mathrm{CH_3}
\end{array}
$$

The Racemization of Biphenyl Derivatives

The racemization of substituted biphenyl derivatives proceeds in the gaseous phase and in solution according to the mechanism of reversible unimolecular reactions. Their rates are found by determining the specific optical rotation, α_t, at measured times, from its initial value of α_0 to its final value of zero. Since $k_1 = k_2$ in this instance, the unimolecular velocity coefficient of the racemization of either reactant is, according to equation (6.19),

$$k_1 = (1/2t) \ln_e(\alpha_0/\alpha_t). \tag{17}$$

The velocity of racemization of 2,2′-dibromo-4,4′-dicarboxybiphenyl in ethanol has been measured at six temperatures in the range $266 \pm 13°K$ by M. M. Harris and R. K. Mitchell (1960) who find, by the method of least squares:

$$k_1 (\text{sec}^{-1}) = 1.07 \times 10^{12} . \exp(-18,900/RT). \tag{18}$$

The error in the energy of activation is ± 500 cal.

The change necessitates the rotation of the plane of one benzene ring relative to the plane of the other. The configuration of the critically activated molecule is generally regarded as that in which the substituent groups are co-planar with the two benzene rings, and the energy of activation as that required to rotate one part of the molecule relative to the other about the biphenyl axis and counter to the repulsions arising from steric hindrance (E. E. Turner and M. M. Harris, 1952). Figure 1 shows a diagram of the critically activated structure of 2,2′-dibromobiphenyl, drawn using conventional covalent radii, and omitting any representation of the hydrogen atoms other than those in positions 6 and 6′. The structure of the two isomers may be visualized by imagining the benzene ring shown on the left hand side to remain the plane of the paper while (1), for one isomer, the hydrogen atom at position 6 on the right hand side is below the plane of the paper, and the bromine atom at position 2 on the right hand side is above it, and (2), for the other isomer, the hydrogen atom on the right is above and the bromine atom on the right below the plane of the paper.

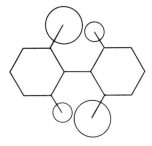

FIG. 1. The planar configuration of 2,2′-dibromobiphenyl.

It is clear from the diagram that the free rotation about the biphenyl axis required for racemization is opposed by, and will not occur without overcoming, the repulsions exerted between the groups in positions 2,2′ and 6,6′. The bending of certain bonds, and the distension of the bond joining the two phenyl radicals can assist in the process. Detailed calculations, made by Westheimer (1947; see also Rieger and Westheimer, 1950) of the critical energy required for the racemization of various derivatives of biphenyl, using experimental values of interatomic and intermolecular constants, gave results in satisfactory agreement with his own experimental values and with the results of other workers in this field, notably Li and Adams (1935); Kuhn and Alrecht (1927), and Turner and Harris (1952). According to Westheimer, the total energy of activation (18 kcal) may be resolved into 7 kcal for the deflections of the C—Br bonds, 6 kcal for intrinsic repulsions, and 5 kcal for all other deformations.

The reactions present kinetic features of more than usual interest. A simple formulation is possible if it is assumed that all the energy is expended in overcoming the hindrance to rotational motion about the biphenyl axis in a direction perpendicular to it. The potential energy of the rotator in this case may be taken as

$$V = \tfrac{1}{2}V^0(1 - \cos 2\theta) = V^0 \sin^2 \theta, \tag{19}$$

where θ is the angle denoting the inclination of one plane relative to the other from a mutually perpendicular position, i.e. θ is taken as zero when the two benzene rings occupy planes are at right angles, and as 90° when they are co-planar. For small displacements from the position of minimum energy there then results a torsional oscillation of frequency

$$\nu = \frac{1}{\pi}\sqrt{\frac{V^0}{2I}}, \tag{20}$$

where I is the moment of inertia of one half of the molecule about its axis of rotation. For 2,2′-diamino-6,6′-dimethyl-biphenyl, I is found to be 648×10^{40} g-cm^2 and for 2,2′-dibromo-4,4′-dicarboxy-biphenyl, $I = 1234 \times 10^{-40}$. If we identify V^0 with the apparent energy of activation, we find ν to have the respective values of $1\cdot57 \times 10^{12}$ and $7\cdot34 \times 10^{11}$ sec^{-1}. Thus $h\nu$ is small compared with kT, and the vibrational partition function of the molecule in its ground state approximates to $kT/h\nu$. The active molecules are those possessing energies equal to or greater than V^0, and these can be regarded as capable of free rotation in a plane at right angles to the biphenyl axis: the rotational partition function for such a motion is $2\pi(2\pi IkT)^{1/2}/h$. The other components of the total partition functions may be regarded as having the same values in the ground and activated states; hence the equilibrium constant is

$$K = \frac{n^*}{n} = \frac{q^*}{q} = \frac{2\pi(2\pi IkT)^{1/2}/h}{kT/hv} \cdot \exp\left(-V^0/kT\right)$$

$$= 2\pi v \left(\frac{2\pi I}{kT}\right)^{1/2} \cdot \exp\left(-v^0/kT\right). \tag{21}$$

The average angular velocity in a given direction is $\bar{\omega} = (kT/2\pi I)^{1/2}$, and the frequency of complete rotation is $\bar{\omega}/2\pi$. Since each complete rotation effects two racemizations, the unimolecular velocity constant becomes (Bauer, 1944)

$$k_1 = 2v \cdot \exp\left(-V^0/kT\right). \tag{22}$$

Thus the calculated values of the first Arrhenius parameter A are 3.14×10^{12} in Kistiakowsky's case, and 1.47×10^{12} sec^{-1} in that of Harris and Mitchell. The former is probably, and the latter certainly, in agreement with experiment.

Equation (22) could more readily have been reached on two basic assumptions, namely, (1) the fractional number of molecules which possess an energy of at least V^0 each, expressible as the sum of two quadratic terms, is $\exp\left(-V^0/kT\right)$, and (2) racemization occurs when the active molecules reach a suitable phase in their internal motions. It is to be emphasized that the frequency v is a property of the normal state of the molecule, rather than of the activated state, as in the derivation of Herzfeld's equation (4.3).

The Effect of Substituents on the Racemization of Biphenyls and Binaphthyls

Harris and her collaborators (Ling and Harris, 1963; A. S. Cook and Harris, 1963; Badar and Harris, 1964; Harris and Mellor, 1961; Harris and Mellor, 1959; D. M. Hall and Harris, 1960; and Leffler and Craybill, 1963) have carried out accurate and extensive kinetic investigations on the effect of substituent groups attached at various positions in the biphenyl and binaphthyl molecules under a variety of solvent and temperature conditions. Only a fraction of their findings can be discussed here.

The Arrhenius parameters obtained by Harris for the racemization of 4,4'-disubstituted 2,2'-di-iodobiphenyls are summarized in Table V. The alternative parameters obtained from equations IV, 36 and IV, 27 are $\Delta^0 H^{\ddagger} = E_A - RT$, and $\Delta^0 S^{\ddagger} = 4 \cdot 557(\log_{10} A - 13 \cdot 2277)$. Since $\log_{10} A$ is found to lie in the region $11 \cdot 9 \pm 0 \cdot 4$, $\Delta^0 S^{\ddagger}$ lies in the region $-6 \cdot 2 \pm 1 \cdot 7$. According to equation (4.26), however, $\Delta^0 S^*$ is zero, provided v is independent of temperature. The differences in the E_A values given in the table, though slight, are significant, and have been critically examined by Harris, who resolves the energy of activation into components due to (1) the energy required to overcome the intrinsic repulsion of the groups in the 2 and 2′ positions when they pass each other, (2) the difference between the internal strains of the solute in the ground and energized states, and (3) the difference between any resonance energy in the two states. In the biphenyl derivatives, there is little doubt that the first of these contributions is the greatest since Harris' observed energies of activation agree reasonably well with the theoretical estimates of Westheimer (1956) and Howlett (1960). Evidence that biphenyl derivatives in both states are strained is provided by X-ray analyses of the crystals, electron diffraction studies on the vapour, dipole moments and ultraviolet absorption spectra. In the activated state, the two benzene rings are not accurately coplanar, and the extra-annular bonds are splayed out of both planes. There is no general agreement concerning the magnitude of the resonance effect. Some of the data seem capable of interpretation without reference to it.

From Table V, it is seen that substitution of amino groups for hydrogen atoms in the 4 and 4′ positions has resulted in an increase in the velocity constant by a factor of 18, which may be resolved into the factors 2·5 arising from a change in A values and 7 arising from a decrease in E_A. The data thus resemble many others already met with, where most of the substitution effects is due to energy changes. According to equations (20) and (22), we

TABLE V

ARRHENIUS CONSTANTS FOR THE RACEMIZATION OF 4,4′-DISUBSTITUTED
2,2′-DI-IODOBIPHENYLS

4,4′-Substituent X	Solvent	$\log_{10} A$ (sec^{-1})	E_A (kcal)
—H	NN-dimethylformamide	11·5	21·0
—COOCH$_3$	NN-dimethylformamide	12·4	22·3
—COOH	NN-dimethylformamide	11·9	21·6
—NH·COOCH$_3$	NN-dimethylformamide	12·0	20·6
—NH$_2$	NN-dimethylformamide	11·9	19·9
—COOH	C$_2$H$_5$OH	11·7	21·3
—COO$^-$	0·1N aqueous NaOH	11·6	21·5
—NH$_3^+$	0·5N aqueous HCl	12·3	22·2

have

$$k_1 = \frac{1}{\pi}\left(\frac{2E_A}{I}\right)^{1/2} . \exp\left(-E_A/RT\right). \tag{23}$$

In the series considered, the magnitude of the moment of inertia, I, is determined chiefly by the iodine atoms, and is but slightly affected by the relatively light substituents in the 4 and 4' positions. If we accept a moment of inertia common to all the solutes, it follows that $\log_{10} A$ should be proportional to $(1/2)\log_{10} E_A$. No such correlation is found. The actual gradients are of a higher order of magnitude, suggesting a more complicated mechanism than that discussed here.

It is clear from Table V that the effect of the solvent on the rate of racemization of the 2,2' di-iode 4,4' dicarboxylic compound is slight. A value of about 21·5 kcal may then be assumed for its racemization in the gas phase, if it could be measured. Comparison with Kistiakowsky's result on the racemization of 2,2' di-amino 6,6' dimethylbiphenyl stresses the fact that the most powerful hindrance to rotation is found when the substituents are in these positions, as could be expected.

Leffler and Graybill (1963) have measured the rates of racemization of numerous 2,2'-6,6' disubstituted biphenyls in a variety of solvents over a range of temperatures. We shall confine the present discussion to their work on the disodium salt of 2,2'-dimethoxy-6,6' dicarboxybiphenyl in water-methanol mixtures of all compositions. The unimolecular constants obtained in pure water may be expressed in either of the forms

$$k_1 = A_1 . \exp\left(-E_A/RT\right) = 3\cdot88 \times 10^{10} . \exp\left(-26{,}011/RT\right), \tag{24}$$

and

$$k_1 = \frac{kT}{h} . \exp\left(-\Delta G^{\ddagger}/RT\right) = \frac{kT}{h} . \exp\left(-29{,}810/RT\right). \tag{4.22}$$

The first two experimental parameters are related as follows to the conventional parameters of Eyring

$$\Delta H^{\ddagger} = E_A - RT, \tag{4.36}$$

$$\Delta S^{\ddagger} = R[\ln A_1 - \ln\left(kTe/h\right)], \tag{4.37}$$

$$\Delta G^{\ddagger} = E_A - RT[\ln A_1 - \ln\left(kT/h\right)]. \tag{4.35}$$

The variation of these parameters with respect to the composition of the solvent is shown in Fig. 2, from which we note that $\Delta G^{\ddagger}$ increases, and therefore the rate of reaction decreases, smoothly and slowly, with the progressive

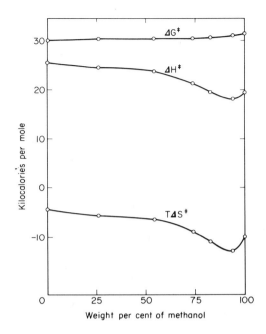

Fig. 2. The free energy, enthalpy and entropy of activation for the racemization of the di-sodium salt of 2,2' dimethoxy-6,6' dicarboxybiphenyl in water-methanol mixtures at 362·85°K (Leffler and Graybill).

addition of methanol to water. The smoothness of this curve, however, conceals the fact that the enthalpy and entropy of activation pass through minima at a mole fraction of water of 0·102. We shall see, in dealing with solvolysis, that a minimum appears at the same composition in the heat of absorption attending the solution of certain solutes. It is probable that in this system also a similar change in the ground state of the solute provides the right explanation. Despite the existence of these minima, it appears that a rough proportionality between the defined entropy and enthalpy of activation persists over a wide range of solvent composition:

$$T\Delta S^{\ddagger} \text{ (kcal-mole}^{-1}) = -33\cdot45 + 1\cdot136\Delta H^{\ddagger}. \qquad (26)$$

Equally extensive information has accumulated on the kinetics of the racemizations of binaphthyl derivatives by Harris and her collaborators (1964), from which we may select for discussion the following representative results (Table VI) on racemizations of molecules of the 1,1'-binaphthyl-8,8'-

TABLE VI

ARRHENIUS PARAMETERS FOR THE RACEMIZATION OF 1,1′-BINAPHTHYL DERIVATIVES

Group in the 8,8′ positions	k_1 at 50°C (sec^{-1})	$\log_{10} A$	E (kcal/mole^{-1})
—H	7.94×10^{-4}	12·1	22·5
$—CH_3$	4.37×10^{-8}	11·3	27·6
—COOH	2.24×10^{-4}	11·3	22·1
$—COO^-$	3.85×10^{-3}	15·2	26·0

disubstituted type such as

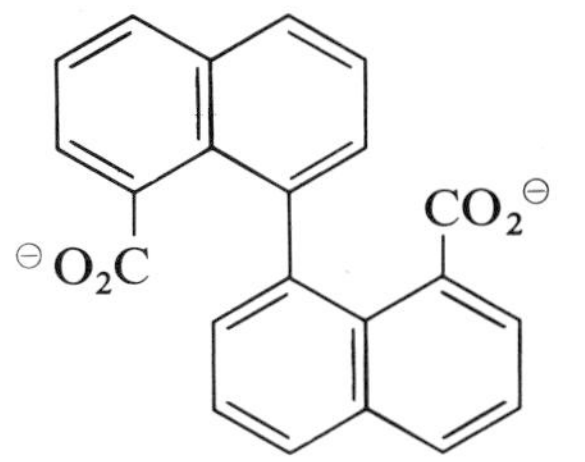

The solvent is NN dimethylformamide, except in the last entry where it is aqueous alkali. The reduction in rate by a factor of over 10^4 caused by the introduction of methyl groups for hydrogen atoms is again due mainly to an increase of some 5 kcal in the apparent energy of activation. Compared with this substituent, the carboxyl radical causes little change in rate. The difference in the energies of the last two reactions listed is probably due in part to the difference in the Coulombic term ε^2/Dr for the critical solute in the two solvents.

When two benzene rings are joined by more than one bridge, it is perhaps natural to expect the energy of activation for the racemization of the compound to be higher than that for the singly bridged compound; and this seems generally to be the case. E_A for the racemization of the following compound in aqueous acidic solution is found to be 27·8 kcal (Ahmed and Hall, 1958):

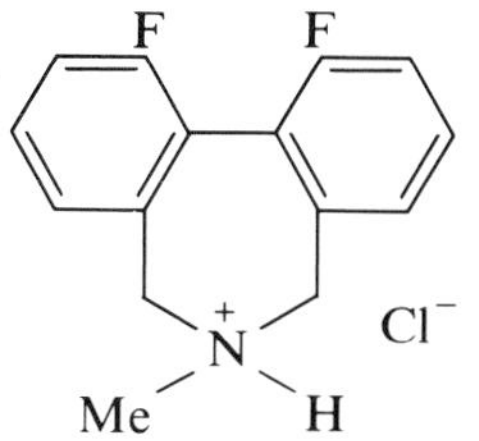

Similarly, for the racemization of 6,7-dihydro-5,5,7,7-tetramethyl-5*H*-dibenzo $[d, f][1,3]$ diarsepinium dibromide in aqueous solution

$$\text{Me}_2\text{As}^+ \quad {}^+\text{AsMe}_2$$
$$\underset{\text{CH}_2}{\diagdown} \qquad 2\text{Br}^-$$

the apparent energy of activation is $25\cdot9 \pm 1\cdot3$ kcal (M. H. Forbes, F. G. Mann, I. T. Millar and E. A. Moelwyn-Hughes, 1963) at a mean temperature of 317°K. The data, however, indicate a value for $-dE_A/dT$ of at least 100 cal/mole-deg. If this result were confirmed by more accurate and extensive kinetic data, it would be of the utmost importance, indicating the operation of a much more complicated means of activation than has hitherto been supposed. Some evidence for such a conclusion has been derived from the magnitude of the gradient $(d \log A/d \log E_A)_T$ afforded by the data of Table V and by earlier results on the mutarotation of monosaccharides.

Mutarotation

Mutarotation consists of a reversible molecular rearrangement attended by a change in optical rotatory power, and differs from racemization in that the final, equilibrated, system retains optical rotation. Due to the prominence given in the literature to mutarotations in aqueous solution, and particularly to the influence of acids and bases on the velocity of reaction, it is frequently overlooked that such changes can take place in non-hydroxylic solvents. The most accurately investigated example is the mutarotation of aluminium benzoylcamphor in carbon tetrachloride solution (Lowry and Traill, 1931). The data of Table VII indicate that the change takes place in consecutive unimolecular stages, according to equation (6.30), the unimolecular constants of which are found to vary with respect to temperature as follows:

$$k_1 = 1\cdot14 \times 10^{12} \exp(-19,000/RT)$$

and

$$k_2 = 2\cdot50 \times 10^{14} \exp(-22,940/RT), \qquad (27)$$

or, on introducing the viscosity

$$(k_1/\eta) = 6\cdot86 \times 10^{15} \exp(-21,375/RT)$$

TABLE VII

THE MUTAROTATION OF $M/100$ ALUMINIUM BENZOYLCAMPHOR IN CARBON TETRACHLORIDE AT 20°C ($\lambda = 5{,}461$ Å)

$$\alpha_t = 1244 \cdot 5 - 867\, e^{-0 \cdot 541t} + 258\, e^{-0 \cdot 115t}$$

t (minutes)	α_t (observed)	α_t (calculated)	Difference (obs. − calc.)
0·9	962	933	+29
1·1	1,031	984	+47
1·7	1,100	1,106	−6
2·1	1,159	1,164	−5
2·3	1,200	1,190	+10
2·8	1,234	1,236	−2
3·1	1,259·5	1,260·5	−1
3·5	1,287·0	1,285·0	+2
4·0	1,301·0	1,306·0	−5
4·5	1,319·0	1,320·5	−1·5
4·9	1,327·0	1,328·0	−1
5·3	1,333·5	1,334·0	−0·5
5·8	1,338·0	1,338·0	0
6·2	1,340·0	1,339·5	+0·5
6·7	1,340·0	1,340·0	0
7·3	1,338·0	1,338·0	0
8·0	1,335·5	1,335·0	+0·5
8·5	1,332·0	1,332·0	0
9·2	1,326·0	1,327·5	−1·5
10·1	1,322·5	1,321·0	+1·5
11·4	1,312·0	1,312·0	0
13·2	1,300·0	1,300·0	0
14·8	1,291·0	1,291·0	0
16·7	1,281·5	1,282·0	−0·5
18·4	1,275·5	1,275·5	0
20·0	1,271·0	1,270·0	+1
23·7	1,261·0	1,261·0	0
29	1,253·5	1,253·5	0
37	1,247·5	1,248·0	−0·5
51	1,245·0	1,245·0	0
∞	1,244·5	—	—

and

$$(k_2/\eta) = 1 \cdot 50 \times 10^{18} \exp\left(-25{,}315/RT\right). \tag{28}$$

The first step in the reaction ($t_{1/2} = 89 \cdot 74$ sec at 20°C) is followed by a slightly slower step requiring an additional energy of 2,375 cal. Galactose in water mutarotates by a similar mechanism (G. F. Smith and T. M. Lowry, 1928).

α-glucose and β-glucose mutarotate in aqueous solution according to first order kinetics. The observed rate constant is given by the equation of general acid-base catalysis

$$k_\alpha + k_\beta = k_0 + k_{H^+}[H^+] + k_{OH^-}[OH^-] + \Sigma k_A[Acid] + \Sigma k_B[Base], \quad (29)$$

where the acids are defined as proton-donors, and the base as proton-acceptors. We are here concerned only with the constant k_0 which refers to the rate of mutarotation in the absence of catalysts, and is sometimes wrongly referred to as the spontaneous, or uncatalysed, rate constant. It is found to have, at a given temperature, the same value when found directly in pure water or indirectly by extrapolating the composite velocity constant obtained in acid or basic media (Table VIII). Mutarotation, as depicted in Fig. 6.2,

TABLE VIII

The Water Constant for the Mutarotation of α-Glucose

Temperature, °K	$k_0 \times 10^5 \sec^{-1}$	
	From catalytic runs*	From water runs†
282·98 ± 0·02	8·87	8·63
293·11 ± 0·01	24·6	24·7
303·14 ± 0·02	64·1	64·3
313·38 ± 0·01	156	159

* P. Johnson and Moelwyn-Hughes, *Trans. Faraday Soc.*, **37**, 289 (1941).
† Kendrew and Moelwyn-Hughes, *Proc. Roy. Soc.*, **A, 176**, 352 (1940).

suggests that the change is intramolecular, but there are two arguments against this view. (1) The puckered hexagon structure of glucose renders such a simple structural change energetically prohibitive. (2) Mutarotation is catalysed by all acids and bases, of which water is one. The real change, therefore, is an essentially bimolecular reaction, which (Christiansen, 1966) may consist of the transfer of the proton indicated here by an asterisk, to the water molecule, leaving a negatively charged carbon atom:

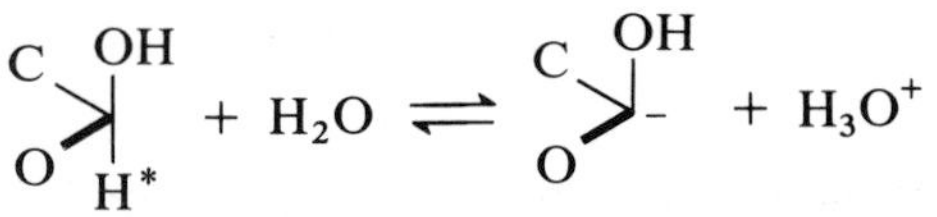

The negatively charged carbon atom is isosteric with the neutral nitrogen atom, and may be assumed to have three valency bonds forming angles near to 120° with respect to one another. Models of the rigid pyranose ring with one carbon atom replaced by a nitrogen atom show that the site on which the —OH group is situated can easily be turned upwards or downwards. Thus the proton which has been removed from the ring may later return to it in either of two positions. According to this mechanism, the energy of activation for the mutarotation of all pyranoses should be nearly the same, as in fact is found to be the case (Table IX). A slightly higher value of $-dE_A/dT$

TABLE IX

KINETIC DATA FOR THE CONVERSION OF α-SUGARS AT 298·1°K

Sugar	$k_\alpha \times 10^{-4}$ sec^{-1}	$E_{A,\alpha}$ cal/g-mol	$J = \left(\dfrac{dE_{A,\alpha}}{dT} \right)_P$
Xylose	8·578	16,245	$-20\cdot4 \pm 8\cdot4$
Mannose	4·572	16,375	$-20\cdot5 \pm 2\cdot8$
Glucose	1·463	16,945	$-19\cdot6 \pm 5\cdot9$
Lactose	1·869	17,225	$-21\cdot9 \pm 5\cdot6$

is given by G. F. Smith and M. C. Smith (1937) for the mutarotation of α-glucose, based on their own experiments and on their analysis of the classical data of Hudson and Dale (1917). Within the limits of experimental error, this gradient is the same for monosaccharides as for disaccharides, which suggests that the activation process for the mutarotation of both types of sugars is located in a group of the same size. This result is in marked contrast with the evidence derived from the decarboxylation reactions of aliphatic and aromatic acids.

The Decomposition of the Benzene–Diazonium Ion in Water

Stoichiometrically, the reaction between benzene diazonium chloride and water is represented as follows:

$$C_6H_5N_2Cl + H_2O \rightarrow C_6H_5OH + H^+ + Cl^- + N_2.$$

Careful analysis shows that the concentration of chloride ion initially present is equivalent to the amount of nitrogen finally formed. Hence, the salt is completely ionized, and the reaction may therefore be written as follows:

$$C_6H_5N_2^+ + H_2O \rightarrow C_6H_5OH + H^+ + N_2.$$

This conclusion (P. Johnson and Moelwyn-Hughes, 1940) differs from an earlier one (Hantzsch, 1900) according to which a percentage of chlorobenzene is produced, which increases with the initial concentration of the diazonium salt. The conclusion, however, is consistent with the fact that diazonium salts of different strong acids decompose at the same rate (Saunders, 1936). There remain two mechanisms for the rate-determining reaction undergone by the phenyldiazonium ion, as explained in Chapters 4 and 9. It may eject a nitrogen molecule, leaving a phenyl ion which rapidly reacts with water, or it may react with numerous water molecules simultaneously, yielding phenol and the hydrated hydrogen ion. The initial presence of more than 5 moles per litre of HCl increases the rate of reaction by 18 per cent, which is comparable to the effect of large concentrations of phenol. There appears to be no evidence that hydrogen ions exert a specific effect.

The influence of temperature on the unimolecular constant has been measured using solutions which were not more than 0·05 molar, at which concentration neither phenol nor sodium chloride has any effect on the reaction rate. The results (Table X) were fitted by the method of least squares to the two constant equation of Arrhenius, which is

$$k_1 = 3\text{·}028 \times 10^{15} \text{. } \exp\left(-27{,}025/RT\right) \tag{30}$$

and to the three-constant equation

$$\ln_e k_1 = B + (J/R)\ln T - E/RT. \tag{31}$$

Two independent methods were used to evaluate J/R, (1) by trial and error, until the values of J/R and E gave the lowest value of $\varepsilon^2 = (k_{1\,\text{obs}} - k_{1\,\text{calc}})^2$, (2) from eight separately determined values of E_A and the least-squares

TABLE X

Temperature, °K	$k_1 \times 10^5 \text{ sec}^{-1}$		
	Observed	By equation (32)	By equation (30)
283·13 ± 0·02	(0·432)	0·386	0·414
288·16 ± 0·03	0·930	0·920	0·958
292·99 ± 0·01	2·01	2·05	2·09
297·80 ± 0·02	4·35	4·41	4·42
303·07 ± 0·02	9·92	9·88	9·77
308·13 ± 0·02	20·7	20·8	20·4
313·20 ± 0·03	42·8	42·6	41·7
317·99 ± 0·02	81·8	81·6	80·2
322·98 ± 0·03	158	157	155
328·20 ± 0·04	301	303	304
333·17 ± 0·05	564	554	563
338·14 ± 0·05	(1070)	991	1030
		$\Sigma\varepsilon^2 = 11$	$\Sigma\varepsilon^2 = 43$

solution of the equation $E_A = JT + E$. The first method gave $J = -38$ cal/mole-deg, and $E = 38{,}831$ cal. The second method gave $J = -32.48$ and $E = 37{,}332$. Accepting the rounded value of $J = -35.76$, so as to make $J/R = 18$, we find

$$\ln k_1 = 157.067 - 18 \ln T - (38{,}200 \pm 750)/RT. \tag{32}$$

The last line gives the sum of the squares of the errors, in order to illustrate the superiority of the more complicated equation. The value of E_A given by the two-constant equation agrees with that of Cain and Nicoll (1902) to within 5 cal. Changing the solvent from water to various alcohols and acids has no detectable influence on the apparent energy of activation (Brandsma 1925; 1930; Pray 1926). In aqueous solution, $dE_A/dT = -35.8$ cal/mole-deg. If the rate of reaction were directly proportional to the viscosity, we would have $E_A = E - B$, and at $310.6°K$, $dE_A/dT = dE/dT - dB/dT = dE/dT + 22.1$ (from Table 3.VIII). On the other hand, if k_1 varied at $1/\eta$, dE/dT would be reduced to -14.

The Decomposition of Highly Polar or Salt-like Molecules

Most of the early kinetic data on the decomposition of highly polar molecules such as diphenyliodonium iodide, triethylsulphonium bromide and many quaternary ammonium halides have proved to be difficult to interpret. It is now realized that no clue is to be expected without ancillary information dealing with, for example, solubility and equilibria. We shall therefore confine ourselves here to one instance in which the kinetic data have been supplemented by static data. A second instance is dealt with in Chapter 16.

The decomposition of triethylsulphonium bromide, $(C_2H_5)_3SBr \rightleftharpoons (C_2H_5)_2S + C_2H_5Br$ was studied kinetically by von Halban (1909, 1911) in various pure solvents. His work has been confirmed and extended by Essex

TABLE XI

THE UNIMOLECULAR DECOMPOSITION OF TRIETHYLSULPHONIUM BROMIDE
IN PURE SOLVENTS

Solvent	k_1 (sec^{-1}) at 25°C	A_1	E_A cal/mole^{-1}
$C_6H_5CH_2OH$	1.15×10^{-11}	2.75×10^{17}	35,990 (32,180)
$C_6H_5NO_2$	6.42×10^{-5}	8.86×10^{16}	$28{,}870 \pm 1{,}500$ (28,290)
$C_2H_2Cl_4$	7.08×10^{-5}	1.38×10^{18}	$30{,}411 \pm 154$ (30,350)

and Gelormini (1920) and by Corran (1927). The kinetic equation is that of a unimolecular decomposition, opposed by a bimolecular combination (equation 6.54). Some of the later results, with those of von Halban in brackets, are summarized in Table XI. The very high values of A_1 immediately suggest that the mechanism is more complicated than that normally expected for a simple unimolecular reaction. This view is in part confirmed by the detailed results found in nitrobenzene solution. This system is discussed further in Chapter 16.

The van't Hoff–Dimroth Equation

If van't Hoff's expression for the chemical potential of a solute in terms of its concentration ($\mu = \mu^0 + kT \ln c$) is valid for concentrations up to the saturated value, s, then $\mu = \mu_{sat} + kT \ln (c/s)$, where μ_{sat} is equal to the chemical potential of the solid or liquid with which the solution is at equilibrium, and is a function of T and P only. Let it be denoted by μ_{cr}. When two such solutes are at equilibrium with each other, it follows that

$$K = \frac{c_B/s_B}{c_A/s_A} = \exp\left[-(\mu_{B,cr} - \mu_{A,cr})/kT\right] \tag{33}$$

is independent of the solvent. This equation (van't Hoff, 1898) has been verified in a number of instances (Dimroth and Mason, 1913; Meyer, 1911), one of which is the reversible system

H   C₆H₅
| |
N   N
/\   /\
N   C—NHC₆H₅   ⇌   N   C—NH₂
‖   ‖   ‖   ‖
N———C—COOR   N———C—COOR

where R stands for the methyl or ethyl group (Dimroth, 1910). Data for the methyl ester in various solvents are quoted in Table XII. Considering the approximate form of the basic equation for the chemical potential, the agreement is satisfactory.

Let us next assume that the rate of either process is proportional to the concentration, so that $k_A c_A = k_B c_B$, and

$$K = \frac{k_A s_A}{k_B s_B} = f(T, P). \tag{34}$$

As in the original derivation of the Arrhenius equation, we may reasonably

TABLE XII

Solvent	Methyl ester		
	c_B/c_A	s_B/s_A	$\dfrac{c_B/s_B}{c_A/s_A}$
$(C_2H_5)_2O$	21·7	53·0	0·40
C_2H_5OH	2·3	7·0	0·33
$C_6H_5CH_3$	1·8	4·3	0·33
C_6H_6	1·02	3·2	0·32
$C_6H_5NO_2$	0·80	2·2	0·36
$CHCl_2$	0·32	1·1	0·32

assume that the numerator and denominator are functions of T and P only and are therefore independent of the nature of the solvent. We find from Dimroth's work (Table XIII) that $k_A s_A$, though not an exact constant, is much less variable than k_A itself.

TABLE XIII

Solvent	k_A at 10°C (min^{-1})	s_A at 10°C (gramme/litre)	$k_A s_A$	$k_A s_A$ (s_A in molar fractions)
CH_3OH	0·00053	218·0	0·116	4·67
C_2H_5OH	0·00103	97.7	0·101	5·85
$C_6H_5CH_2OH$	0·00110	90·0	0·099	9·12
$(CH_3)_2CO$	0·00527	56·5	0·298	21·7
$CHCl_3$	0·0211	8·8	0·186	14·8
$CH_3COOC_2H_5$	0·0267	12·0	0·320	32·7
$C_6H_5NO_2$	0·0460	6·5	0·299	31·3
$C_2H_5NO_2$	0·0550	3·2	0·176	14·4

The important conclusion to be drawn is that the instantaneous rate, $-dc/dt$, of chemical change is governed by the concentration, rather than the activity, of the reactant. This, as we have seen (Chapter 4) is one of the basic principles of chemical kinetics, and was successfully used by Brönsted, Bjerrum and Christiansen in their treatment of ionic reactions.

Doubt remains as to the validity of applying, without verification, van't Hoff's expression for the chemical potential of a solute in such concentrated solutions (*ca.* 0·5 molar) as those used by Dimroth. A better, but still inadequate, expression is given in terms of the mole fraction, x, of solute ($\mu = \mu^0 + kT \ln x$), where μ^0 is clearly the chemical potential of the pure liquid solute. K now becomes x_B/x_A and is still independent of the solvent. If, as before, we use the kinetic expression $K = k_A/k_B$, we see that

$$\frac{k_A x_A}{k_B x_B} = f(T, P). \tag{35}$$

Arguing as before, we may anticipate that, for a given reaction in a variety of solvents at constant temperature and pressure,

$$k_A x_A = \text{constant}. \tag{36}$$

The conversion has led to little improvement in Dimroth's data of Table XIII. A better test, however, is afforded by the work of H. Eyring and J. F. Daniels (1930) on the rate of decomposition of dinitrogen pentoxide in the saturated gaseous phase and in two solvents. Their results are given in the first three columns of Table XIV. The product $k_A s_A$ is seen to be about 1,000 times as great in carbon tetrachloride as in the gas phase. The fifth column contains rough estimates of the saturation concentrations in terms of molar fractions. $k_A x_A$ is seen to have the same value in these two media.

TABLE XIV

THE DECOMPOSITION OF DINITROGEN PENTOXIDE IN VARIOUS MEDIA

Medium	$k_A \times 10^5$ (sec^{-1})	s_A (mole/litre)	$k_A s_A \times 10^5$	x_A (mole fraction)	$k_A x_A \times 10^5$
Gas phase	0·79	0·0102	0·008	1	0·79
CH_3NO_2	1·35	4·38	5·91	0·23	0·31
CCl_4	1·83	4·78	8·75	0·45	0·82

When a reaction conforms to the revised form of Dimroth's relationship ($k_A x_A = \text{constant}$) the sum of the energy of activation for a given reaction and the heat of solution of the reactant should also be a constant independent of the nature of the solvent:

$$E_A + \Delta H_{\text{diss}} = \text{constant}. \tag{37}$$

A striking example is afforded by the work of Freundlich and Richards (1912) on the cyclization of ε-chloramylamine in water and in tetrachlorethane:

In water: $\Delta H_{\text{diss}} = \quad 450.$ $E_A = 20,670.$ Sum $= 21,120$ cal.

In $C_2H_2Cl_4$: $\Delta H_{\text{diss}} = 5,200.$ $E_A = 14,420.$ Sum $= 19,620$ cal. (38)

Here the main influence of the solvent is to alter the enthalpy of the solute in the ground state.

The Additive Effect of Substituents

If the energy of activation consists partly of the energy required to overcome steric hindrance, it is natural to expect the introduction of two substituents into the reacting molecule to lead to an additive effect.

A clear demonstration of the additive effect of substituents on the kinetics of unimolecular reactions is to be found in the work of A. W. Chapman (1936) on the Beckmann transformation of ketoximes in carbon tetrachloride solution:

$$R_1 - \underset{\underset{N-OR_3}{\|}}{C} - R_2 \quad \rightarrow \quad \left[\underset{R_1 - N}{\overset{R_2}{\underset{\|}{C}}}_{\diagdown OR_3} \right] \quad \rightarrow \quad R_2 - \underset{R_1 - N - R_3}{\overset{C=O}{|}}$$

Let us confine our attention to his results on the picryl ethers, for which R_3 is $2:4:6\text{-}C_6H_2(NO_2)_3$. When R_2 is the *p*-nitrophenyl group, the oxime is not readily isomerized; when R_2 is the *p*-anisyl radical, a very rapid rearrangement takes place. Measurable velocities obtain with other substituents, and the results show that substitutions into the groups R_1 and R_2 make additive contributions to the energy of activation. Some figures illustrating Chapman's conclusion are to be found in Table XV, in which the energies have been corrected for the viscosity of the solvent.

TABLE XV

The Effect of Substitution on the Energy of Activation of the Beckmann Isomerism of Certain Ketoxime Picryl Ethers in Carbon Tetrachloride Solution

R_1	R_2	E (kcal/gramme-mole)	$\Delta E = E_s - E_u$	
C_6H_5	C_6H_5	29·2	0	
C_6H_5	$p\text{-}C_6H_4Cl$	30·8	$+1\cdot6$	$+2\cdot2$
$p\text{-}C_6H_4Cl$	C_6H_5	29·8	$+0\cdot6$	
$p\text{-}C_6H_4Cl$	$p\text{-}C_6H_4Cl$	31·5	$+2\cdot3$	
C_6H_5	$p\text{-}C_6H_4CH_3$	26·5	$-2\cdot7$	$-3\cdot6$
$p\text{-}C_6H_4CH_3$	C_6H_5	28·3	$-0\cdot9$	
$p\text{-}C_6H_4CH_3$	$p\text{-}C_6H_4CH_3$	25·8	$-3\cdot6$	

A General Statistical Treatment of Unimolecular Reactions in Gases

Provided the rate at which molecules undergo chemical change is insufficient to invalidate the Boltzmann law, we may write for the number, N_a, of molecules possessing energy ε_a each, the general expression

$$N_a = N\frac{\exp\left(-\varepsilon_a/kT\right)}{\Sigma\exp\left(-\varepsilon_i/kT\right)} \tag{39}$$

where N is the total number of molecules, k is Boltzmann's constant, and the summation in the denominator (which is the partition function) is to be taken over all energy states. The kind of energy is immaterial. Denoting by v_a the probability per second that a molecule with energy ε_a shall react, the instantaneous rate of chemical change is

$$-\frac{dN}{dt} = \Sigma - \frac{dN_a}{dt} = N\frac{\Sigma v_a\exp\left(-\varepsilon_a/kT\right)}{\Sigma\exp\left(-\varepsilon_i/kT\right)}.$$

The summation in the numerator is to be taken over all the molecules that react. Then the unimolecular velocity constant is

$$k_1 = -\frac{1}{N}\frac{dN}{dt} = \frac{\Sigma v_a\exp\left(-\varepsilon_a/kT\right)}{\Sigma\exp\left(-\varepsilon_i/kT\right)}.$$

The apparent energy of activation is

$$\varepsilon_A = kT^2\frac{d\ln k_1}{dT} = \frac{\Sigma v_a\varepsilon_a\exp\left(-\varepsilon_A/kT\right)}{\Sigma v_A\exp\left(-\varepsilon_A/kT\right)} - \frac{\Sigma\varepsilon_i\exp\left(-\varepsilon_i/kT\right)}{\Sigma\exp\left(-\varepsilon_i/kT\right)}. \tag{40}$$

The first term on the right is recognized as the average energy of the active molecules, and the second term as the average energy of all the molecules in the system (R. C. Tolman):

$$\varepsilon_A = \bar{\varepsilon}_a - \bar{\varepsilon}_i$$

These we may denote by the symbols ε^* and $\bar{\varepsilon}$.

$$\varepsilon_A = \varepsilon^* - \bar{\varepsilon}. \tag{41}$$

Since the average energy, $\bar{\varepsilon}$, of all stable molecules is a function of temperature, it follows that the apparent energy of activation, ε_A, must also be a function of temperature, except in the improbable event of ε^* and $\bar{\varepsilon}$ responding identically to T.

Let us consider the case where ε^* consists of a temperature-independent energy, ε_0, and the equipartition energy, kT, associated with a single classical oscillator, and when $\bar{\varepsilon}$ is made up entirely of the equipartition contributions from s classical oscillators. Then

$$\varepsilon^* = \varepsilon_0 + kT$$

$$\bar{\varepsilon} = skT$$

and

$$\varepsilon_A = \varepsilon_0 - (s-1)kT \tag{42}$$

as derived from Berthoud's equation (53). The apparent energy of activation is now seen to decrease linearly with respect to temperature because, although an energy height, $\varepsilon_0 + kT$, must be reached by all the molecules that react, normal molecules spring to it from platforms of varying heights, determined by the temperature. On using the Arrhenius equation in its general form (equation 1.21) and integrating, we find that

$$\ln k_1 = \text{constant} - (s - 1)\ln T - \varepsilon_0/kT. \tag{43}$$

In deriving equation (41), it was assumed that all v_a terms were independent of temperature. When this restriction is removed, we obtain (La Mer, 1933)*

$$\varepsilon_A = \bar{\varepsilon}_a - \bar{\varepsilon}_i + kT \frac{\Sigma(dv_a/d\ln T)\cdot\exp(-\varepsilon_a/kT)}{\Sigma v_a \exp(-\varepsilon_a/kT)}. \tag{44}$$

Just as we have regarded the first term of equation (40) as the average value of the energies of the active molecules, we may regard the last term of equation (44) as the average value of $d\ln v_a/d\ln T$, so that

$$\varepsilon_A = \bar{\varepsilon}_a - \bar{\varepsilon}_i + kT\overline{\left(\frac{d\ln v_a}{d\ln T}\right)}. \tag{45}$$

In solution kinetics, that last term may be an important one, as is obvious if the average value of v_a depends on the viscosity of the solvent.

The Distribution of Energy among a System of Simple Harmonic Oscillators

All theories of unimolecular reactions require a knowledge of the manner in which the energy within a molecule is distributed throughout its structure. When the molecular energy is made up of simple harmonic vibrations, both classical and quantal theories provide appropriate forms of the distribution law, to which reference has been made in earlier chapters. We shall here derive the classical law only.

When the energy per molecule is expressible as a sum of terms which are quadratic with respect to the positional and momental variables, the distribution law is independent of whether these terms represent kinetic energy or potential energy. If there are equal numbers, s, of potential energy terms, in Hooke's form $\frac{1}{2}fq^2$, and of kinetic terms, $(1/2m)p^2$, the energy of one molecule

---

* Equation (45), derived statistically, coincides with equation (4.25), derived in terms of thermodynamic variables.

is

$$\varepsilon = \tfrac{1}{2}f_1 q_1^2 + \tfrac{1}{2}f_2 q_2^2 + \cdots + \tfrac{1}{2}f_s q_s^2$$
$$+ (1/2m_1)p_1^2 + (1/2m_2)p_2^2 + \cdots + (1/2m_s)p_s^2$$
$$= \tfrac{1}{2}\sum^{s} f_i q_i^2 + \tfrac{1}{2}\sum^{s} (p_i^2/m_i), \tag{46}$$

and its complete motion is that of s independent harmonic oscillators, the individual vibration frequencies of which are given by the equation:

$$v_i = (1/2\pi)\sqrt{(f_i/m_i)}.$$

Then,

$$\varepsilon = \tfrac{1}{2}\sum^{s} (2\pi v_i)^2 m_i q_i^2 + \tfrac{1}{2}\sum^{s} (p_i^2/m_i).$$

In order to apply the general distribution law,

$$\frac{dN}{N} = \frac{\exp(-\varepsilon/kT)\,dq_1 \ldots dp_s}{\int_{-\infty}^{\infty}\ldots\int \exp(-\varepsilon/kT)\,dq_1 \ldots dp_s}, \tag{47}$$

let us first make the substitutions

$$w_i = (2m_i)^{1/2}\pi v_i q_i \quad \text{and} \quad z_i = (2m_i)^{-1/2} p_i, \tag{48}$$

so that

$$\varepsilon = \sum^{s} w_i^2 + \sum^{s} z_i^2 \quad \text{and} \quad dq_i\,dp_i = (1/\pi v_i)\,dw_i\,dz_i.$$

The element of volume in $2s$-dimensional space is thus

$$dq_1 \ldots dp_s = \prod^{s} (1/\pi v_i)\,dw_1 \ldots dw_s\,dz_1 \ldots dz_s.$$

Let us next substitute

$$\varepsilon = R^2. \tag{49}$$

Because ε consists of $2s$ squared terms, R may be regarded as the radius of a sphere of $2s$ dimensions. If the volume of this hyper-sphere is V, the volume element, previously denoted by $dq_1 \ldots dp_s$, is simply dV.

Now the volume of a sphere of $2s$ dimensions is*

$$V = \pi^s R^{2s}/s!, \tag{50}$$

or, making use of equation (49),

$$V = \pi^s \varepsilon^s/s!.$$

---

* This equation may be regarded as a generalization from the known volumes of spheres of small dimensions, e.g. πr^2 and $\tfrac{4}{3}r^3$ for spheres of 2 and 3 dimensions.

Hence

$$dV = \frac{\pi^s}{(s-1)!} \varepsilon^{s-1} \, d\varepsilon,$$

and

$$dq_1 \ldots dp_s = \frac{\prod\limits^{s}(1/v_i)}{(s-1)!} \varepsilon^{s-1} \, d\varepsilon.$$

The numerator of equation (47) is thus

$$\frac{\prod\limits^{s}(1/v_i)}{(s-1)!} \exp\left(-\varepsilon/kT\right)\varepsilon^{s-1} \, d\varepsilon.$$

The denominator is this expression integrated over all values of ε from zero to infinity. By means of the standard equation

$$\int_0^{\infty} e^{-ax} \cdot x^n \, dx = \frac{n!}{a^{n+1}} \tag{51}$$

it is found to be $\prod\limits^{s}(kT/v_i)$. Hence

$$\frac{dN}{N} = \frac{\exp\left(-\varepsilon/kT\right)\varepsilon^{s-1} \, d\varepsilon}{(kT)^s(s-1)!}, \tag{52}$$

which is the desired expression for the fraction of the total number of molecules which possess an energy lying within the limits ε and $\varepsilon + d\varepsilon$, or the probability that any molecule taken at random shall be found to possess an energy within these limits.

The fraction of the total number of molecules which have an energy of at least ε_i, or the probability, P, that any molecule taken at random shall be found to have at least this amount of energy, is obtained by integrating this expression with respect to the energy, from the lower limit, ε_i, to infinity. The result is readily found to be

$$P(\varepsilon \geq \varepsilon_i) = \exp\left(-\varepsilon_i/kT\right)\left[\frac{(\varepsilon_i/kT)^{s-1}}{(s-1)!} + \frac{(\varepsilon_i/kT)^{s-2}}{(s-2)!} + \ldots + 1\right]. \tag{53}$$

There is a slight dynamical inconsistency in the treatment, because we have expressed the energy as the sum of the energy of s independent oscillators, and have later regarded the whole energy of the molecule as pooled among the oscillators. Pooling could clearly not happen if the oscillators vibrated independently. For this reason, the system considered is sometimes referred to as a system of s feebly coupled oscillators, by which is implied a fictitious coupling energy of zero.

It will be observed that the term inside the square brackets is the expansion of the expression $e^{+s_i/kT}$, stopping at the sth term. When $s = 1$, we have

$$\frac{dN_{\varepsilon \geq \varepsilon_i}}{N} = \exp(-\varepsilon_i/kT),$$

When $s = 2$, we have

$$\frac{dN_{\varepsilon \geq \varepsilon_i}}{N} = \exp(-\varepsilon_i/kT)\left(\frac{\varepsilon_i}{kT} + 1\right).$$

The ratio of the leading term in the square brackets to the next term is seen to be $\varepsilon_i/(s - 1)kT$. When this ratio is large, i.e. when the minimum energy, ε_i, greatly exceeds $(s - 1)kT$, the leading term alone suffices as an approximation, giving us for the fractional number of activated molecules

$$\frac{N_a}{N} = \frac{dN_{\varepsilon \geq \varepsilon_i}}{N} = \exp(-\varepsilon_i/kT)\frac{(\varepsilon_i/kT)^{s-1}}{(s-1)!}. \tag{54}$$

The simplest theories of unimolecular and bimolecular reactions are based on the assumption that the velocity constants are directly proportional to this ratio, the proportionality constant being a mean vibration frequency in

TABLE XVI

$$\log_{10} k_1 \ (\text{sec}^{-1}) = a - b \log_{10} T - c/T$$

Reaction	a	b	c	ε_i (kcal/mole)	s	$\bar{\nu}$ (sec^{-1})
Decomposition of N_2O_5 in the gas phase	66·766	18·053	7920·87	36·24	19	7.45×10^5
Conversion of α- to β-glucose in water	37·814	10	5039·34	23·06	11	5.71×10^3
Decomposition of $C_6H_5N\equiv N^+$ in water	68·210	18	8351	38·20	19	8.05×10^6
Decomposition of CCl_3COOH in water	48·13	10·07	9378·5	42·91	11	2.22×10^{11}
Methanolysis of tertiary C_4H_9Cl	76·402	21·905	8424·7	38·55	23	1.36×10^3
Methanolysis of tertiary C_4H_9Br	69·927	19·354	7899·8	36·15	20	1.13×10^5
Methanolysis of tertiary C_4H_9I	64·998	17·981	7277·6	33·30	19	5.94×10^4

the former and a standard collision frequency in the latter:

$$k_1 = \bar{v} \cdot \exp\left(-\varepsilon_i/kT\right)\frac{(\varepsilon_i/kT)^{s-1}}{(s-1)!}, \tag{55}$$

$$k_2 = Z^0 \cdot \exp\left(-\varepsilon_i/kT\right)\frac{(\varepsilon_i/kT)^{s-1}}{(s-1)!}. \tag{56}$$

In either case, the rate constant may be reproduced by equations of the form

$$\log_{10} k_n = a - b \log_{10} T - c/T, \tag{57}$$

where a, b and c are empirical constants repectively proportional to $\bar{v}$ or Z^0 and to s and ε_i. Some numerical values for reactions of the first kinetic order are given in Table XVI.

More Elaborate Theories

In the theory of unimolecular reactions developed by O. K. Rice and Ramsperger (1927) and by Kassel (1928, 1932), a distinction is drawn between an energized molecule, possessing sufficient energy ε_i to decompose, and an activated molecule possessing energy ε_m in a given bond. From a statistical expression derived for the probability that a given oscillator has m identical quanta of vibrational energy hv when the total number, j, of such quanta are distributed among s identical oscillators, the expression derived for the first order constant in gaseous systems at relatively high pressures is $k_1 = \bar{v} \cdot \exp\left(-mhv/kT\right)$. At relatively low pressures, however, there results a complicated dependence of the half-life on the pressure, which has been quantitatively confirmed in the pyrolysis of ethane (M. C. Lin and K. J. Laidler, 1968) and in numerous other recently-investigated decompositions (Laidler, 1965). Because the low pressure conditions attainable in gaseous systems has no obvious parallel in solutions, we shall not further pursue the subject here.

Rather than concentrate on the distribution of energy within a group of independent oscillators, N. B. Slater (1959) concentrates attention on a single bond, and identifies most of the energy of activation with the energy required to stretch it by a given distance. The restoring force of the bond itself is alternately opposed and reinforced by the co-operation of neighbouring forces until, when two or more vibrations come into phase, their super-position gives rise to the considerable bond distension required for its rupture.

In dealing with unimolecular reactions in solution, J. C. Kendrew and E. A. Moelwyn-Hughes (1940) define a mean frequency as follows in terms of the number of oscillators and the various frequencies of the classical oscillators in the ground (q) and activated (s) states:

$$\bar{v} = \left(\frac{\prod\limits^{q} v_q}{\prod\limits_{s} v_s}\right)(q - s)^{-1}. \tag{58}$$

The unimolecular constant can then be expressed as follows:

$$k_1 = k_1 v\left(\frac{h\bar{v}}{kT}\right)^{q-s} . \exp\left(-\varepsilon/kT\right). \tag{59}$$

It is not possible to evaluate all the terms in this equation without assuming that $v = \bar{v}$. On doing this we obtain the estimates given in Table XVII.

TABLE XVII

Reactant in aqueous solution	ε cal/gramme-mol	$\bar{v} \times 10^{-13}$ sec^{-1}	$q - s$ obs.	$E_0/h\bar{v}$
α-Glucose	23,055	0·66	10 ± 3	36·7
Tribromacetate ion	39,610	5·25	10 ± 5	8·0
Trichloracetate ion	42,910	5·70	10 ± 5	7·8
Methyl bromide	46,820	1·66	34 ± 4	29·7
2:4:6-Trinitrobenzoate ion	52,300	2·95	23 ± 5	18·7

Columns 2 and 3 contain the experimental values of the energy of activation and effective mean frequency. In the last column is found the number of quanta, $h\bar{v}$, which will fully account for ε. We infer that in mutarotation each oscillator contributes many quanta. In the other cases, the figures in columns 4 and 5 tally. $q - s$ is about 21 for trinitrobenzoic acid and about 9 for trichloracetic acid; one is naturally inclined to attribute the difference of 12 to the three nitro groups, which may claim 4 each. When a molecule of trinitrobenzoic acid decomposes, therefore, the whole molecule seems to contribute to the localized ejection of the three atoms.

Unimolecular Reactions Controlled by Rotational Relaxation

Experimental work on the heat capacity and infra-red spectrum of chloroform indicate that the only effective internal rotation of the molecule is that taking place round the carbon–hydrogen bond (Chapter 1), and it is worth considering whether the rate at which such a molecule undergoes chemical change may be influenced or even determined by such a motion. Let us first combine Smoluchowski's expression for the bimolecular encounter constant of like spherical molecules,

$$k_2 = \frac{8kT}{3\eta} \cdot \exp - (E_2/RT) \tag{4.19}$$

with the approximate equation of Fuoss for the equilibrium constant

$$K = \frac{k_1}{k_2} = \frac{3}{4\pi\sigma^3} \cdot \exp(-\Delta E/RT), \tag{3.96}$$

thus obtaining for the unimolecular rate constant

$$k_1 = \frac{kT}{4\pi r^3 \eta} \cdot \exp(-E_1/RT), \tag{60}$$

where $E_1 = E_2 + \Delta E$. The pre-exponential term is recognized as the reciprocal of Debye's expression for the time of rotational relaxation of a spherical molecule of radius r:

$$\tau = \frac{4\pi r^3 \eta}{kT}. \tag{61}$$

Equation (60) is one of a number which have been applied to unimolecular reactions in solution, to explain the magnitude of k_1 and particularly the temperature variation of the apparent energy of activation (Moelwyn-Hughes, 1938). Debye has stressed that η of equation (61) is an effective viscosity, differing from the microscopic viscosity. For reactions proceeding according to such a mechanism, we have $E_A = E_1 + RT + B$. If we further employ equation (5.33), we find that

$$k_1 = \frac{kTe}{4\pi r_3 b} \cdot \exp(-E_A/RT).$$

For the ionization of water at 25°C, we have $\sigma = 2\cdot685$ Å, and $b = 1\cdot006 \times 10^{-5}$, so that $k_1 = 3\cdot66 \times 10^{14} \cdot \exp(-E_A/RT)$.

The mean rotational frequency, according to equation (61), is

$$v = \frac{kT}{4\pi r^3 \eta}. \tag{62}$$

By an electron spin resonance technique, the rate of rotation and the apparent energy of activation have been measured for the rotation of the nitroxide radical in toluene and decane solutions at about 200°K with the results: $v = 2.45 \times 10^{12} \sec^{-1}$ and $E_A = 2.5$ kcal in toluene solution, and $v = 270 \times 10^{12} \sec^{-1}$ and $E_A = 4.9$ kcal in decane solution (A. L. Buchachenko, A. M. Wasserman and A. L. Kovarskii, 1969).

A Possible Interpretation of High Values of the First Arrhenius Parameter

The first-order constant, k_1, governing the decomposition of benzene-diazonium chloride in a variety of hydroxylic solvents, including water can be reproduced over a small temperature range by the integrated form of the Arrhenius equation

$$k_1 = Z_{obs} \cdot e^{-E_A/RT} = 5.67 \times 10^{15} e^{-E_A/RT} \tag{63}$$

The pre-exponential term is about 1,000 times greater than the value considered most probable for first-order reactions by Polanyi and Wigner (*Z. physikal. Chem.*, *Haber-Band*, 439 (1928)) in a statistical survey. There are many explanations for such high values. We shall consider here one which is fairly general, though seldom applied (Moelwyn-Hughes, *Trans. Faraday Soc.*, **34**, 91 (1937). As in previous studies, we shall first resolve the free Gibbs energy of activation into non-electrostatic and electrostatic components, E_0 and E_e, regarding the former, and a theoretical frequency term, Z, as independent of temperature, so that

$$k_1 = Z e^{-E_0 RT} \cdot e^{-E_e/RT} \tag{64}$$

In the example considered each solute molecule, of dipole moment μ_A is attracted by c solvent molecules, each of moment μ_S at an average separation of r in the unreactive system and of r_* in the reactive system, so that

$$E_e = \frac{-c\mu_A\mu_S}{r_*^3} - \left(-\frac{c\mu_A\mu_S}{r^3}\right) = \frac{c\mu_A\mu_S}{r^3}\left[1 - \left(\frac{r}{r_*}\right)^3\right] \tag{65}$$

On equating equations 63 and 64, we have

$$Z_{obs} \cdot e^{-E_A/kT} = Z \cdot e^{-E_0/kT} \cdot e^{-E_e/kT}, \tag{66}$$

On using equation 65, we have

$$Z_{obs} \exp\left(-E_A/kT\right) = Z \exp\left(-E_0/kT\right) \cdot \exp\left(-(c\mu_A\mu_B/r^2kT)\left[1 - \left(\frac{r}{r_*}\right)^3\right]\right) \tag{67}$$

In order to obtain the experimental enthalpy of activation, we use the general expression of $E_A = kT^2(d \ln k_1/dT)_P$ and assume that r_* is independent of temperature, while r, the intermolecular separation of the unreactive system is not. We then find

$$E_A = E_0 + \frac{c\mu_A\mu_S}{r^3}\left[1 - \left(\frac{r}{r_*}\right)^3\right] + \frac{3c\mu_A\mu_S}{r^3}\left(\frac{d \ln r}{d \ln T}\right)_P \qquad (68)$$

From equation 67, it follows that

$$\ln \frac{Z_{obs}}{Z} = \frac{3c\mu_A\mu_S}{r^3kT}\left(\frac{d \ln r}{d \ln T}\right)_P,$$

which has been shown to provide a qualitatively satisfactory interpretation of the high value of the pre-exponential term of the Arrhenius equation.

REFERENCES

Ahmed and Hall, *Trans. Chem. Soc.*, 3043 (1958).
Auerbach, F. H., Verhoek and A. L. Henne, *J. Amer. Chem. Soc.*, **72**, 299 (1950).
Badar and Harris, *Chem. and Ind.*, 1426 (1964).
Bauer, *Cahiers de Physique*, **20**, 1 (1944).
Belton, Griffith and McKeown, *Trans. Chem. Soc.*, **129**, 3153 (1926).
Bennett, G. M., *Trans. Faraday Soc.*, **37**, 794 (1941).
Benson and Axworthy, *J. Chem. Physics*, **26**, 1718 (1957).
Bernouilli and Jakubowicz, *Helv. Chim. Acta*, **4**, 1018, (1921).
Bowen, E. J., E. A. Moelwyn-Hughes and C. N. Hinshelwood, *Proc. Roy. Soc.*, A, **134**, 211 (1931).
Brandsma, *Thesis*, Delft (1925); *Tables Annuelles* (1930).
Branton, G. R., H. M. Frey, D. C. Montague and I. D. R. Stevens, *Trans. Faraday Soc.*, **62**, 659 (1966).
Buchachenko, A. L., A. M. Wasserman and A. L. Kovarskii, *J. Chem. Kinetics*, **1**, 361 (1969).
Cain and Nicoll, *Trans. Chem. Soc.*, **81**, 1412 (1902).
Chapman, A. W., *Trans. Chem. Soc.*, 448 (1936).
Chapman, D. L., and H. E. Clarke, *Trans. Chem. Soc.*, **93**, 1638 (1908).
Christiansen, *J. Colloid and Interface Science*, **22**, 1 (1966).
Christiansen, *Z. physikal Chem.*, **103**, 91 (1923).
Clark, L. W., *J. Amer. Chem. Soc.*, **77**, 3130 (1955); *J. Phys. Chem.*, **64**, 692 (1960).
Clement, *Ann. Physik*, **14**, 341 (1904).
Conant and Carlson, *J. Amer. Chem. Soc.*, **51**, 3464 (1929).
Cooke, A. S., and Harris, *Trans. Chem. Soc.*, 2365 (1963).
Corran, *Trans. Faraday Soc.*, **23**, 605 (1927).
Daniels, F., and E. H. Johnston, *J. Amer. Chem. Soc.*, **43**, 53 (1921).
Davies, A. G., Mansel Davies and M. Stoll, *Helv. Chim. Acta*, **37**, 1351 (1954).
Dimroth, *Annalen*, **377**, 127 (1910).
Dimroth and Mason, *Annalen*, **399**, 91 (1913).
Dinglinger and Schröer, *Z. physikal Chem.*, A, **179**, 401 (1937).
Essex and Gelormini, *J. Amer. Chem. Soc.*, **48**, 882 (1920).
Eyring, H., and J. F. Daniels, *J. Amer. Chem. Soc.*, **52**, 1472 (1930).

Fairclough, *Trans. Chem. Soc.*, 1186 (1938).
Forbes, M. H., F. G. Mann, I. T. Millar and E. A. Moelwyn-Hughes, *Trans. Chem. Soc.*, 2833 (1963).
Freundlich and Richards, *Z. physikal Chem.*, **79**, 681 (1912).
Garrett and W. C. McC. Lewis, *J. Amer. Chem. Soc.*, 1091 (1923).
Gelles, *J. Amer. Chem. Soc.*, **75**, 6199 (1953).
Hall, D. M., and Harris, *Trans. Chem. Soc.*, 490 (1960).
Hall, G. A., and F. H. Verhoek, *J. Amer. Chem. Soc.*, **69**, 613 (1947).
Hantzsch, *Ber.*, **33**, 2534 (1900).
Harris, M. M., and R. K. Mitchell, *Trans. Chem. Soc.*, 1905 (1960).
Harris and Mellor, *Chem. and Ind.*, 557 (1961); 949 (1959).
Hinshelwood, C. N., and E. A. Moelwyn-Hughes, *Proc. Roy. Soc.*, A, **131**, 186 (1931).
Hinshelwood, C. N., and H. W. Thompson, *Proc. Roy. Soc.*, A, **113**, 221 (1926).
Howlett, *Trans. Chem. Soc.*, 1055 (1960).
Hudson and Dale, *J. Amer. Chem. Soc.*, **39**, 320 (1917).
Jahn, *Z. anorg. Chem.*, **48**, 260 (1906).
Jakubowicz, *Helv. Chim. Acta*, **4**, 1018 (1921).
Johnson, P., and E. A. Moelwyn-Hughes, *Proc. Roy. Soc.*, A, **175**, 118 (1940).
Kappanna, *Z. physikal Chem.*, A, **158**, 355 (1932).
Kassel, *J. Phys. Chem.*, **32**, 1065 (1928); *The Kinetics of Homogeneous Gas Reactions*, Reinhold, New York, (1932).
Kendrew, J. C., and E. A. Moelwyn-Hughes, *Proc. Roy. Soc.* A, **176**, 352 (1940).
Kistiakowsky and W. R. Smith, *J. Amer. Chem. Soc.*, **58**, 1042 (1936).
Kuhn and Albrecht, *Ann.*, **458**, 221 (1927).
Laidler, *Chemical Kinetics*, McGraw-Hill, New York, (1965).
La Mer, *J. Chem. Physics*, **1**, 289 (1933).
Leffler and Graybill, *J. Phys. Chem.*, **63**, 1457 (1963).
Lewis, W. C. McC., *Trans. Chem. Soc.*, **113**, 471 (1918).
Li and Adams, *J. Amer. Chem. Soc.*, **57**, 1565 (1935).
Lin, M. C., and K. J. Laidler, *Trans. Faraday Soc.*, **64**, 79 (1968).
Lindemann, *Trans. Faraday Soc.*, **17**, 599 (1922).
Ling, C. K., and M. M. Harris, *Trans. Chem. Soc.*, 1825 (1964).
Lowry and Traill, *Proc. Roy. Soc.*, A, **132**, 416 (1931).
McKenney and Laidler, *Can. J. Chem.*, **40**, 539 (1962).
Mann, F. G.: See Forbes *et alia*.
Mellor, *Chem. and Ind.*, 949 (1959).
Meyer, *Annalen*, **380**, 231 (1911).
Moelwyn-Hughes, E. A., *Proc. Roy. Soc.*, A, **164**, 295 (1938).
Ogg, *J. Chem. Physics*, **15**, 337 (1947).
Perman and Greaves, *Proc. Roy. Soc.*, A, **80**, 353 (1908).
Pray, *J. Phys. Chem.*, **30**, 1477 (1926).
Rieger and Westheimer, *J. Amer. Chem. Soc.*, **72**, 19 (1950).
Rice and Herzfeld, *J. Amer. Chem. Soc.*, **56**, 284 (1934).
Rice, O. K., and Ramsperger, *J. Amer. Chem. Soc.*, **49**, 1617 (1927).
Reisenfeld and Wassmuth, *Z. physikal Chem.*, A, **143**, 397 (1929).
Ruzicka, L., W. Brugger, M. Pfeiffer, H. Schinz and M. Stoll, *Helv. Chim. Acta*, **9**, 499 (1926).
Salomon, *Helv. Chim. Acta*, **19**, 743 (1936).
Salomon, *Trans. Faraday Soc.*, **32**, 153 (1936).

Saunders, *The Aromatic Diazo-Compounds and their Technical Applications*, p. 69, Arnold, London, (1936).

Slater, N. B., *The Theory of Unimolecular Reactions*, Methuen, London, (1959).

Smith, D. F., *J. Amer. Chem. Soc.*, **49**, 43 (1927).

Smith, G. F., and T. M. Lawry, *Trans. Chem. Soc.*, 666 (1928).

Smith, J. H., and Daniels, *J. Amer. Chem. Soc.*, **55**, 922 (1933).

Sprenger, *Proc. Nat. Acad. Sci.*, **16**, 129 (1930).

Tolman, R. C., *Statistical Mechanics*, Chemical Catalogue Co., New York, (1927).

Trautz, *Z. anorg. Chem.*, **91**, 1 (1916).

Trivich and Verhoek, *J. Amer. Chem. Soc.*, **65**, 1919 (1943).

Turner, E. E., and M. M. Harris, *Organic Chemistry*, Longmans, (1952).

van't Hoff, *Lectures on Theoretical and Physical Chemistry*, **1**, 221 (1898).

von Halban, *Dissertation*, Würzburg (1909); *Z. physikal Chem.*, **67**, 129 (1909); *ibid.*, **77**, 719 (1911).

Warburg, *Ann. Physik*, **9**, 286 (1902).

Westheimer, *J. Chem. Physics*, **15**, 252 (1947).

Westheimer, *Steric Effects in Organic Chemistry*, Wiley, New York, (1956).

Wiig, *J. Phys. Chem.*, **32**, 961 (1928).

11

CATALYSED REACTIONS

A substance which hastens a chemical reaction while retaining its identity is termed a catalyst. It can act in various ways. In some gaseous systems, its role is to generate active centres which then propagate reaction chains. More usually, it lowers the energy barrier required for reaction to take place, and thus increases the concentration of suitably activated molecules.

In the presence of the catalyst, the uncatalysed and the catalysed reactions proceed concurrently, so that an observed first-order rate constant can be resolved as follows:

$$k = k_0 + k_c \, [\text{catalyst}]^n. \tag{1}$$

If successive additions of the catalyst bring about proportionate increases in k, n is unity, and the catalytic coefficient, k_c, becomes effectively a bimolecular constant:

$$k = k_0 + k_2 n_c. \tag{2}$$

By plotting $\ln k_0$ and $\ln k_2$ separately against $1/T$, we derive the apparent energies of activation of the uncatalysed and the catalysed reactions. It is clearly pointless to plot $\ln k$ against $1/T$. When small amounts of catalyst produce a large effect on the velocity, k_0 may be ignored in comparison with k_2. This is the case with the decomposition of hydrogen peroxide; in the presence of minute quantities of certain catalysts the unimolecular velocity constant is millions of times greater than the constant for the uncatalysed decomposition. Occasionally, as with the hydrolysis of casein, the spontaneous reaction has never been detected under any conditions, and k_0 accordingly does not enter. On the other hand, the two constants of the reaction may be commensurate: when this is so, a careful analysis of varying amounts of the catalysts is essential at all temperatures. A case in point is the decomposition of acetonedicarboxylic acid in aqueous solution: k_0 at 60°C is $5.48 \times 10^{-2} \sec^{-1}$; the addition of 0.0165 gramme-mole of aniline per litre gives $k = 11.9 \times 10^{-2}$. k_c thus becomes 3.89 litres/gramme-mole-sec. On analysing the remaining data of Wiig in the same way, we find an energy of activation of about 13,900 cal for the catalysed reaction, which is 9,300 cal less than the value for the uncatalysed reaction. Aniline depresses by a similar amount (8,700 cal) the energy required for the decomposition of trichloracetic acid. Further examples are furnished in Table I, from which we see that catalysts are as varied as the reactions which they facilitate and the media in which the chemical changes take place. Solute and solvent may both act as catalysts.

TABLE I

Reference	Reaction	Catalyst	E_A
* † ‡ §	The decomposition of hydrogen peroxide in aqueous solution	None Iodide ion Colloidal platinum Liver catalase	18,000 13,500 11,700 5,500
‖	The decomposition of acetone-dicarboxylic acid in aqueous solution	None Aniline	23,200 13,900
¶ **	The hydrolysis of sucrose in aqueous solution	Hydrogen ion Saccharase	25,560 8,700
†† ‡‡	The hydrolysis of casein in aqueous solution	Hydrochloric acid Trypsin-kinase	20,600 14,400
§§	The decomposition of triethyl-sulphonium bromide in acetone solution	None 4 per cent water	33,500 30,700
‖ ‖	The decomposition of trinitrobenzoic acid in nitrobenzene solution	None Adventitious impurity, probably water	35,000 21,700
¶¶ ***	The decomposition of trichloracetic acid	Water (solvent) Aniline (solvent)	37,050 28,350
†††	The Beckmann rearrangement of the picryl ether of benzophenone oxime in carbon tetrachloride solution	None Nitromethane	30,250 23,800
‡‡‡	The decomposition of ethylene iodide in the gas phase and in carbon tetrachloride solution	None Iodine (atomic)	37,000 12,500
§§§	The decomposition of diethylether in the gas phase	None Iodine (molecular)	53,500 34,300

REFERENCES

* Possibly some decomposition upon dust particles. Pana, *Trans. Faraday Soc.*, **24**, 486 (1928).

† Walton, *Z. physikal Chem.*, **47**, 185 (1904).

‡ Bredig and von Berneck, *ibid.*, **31**, 258 (1899).

§ Williams, J., *J. General Physiol.*, **11**, 309 (1928).

‖ Wiig, *J. Physical Chem.*, **32**, 961 (1928).

¶ Lamble and Lewis, *Trans. Chem. Soc.*, **105**, 2330 (1914).

** Nelson and Bloomfield, *J. Amer. Chem. Soc.*, **46**, 1025 (1924).

†† Nasset and Greenberg, *ibid.*, **51**, 836 (1929).

‡‡ Moelwyn-Hughes, Pace and Lewis, *J. General Physiol.*, **13**, 323 (1930).

§§ von Halban, *Z. physikal Chem.*, **67**, 129 (1909).

‖ ‖ Moelwyn-Hughes and Hinshelwood, *Proc. Roy. Soc.*, A, **131**, 186 (1931).

¶¶ Kappanna, *Z. physikal Chem.*, A, **158**, 355 (1932); P. Johnson and Moelwyn-Hughes, *Proc. Roy. Soc.*, A, **175**, 118 (1940).

*** Goldschmidt and Bräuer, *Berichte*, **39**, 109 (1906).

††† A. W. Chapman, *Trans. Chem. Soc.*, 1550 (1934).

‡‡‡ Polissar, *J. Amer. Chem. Soc.*, **52**, 956 (1930); Arnold and Kistiakowsky, *J. Chem. Physics*, **1**, 166 (1933). See erratum added later.

§§§ Clusius and Hinshelwood, *Proc. Roy. Soc.*, A, **128**, 82 (1930).

§, **, ‡‡ These figures refer to the optimum *p*H.

There are three criteria of homogeneous catalysts. The first criterion is a reduction by unity of the apparent order of reaction. The second is a reduction in the apparent energy of activation. The third criterion, which is fairly general though not universal, is that the catalytic coefficient belongs to the category of normal reactions, for which there is unit efficiency of activating collisions involving no internal degrees of freedom.

The examples of catalysis selected for study in this chapter are those in which the uncatalysed change is a unimolecular one and for which the catalytic integer is 1. We then have

$$-dn_A/dt = k_1 n_A + k_2 n_A n_C, \tag{3}$$

where n_A is the concentration of the substance undergoing change, and n_C is the concentration of the catalyst. The observed first-order constant is thus linearly related to the concentration of catalyst:

$$k = k_1 + k_2 n_c. \tag{4}$$

These equations apply in the first instance to catalysis in dilute solutions, to which our main attention is directed.

Some Simple Catalytic Mechanisms

For the purpose of a preliminary discussion, we shall assume that the rate of chemical reaction is determined principally by the magnitude of the energy of activation, and that a catalyst acts by lowering it. This reduction can be effected in various ways; for example:

(i) By Changing the Distribution of Internal Energy

Let us assume that the real energy of activation, E_0, is the same for the catalysed as for the uncatalysed reaction, but that, in the two activated complexes, it is distributed in different ways, according to the number, F_c and F_u, of classical oscillators, among which the energy is distributed. Then, for the uncatalysed and catalysed reactions,

$$E_u = E_0 + s_u RT$$

and

$$E_c = E_0 + s_c RT.$$

Then

$$s_u - s_c = (E_u - E_c)/RT. \tag{5}$$

Applied to the decomposition of hydrogen peroxide in aqueous solution, uncatalysed and catalysed by the iodide ion, we see, from Table I, that $s_u - s_c$ is 9. The presence of the ion has thus made it possible for a small energy, delivered in the right spot, to do the work of a larger energy when distributed throughout the molecule. The value of $S_u - S_c$ similarly obtained for the decomposition of diethylether is 10.

(ii) By Bringing the Reactants Closer Together

In the reaction between the persulphate and iodide ions, the energy of activation includes the Coulombic energy $2\varepsilon^2/Da$ required to bring them to a distance a apart. ε is the charge on the proton, and D the permittivity. In the presence of a potassium ion placed midway between them, the work required to bring the reactant ions to within the same distance is $-4\varepsilon^2/Da$. Thus the energy of activation is reduced by $6\varepsilon^2/Da$, which, with $a = 6$ Å in water at 25°C, is 4,230 cal. Such a mechanism has been established for this reaction (equation 7.19b) and for several other reactions involving bridged complexes.

(iii) By Breaking a Bond in One of the Reactants

The reduction of the iodate ion by molecular hydrogen in neutral, homogeneous, aqueous solutions, $IO_3^- + 3H_2 \rightarrow I^- + 3H_2O$, is too slow to be measured. In the presence of various cations, however, it proceeds rapidly, due to the breaking, by the cation, of the very firm bond in the hydrogen molecule. The rate-determining step in the presence of the cupric ion is $Cu^{2+} + H_2 \rightarrow CuH^+ + H^+$. This is followed rapidly by the oxidation proper, and by the regeneration of the catalyst (Calvin and Halpern, 1959).

(iv) BY ALTERING THE STRUCTURE OF ONE OF THE REACTANTS

The reaction between iodine and acetone in aqueous solutions of strong acids is a simple example of this mechanism. The role of the hydrogen ion is to change the structure of the acetone molecule from the ketonic to the enolic form. The subsequent reactions include the rapid iodination of acetone and the regeneration of the catalyst.

General and Specific Catalyses

Certain reactions, such as mutarotations in non-planar solvents, are catalysed to nearly the same extent by various organic solutes which are in no way structurally related to one another. Such catalyses may be termed general. On the other hand, there are reactions, particularly in aqueous solution, whose rates are sensitive to one specific catalyst only. Enzyme reactions are of this type. The catalytic response of some substrates is limited to a specific ion, such as the cyanide or the hydroxyl ion. These also are examples of specific catalysis. Finally, we have the widely studied set of reactions which respond not only to hydrogen ions and hydroxyl ions but to acids and bases generally. We shall here examine a few instances from each category (E. K. Rideal and H. S. Taylor, 1926; G.-M. Schwab, 1937; H. Schmid, 1940; R. P. Bell, 1941; P. G. Ashmore, 1963; E. L. King, 1955).

Catalysis in Carbon Tetrachloride Solution

(i) THE MUTAROTATION OF BERYLLIUM BENZOYLCAMPHOR

In pure carbon tetrachloride, beryllium benzoylcamphor undergoes mutarotation unimolecularly at a slow rate, indicating an apparent energy of activation of at least 25,000 calories per mole. Rapid mutarotation takes place when small amounts of pyridine, ethanol, *p*-cresol or acetone are added, and the first-order rate constants are proportional to their concentrations, since the rate of the uncatalysed reaction is negligibly small in comparison. Some of the results obtained by Lowry and Traill (1931) are summarized in Table II. In striking contrast to the work on mutarotation

TABLE II

CATALYTIC COEFFICIENTS (LITRES/GRAMME-MOL-SEC) MUTAROTATION OF
BERYLLIUM BENZOYLCAMPHOR

Pyridine in CCl_4 at 20°	0·132
Pyridine in CCl_4 at 25°	0·227
Pyridine in CCl_4 at 30°	0·385
Pyridine in $CHCl_3$ at 25°	0·100
Alcohol in CCl_4 at 20°	0·231†
Alcohol in CCl_4 at 25°	0·370†
Alcohol in CCl_4 at 30°	0·658†
Alcohol in $CHCl_3$ at 25°	0·204†
p-Cresol in CCl_4 at 25°	0·318†
p-Cresol in $CHCl_3$ at 25°	0·103†
Acetone in CCl_4 at 25°	0·015†
CCl_1 and $CHCl_3$ at 25°	Negligible in comparison

REFERENCE

† Extrapolated to infinite dilution.

in aqueous solution (Table XI), the catalytic coefficients in carbon tetrachloride solution are nearly the same for the weak acid, cresol, and the
weak base, pyridine. Traill (1932) (see also Moelwyn-Hughes (1932)) has
analysed the results in terms of the collision theory of Trautz and Lewis.
The catalytic coefficients vary with respect to temperature as follows:

Pyridine as catalyst: $k_2 = 2.50 \times 10^{11} \cdot \exp(-18,850/RT)$,
Ethanol as catalyst: $k_2 = 1.91 \times 10^{11} \cdot \exp(-18,410/RT)$.

The pre-exponential terms are in reasonable agreement with the simple
collision equation. Both catalysts have reduced the energy of activation
by at least 6,000 cal. It is strange that chloroform as a solute is catalytically
inefficient. The fact that a complex between it and the reactant is formed
may be a clue to its special role. Burgess and Lowry (1924) have isolated it
in the form of a crystalline solvate containing two molecules of solvent to
one of solute. From the value of its optical rotation, the following symmetrical
structure has been ascribed to it.

Since the racemization of an optically active compound usually proceeds through the formation of an intermediate complex of higher symmetry, the complex thus formulated provides an ideal mechanism for the reaction. From a knowledge of all the specific rotations involved, Lowry and Traill have shown that 46 per cent of the α form and 64 per cent of the β form are combined with the solvent. Experiments prove that these figures account quantitatively for the lessened catalytic activity of pyridine in chloroform, as compared with solutions in carbon tetrachloride, where the whole of the solute is apparently free to interact with the catalyst instead of being bound in part to the solvent. These investigations have accordingly given us a description—possibly as clear and precise as it is possible to expect— of the definite part played by a well-defined intermediate compound in the mechanism of chemical change in solution.

(ii) The Beckmann Transformation

A. W. Chapman's investigation of the kinetics of the Beckmann rearrangement of numerous ketoximes in inert solvents and under the influence of various catalysts takes the subject a whole stage forward by demonstrating a direct field effect in catalysis.

In carbon tetrachloride solution, the picryl ether of benzophenone oxime undergoes the Beckmann rearrangement

$$C_6H_5-\underset{\underset{N-O-C_6H_2(NO_2)_3}{\|}}{C}-C_6H_5 \rightarrow C_6H_5-\underset{\underset{C_6H_5-N}{\|}}{C}-O-C_6H_2(NO_2)_3$$

unimolecularly, but the velocity coefficient is found to be linearly related to the concentration of catalyst. This Chapman (1934) interprets as evidence that, in addition to the unimolecular change proper there is a concomitant change due to catalysis by the oxime itself. Some of the results obtained are reproduced in Table III. The apparent energy of activation for the uni-

TABLE III

Unimolecular Velocity Constants and Catalytic Coefficients for the Beckmann Transformation in Carbon Tetrachloride Solution

$t°C$	$k_1 \times 10^5$	$k_2 \times 10^5$
80·5	4·40	67
85·9	7·90	—
92·8	16·8	190
100·0	34·1	392

molecular reaction is 27,880 cal, and its rate is such as to indicate the necessity of a small number of internal degrees of freedom. The energy of activation for the catalytic process is 22,620 cal, and the velocity of chemical change is, within a factor of 2, equal to that predicted by the simple collision theory. Non-polar solutes, such as hexane, have no influence on the rate of transformation; other solutes affect it with an efficiency which increases as the dipole moment (Fig. 1). The slight curvature of some of the lines, and the

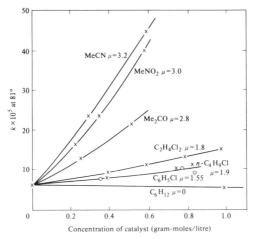

FIG. 1. The catalysis of the Beckmann change by a variety of solutes.

apparently inhibitive effect of hexane are probably to be attributed to the influence of dilution. The relation between the dipole moments of the catalysts and the energies of activation of the catalytic transformation is shown in Table IV. Within the limits of error, the catalysts affect the reaction by lowering the energy of activation, and are otherwise equally efficient. The figure in brackets is merely a rough kinetic estimate of the dipole moment of

TABLE IV

MOLECULAR STATISTICS OF THE CATALYSED BECKMANN CHANGE IN CARBON TETRACHLORIDE

Catalyst	$\mu \times 10^{18}$ e.s.u.	E_2 (cal/gramme-mole)	P
The reactant	(5·4)	22,620	0·43
CH_3NO_2	3·0	23,800	0·53
$C_2H_4Cl_2$	1·8	24,400	0·26

(After A. W. Chapman, *Trans. Chem. Soc.*, 1550, (1934).)

the oxime, consistent with the evident experimental value of 50 cal per gramme-mole per debye for $dE/d\mu$. It is a useful figure because it enables us to ascribe a value of about 10, in the usual units, for $dE/d(\mu_A\mu_C)$. The Beckmann change is clearly facilitated by the presence of even weak electrostatic fields. We must not, however, conclude that only a solute molecule possessing a dipole moment will act as a positive catalyst. Chapman has shown that non-polar molecules may act as catalysts provided the two (internally compensating) dipoles which they contain are sufficiently far apart not to cancel out each other's external fields. Thus, *trans*-dichlorethylene has no catalytic effect, but *p*-dichlorobenzene and *trans*-1:4-dibrom*cyclo*hexane are catalytically efficient, the latter being almost as good as ethylene dichloride.

(iii) Discussion

The inefficiency of chloroform as a catalyst in the mutarotation of beryllium benzoylcamphor may be understood if the solute–solvent complex formed in carbon tetrachloride solution has the symmetrical structure ascribed by Burgess and Lowry, in which the fields exerted by the two solvate molecules probably may cancel out.

In the two chemically different but kinetically similar reactions discussed here, the catalysts have brought about reductions of from 5 to 6 kcal in the energies of activation. How the mere presence of polar molecules can effect this is by no means clear. In some way which we do not understand, the transfer of energy which is always necessary during activation seems to be invariably facilitated by the presence of an electrostatic field.

The Inversion of Cane Sugar; Specific Catalysis by Hydrogen Ion

Reference has already been made to this parent of all kinetic studies. At constant temperature, pressure and acid concentration, the rate of hydrolysis as measured by polarimetric, analytic, dilatometric and calorimetric techniques is unimolecular with respect to sucrose. The observed first-order constant increases as the concentration of hydrogen ion increases, though not exactly in direct proportion, giving a catalytic coefficient $k_c = k/c_{H^+}$ which increases slightly with increase in c_{H^+} and in the concentration of sugar. The rate of hydrolysis is unaffected by unionized acids or by ions other than hydrogen ions, i.e. we are dealing with specific hydrogen–ion catalysis. The corresponding catalytic coefficient using deuterium instead of hydrogen, k/c_{D^+}, is greater than k/c_{H^+} by a factor of 1·80 at 18·71°C and 1·55 at 37·13°C (Moelwyn-Hughes and Bonhoeffer, 1934), in marked contrast to the ratio of 0·64 (25°C) found for the mutarotation of glucose (Moelwyn-Hughes, Klar and Bonhoeffer, 1934). The latter reaction is known to be subject to general catalysis. It therefore seems that experimental ratios k_{D^+}/k_{H^+} which are less than unity indicate general acid-base catalysis, and that ratios greater than unity indicate specific hydrogen-ion catalysis. The hydrolysis of sucrose has long been regarded as proceeding by the following mechanism:

$$S + H^+ \underset{k_1}{\overset{k_2}{\rightleftarrows}} SH^+ \overset{k_3}{\rightarrow} G + F + H^+.$$

The equilibrium between hydrogen ion and sucrose is assumed to be rapidly established. The rate-determining step is that of the breakdown of the complex (see contributions by Bell, Bonhoeffer, Pedersen and Wynne-Jones, (1937)). In dilute solutions, the catalytic coefficient thus becomes

$$k_c = -\frac{1}{c_S c_{H^+}}\frac{dc_S}{dt} = \frac{k_2 k_3}{k_1 + k_3}. \tag{1.4}$$

Since $k_1 \gg k_3$,

$$k_c = (k_2/k_1)k_3 = Kk_3, \tag{6}$$

where K is the thermodynamic equilibrium constant governing the formation and dissociation of the complex at infinite dilution and k_3 is the unimolecular constant governing its conversion into products. In terms of activity coefficients, we have, at real concentrations,

$$k_c = k_3 K^0 \frac{\gamma_S \gamma_{H^+}}{\gamma_{SH^+}}$$

$$= k_c^0 F, \tag{7}$$

where k_c^0 is the catalytic coefficient extrapolated to zero concentrations, and F, the kinetic activity factor, depends on the concentrations of both the hydrogen ion and the sugar. In solutions with a fixed concentration of sucrose at 0°C, for example, it is found that

$$F = \exp(0·2393 c_{H^+}). \tag{8}$$

The numerical constant determined by Leininger and M. Kilpatrick (1938) from their own data with hydrochloric acid at concentrations up to 5·8M is 0·2394, and that calculated by them from the data of Duboux (1938) is 0·2392. The activity coefficient, γ_S, of sucrose in water is known with precision from the vapour pressure data. A comparison with the kinetic results makes it clear that the ratio of the activity coefficients of the two cations is far from unity, as would probably be the case if they were structurally similar. Later reference is made to other, slightly less empirical, attempts to represent the rate of hydrolysis in concentrated acid solutions at a single temperature.

We shall here deal with the rate of hydrolysis in dilute acid solutions at different temperatures. A three-constant equation is insufficient to reproduce the experimental values of the catalytic coefficient, but the 4-constant equation (D. N. Glew, Private communication),

$$\log_{10} k_c = 1308 \cdot 6014 - 513 \cdot 0589 \log_{10} T + 0 \cdot 33257T - 42,352 \cdot 60/T \quad (9)$$

provides the best fit (see Table V), with a standard deviation of

$$\sigma = \left\{ \frac{\Sigma[\delta \log_{10} k_c]^2}{7} \right\}^{1/2} = \pm 0 \cdot 0059.$$

The apparent energy of activation, in cal per mole, is consequently

$$E_A = 193,789 - 1,019 \cdot 53T + 1 \cdot 52171T^2. \quad (10)$$

The value of $d^2 E_A/dT^2 = 3 \cdot 04342$ is significant at a 31·3 per cent probability level. E_A itself passes through a minimum value at 61·85°C.

TABLE V

CATALYTIC COEFFICIENTS FOR THE HYDROLYSIS OF SUCROSE (5 PER CENT BY WEIGHT) IN APPROXIMATELY N/5 AQUEOUS HYDROCHLORIC ACID

$t°C$	$\log_{10} k_c$(litre/mole^{-1}-sec^{-1}) observed	calculated	$\delta \log_{10} k_c$ obs − calc
15·45	−4·4461	−4·4478	+0·0017
19·40	−4·1864	−4·1817	−0·0047
23·09	−3·9393	−3·9441	+0·0048
25·00	—	−3·8243	—
27·26	−3·6882	−3·6873	−0·0009
31·08	−3·4647	−3·4620	−0·0025
35·98	−3·1884	−3·1861	−0·0023
41·00	−2·9066	−2·9166	+0·0100
46·10	−2·6615	−2·6551	−0·0064
51·15	−2·4078	−2·4068	−0·0010
57·10	−2·1249	−2·1262	+0·0013

Further work on the temperature effect has been carried out by Hitchcock and Dugan (1935), Sturtevant (1937), Heidt and Purves (1938), Pearce and Thomas (1938) and by Leininger and Kilpatrick (1938), whose results are shown in Fig. 2. In this work, the dilatometer method was used, and each temperature was kept constant to within ±0·005°. The results show that E_A falls as T is raised, though not so rapidly as in the earlier and less accurate work, and that dE_A/dT becomes less negative as the temperature is raised. There are clear indications of minima in E_A at temperatures slightly above

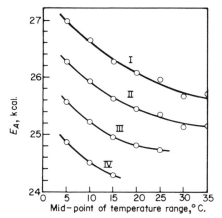

FIG. 2. Effect of temperature on energy of activation: I, 0 molar HCl; II, 1 molar HCl; III, 2 molar HCl; IV, 3 molar HCl. From "Kinetics of Reactions in Solution," 2nd ed., reproduced by permission of The Clarendon Press, Oxford.

those covered. From the catalytic coefficients extrapolated to zero concentration of acid, the following approximate expression holds:

$$E_A (\text{cal/mole}^{-1}) = 128,725 - 647T - 1.014T^2, \tag{11}$$

from which Table VI has been constructed. The temperature at which E_A

TABLE VI

APPARENT ENERGIES OF ACTIVATION FOR THE
HYDROLYSIS OF SUCROSE

$t°C$	E_A	dE_A/dT	dB/dT
0	27,653	−93	−53
25	25,960	−42	−32
45·85	25,518	0	—
50	25,535	+8	−12

attains its minimum value is lower than that found in the earlier work, and depends on the composition of the solutions. With some certainty, however, it may be said that dE_A/dT for this reaction becomes zero at $53 \pm 8°C$. If, as Ölander (1929) suggested, the catalytic coefficient varies inversely as the viscosity, part of the term $-dE_A/dT$ may be attributed to the variation of the viscous energy, B (Table 5.V) with respect to temperature. The results summarized in Table VI suggest that, at least in the neighbourhood of 25°C, the temperature variation of E_A is due mainly to the temperature variation of the fluidity. The composite nature of the catalytic coefficient (equation 4.5) must not be overlooked in this connection, or the possibility that there may be at least two concurrent mechanisms of hydrolysis.

The Reaction of Ethyl Diazoacetate with Water: A Further Instance of Specific Catalysis by Hydrogen Ion

The reaction between ethyl diazoacetate and water entails the evolution of nitrogen, which makes its rate conveniently measurable nanometrically:

$$N_2CH \cdot COOC_2H_5 + H_2O \rightarrow HOCH_2 \cdot COOC_2H_5 + N_2.$$

Hydrogen ion is its exclusive catalyst. At 25°C the catalytic coefficient is 827 times as great as that for sucrose hydrolysis. Since the discovery of a direct proportionality† between rate of reaction and concentration of hydrogen ion, the reaction has frequently been used as a kinetic method for the determination of pH. Fraenkel's (1907) classical researches, however, leave certain points in obscurity. In the first place, a number of strong acids, including HCl, attack the ester molecule effecting substitution, according to the scheme

$$N_2CH \cdot COOC_2H_5 + HX \rightarrow XCH_2 \cdot COOC_2H_5 + N_2.$$

On account of the partial removal of catalyst, the velocity of elimination of nitrogen falls off at a greater rate than that given by the unimolecular law. In the second place, strong acids which do not appreciably attack the ester by the substitution mechanism and which consequently give good unimolecular constants, yield catalytic coefficients which vary somewhat with the acid concentration and the nature of the anion. At 283·16°K, for example, using nitric acid, the unimolecular velocity constant is given by the empirical relation‡

$$k_1 = k_0 + k_2 c_{H^+} + k_3 c_{H^+}^2, \tag{11}$$

where c_{H^+} is the concentration of hydrogen ion, in gramme moles/litre, and k_1 is the unimolecular velocity coefficient, in reciprocal seconds. The specific constants k_0, k_2 and k_3 under these conditions have the values 0, 0·124 and 0·525 (Fig. 3).

The conditions of temperature and the nature of the acid necessary for uncomplicated catalysis must therefore be carefully selected. They are found to correspond to dilute solutions of nitric acid above 10°C, in which region

---

† Fraenkel (1907).
‡ Moelwyn-Hughes and Johnson (1941).

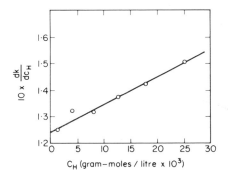

FIG. 3. Hydrogen-ion catalysis of the decomposition of ethyl diazoacetate.

there is no detectable substitution, and the constant k_3 is zero. Some values of the unimolecular velocity coefficients are given in Table VII, from which the catalytic coefficient, k_2, has been determined by the method of least squares. An analysis of the dependence of $\ln k_2$ on temperature, shows that either of the equations

$$\ln k_2 = 44\cdot6801 - 4\cdot615 \ln T - 18,890/RT$$

and

$$\ln k_H = 29\cdot0968 - 17,500/RT$$

satisfactorily reproduces the facts. If anything, the latter equation is arithmetically slightly superior, though physically less acceptable. In order to account for the rate of this reaction it is found necessary to postulate the sharing of the activation energy among a small number of degrees of freedom, the first estimate of which* was 4.

The specific role of the anion suggests that the chloride ion can efficiently compete with water but that the nitrate ion is less successful.

The value of E_A given by Fraenkel is 17,560 calories, though his results make possible a value as low as 17,480 (Bredig and Fraenkel, 1905).

Using picric acid as catalyst, the catalytic coefficient in deuterium oxide, D_2O, is greater than that in ordinary water, and the ratio k_{D_2O}/k_{H_2O} decreases with an increase in temperature; (k_{D_2O}/k_{H_2O}) is 3·27 at 15°C and 2·94 at 35°C (P. Gross, H. Steiner and F. Krauss, 1937; O. Halpern and P. Gross, 1935). These findings are analogous to those found for the hydrolysis of sucrose, indicating a similar mechanism of reaction. Equilibrium between

---

* *Phil. Mag.*, **14**, 112, (1932).

TABLE VII
UNIMOLECULAR VELOCITY COEFFICIENTS FOR THE DECOMPOSITION OF ETHYL DIAZO-
ACETATE IN AQUEOUS NITRIC ACID SOLUTION

Temperature (°K)	$c_H \times 10^4$	$c_E \times 10^2$	$k_1 \times 10^4$ (sec^{-1}) observed	calculated
$273 \cdot 22 \pm 0 \cdot 02$	53·67	3·76	2·190	2·194
$273 \cdot 23 \pm 0 \cdot 01$	71·61	1·56	2·982	2·969
$273 \cdot 29 \pm 0 \cdot 03$	107·4	3·09	4·509	4·519
$273 \cdot 31 \pm 0 \cdot 02$	214·8	1·90	9·168	9·166
$273 \cdot 29 \pm 0 \cdot 03$	286·5	1·31	12·78	12·27
$278 \cdot 11 \pm 0 \cdot 04$	17·90	3·94	1·296	1·311
$278 \cdot 11 \pm 0 \cdot 04$	35·80	3·24	2·641	2·71
$278 \cdot 11 \pm 0 \cdot 04$	107·4	1·67	8·45	8·31
$278 \cdot 11 \pm 0 \cdot 04$	214·9	0·95	16·65	16·71
$283 \cdot 16 \pm 0 \cdot 03$	26·85	3·22	3·351	3·359
$283 \cdot 16 \pm 0 \cdot 03$	53·69	2·72	6·907	6·902
$283 \cdot 17 \pm 0 \cdot 03$	107·4	2·24	13·98	13·99
$283 \cdot 16 \pm 0 \cdot 05$	143·2	0·92	18·90	18·71
$283 \cdot 16 \pm 0 \cdot 03$	214·7	1·60, 2·49	29·08	28·15
$283 \cdot 16 \pm 0 \cdot 03$	286·5	1·38	39·88	37·64
$288 \cdot 00 \pm 0 \cdot 01$	13·46	3·62	2·619	2·675
$288 \cdot 00 \pm 0 \cdot 01$	26·83	3·62	5·837	5·785
$288 \cdot 00 \pm 0 \cdot 01$	53·65	2·96	11·99	12·00
$288 \cdot 00 \pm 0 \cdot 02$	107·3	1·28	24·41	24·43
$293 \cdot 39 \pm 0 \cdot 01$	13·45	4·19	4·796	4·92
$293 \cdot 39 \pm 0 \cdot 01$	26·80	1·50, 3·26	10·24	10·11
$293 \cdot 39 \pm 0 \cdot 01$	53·60	3·10	20·57	20·54
$293 \cdot 39 \pm 0 \cdot 01$	107·2	1·90	41·37	41·41
$298 \cdot 11 \pm 0 \cdot 01$	13·40	3·84, 2·80	8·03	8·19
$298 \cdot 11 \pm 0 \cdot 01$	26·78	1·89	16·9	16·87
$298 \cdot 11 \pm 0 \cdot 01$	35·70	1·93	22·7	22·66
$298 \cdot 11 \pm 0 \cdot 01$	53·56	1·16	34·4	34·25
$298 \cdot 11 \pm 0 \cdot 01$	71·39	0·671, 2·98, 3·36	44·4	45·82
$298 \cdot 11 \pm 0 \cdot 01$	107·1	1·49, 1·19	68·9	68·99
$303 \cdot 13 \pm 0 \cdot 03$	5·350	3·48	3·625	3·784
$303 \cdot 13 \pm 0 \cdot 04$	13·42	2·78	12·43	12·32
$303 \cdot 13 \pm 0 \cdot 04$	26·75	2·23	26·51	26·42
$303 \cdot 13 \pm 0 \cdot 04$	53·50	1·76	54·64	54·74
$308 \cdot 33 \pm 0 \cdot 03$	2·671	4·94	2·293	2·33
$308 \cdot 33 \pm 0 \cdot 03$	5·342	4·36	6·705	6·77
$308 \cdot 33 \pm 0 \cdot 03$	7·118	2·17	9·835	9·72
$308 \cdot 33 \pm 0 \cdot 03$	26·69	2·04	42·23	42·25
$313 \cdot 20 \pm 0 \cdot 04$	2·665	3·02	3·785	3·323
$313 \cdot 20 \pm 0 \cdot 04$	5·325	2·72	10·23	10·36
$313 \cdot 18 \pm 0 \cdot 05$	13·36	1·92	30·74	31·63
$313 \cdot 20 \pm 0 \cdot 04$	17·76	1·27	43·76	43·28
$313 \cdot 20 \pm 0 \cdot 04$	26·65	1·91	66·92	66·80

the ester and the hydrogen ion is rapidly established:

$$\bar{N}{=}\overset{+}{N}{-}CHCOOC_2H_5 + H_3O^+ \rightleftarrows N{\equiv}\overset{+}{N}{-}CH_2COOC_2H_5 + H_2O;$$
$$c_E c_{H^+} c_C$$

$$K = c_C/c_E c_{H^+}.$$

There follow two rate-determining steps when the complex reacts simultaneously with a water molecule and with the anion of the acid:

$$N{\equiv}\overset{+}{N}{-}CH_2COOC_2H_5 + H_2O \rightarrow N_2 + CCH_2OHCOOC_2H_5 + H^+ \quad k_2'$$

$$N{\equiv}\overset{+}{N}{-}CH_2COOC_2H_5 + X^- \rightarrow N_2 + CH_2XCOOC_2H_5 \quad\quad\quad k_3'$$

The net rate of reaction is thus

$$-\frac{dc_E}{dt} = k_2'c_C + k_3'c_C c_X$$

$$= k_2'K c_E c_{H^+} + k_3'K c_E c_{H^+}c_X$$

$$= k_2 c_E c_{H^+} + k_3 c_E c_{H^+}c_X.$$

Electrical neutrality demands that the concentration of hydrogen ion at any time is equal to that of the anion, X^-. Hence, since

$$k_1 = -(1/c_E)(dc_E/dt)$$

$$k_1 = k_2 c_{H^+} + k_3 c_{H^+}^2 \tag{12}$$

which, as W. J. Albery and R. P. Bell (1961) have shown, offers an explanation of the experimental equation of P. Johnson and Moelwyn-Hughes.

In order to estimate the relative extent of these two reactions, it may be assumed that the rate of elimination of nitrogen is due entirely to reaction between the ester and hydrogen ion. The fall off in the observed rate is then proportional to the concentration of ethylmonochloracetate formed. Johnson and Moelwyn-Hughes thus found the fraction of the change due to the substitution reaction to be 19·7 per cent for HCl at 25°C, and 0·5 per cent for HNO₃ at the same temperature. Albery and Bell, by a combination of thermometric and analytical methods, find values of from 6·1 to 7·7 for HCl at 0°C. The fraction considered is thus sensitive to temperature and to the nature of the anion.

Equation (12) can be expressed in terms of the initial concentrations, a and c, of ethyldiazoacetate and catalysing acid respectively. Let x be the decrease in the concentration of ester due to the formation of ethylglycollate, and y be the decrease due to the formation of ethylchloracetate. The total rate of reaction becomes

$$-\frac{d(a - x - y)}{dt} = k_2(a - x - y)(c - y) + k_3(a - x - y)(c - y)^2,$$

and the rates of formation of the two organic products are

$$dx/dt = k_2(a - x - y)(c - y)$$

and

$$dy/dt = k_3(a - x - y)(c - y)^2.$$

At all stages of the reaction, therefore,

$$dy/dx = (k_3/k_2)(c - y),$$

or, after integrating with the usual initial conditions,

$$k_3/k_2 = (1/x) \ln [c/(c - y)]. \tag{13}$$

Thus the ratio of the two rate constants may be found by chemical analysis of the organic products, as in the treatment leading to equation (9.17). Values of k_3 for numerous elementary ions have been measured in this way (Albery, J. E. C. Hutchins, R. M. Hyde and R. H. Johnson, 1962).

Reference is made later to the catalytic hydrolysis of sucrose and the catalytic decomposition of ethyldiazoacetate in concentrated solutions of acids.

The Decomposition of Diacetone-Alcohol; Catalysis by Hydroxyl Ions

The proportionality of the rate of decomposition of diacetone-alcohol to the concentration of hydroxyl ion in aqueous solution,

$$(CH_3)_2C(OH)CH_2COCH_3 + OH^- \rightarrow 2CH_3COCH_3 + OH^-,$$

TABLE VIII

CATALYTIC COEFFICIENTS FOR THE DECOMPOSITION OF
DIACETONE-ALCOHOL AT 25°C

[NaOH] $\times 10^3$ (gramme-mole/litre)	$k \times 10^3$ (min^{-1}; log$_{10}$)	$k/$[NaOH]
5	1·01	0·202
10	2·03	0·203
20	4·08	0·204
40	8·33	0·208
100	20·80	0·208

was established by French (1929), (see also Akerlöf (1926)) whose unimolecular constants are shown in Table VIII. Expressed in the customary units, the catalytic coefficient at this temperature is thus 7.87×10^{-3}. Murphy (1931) measured the temperature effect with concentrations of alkalies varying between 0.01 and 0.10 molar. The mean of eight values of k_{30}/k_{20} is 2.57 and of k_{35}/k_{25} is 2.67, which give us, respectively, 17,918 and 18,122 cal as the apparent energy of activation. Within the limits of error we may take E_A to be $18,020 \pm 100$ cal at 27.5°C. On combining this with French's catalytic coefficient we obtain

$$k_2 = 1.31 \times 10^{11} \times \exp(-18,020/RT),$$

from which at a glance we recognize a reaction proceeding with the normal velocity.

TABLE IX

APPARENT ENERGIES OF ACTIVATION (E_A) AND CORRECTED VALUES
($E = E_A + B$) FOR THE CATALYTIC DECOMPOSITION OF DIACETONE-
ALCOHOL BY HYDROXYL IONS IN WATER
(' for 15° interval, " for 20°, others for 10°)

Mid-point of temp. interval, °C	A moles litre/min	E_A	E
4.89	10.92	15,850	20,630
7.49	11.13'	16,100'	20,760
9.96	11.22"	16,230"	20,780
10.42	11.42	16,500	21,030
12.89	11.46'	16,550'	20,980
14.85	11.52	16,620	20,970
15.43	11.56"	16,680"	21,010
17.39	11.63	16,770'	21,030
19.84	11.75"	16,920"	21,140
19.99	11.69	16,850	21,020
22.44	11.83'	17,040'	21,130
24.91	11.97	17,250	21,260
24.97	11.84"	17,040"	21,050
27.44	11.93	17,180'	21,120
29.96	11.96"	17,220"	21,090
29.98	11.98	17,240	21,110
32.50	12.00'	17,270'	21,080
34.95	11.94	17,190	20,940
35.05	12.02"	17,290"	21,040
37.50	11.98'	17,250'	20,940
39.93	11.78"	16,960"	20,600
40.04	12.05	17,350	20,990
42.47	11.76'	16,930'	20,530
44.99	11.61	16,720	20,288

In a careful repetition of the experiments (La Mer and M. L. Miller, 1935) an increase in E_A with a rise in temperature has been found. The point-to-point values of E_A and A calculated by La Mer and Miller are reproduced in Table IX, to which have been added the corrected energies $E = (E_A + B)$. It will be observed that E is less variable than E_A, and, from 5 to 40°C, has an average value of 20,990 ± 360 cal. Inspection of the E_A figures shows that two of them may differ at the same temperature by 210 cal. A further error of 150 cal in $RT^2(d \ln \eta/dT)_P$ is not an unreasonable one to allow when we consider the effect of the various solutes on the viscosity. Within the limits of error, therefore, we conclude that E is constant. The presence of methyl alcohol diminishes B and increases E_A, so that the introduction of the viscosity correction here, as in other cases, seems to explain at least one of the factors in the solvent effect. In cases such as the present one, however, kinetic work has far outstripped our knowledge of the static properties of the system. As far as one is able to apply equation (5.2), there results, as La Mer and Miller observe, a good correlation.

A Comparison of Catalysis by Hydrogen Ion and by Hydroxyl Ion

It has long been known that hydroxyl ions are more efficient catalysts for many reactions than are hydrogen ions. According to the mechanism of catalysis which was shown in the preceding sections to be most commonly at work in solutions, we should anticipate a lower energy of activation for the basic catalysis. The anticipation is fully realised, as the results in Table X

TABLE X

	E_A		
Reaction	Catalysed by hydrogen ion	Catalysed by hydroxyl ion	Difference
Decomposition of nitrosotri-acetonamine	22,100	16,210	5,890
Hydrolysis of methyl acetate	16,920	11,220	5,700 ⎫ 5,660
Hydrolysis of ethyl acetate	16,830	11,210	5,620 ⎭
Conversion of hexamethylpara-osaniline hydrochloride from carbinol to quinonoid form†	17,770 18,910 19,460	11,260 12,950 13,750	6,510 ⎫ 5,960 ⎬ 6,060 5,710 ⎭
Hydrolysis of acetamide	19,210	13,320	5,890 ⎫ 6,230
Hydrolysis of propionamide	19,490	12,930	6,560 ⎭

† This reaction is probably not simple, the critical increments falling with rise in temperature. Nevertheless $(E_H - E_{OH})$ remains roughly constant at all temperatures. The reaction was measured by Biddle and Porter (1915).

indicate. The data, which refer to aqueous solutions, point to an approximately fixed value of 5,980 cal for $(E_H - E_{OH})$ for all these reactions. At 25°C, the ratio k_{OH}/k_H of the catalytic coefficients is in the case of the mutarotation of glucose (Table XI) is 2.91×10^4. Assuming the Arrhenius A terms to be the same for basic and acidic catalysis, we have

$$\Delta E = RT \ln (k_{OH}/k_H) = 6,090 \text{ cal}.$$

The apparent energy of activation for the hydrogen-ion catalysed enolisation of acetone is $20,630 \pm 50$ cal. The apparent energy of activation for the hydroxyl-ion catalysis is thus anticipated to be 14,540,* in rough agreement with that found in later experiments, which is 13,400 cal (R. P. Bell and H. C. Longuet-Higgins, 1946). The rate-determining step is considered to be the transfer of a proton from the acetone molecule to the hydroxyl ion. This is followed by rapid reactions, eventually yielding iodoform or bromoform, as described in greater detail in a later passage.

The approximate constancy of $E_H - E_{OH}$ may be used to determine the variation, with respect to temperature of the pH corresponding to the minimum velocity. From Wijs's relation (14.3), we have

$$[H^+]_i = \sqrt{\frac{k_{OH} K_W}{k_H}},$$

whence

$$\frac{d(pH)_i}{dT} = \frac{E_H - E_{OH} - \Delta H_W}{2 \times 2.303 \times RT^2}. \tag{14}$$

ΔH_W is the heat of ionization of water (13,450 cal), hence the numerator becomes roughly $-7,450$ cal. This means that the pH corresponding to the minimum velocity has a negative temperature coefficient, which is nearly the same for all hydrolytic reactions. A rise in temperature from 20 to 30°C reduces $(pH)_i$ by nearly 0.1.

The Mutarotation of Glucose; Catalysis by Acids and Bases in General

By mutarotation is meant the conversion of the α-form or the β-form of an optically active substance to the equilibrium mixture of the α- and β-forms (see Fig. 6.2). We have, for example,

$$\underset{a-x}{\alpha\text{-Glucose}} \underset{k_\beta}{\overset{k_\alpha}{\rightleftarrows}} \underset{b+x}{\beta\text{-glucose}}$$

---

* *Kinetics of Reactions in Solution*, 1st Ed., p. 253 (1933).

and the observed first-order constant is

$$k = k_\alpha + k_\beta = \frac{1}{t} \ln \frac{x_e}{x_e - x}. \tag{6.16}$$

The rate constants k_α and k_β in the pure solvent may be denoted by k_α^0 and k_β^0. The addition of neutral, acidic or basic solutes increases the rate of reaction, according to the equation for multiple catalysis

$$k_\alpha = k_\alpha^0 + \Sigma k_i c_i. \tag{6.91}$$

k_i is the catalytic coefficient for the solute of species i, and c_i is its concentration. When the catalysts are only water molecules, hydrogen ions and hydroxyl ions, for example, we have

$$\begin{aligned} k &= k_0^\alpha + k_{H^+}^\alpha [H^+] + k_{OH^-}^\alpha [OH^-] \\ &\quad + k_0^\beta + k_{H^+}^\beta [H^+] + k_{OH^-}^\beta [OH^-] \\ &= (k_0^\alpha + k_0^\beta) + (k_{H^+}^\alpha + k_{OH^-}^\beta)[H^+] + (k_{OH^-}^\alpha + k_{OH^-}^\beta)[OH^-] \\ &= k_0 + k_{H^+}[H^+] + k_{OH^-}[OH^-]. \end{aligned} \tag{15}$$

Each of the last three constants is seen to be a composite term. To evaluate the rate constants separately, we require a knowledge of the equilibrium constant $K = k_\alpha/k_\beta$. Then

$$k_\alpha = k\frac{K}{K + 1},$$

$$k_\beta = k\frac{1}{K + 1}, \tag{16}$$

and

$$E_{A,\alpha} = E_A + \Delta H \frac{1}{K + 1},$$

$$E_{A,\beta} = E_A - \Delta H \frac{K}{K + 1}. \tag{17}$$

where ΔH is the heat absorbed in the conversion of the α- to the β-form. When this is positive, as with lactose, or negative, as with xylose, the catalytic coefficient for a given ion and the corresponding energy of activation are different for the two forms of the sugar. Fortunately, in the case of glucose, ΔH is zero in the range 273 to 318°K (Nelson and Beagle, 1919; Kendrew and Moelwyn-Hughes, 1940), and $K = 0.5745 \pm 0.0045$, in good agreement with the statistical estimate of Christiansen (1966), which is $K = \frac{4}{7} = 0.5714$.

Some of the catalytic constants for the water molecule, the hydrogen ion and the hydroxyl ion towards the mutarotation of a-glucose are given in Table XI. They have been obtained polarimetrically, and, with similar data

TABLE XI

k_0 (sec^{-1}) AND k_{H^+} AND k_{OH^-} (LITRE/MOLE-SEC) FOR THE MUTAROTATION OF d-GLUCOSE AT 25°C

$k_0 \times 10^4$	$k_{H^+} \times 10^3$	$k_{OH^-} \times 10^{-2}$
3·68*	9·98*	3·76*
3·99†	12·67†	3·60†
3·94‡	12·80§	3·79‖

REFERENCES

* Hudson, *J. Amer. Chem. Soc.*, **29**, 1571 (1907).
† Kuhn and Jacob, *Z. Physikal Chem.*, **113**, 389 (1924).
‡ Kendrew and Moelwyn-Hughes, *Proc. Roy. Soc.*, A, **176**, 352 (1940).
§ Hudson and Dale, *J. Amer. Chem. Soc.*, **39**, 320 (1917); Kilpatrick and Kilpatrick, *J. Amer. Chem. Soc.*, **53**, 3698 (1931): Moelwyn-Hughes, *Z. physikal Chem.*, B, **26**, 272 (1934); P. Johnson and Moelwyn-Hughes, *Trans. Faraday Soc.*, **37**, 289 (1941).
‖ Lowry and Traill, *Proc. Roy. Soc.*, A, **132**, 398 (1931): extrapolated.

enabled Lowry and G. F. Smith (1927) to advance the theory of general acid-base catalysis. Independently and almost simultaneously, Brönsted and Guggenheim (1927) proposed a very similar theory, based on dilatometric evidence. A thorough re-examination of the theory of acid-base catalysis, based on new experimental material has been carried out by H. Schmid and G. Bauer (1965, 1966). It includes values for the apparent energies of activation associated with many catalytic coefficients governing the mutarotation of glucose. (See also reference to Table XI). The correlations established by Brönsted between rate constants and ionization constants are discussed in Chapter 15 (equations 78–85).

A Comparison of Catalytic Hydrolysis and Deuterolysis

Reactions of deuterium atoms and molecules in the gas phase are usually slower than the corresponding reactions of hydrogen.* A similar state of affairs prevails in certain solutions. The uncatalysed mutarotation of glucose in water, for example, is roughly 3 times as fast as the corresponding change

---

* *Ann. Reports Chem. Soc.*, **33**, 86, (1937).

in liquid deuterium oxide, and the catalytic efficiency of the deuterated deuterium ion (D_3O^+) towards the same reaction is about $\frac{3}{2}$ times as great as that of the hydrated hydrogen ion (H_3O^+). As a rule, however, the position in solution is reversed, i.e. the reaction involving D is faster than that involving H. This is true not only for those cases listed in Table XII, which summarizes

TABLE XII

THE COMPARATIVE ENERGETICS OF HYDROLYSIS AND DEUTEROLYSIS

Reference	Reaction	Catalysts	k_D/k_H at 25°C	$E_D - E_H$ cal/gm-mole
*	Mutarotation of α-d-glucose	H_2O and D_2O	...	405
†	Mutarotation of α-d-glucose	H_3O^+ and D_3O^+	0·64	1,250
‡	Hydrolysis of sucrose	H_3O^+ and D_3O^+	1.76	−1,450
§	Hydrolysis of methyl acetate	H_3O^+ and D_3O^+	1·68	−1,200
¶	Hydrolysis of ethyl orthoformate	H_3O and D_3O^+	2·50	−1,400
‖	Hydrolysis of ethyl diazoacetate	H_3O^+ and D_3O^+	3·11	−940

REFERENCES

* Moelwyn-Hughes, Klar, and Bonhoeffer, *Z. physikal Chem.*, **A, 169**, 113, (1934).
† Moelwyn-Hughes, *ibid*, **B, 26**, 272, (1934).
‡ Moelwyn-Hughes and Bonhoeffer, *Naturwiss.*, **11**, 174, (1934); also ref. 2.
§ Hornel and J. A. V. Butler, *Trans. Chem. Soc.*, 1361, (1936).
¶ Brescia and La Mer, *J. Amer. Chem. Soc.*, **62**, 612, (1940).
‖ O. Halpern and P. Gross, *J. Chem. Physics*, **3**, 454 (1935).

the information available in 1947 on the changes in activation energy, but for very numerous reactions which have since been measured at single temperatures only. The contrast between the isotopic effects in mutarotation and hydrolysis is rather remarkable. Moelwyn-Hughes and Bonhoeffer (1934) have suggested that mutarotation and hydrolysis differ in that the rate of the former is governed by the rate of activating collisions between ion and sugar, whereas the rate of the latter is governed by the rate of decomposition of an ion–sugar complex. According to the familiar scheme

$$S + H^+ \underset{k_1}{\overset{k_2}{\rightleftarrows}} SH^+ \overset{k_3}{\rightarrow} G + F + H^+,$$

the general expression for the catalytic coefficient, k_c, becomes

$$k_c = \frac{k_2 k_3}{k_1 + k_3},$$

which reduces to k_2 when $k_3 \gg k_1$, as in mutarotation, and to $k_3(k_2 k_1) = k_3/K$ when $k_1 \gg k_3$, as in hydrolysis. They consequently attribute the enhanced rate of hydrolysis in deuterium oxide to the greater stability of the complex, as reflected in a lower value of its dissociation constant (cf. Halpern, 1935; R. P. Bell, 1936; Hornel and Butler, 1936; Reitz, 1939, Wynne-Jones, 1935; F. A. Long and J. Bigeleisen, 1959; W. J. Albery, 1967).

The Role of Hydrogen, Deuterium and Tritium in the Alkaline Bromination of Ketones

Acetophenone-d_3 reacts as follows in alkaline solutions of bromine:

$$C_6H_5COCD_3 + 3BrO^- \rightarrow C_6H_5COO^- + CDBr_3 + 2DO^-.$$

The rate-determining step is

$$C_6H_5COCD_3 + HO^- \rightarrow C_6H_5COCD_2^- + HOD,$$

followed by the rapid introduction of three bromine atoms per molecule. Denoting by A the acetophenone molecule, and by $k_{OH^-}^D$ the catalytic co-efficient, the bimolecular rate constant is given by the equation

$$-d[A]/dt = k_{OH^-}^D [A][OH^-].$$

The Arrhenius parameters for the three reactions are as follows (J. R. Jones, 1969):

Ion	E_A (kcal.)	$\log_{10} A$(litre-mole^{-1}-sec^{-1})	Rate constant
HO$^-$	12·85	8·82	$k_{OH^-}^D$
DO$^-$	14·60	9·24	$k_{OD^-}^D$
TO$^-$	15·35	8·99	$k_{OT^-}^D$

Jones gives a valuable summary of kindred isotopic comparisons.

Elementary Theory of the Isotope Effect in Gaseous Systems

The gain in potential energy when a molecule RH dissociates into the radical R and the atom H from the hypothetical, motionless state of zero energy at the absolute zero of temperature is the same as the corresponding gain attending the dissociation of RD into R and D because the energy change in each case is unaffected by the difference in the nuclei, and is determined solely by the electronic states of R, H and D. The observed energies of dissociation of RH and RD, however, differ, since experiments reveal only the difference between the energies of the products of dissociation and the energies of the molecules in the ground state, where they are not motionless, but retain a residual energy $(1/2)hv$ each (see Fig. 2.1). The observed difference in the energies of dissociation is thus the difference in the residual (sometimes called zero-point) energies. From Fig. 4, we see that $\varepsilon_s = E_H + (1/2)hv_H = E_D + (1/2)hv_D$, so that

$$E_D - E_H = \tfrac{1}{2}h(v_H - v_D). \tag{16}$$

Because the two potential energy curves are identical, the restoring force constants are the same in RH and RD. The reduced masses, μ, however, differ, and consequently the harmonic vibration frequencies stand in the ratio

$$\frac{v_D}{v_H} = \left(\frac{\mu_H}{\mu_D}\right)^{1/2}. \tag{17}$$

Because the ionization potentials of the hydrogen and deuterium atoms are equal, a similar energy difference is found for the dissociation of RH and RD into $R^- + H^+$ and $R^- + D^+$ respectively. Equal energy differences, but of smaller magnitude, prevail for the distention, or partial dissociation of the homopolar bonds and for the distention, or partial ionization, of the polar bonds. Figure 4 shows that the $R{-}D$ and $R^-{-}D^+$ bonds are firmer than the $R{-}H$ and $R^-{-}H^+$ bonds.

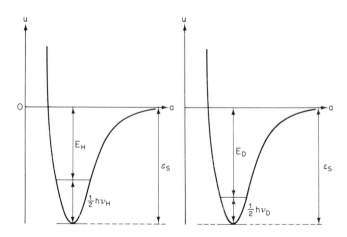

FIG. 4. The effect of residual (zero-point) energy on the observed energy of dissociation or ionization of isotopic diatomic molecules in the gaseous phase.

Quantitative Treatment of the Kinetic Isotope Effect, Applied to the Gaseous Reactions $Br + H_2 \rightarrow BrH + H$ and $Br + D_2 \rightarrow BrD + D$

The equations of the preceding paragraph are valid only for gaseous diatomic molecules at the absolute zero of temperature. When R is an atomic group rather than a single atom, and the system is at a temperature, T, more elaborate equations naturally emerge. Their nature may be realized by considering the application of statistical mechanics to the reaction between the bromine atom and the hydrogen molecule, the bimolecular velocity constant for which, as derived by the method of Pelzer and Wigner (1932), is

$$k_2 = \frac{1}{4}\left(\frac{h}{2\pi}\right)^2\left(\frac{2\pi}{kT}\right)^{1/2}\left(\frac{m_{Br} + 2m_H}{2m_{Br}m_H}\right)^{3/2}\frac{I_{BrHH}}{I_{HH}}\frac{\sigma_{HH}}{\sigma_{BrHH}}\frac{\sinh \beta v_{HH}}{\sinh \beta v_s . \sinh^2 \beta v_\phi}$$
$$\times \exp(-\varepsilon_s/kT), \qquad (4.34)$$

where $\beta = h/2kT$, and the remaining terms have been defined (equation 4.38). The model of the activated complex $Br \cdots H \cdots H$ underlying the treatment is a linear arrangement of the three atoms, allowing the triatomic system to possess four internal modes of vibration, two of which (v_ϕ) are doubly degenerate. The breaking up of the complex may be treated as a linear translation of the atom H relative to the group $Br \cdots H$ (Hirschfelder, Eyring and Topley, 1936) or as a weak vibration (Moelwyn-Hughes, 1936; 1937), real or imaginary, depending on the curvature of the potential energy contour at the saddle point. Recent improved calculations on the shape of the potential energy mountain for the $H \cdots H \cdots H$ system by R. N. Porter and K. Marplus (1964) yield a curvature at the saddle point similar to that found for the $Br \cdots H \cdots H$ system (Figs. 4.4 and 4.5). As stated, equation (4.34) can be derived irrespective of whether the reaction path at the col is a hump or a dip or just flat. Since ε_s is the same for the systems BrHH and BrDD, we see that the ratio of the bimolecular velocity constants for the isotopic pair is simply

$$\frac{k_D}{k_H} = 0.372\left(\frac{\sinh \beta v_{DD}}{\sinh \beta v_{HH}}\right)\left(\frac{\sinh \beta v_{\phi,H}}{\sinh \beta v_{\phi,D}}\right)^2\left(\frac{\sinh \beta v_{s,H}}{\sinh \beta v_{s,D}}\right). \qquad (18)$$

The numerical term has been evaluated in terms of the masses and moments of inertia, and in this instance by taking advantage of the fact that the internuclear distances Br—H and H—H in the activated complex are equal. It is approximately $(m_H/m_D)^{3/2} = 1/2^{3/2} = 0.354$. Each vibrational partition function may be re-written as follows:

$$\sinh(hv/2kT) = \tfrac{1}{2}[1 - \exp(-hv/kT)] . \exp(hv/2kT). \qquad (19)$$

From the known magnitudes of v_{HH} and v_{DD}, the term in the first bracket of equation (18) reduces to $\exp\left[-(h\nu_{HH} - h\nu_{DD})/2kT)\right]$, correct to 1 in 1,000. Only approximate values can be given for the remaining vibration frequencies. If each quantum $h\nu$ is considerably less than kT, we can use the approximation $\sinh(h\nu/2kT) = h\nu/2kT$, and thus obtain the ratio

$$\frac{k_D}{k_H} = 0.372\left(\frac{\nu_{\phi,H}}{\nu_{\phi,D}}\right)^2 \cdot \left(\frac{\nu_{s,H}}{\nu_{s,D}}\right) \cdot \exp\left[-(h\nu_{HH} - h\nu_{DD})/2kT\right]. \qquad (20)$$

A further approximation is possible if we assume that the vibration frequencies vary inversely as the square-root of the masses of the two light atoms (equation 17). The product of the frequency ratios then becomes $2\sqrt{2}$, and, after inserting the spectroscopic values of the vibration frequencies of the diatomic molecules, we have the theoretical ratio

$$\frac{k_D}{k_H} = 1.05 \cdot \exp(-1,878/RT), \qquad (21)$$

in substantial agreement with the experimental equation given by F. Bach, K. F. Bonhoeffer and E. A. Moelwyn-Hughes (1935);

$$\frac{k_D}{k_H} = 1.21 \cdot \exp\left[-(2,130 \pm 1,000)/RT\right].$$

D. Britton and R. M. Cole (1961), have studied the same pair of reactions by a shock wave method, which enabled them to extend the temperature of investigation to $1,700°K$, and to obtain the more precise experimental ratio

$$\frac{k_D}{k_H} = 1.285 \cdot \exp\left[-(1,380 \pm 137)/RT\right]. \qquad (22)$$

Real and Imaginary Vibration within the Complex

Let us consider the expression for the equilibrium constant governing the reaction between an atom A and a diatomic molecule BC on the one hand and a linear triatomic molecule, ABC, on the other hand. Expressed in the units of cc per molecule, it is

$$K = \frac{h^3}{(2\pi kT)^{3/2}}\left[\frac{m_A + m_B + m_C}{m_A(m_B + m_C)}\right]^{3/2} \frac{I_{ABC}}{I_{BC}} \cdot \frac{\sigma_{BC}}{\sigma_{ABC}} \frac{2\sinh(h\nu_{AB}/2kT)}{\prod^4 2\sinh(h\nu/2kT)}$$
$$\times \exp(-\varepsilon_s/kT). \qquad (23)$$

The vibrational motions of which the complex is capable may again be denoted by ν_s, ν_a, and ν_ϕ (reckoned twice). The linear complex formed in reactive systems is manifestly unstable, and the rate of its decomposition is the product of K and a probability per second that it decomposes. If this probability is independent of temperature, the apparent energy of activation becomes

$$\varepsilon_A^r = -(3/2)kT + (1/2)hv_s \coth \beta v_s + (1/2)hv_a \coth \beta v_a + hv_\phi \coth \beta v_\phi$$
$$- (1/2)hv_{AB} \coth \beta v_{AB} + \varepsilon_s. \tag{24}$$

The superscript, r, is to remind us that we are treating all the vibrations as real. One of the internal frequencies, however, is imaginary (L. Farkas and E. Wigner, 1936). Let it be denoted by $v_x = iv_a$. The equilibrium constant is thus imaginary, but the rate constant is real. It follows that

$$\varepsilon_A^i = -(3/2)kT + (1/2)hv_s \coth \beta v_s + (1/2)hv_a \cot \beta v_a + hv_\phi \coth \beta v_\phi$$
$$- (1/2)hv_{AB} \coth \beta v_{AB} + \varepsilon_s. \tag{25}$$

The difference is

$$\varepsilon_A^r - \varepsilon_A^i = (1/2)hv_a(\coth \beta v_a - \cot \beta v_a)$$
$$= kT(\beta v_a \coth \beta v_a - \beta v_a \cot \beta v_a). \tag{26}$$

It is generally assumed that βv_a is small enough to justify our limiting the following expansions to the first two terms in each series:

$$\beta v_A \coth \beta v_A = 1 + \tfrac{1}{3}(\beta v_A)^2;$$
$$\beta v_A \cot \beta v_A = 1 - \tfrac{1}{3}(\beta v_A)^2.$$

Then

$$\varepsilon_A^r - \varepsilon_A^i = kT \cdot \tfrac{2}{3}(\beta v_A)^2 = \frac{2}{3}kT\left(\frac{hv_A}{2kT}\right)^2$$

or

$$\varepsilon_A^i - \varepsilon_A^r = -\frac{2}{3}kT\left(\frac{hv_A}{2kT}\right)^2. \tag{27}$$

The presence of an imaginary frequency thus reduces the apparent energy of activation, and causes an increase in the rate of reaction. It also introduces into the heat capacity of activation the small positive term $(k/6)(hv_a/kT)^2$.

The ratio of the rate constants, from equation (23), is seen to be

$$R = \frac{k_i}{k_r} = \frac{\sinh (hv_a/2kT)}{\sin (hv_a/2kT)}. \tag{28}$$

If we approximate the numerator only (to $hv_a/2kT$), the ratio becomes approximately*

$$R = \frac{hv_a/2kT}{\sin(hv_a/2kT)} = \frac{hv_a/2kT}{(hv_a/2kT) - (1/6)(hv_a/2kT)^3} \tag{29}$$

$$= \left[1 - \frac{1}{6}\left(\frac{hv_a}{2kT}\right)^2\right]^{-1} \doteq 1 + \frac{1}{6}(hv_a/2kT)^2 \tag{30}$$

as given by Wigner (1932). When, however, both numerator and denominator are expanded, the approximate ratio is

$$R = 1 + (1/3)\left(\frac{hv_a}{2kT}\right)^2. \tag{31}$$

On differentiating with respect to T and multiplying by kT^2, we again obtain equation (27). For the reaction studied here, it has been shown that hv_a/kT at the mean temperature of experiment is 0·29. Hence $\varepsilon_A^i - \varepsilon_A^r = -7·7$ cal/mole, and the difference between the computed heat capacities of the reaction is $C^i - C^r = 0·014R$ cal/mole-deg. It would thus be impossible from experimental data on the temperature variation of the activation energy of this reaction to decide whether the breaking frequency is real or imaginary.

It may be argued that, since the breaking frequency in these triatomic systems is known to be imaginary, the equations derived here are unnecessary. Such, however, is not the case. They are helpful in understanding the kinetics of reactions involving quantal tunnelling.

Application of the Theory of the Isotope Effect to Reactions in Solution

In applying the foregoing considerations to reactions in solution, we must, in any elementary treatment, ignore the difference between the solute–solvent interaction energies in the ground and the activated states. How serious such an omission is depends on the extent of the distension of the RH and RD bonds during chemical reaction. If the rate of catalysis is determined by the rate at which a proton or a deuteron is transferred from the parent acids $R^- \cdots H^+$ and $R^- \cdots D^+$ to a common substrate, as is thought to be the case in general acid–base catalysis, we may adopt expressions similar to those derived for gaseous systems, and thereby attribute most of the isotopic effect to the difference in the residual energies of the two acids. By combining

---

* For small values of x, $\sin x = x - \dfrac{x^3}{3!} + \dfrac{x^5}{5!} - \cdots$

equations (16) and (17), we then have

$$E_A(D) - E_A(H) = \frac{1}{2}h\nu_H\left[1 - \left(\frac{\mu_H}{\mu_D}\right)^{1/2}\right].\tag{32}$$

When the proton or deuteron vibrates with respect to a massive partner, we may replace the reduced masses, μ, by the nuclear masses, m. Let us apply this equation to the ionization of the bonds —O—H and —O—D forming part of, say, the glucose and deuterated glucose molecules respectively. The vibration frequency given by Raman spectroscopy for the —O—H bond is $3,650$ cm^{-1}; hence $(1/2)h\nu_H = 5,215$ cal/mole, and $E_A(D) - E_A(H)$ becomes $1,527$ cal, which lies near to the experimental value of $1,250$ cal (Table XII) but is not directly comparable with it. If the same argument is applied to the ionization equilibria of H_2O and D_2O, a similar energy difference is to be expected. Actually, $\Delta H(H_2O)$ and $\Delta H(D_2O)$ are equal, and the ionic product ratio $K(D_2O)/K(H_2O)$ is in fact $0\cdot195$ rather than $0\cdot076$ as would be the case if the entropy changes were equal. Qualitatively, nevertheless, the general conclusion that D_2O is the more stable acid is a sound one (R. P. Bell, 1959). The rate ratio for the hydrogen-tritium catalysed reactions is greater than that for the hydrogen-deuterium catalytic ratio, and the apparent energies of activation in kcal for the base-catalysed enolization of acetone are in the expected sequence: $12\cdot4(OH^-)$, $14\cdot5(OD^-)$, $15\cdot3(OT^-)$ (J. R. Jones, 1965).

The Tunnel Effect

It is one of the consequences of wave mechanics that particles initially on one side of a potential energy barrier can be found on the other side, though not possessing, on classical grounds, sufficient energy to surmount it. Passage through the barrier, referred to as tunnelling, is greatest for light particles, such as photons and electrons, but can be a significant feature of the motion of a proton or deuteron, so that the phenomenon appears in chemical reactions wherein their transfer plays a part. To estimate the magnitude of its effect, we proceed as follows.

The exact form of the barrier is not known. If it is assumed to be parabolic, as in Fig. 5, its permeability, i.e. the fraction of the number of particles that approach it with kinetic energy, E and are transmitted, is given by Mott and Sneddon (1948) as

$$\kappa = \exp -\left\{\frac{2\pi^2a}{h}[2m(V_0 - E)]^{1/2}\right\}.\tag{33}$$

where a is half the width of the barrier, V_0 its height; m is the mass and E is

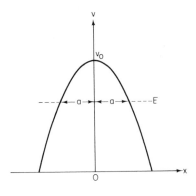

FIG. 5.

the kinetic energy of the penetrating particle. The change in potential energy around the vertex is $V - V_0 = -(1/2)Kx^2$, and the consequent motion is that of a harmonic oscillator with an imaginary frequency $v_i = (1/2\pi) \times (-K/m)^{1/2}$. At the point of penetration, $V_0 - E = (1/2)Ka^2$. On eliminating K and a, we see that

$$\kappa = \exp -\frac{2\pi(V^0 - E)}{hv} \tag{34}$$

where v is a real frequency given by $v = -iv_i$. The permeability for a given particle thus decreases as the height of the barrier increases, increases as the kinetic energy of the particle increases, and becomes unity when $V_0 = E$. For a given barrier height and a fixed energy, light particles penetrate more readily than heavy particles.

Bell (1959) treating the problem by a different method, derives the following expression for the permeability:

$$G = \{1 + \exp[2\pi(V_0 - E)/hv]\}^{-1}. \tag{35}$$

According to this equation, G is small for high barriers and heavy particles, becomes $\frac{1}{2}$ when $E = V_0$, and tends towards unity as $E - V$ becomes infinite. At temperatures other than the absolute zero, allowance must be made for the transmission of particles possessing a range of energies. The fractional number of particles of an ideal gas colliding normally with a surface while possessing energies in the range between E and $(E + dE)$ is

$$\frac{dZ}{Z} = \exp(-E/kT)\frac{dE}{kT}.$$

This is the classical law of the distribution of energies among molecules when each molecule possesses an energy expressible as the sum of two quadratic terms. The relative kinetic energy of a molecule in such a linear system, however, is given by one quadratic term only; it is because the collision frequency is proportional to the relative velocity that the second quadratic term appears. According to classical theory, the fraction dZ/Z, integrated from $E = V_0$ to $E = \infty$, is $e^{-V_0/kT}$, and to this the rate of reaction is proportional. If tunnelling occurs, however, we must allow for the fact that G is a function of E. Clearly

$$R = \frac{\text{rate of reaction during tunnelling}}{\text{classical rate of reaction}} = \frac{\displaystyle\int_0^\infty \exp\left(-E/kT\right).G(E)\frac{dE}{kT}}{\displaystyle\int_{V_0}^\infty \exp\left(-E/kT\right)\frac{dE}{kT}}$$

$$= \frac{\exp\left(V_0/kT\right)}{kT}\int_0^\infty \exp\left(-E/kT\right).G(E)\,dE.$$

Evaluation of the integral depends on the magnitude of $h\nu/2kT$. When this ratio is much less than π, we have (Biegeleisen, (1958) cited by Bell)

$$R = \frac{h\nu/2kT}{\sin\left(h\nu/2kT\right)}. \tag{29}$$

On expansion of the denominator, the approximate ratio becomes

$$R = 1 + \frac{1}{6}\left(\frac{h\nu}{2kT}\right)^2. \tag{30}$$

It will be observed that this equation is identical with Wigner's equation (30), which means that the quantal treatment of the reactive complex, allowing for one imaginary vibration frequency, is equivalent to the classical treatment plus a quantal correction for tunnelling.

The effect of the tunnelling is seen to lower the energy of activation. On using equation (29),

$$E_A\text{(tun)} - E_A\text{(class)} = kT^2\left(\frac{d\ln R}{dT}\right)_V = kT\left[\frac{h\nu}{2kT}\cot\left(\frac{h\nu}{2kT}\right) - 1\right].$$

For low values of $h\nu/2kT$,

$$E_A\text{(tun)} - E_A\text{(class)} = -\frac{kT}{3}\left(\frac{h\nu}{2kT}\right)^2.$$

A closer approximation, found by adopting equation (28), is

$$E_A\text{(tun)} - E_A\text{(class)} = -\frac{2}{3}kT\left(\frac{h\nu}{2kT}\right)^2. \tag{27}$$

Tunnelling contributes a positive heat capacity to the energy of activation:

$$\frac{dE_A(\text{tun})}{dT} - \frac{dE_A(\text{class})}{dT} = \frac{2}{3}k\left(\frac{hv}{2kT}\right)^2. \qquad (36)$$

Two of the methods used to study the tunnelling effect are (1) the measurement of the rates of proton–transfer reactions at low temperature, where the effect of tunnelling is most marked, and (2) the measurement of the rates of pairs of isotopic ion–transfer reactions.

(1) Caldin and Kasparian (1965) have measured the rates of several proton–transfer reactions in ethanolic solutions at low temperatures. A typical instance is

$$C_6H_2(NO_2)_3CH_2^- + HF \rightarrow C_6H_2(NO_2)_3CH_3 + F^-.$$

From the data cited in Table XIII, we see that, as the temperature is raised, the apparent energy of activation increases. If we assume the rate of increase to be constant, we have the approximate relation $E_A = 6,315 + 8.35RT$.

TABLE XIII

REACTION OF THE 2:4:6 TRINITROBENZYL ION WITH HYDROGEN
FLUORIDE IN ETHANOLIC SOLUTION (k_2 IN LITRE/MOLE-SEC)

$t°C$	$\log_{10} k_2$	Mean $T °K$	E_A(cal/mole)	dE_A/dT
25	3·173	285·65	11,060	
0	2·431			
				+ 16·6
− 59·92	0·045	198·28	9,610	
− 89·82	− 1·565			

Analysis of numerous data in the light of quantal tunnelling through parabolic barriers yields barrier thicknesses (2a) at the base of from 1·22 Å to 1·66 Å which seem reasonable in themselves, though not yet satisfactorily correlated with the lengths of the relevant bonds in the various bases.

The difficulty in accepting these numerical conclusions unreservedly arises from the fact that, when the temperature of a reacting system is altered, most properties of the solvent and the reactants are also altered. The viscosity, for example, is temperature-dependent, and so, consequently, is the rate of rotation of the solute. Moreover, as G. J. Hills, P. J. Ovenden and D. R. Whitehouse (1965) have emphasized, theoretical equations such as equation (27) relate to energies of activation, whereas experimental values relate to enthalpies of activation.

(2) These observations apply with equal force to conclusions drawn from isotopic investigations of ion–transfer reaction, which have been critically reviewed by R. P. Bell (1959). Knowledge of the linear vibration frequencies of the bonds R—H and R—D is insufficient to formulate a reasonable reaction mechanism and must be supplemented, for example, by information on bending frequencies, as was attempted in deriving equation (21).

While experimental information of the kinetics of those reactions in solution which depend on tunnelling is rapidly advancing, their exact interpretation, as in most other sections of this wide field, is fraught with difficulties.

Reactions Catalysed by Hydrogen Ions in Concentrated Acid Solutions

In attempts to gain details of the rate-determining step in reactions catalysed by hydrogen ions, experiments have been carried out on hydrogen–ion catalysis in very concentrated aqueous solutions of inorganic acids. The equilibrium established between a monohydrated proton and a neutral or basic solute, S, may be expressed as follows:

$$S + H_3O^+ \rightleftarrows SH^+ + H_2O,$$

which is a considerable simplification of the completer equation

$$[S,mH_2O] + [H,nH_2O]^+ \rightleftarrows [SH \cdot (m - x)H_2O]^+ + (n + x)H_2O.$$

The dissociation constant of the acid SH^+, expressed in terms of activities, is

$$K' = \frac{a_{H_3O^+} \cdot a_S}{a_{SH^+} \cdot a_{H_2O}}. \tag{37}$$

If the activity of the water is assumed to be independent of the composition of the solution, we may define a revised dissociation constant as

$$K = \frac{a_{H_3O^+} \cdot a_S}{a_{SH^+}} = a_{H_3O^+} \left(\frac{\gamma_S}{\gamma_{SH^+}} \right) \frac{[S]}{[SH^+]} \tag{38}$$

where the γ terms are activity coefficients, and the square-bracketed terms are concentrations. The instantaneous rate of reaction is proportional to SH^+:

$$-\frac{d[S]}{dt} = k_3[SH^+] = \frac{k_3}{K} \cdot a_{H_3O^+} \left(\frac{\gamma_S}{\gamma_{SH^+}} \right) [S],$$

and the first-order rate constant is

$$k_1 = -\frac{1}{[S]} \frac{d[S]}{dt} = \frac{k_3}{K} \cdot a_{H_3O^+} \left(\frac{\gamma_S}{\gamma_{SH^+}} \right) = \frac{k_3}{K} [H^+] \left(\frac{\gamma_H + \gamma_S}{\gamma_{SH^+}} \right) = \frac{k_3}{K} \cdot h_0. \tag{39}$$

The formulation is that of the Brönsted–Bjerrum theory (equation 7), giving a catalytic coefficient, $k_1/[H^+]$, which is proportional to the kinetic activity factor $\gamma_{H^+}\gamma_S/\gamma_{SH^+}$. This factor, as we have seen (Chapter 7) is readily evaluated theoretically for aqueous solutions containing electrolytes at a concentration of about 1 millimole per litre. Unfortunately, there is as yet no simple and satisfactory theory of electrolytes in concentrated aqueous solutions. It has been found, however, that the term denoted by h_0 can be estimated colourimetrically, and has a common value for a variety of bases. We can then write

$$\log_{10} k_1 = \log_{10}(k_3/K) + \log_{10} h_0 = \log_{10}(k_3/K) - H_0 \qquad (40)$$

which defines the "acidity function" H_0 (see Long and Paul, 1957).

The reaction between ethyl diazoacetate and water, catalysed by strong acids in concentrated solutions provides an example of a linear relationship between $\log_{10} k_1$ and H_0 (Fig. 6, after Albery and Bell, 1961). Its mechanism has already been established as a simultaneous attack of the ester–hydrogen ion complex with water and the anions of the catalysing acids rather than as the slow formation of a carbonium ion which subsequently reacts rapidly with these solutes.

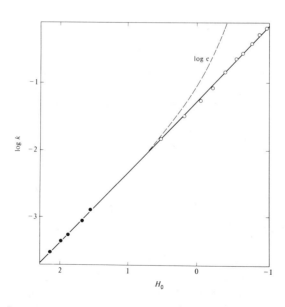

FIG. 6. The first-order rate constant for the hydrogen–ion catalysis of ethyl diazoacetate in aqueous solution. Open circles, W. J. Albery and R. P. Bell, *Trans. Faraday Soc.*, **57**, 1942 (1961). Full circles, E. A. Moelwyn-Hughes and P. Johnson, *ibid*, **37**, 1 (1941).

In an aqueous solution containing 2·5 moles of hydrochloric acid per litre there are, approximately, only 10 water molecules per ion. Their state differs widely from that of ordinary water molecules. The activity of water in acid solutions is known, from vapour pressure data, to fall as the concentration of acid increases; and it is false to assume, as in the formulation of equation (38) that a_{H_2O} is constant.

The viscosity of 2·5 molar aqueous hydrochloric acid at 25°C is 13·8 per cent higher than that of water—a factor which is generally ignored when attempts are made to relate catalytic rate constants to functions of acidity only.

Catalytic Hydrolysis by Hydrogen Ion According to Two Mechanisms

The catalytic hydrolysis of esters by hydrogen ions is conventionally written as follows: $R'COOR + H_2O \xrightarrow{H^+} R'COOH + ROH$. Work with the heavy isotope of oxygen, however, has shown that the hydrolysis can take place in two ways, according as the link cleft is the acyl-oxygen or the alkyl-oxygen link. In the hydrolysis of methyl acetate, there is acyl cleavage:

$$H_3C-\overset{\overset{\displaystyle O}{\|}}{C}\overset{|}{-}O-CH_3,$$
$$HO\overset{|}{-}H$$

as proved by the fact that the oxygen atom appearing in the alcohol derives from the ester (I. Roberts and Urey, 1938; Datta, Day and Ingold, 1939). In the hydrolysis of tertiarybutyl benzoate (Cohen and Schneider, 1941) and tertiarybutyl acetate (C. A. Bunton and J. L. Wood, 1955), cleavage is in the alkyl–oxygen link:

$$H_3C-\overset{\overset{\displaystyle O}{\|}}{C}-O\overset{|}{-}C_4H_9$$
$$H\overset{|}{-}OH$$

as proved by the fact that the (tagged) oxygen atom appearing in the alcohol derives from the water.

A thorough investigation of the hydrolysis of tertiarybutyl acetate catalysed by hydrogen ions in water containing $H_2^{18}O$ (K. R. Adam, I. Lauder and V. R. Stimson, 1962), has shown, however, by mass-spectrometric and other methods of analysis that both types of hydrolysis proceed simultaneously at all temperatures between 25° and 85°C. The mechanisms by means of which these facts are interpreted are as follows.

ALKYL-OXYGEN CLEAVAGE:

$$R'-\overset{\overset{\textstyle O}{\|}}{C}-OR + H^+ \rightleftarrows R'-\overset{\overset{\textstyle OH^+}{\|}}{C}-OR; \qquad \text{fast}$$

$$R'-\overset{\overset{\textstyle OH^+}{\|}}{\underset{\underset{\textstyle HO-H}{|}}{C}}-OR \rightarrow R'-\overset{\overset{\textstyle OH^+}{\|}}{C}-OH + ROH; \quad k_2$$

$$R'-\overset{\overset{\textstyle OH^+}{\|}}{C}-OH \rightarrow R'-\overset{\overset{\textstyle O}{\|}}{C}-OH + H^+; \qquad \text{fast.}$$

The rate-determining step is the pseudo-bimolecular reaction between the ester-hydrogen-ion complex and a water molecule.

TOTAL CLEAVAGE

It is assumed that the same equilibrium between ester and hydrogen–ion is rapidly set up, and that the rate-determining step is the pseudo-unimolecular ionization

$$R'-\overset{\overset{\textstyle OH^+}{\|}}{C}-OR \rightarrow R-\overset{\overset{\textstyle O}{\|}}{C}-OH + R^+; \quad k_1$$

$$R^+ + H_2O \rightarrow ROH + H^+; \qquad \text{fast.}$$

When the two kinds of hydrolysis proceed concurrently, we have a composite catalytic coefficient $k_c = k_{uni\,obs}/c_{H^+} = k_1 + k_2$. The relative contribution from each mechanism is given by the relation

$$\text{Fraction of alkyl–oxygen fission} = \frac{[RO^*H]}{[RO^*H] + [ROH]} = \frac{k_1}{k_1 + k_2}.$$

Stimson, by measuring $(k_1 + k_2)$ and the isotopic ratio over a range of temperatures, has been able unambiguously to assign reliable values to the two catalytic constants and their Arrhenius parameters. A summary of his values is given in Table XIV, which includes, for purposes of comparison, the parameters found at 25°C for the catalytic hydrolysis of methyl acetate (McKinley-McKee and Moelwyn-Hughes, 1952). It will be observed that the Arrhenius parameters for the acyl-oxygen fission for methyl acetate and tertiarybutyl acetate are virtually identical. The composite E_A has been resolved into two apparent energies of activation, one of which is slightly higher and the other of which is very much lower than the composite value. These results confirm van't Hoff's (1896) observation that, in the acid

TABLE XIV

CATALYTIC HYDROLYSIS OF TERTIARYBUTYL ACETATE IN WATER
(DATA OF STIMSON)

		$\log_{10} A$	E_A (cal/mole)
Composite catalytic hydrolysis	$k_1 + k_2$	17·56	26,845 $\pm$ 55
Uncatalysed (water) hydrolysis	k_0 (sec^{-1})	12·3	26,800
Alkyl-oxygen fission	k_1	16·1	27,500
Acyl-oxygen fission	k_2	7·9	17,300
Hydrolysis of methyl acetate	k_2	7·57	17,537 $\pm$ 65

hydrolysis of esters, the rate of reaction is sensitive to the nature of R' but insensitive to the nature of R. They also furnish yet another example of the so-called compensatory effect caused by simultaneous changes of A and E_A in the same direction.

Hydrolysis of Esters in Mixed Solvents

The catalysis by acids of the hydrolysis of aliphatic esters in acetone-water mixtures has been carefully investigated by Tommila and A. Hella (1954) at various temperatures. As the results in general resemble those found during uncatalysed hydrolyses in mixed solvents (Chapter 12), we shall be content with reproducing Fig. 7, which shows, as in the work of Winstein and of Arnett, the remarkable maximum in the rate in acetone-rich solvents.

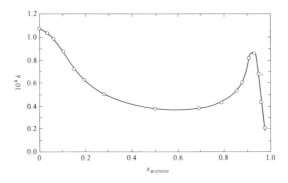

FIG. 7. The catalytic coefficient for the acid hydrolysis of ethyl acetate as a function of the mole-fraction of acetone at 25°C. (After Tommila and Hella.)

Homogeneous Catalysis by Metallic Cations

It is well known that hydrogen at metal surfaces is a powerful reducing agent. It can also act in aqueous solution in the presence of various cations (Calvin 1938). Let us consider, as an example, the reduction of $Cr_2O_7^{2-}$, IO_3^- and Ce^{4+} by H_2 in aqueous solutions containing cupric salts (Halpern 1956). The rate of consumption of hydrogen is found to be independent of the concentration of the reducible substrate and to be inhibited by hydrogen ions according to the equation

$$-\frac{d[H_2]}{dt} = \frac{k_1[H_2][Cu^{2+}]^2}{[Cu^{2+}] + (k_{-1}/k_2)[H^+]}. \tag{41}$$

The mechanism proposed consists of the following reactions:

$$Cu^{2+} + H_2 \underset{k_{-1}}{\overset{k_1}{\rightleftarrows}} CuH^+ + H^+ ;$$

$$CuH^+ + Cu^{2+} \overset{k_2}{\rightarrow} 2Cu^+ + H^+ ;$$

$$2Cu^+ + Substrate \rightarrow 2Cu^{2+} + Products; \quad \text{(rapid reaction)}.$$

Application of the stationary state hypothesis to determine the concentration of CuH^+ accounts for the experimental equation. The simpler equation which holds at low concentrations of hydrogen ions is

$$-d[H_2]/dt = k_1[H_2][Cu^{2+}].$$

Under these conditions, the rate-determining step is the bimolecular reaction between dissolved hydrogen molecules and cupric ions, the rate constant for which varies with temperature according to the equation k_1 (litre/mole^{-1}-sec^{-1}) $= 1.54 \times 10^{10} . \exp(-26,600/RT)$. The pre-exponential term has the normal value. We may regard the cupric ion as having effected a heterolytic cleavage of the hydrogen molecule: $H_2 \rightarrow H^- \cdots H^+$. Silver ions behave in the same way (A. J. Chalk, J. Halpern and A. C. Harkness, 1959), the first step being $Ag^+ + H_2 \rightarrow AgH + H^+$. This may be compared with the rate-determining step in the reaction between the silver ion and the methyl halides, which is $Ag^+ + CH_3X \rightarrow AgX + CH_3^+$, accompanied by $CH_3^+ + H_2O \rightarrow CH_3OH + H^+$.

REFERENCES

Adam, K. R., J. Lander, and V. R. Stimson, *Aust. J. Chem.*, **15**, 467 (1962).
Akerlöf, *J. Amer. Chem. Soc.*, **48**, 3046 (1926).
Albery, W. J., *Trans. Faraday Soc.*, **63**, 200 (1967).
Albery, W. J. and R. P. Bell, *Trans. Faraday Soc.*, **57**, 1942 (1961).

Albery, W. J., J. E. C. Hutchins, R. M. Hyde and R. H. Johnson, *Trans. Chem. Soc.*, **B**, 219 (1962).

Ashmore, P. G., *Catalysis and Inhibition of Chemical Reactions*, Butterworths, London (1963).

Bach, F., K. F. Bonhoeffer and E. A. Moelwyn-Hughes, *Z. physikal Chem.*, **27B**, 71 (1935).

Bell, *Trans. Faraday Soc.*, **55**, 1 (1959).

Bell, R. P., *Acid-Base Catalysis*, Oxford (1941).

Bell, R. P., *Proc. Roy. Soc.*, **A**, **154**, 297 (1936).

Bell, R. P., *The Proton in Chemistry*, Methuen, London (1959).

Bell, R. P. and H. C. Longuet-Higgins, *Trans. Chem. Soc.*, 636 (1946).

Bell, R. P., Bonhoeffer, Pedersen and Wynne-Jones, *Faraday Soc. Diss.*, (1937).

Biddle and Porter, *J. Amer. Chem. Soc.*, **37**, 1571 (1915).

Biegeleisen (1958); cited by Bell.

Bredig and Fraenkel, *Z. Electrochem*, **11**, 515 (1905).

Britton, D. and R. M. Cole, *J. Phys. Chem.*, **65**, 1302 (1961).

Brönsted and Guggenheim, *J. Amer. Chem. Soc.*, **49**, 2554 (1927).

Bunton, C. A. and J. L. Wood, *Trans. Chem. Soc.*, 1522 (1955).

Burgess and Lowry, *Trans. Chem. Soc.*, **125**, 2081 (1924).

Caldin and Kasparian, *Faraday Soc. Diss.*, **39**, 25 (1965).

Calvin, *Trans. Faraday Soc.*, **34**, 1181 (1938).

Chalk, A. J., J. Halpern and A. C. Harkness, *J. Amer. Chem. Soc.*, **81**, 5854 (1959).

Chapman, *Trans. Chem. Soc.*, 1550 (1934).

Christiansen, *J. Colloid and Interface Science*, **22**, 1 (1966).

Cohen and Schneider, *J. Amer. Chem. Soc.*, **63**, 3382 (1941).

Datta, Day and Ingold, *Trans. Chem. Soc.*, 838 (1939).

Duboux, *Helv. Chim. Acta*, **21**, 236 (1938).

Farkas, L. and E. Wigner, *Trans. Faraday Soc.*, **32**, 1 (1936).

Fraenkel, *Z. physikal Chem.*, **A**, **60**, 207 (1907).

French, *J. Amer. Chem. Soc.*, **51**, 3215 (1929).

Gross, P., H. Steiner and F. Krauss, *Trans. Faraday Soc.*, **34**, 351 (1937).

Halpern, O., *J. Chem. Physics*, **3**, 459 (1935).

Halpern, J., *Quart. Rev.*, **10**, 463 (1956).

Halpern, O. and P. Gross, *J. Chem. Physics*, **3**, 454 (1935).

Heidt and Purves, *ibid.*, **60**, 1206 (1938).

Hella, A., *Ann. Acad. Sci. Fennicae*, II, 53 (1954).

Hills, G. J., P. J. Ovenden and D. R. Whitehouse, *Faraday Soc. Diss.*, **39**, 207 (1965).

Hirschfelder, Eyring and Topley, *J. Chem. Physics*, **4**, 170 (1936).

Hitchcock and Dugan, *J. Phys. Chem.*, **39**, 1177 (1935).

Hornel and Butler, *Trans. Chem. Soc.*, 1361, (1936).

Jones, J. R., *Trans. Faraday Soc.*, **61**, 95 (1965).

Jones, J. R., *Trans. Faraday Soc.*, **65**, 2138 (1969).

Kendrew and E. A. Moelwyn-Hughes, *Proc. Roy. Soc.*, **A**, **176**, 352 (1940).

King, E. L., *Catalysis*, II, 337 (Ed. Emmett), Reinhold, New York (1955).

La Mer and M. L. Miller, *J. Amer. Chem. Soc.*, **57**, 2674 (1935).

Leininger and M. Kilpatrick, *J. Amer. Chem. Soc.*, **60**, 2891 (1938).

Long, F. A. and J. Biegeleisen, *Trans. Faraday Soc.*, **55**, 2077 (1959).

Long and Paul, *Chem. Rev.*, **57**, 1 (1957).

Lowry and G. F. Smith, *Trans. Chem. Soc.*, 2539 (1927).

Lowry and Traill, *Proc. Roy. Soc.*, **A**, **132**, 398 (1931).

McKinley-McKee and Moelwyn-Hughes, *Trans. Faraday Soc.*, **48**, 247 (1952).

Moelwyn-Hughes, E. A., *Ann. Rep.*, **32**, 90 (1936); **33**, 86 (1937).

Moelwyn-Hughes, E. A., *Nature*, **129**, 316 (1932).

Moelwyn-Hughes, E. A. and Bonhoeffer, *Z. Electrochem.*, **40**, 469 (1934).

Moelwyn-Hughes, E. A. and Bonhoeffer, *Naturwiss*, **11**, 174 (1934); *Z. physikal Chem.*, **B**, **20**, 272 (1934).

Moelwyn-Hughes and Johnson, *Trans. Faraday Soc.*, **37**, 1 (1941).

Moelwyn-Hughes, E. A. Klar and Bonhoeffer, *Z. physikal Chem.*, **A**, **169**, 113 (1934).

Mott and Sneddon, *Wave Mechanics and its Applications*, Oxford (1948).

Murphy, *J. Amer. Chem. Soc.*, **53**, 977 (1931).

Nelson and Beagle, *J. Amer. Chem. Soc.*, **41**, 559 (1919).

Ölander, *Z. physikal Chem.*, **144**, 118 (1929).

Pearce and Thomas, *J. Phys. Chem.*, **42**, 455 (1938).

Pelzer and Wigner, *Z. physikal Chem.*, **15B**, 445 (1932).

Porter, R. N. and K. Marplus, *J. Chem. Physics*, **40**, 1105 (1964).

Reitz, *Z. physikal Chem.*, **A**, **183**, 371 (1939).

Rideal, E. K. and H. S. Taylor, *Catalysis in Theory and Practice*, 2nd ed., Macmillan (1926).

Roberts, I. and Urey, *J. Amer. Chem. Soc.*, **60**, 1391 (1938).

Schmid, H., *Zwischenreaktionen, Handbuch der Katalyse*, **2**, 1 (1940).

Schmid, H. and G. Bauer, *Angew. Chem.*, **77**, 973 (1965).

Schmid, H. and G. Bauer, *Monatsh.*, **97**, 168 (1966), and 9 previous papers in the same series.

Schwab, G. M., (*trans.* H. S. Taylor and R. Spence), *Catalysis from the Standpoint of Chemical Kinetics*, Macmillan (1937).

Sturtevant, *J. Amer. Chem. Soc.*, **59**, 1528 (1937).

Traill, *Phil. Mag.*, **13**, 225 (1932).

Wigner, *Z. physikal Chem.*, **B19**, 203 (1932).

Wynne-Jones, *Chem. Rev.*, **17**, 115 (1935).

van't Hoff, *Studies in Chemical Dynamics*, p. 119 (1896).

12

HYDROLYSIS AND OTHER SOLVOLYSES

Hydrolysis among chemical reactions is like water among solvents—the most frequently encountered and the least understood. The text books define hydrolysis as a reaction in which water takes part. According to this definition, the water-gas reaction ($CO + H_2O \rightarrow CO + H_2$) would qualify as an example of hydrolysis. Its mechanism in the gaseous phase, however, though still not elucidated, almost certainly entails atoms and radicals, like H and $\tilde{H}O_2$, which are known to take part in similar gas reactions. We shall here define hydrolysis as a reaction between a solute in aqueous solution and the water molecules surrounding it.

The Hydrolysis of Sugars and Glucosides

The earliest kinetically investigated instance of hydrolysis is the so-called inversion of cane sugar or sucrose (Wilhelmy, 1850), conventionally written as

$$C_{12}H_{22}O_{11} + H_2O \xrightarrow{H^+} C_6H_{12}O_6 + C_6H_{12}O_6.$$

sucrose glucose fructose

The reaction has been generally followed polarimetrically, since the specific optical rotatory power of sucrose is positive and the sum of the specific optical rotatory powers of glucose and fructose is negative. During the course of hydrolysis, therefore, the optical rotation of any given solution is inverted in sign, which explains the term 'inversion'. Early work in enzyme chemistry based the unit of enzyme activity on the time required, under stipulated conditions of temperature and pH, for a solution to reach the point of zero optical rotation. The somewhat elaborate transformation of units has enabled later workers in the field of enzyme chemistry to deal with the number of sucrose molecules hydrolysed in unit time by one enzyme molecule. Accurate data on the rate of inversion have also been obtained dilatometrically.

As indicated by the symbol H^+ over the arrow in this equation, the hydrolysis is catalysed by hydrogen ions. The reaction is the first known instance of homogeneous catalysis and probably the best known example of specific hydrogen-ion catalysis. Acids in the general sense of the term do not affect the rate, which, in dilute solution, is given simply by the equation

$$-\frac{dc_S}{dt} = k_2 c_H c_S, \tag{1}$$

where c_H and c_S are, respectively, the concentrations of hydrogen ion and sucrose. There is no catalysis by the solvent.* Since, during the course of a run, $k_2 c_H$ is constant, the reaction is ostensibly of the first order,

$$-\frac{dc_S}{dt} = k_1 c_S, \tag{2}$$

where the catalytic coefficient is

$$k_2 = k_1/c_H. \tag{3}$$

The mechanism of the reaction has long been regarded† as the rapid equilibration of the sugar molecule with the hydrogen ion to form a complex

$$C_{12}H_{22}O_{11} + H^+ \rightleftarrows (C_{12}H_{22}O_{11}H)^+,$$

followed by the rate-determining reaction of the complex with water

$$(C_{12}H_{22}O_{11}, H)^+ + H_2O \rightarrow C_6H_{12}O_6 + C_6H_{12}O_6 + H^+ \text{ (slow)}.$$

In the presence of strong acids like HCl, sucrose may be regarded as an uncharged solute, since its ionization constant has the order of magnitude of 10^{-13} gramme ions/litre (Kullgren, 1902; Smolenski and Kozlowski, 1934).

W. C. McC. Lewis and his collaborators (C. M. Jones and Lewis, 1920; Moran and Lewis, 1922) showed that the rate of reaction at a constant temperature was directly proportional to the activity of the hydrogen ion, as determined electrometrically. Scatchard (1921) argued that the instantaneous rate of reaction must be given by an equation of the form

$$-\frac{dc_S}{dt} = k a_w^{n_w} a_H^{n_H} c_S, \tag{4}$$

---

* Sucrose hydrolyses slowly in pure water, but the reaction is due to catalysis by the hydrogen ion inevitably present.

† Reference to the first few hundred publications, out of many thousands, are given by Caldwell, *British Association Reports*, p. 351 (1906).

where the a terms are activities, and the n's are integers. W. C. McC. Lewis and his collaborators (C. M. Jones and Lewis, 1920; Moran and Lewis, 1922) showed that n_H was unity, and Scatchard, from data on the partial pressure of water vapour over solutions of cane sugar in the presence of acids, evaluated n_W as 6 ± 1. Accordingly, the hydrolysis of sucrose catalysed by hydrogen ion in aqueous solution may be regarded as a reaction of the first order with respect to cane sugar and hydrogen ion and of the order 6 ± 1 with respect to water. Scatchard suggested that the sugar which reacted was a hexahydrate; and it is relevant to note that the osmotic pressure of cane sugar solutions, up to the highest limit measured, can be simply calculated by assuming that the number of effective solvent molecules in solution is the total number of water molecules, less six times the number of sucrose molecules (Callendar, 1908; Findlay, 1913).

We turn next to the effect of temperature on the catalytic coefficients, k_2 for the hydrolysis of pyranosides and furanosides generally. For sucrose in the neighbourhood of 25°C, we have (Urech, 1884; Spohr, 1888; Arrhenius, 1889; Lamble and W. C. McC. Lewis, 1915; Moelwyn-Hughes, 1929)

$$k_2 \text{ (litre/mole-sec)} = 8 \cdot 89 \times 10^{14} . \exp(-25,560/RT),$$

from the first term of which it is clear that no simple collision theory can account for the rate of reaction unless the energy of activation is distributed among a fairly large number, s, of degrees of freedom. To estimate this number on the basis of an equation of the form

$$k_c = Z^0 \left(\frac{E}{RT} \right)^{s-1} \frac{\exp(-E/RT)}{(s-1)!}, \tag{4.45}$$

it was assumed that with a comparable pair of substrates, such as α-methyl glucoside and β-methyl glucoside, the Z^0 and s values were the same for the two isomers, so that

$$\frac{k_\alpha}{k_\beta} = \exp\left[-(E_\alpha - E_\beta)/RT\right] \left(\frac{E_\alpha}{E_\beta} \right)^{s-1},$$

from which s may be evaluated. The mean value of $(s-1)$ so found from fourteen different pairs of carbohydrates was 45 ± 4. If this theory is right, it follows that, provided Z^0 is independent of temperature, the Arrhenius energy of activation should decrease linearly with respect to the absolute temperature:

$$E_A = E_0 - (s-1)RT. \tag{5}$$

Experiment shows (Moelwyn-Hughes, 1934) that such a linear relationship is invariably obeyed for hydrolyses. In the case of the inversion of sucrose, for example, it is found that

$$E_A \text{ (cal/mole)} = 48{,}878 - (40.179 \pm 5.35)RT. \tag{6}$$

The experimental value of dE_A/dT thus confirms the calculated value. A more accurate expression has been derived in Chapter 11, equation (10).

The Hydrolysis of Methyl Chloride, Bromide and Iodide

Sucrose is manifestly a complicated molecule, which, moreover, hydrolyses only in the presence of hydrogen ion; and the questions arise as to how far the high values of $-dE_A/dT$ found for the hydrolysis of glykosides are specific to changes suffered by large molecules, or to catalysis. A study of the hydrolysis of the methyl halides provides a direct answer. The reactions $CH_3X + H_2O \rightarrow CH_3OH + H^+ + X^-$ were first examined (E. A. Moelwyn-Hughes, 1938) by the sealed tube technique, allowing as little vapour space as was compatible with sealing. Their rates, measured by titrating the halogen ions formed, followed the pseudo-unimolecular law at all temperatures between 290 and 384°K. The reactions are free from complications, and go to completion. The rate constants are almost unaffected by changes in the initial concentration of solute, and were found to vary with temperature according to the empirical equations

Methyl chloride: $\log_{10} k_1 = 110 \cdot 223 - 33 \cdot 559 \log_{10} T - 10{,}403/T$; (7)

Methyl bromide: $\log_{10} k_1 = 112 \cdot 656 - 34 \cdot 259 \log_{10} T - 10{,}236/T$; (8)

Methyl iodide: $\log_{10} k_1 = 111 \cdot 859 - 33 \cdot 821 \log_{10} T - 10{,}534/T$. (9)

There is thus no doubt that the phenomenon considered appears with simple solutes and is not dependent on catalysis (E. A. Moelwyn-Hughes, 1938). It is in fact, as later work has shown, fairly common, and is not restricted to hydrolysis.

The work was repeated using vapour-free reaction vessels, and supplementing the analytical with an electrometric technique (E. A. Moelwyn-Hughes, 1953).

The newly found rate constants were ten times as accurate as, and uniformly greater than, the earlier ones by from 7 to 18 per cent, and suggested that E_A passed through slight minima at high temperatures. The results of a specimen run, followed for over a month, are given in Table I. The best plot of $\ln(a - x)$ against t gives $k_1 = 2 \cdot 430 \times 10^{-7}$. From a large-scale graph of the concentration as a function of time, the quarter-life is found to be 19,850 min, and the half-life 46,960 min, giving values of $2 \cdot 460$ and $2 \cdot 414 \times 10^{-7} \, \text{s}^{-1}$ for k_1. The adopted value is $2 \cdot 43 \times 10^{-7}$. The calculated values of x are those

TABLE I

METHYL CHLORIDE IN WATER AT 313·26°K

| $t \times 10^{-3}$ (min) | Millimoles per litre of solution | | | | $k_1 \times 10^7$ (s^{-1}) |
	[Cl$^-$]	[H$^+$]	Mean observed x	Calculated x	
0	0	0	0	0	—
6·80	4·22	4·42	4·32	4·05	(2·60)
12·62	7·02	7·33	7·18	7·20	2·417
19·75	10·40	10·96	10·68	10·74	2·415
24·10	12·44	12·74	12·59	12·71	2·403
31·30	15·46	15·67	15·57	15·73	2·400
37·06	18·00	17·78	17·89	17·91	2·427
41·34	19·44	19·50	19·47	19·42	2·438
47·10	21·64	21·38	21·51	21·32	2·461
54·21	23·40	23·44	23·42	23·45	2·426
∞	42·92	—	42·92	42·92	
				average	2·423

reproduced by the equation:

$$x/a = 1 - \exp(-k_1 t), \tag{6.3}$$

using the adopted value of k_1.

TABLE II

KINETIC DATA ON THE HYDROLYSIS OF METHYL BROMIDE IN WATER (HEPPOLETTE AND ROBERTSON, 1959)

T(°K)	k_1 (sec^{-1})	n
308·16	$(1·658 \pm 0·008) \times 10^{-6}$	4
313·15	$(3·213 \pm 0·009) \times 10^{-5}$	3
323·17	$(1·114 \pm 0·003) \times 10^{-5}$	4
328·11	$(2·006 \pm 0·006) \times 10^{-5}$	4
333·16	$(3·512 \pm 0·005) \times 10^{-5}$	3
338·19	$(6·050 \pm 0·004) \times 10^{-5}$	3
343·16	$(1·014 \pm 0·002) \times 10^{-4}$	4
350·27	$(2·065 \pm 0·004) \times 10^{-4}$	3
353·16	$(2·713 \pm 0·002) \times 10^{-4}$	2
358·17	$(4·323 \pm 0·022) \times 10^{-4}$	4
363·10	$(6·700 \pm 0·015) \times 10^{-4}$	2
368·15	$(1·033 \pm 0·002) \times 10^{-3}$	2
373·15	$(1·585 \pm 0·004) \times 10^{-3}$	3

A still more precise investigation of these three reactions was carried out by Heppolette and R. E. Robertson (1959) who measured the rates from electrical conductivities in solutions containing initially from 2 to 5 millimoles of solute per litre. A specimen set of their results is given in Table II. The temperatures were controlled to within 0·003°, and the velocity constants were calculated by Guggenheim's method (equation 6.10). The errors shown are the maximum deviations of k_1 from the average value of n graphical rate plots. Their data are summarized in the equations

Methyl chloride: $\log_{10} k_1 = 85 \cdot 3556 - 25 \cdot 1185 \log_{10} T - 9192 \cdot 18/T$, (10)

Methyl bromide: $\log_{10} k_1 = 77 \cdot 6800 - 22 \cdot 3771 \log_{10} T - 8557 \cdot 10/T$, (11)

Methyl iodide: $\log_{10} k_1 = 93 \cdot 1459 - 27 \cdot 4294 \log_{10} T - 9661 \cdot 27/T$. (12)

It will be observed that the coefficients of $\log_{10} T$ and $1/T$ are lower and slightly more variable, than the earlier estimates, and that, within the limits of error, the apparent energies of activation decrease linearly as the temperature is raised. From these constants are derived the kinetic parameters shown in Table III.

TABLE III

KINETIC CONSTANTS FOR THE HYDROLYSIS OF THE METHYL HALIDES

Halide	$k_1 \times 10^8$ (sec^{-1}) at 298·16°K	E_0 (cal/mole)	dE_A/dT	E_A at 298·16°K
CH$_3$F	0·074	47,900	−67·02	27,910
CH$_3$Cl	2·353	42,060	−49·914	27,178
CH$_3$Br	40·69	39,154	−44·364	25,926
CH$_3$I	7·418	44,206	−54·506	27,955

The Kinetics of the Hydrolysis of Methyl Fluoride

The pseudo-unimolecular velocity constant governing the hydrolysis of methyl fluoride in water (D. N. Glew and E. A. Moelwyn-Hughes, 1952), as calculated from the usual equation

$$\bar{k}t = \ln [a/(a - x)],$$	(12)

decreases during the course of the run—an effect which suggests retardation by the products of reaction. That this is the likely cause of the drift is clear from Table IV. Typical strong electrolytes like perchloric acid and potassium perchlorate have no effect on the rate of reaction, but the fluoride ion, hydrofluoric acid and methanol reduce the rate. The apparent positive effect of

TABLE IV

THE INFLUENCE OF ADDITIONAL SOLUTES

AT 393·1°K

$[CH_3F]_0 = 25$ to 30 millimoles/litre

additional solute	concentration (millimoles/litre)	$10^5 \bar{k}_0$ (sec^{-1})
none	0	1·95 ± 0·04
$HClO_4$	30	1·95 ± 0·06
$KClO_4$	40	1·95 ± 0·04
NaF	40	1·92 ± 0·04
NaF	100	1·84 ± 0·04
CH_3OH	30	1·90 ± 0·05
HF	13·5	1·86 ± 0·04
KBr	40	2·27 ± 0·04

alkali bromides is due to bimolecular substitution, as indicated in Chapter 8. Hydrofluoric acid, which is only slightly ionized, exerts by far the greatest retarding effect. The simplest kinetic scheme required to incorporate the observed effects is obtained by superimposing on the usual mechanism a secondary reaction, which is the reversible bimolecular deactivation of the intermediate complex. In a system containing initially a concentration c of hydrofluoric acid, we have the scheme

$$\underset{a-y}{CH_3F} \underset{k_4}{\overset{k_1}{\rightleftarrows}} \underset{y}{CH_3F^*} \overset{k_3}{\rightarrow} \underset{x}{CH_3OH} + \underset{c+x}{HF}$$

$$(+HF) \, k_2 \downarrow \uparrow k_5$$

$$\underset{y}{CH_3F} + \underset{c+x}{HF}.$$

The stationary concentration of the activated complex is obtained from the relation

$$dy/dt = k_1(a - x) - (k_3 + k_4)y + (k_2 - k_5)(c + x)y = 0, \qquad (13)$$

and the rate of formation of the product becomes

$$\frac{dx}{dt} = k_3 y = \frac{k_1 k_3(a - x)}{k_3 + k_4 + (k_2 - k_5)(c + x)},$$

which gives on integration

$$t = \left[\frac{k_3 + k_4 + (k_2 - k_5)(c + a)}{k_1 k_3} \right] \ln_e \left(\frac{a}{a - x} \right) - \frac{(k_2 - k_5)}{k_1 k_3} x. \qquad (14)$$

On eliminating t from equations (12) and (14), we obtain an expression for $\bar{k}$

in terms of x:

$$\frac{1}{k} = \frac{k_3 + k_4 + (k_2 - k_5)c}{k_1 k_3} + \frac{(k_2 - k_5)a}{k_1 k_3}\left[1 - \frac{x/a}{\ln_e (a/a - x)}\right]. \quad (15)$$

If $1/\bar{k}$ is plotted against the bracketed function of (x/a), the gradient is seen to be $(k_2 - k_5)a/k_1 k_3$ and the intercept $1/\bar{k}_0$, where

$$\bar{k}_0 = \frac{k_1 k_3}{k_3 + k_4 + (k_2 - k_5)c}. \quad (16)$$

These equations have been found adequate to reproduce the experimental results at all stages of the reaction, and at all concentrations of products. Similar equations have been derived for other systems (Hinshelwood and Prichard, 1925; Moelwyn-Hughes, 1933).

The variation with respect to temperature of the corrected pseudo-unimolecular constant for the hydrolysis of methyl fluoride are reproduced by the equation

$$\log_{10} k_1 \text{ (sec}^{-1}) = 109.435 - 33.729 \log_{10} T - 10{,}467/T. \quad (17)$$

Kinetic constants derived from this equation have been included in Table III.

The Absolute Enthalpy of a Critically Activated Complex in Aqueous Solution

The absolute heat content of a reactive complex in solution is the sum of its absolute value in the ground state and the enthalpy of activation:

$$H^* = H^0 + E_A. \quad (18)$$

This equation we shall now apply to aqueous methyl bromide. H^0 for the gas is found by combining the calorimetric data for the liquid (Egan and Kemp, 1938) with the spectroscopically evaluated heat content of the gas. The decrease in heat content attending dissolution of the gas is obtained (Table III) from Henry's constant, corrected for gas imperfection:

$$\Delta^0 H \text{ (gas} \rightarrow \text{aq soln)} = -19{,}461 + 44.22T. \quad (19)$$

By addition, we obtain the heat content of the solute in its reference state:

$$H^0 = -11{,}470 + 54.57T. \quad (20)$$

The energy of activation, derived from Robertson's data (Table II), is

$$E_A = 39{,}154 - 44.47T. \quad (21)$$

TABLE V

STATIC AND KINETIC DATA ON AQUEOUS METHYL BROMIDE

$t°C$	$-273·15$	0	25	50	75
H^0 (gas)	8,539	10,828	11,077	11,337	11,602
ΔH^0 (gas → aq soln)	$-19,461$	$-7,381$	$-6,281$	$-5,171$	$-4,071$
H^0 (aq)	$-10,922$	3,447	4,796	6,166	7,531
E_A	39,154	27,007	25,895	24,784	23,632
H^0 (aq) $+ E_A$	28,232	30,454	30,691	30,950	31,203

Hence

$$H* = 27,684 + 10·10T. \tag{22}$$

The slightly higher value of H_0^x given in Table V results from a different method of extrapolation. The heat content of the activated solute is seen to increase linearly with respect to temperature; $dH*/dT = 10·10 \pm 0·5$ cal/mole^{-1}-deg^{-1}, which, possibly by accident, is near to C_P^0 (gas) $= 10·17$ at 25°C. The important conclusion is that H^0 and E_A vary linearly with respect to temperature, giving gradients that are nearly equal in magnitude but opposite in sign. The absolute enthalpy of the critical complex thus becomes, very nearly, constant, as shown in Fig. 1. The only assumption underlying these calculations (E. A. Moelwyn-Hughes, 1953) is that the break-down frequency of equation (4.22) is independent of temperature.

It seems clear that most of the decrease in E_A with increasing temperature can be attributed to the increase in H^0.

The Hydrolysis of Tertiary Butyl Chloride in Pure Water

The reaction between tertiary butyl chloride and water, unadulterated by any other solute or solvent, is accurately represented by the equation $(CH_3)_3Cl + H_2O \rightarrow (CH_3)_3COH + H^+ + Cl^-$. Its rate of hydrolysis was first measured by means of electrical conductivity (E. Tommila, M. Tiilikainen and A. Voipio, 1955) in the temperature range 10–25°C. Winstein and Fainberg (1957) measured the rate of reaction electrometrically, using a glass electrode at 0 and 25°C. E. A. Moelwyn-Hughes, R. E. Robertson and S. Sugamori (1965) used the conductometric method at intervals of 1 degree between 1·008 and 20·013°C. Some of these results are compared in Table VI. The apparent energy of activation decreases with a rise in temperature according to the linear equation

$$E_A = 47,000 - 81·331T, \tag{23}$$

TABLE VI

KINETIC DATA ON THE HYDROLYSIS OF TERTIARY BUTYL CHLORIDE IN PURE WATER

$k_1 \times 10^2$ at 25°C (sec^{-1})	E_A at 12·5°C (cal/mole)	$-dE_A/dT$ (cal/mole-deg)	Investigators
3·03	23,340	70	Tommila *et al.* (1955)
2·70	23,400	—	Winstein *et al.* (1957)
3·12	24,167	81·3	Robertson *et al.* (1965)

as shown in Fig. 2. The dots represent experimental values, and the line is drawn from this equation.

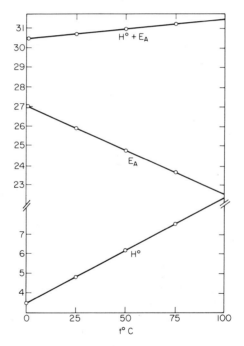

FIG. 1. The partial molar heat content, H^0, of aqueous methyl bromide, and the apparent energy of activation, E_A, for its hydrolysis (kcal/mole).

Further Instances of Simple Hydrolyses in Pure Water

Data relating to the uncatalysed hydrolysis of various solutes in pure water have been summarized in Table VII in terms of the constants, *a*, *b* and *c* of the

TABLE VII

KINETIC CONSTANTS FOR SOME UNCATALYSED HYDROLYSES

$$\log_{10} k_1 \, (\text{sec}^{-1}) = a - b \log_{10} T - c/T$$

Solute	Reference	a	b	c
CH_3F	*	109·435	33·729	10,467
CH_3Cl	†	83·356	25·119	9,192
CH_3Br	†	77·680	22·377	8,557
CH_3I	†	93·146	27·429	9,661
CH_2Cl_2	‡	98·441	29·661	10,597
CH_3NO_3	§	75·324	21·432	9,229
$C_6H_5CH_2Cl$	‖	67·404	19·309	7,306
C_3H_5Cl	‖	81·916	24·189	8,345
C_3H_5Br	‖	95·813	28·727	8,794
C_3H_5I	‖	78·289	22·560	8,304
$C_6H_5SO_3CH_3$	¶	57·219	15·819	6,860
$C_6H_5SO_3C_2H_5$	¶	59·218	16·376	7,063
$C_6H_5SO_3C_3H_7$	¶	70·673	19·742	7,455
iso $C_6H_5SO_3C_3H_7$	¶	53·953	14·740	6,734
$C_6H_5N_2^+$	**	45·603	18·00	8,349
$(CH_3CO)_2O$	††	126·362	41·77	7,619
$(CH_3)_3CCl$	‡‡	134·490	40·928	10,359
$[Co(NH_3)_5Cl]^{3+}$	§§	84·487	25·05	8,429

REFERENCES

 * Glew and Moelwyn-Hughes, *Proc. Roy. Soc.* **A 211**, 254 (1952).

 † Heppolette and Robertson, *Proc. Roy. Soc.*, **A 252**, 273 (1959).

 ‡ Fells and Moelwyn-Hughes, *Trans. Chem. Soc.*, 1326 (1959).

 § McKinley-McKee and Moelwyn-Hughes, *Trans. Faraday Soc.*, **48**, 247 (1952).

 ‖ Robertson and Scott, *Trans. Chem. Soc.*, 1956 (1961).

 ¶ Robertson, *Can. J. Chem.*, **35**, 613 (1957).

** Moelwyn-Hughes and Johnson, *Trans. Faraday Soc.*, **36**, 948 (1940).

†† Gold, *Trans. Faraday Soc.*, **44**, 506 (1948).

‡‡ Moelwyn-Hughes, Robertson and Sugamori, *Trans. Chem. Soc.*, 1965 (1965).

§§ S. C. Chan, *Trans. Chem. Soc.*, 291 (1967).

empirical equation

$$\log_{10} k_1 = a - b \log_{10} T - c/T. \tag{24}$$

The coefficient of $\log_{10} T$ is invariably negative, and shows no regular dependence on the size of the solute. If the break-down frequency of equation (4.22) is independent of temperature,

$$E_A = \Delta^0 H^x,$$

and

$$\left(\frac{dE_A}{dT}\right)_P = \frac{d\Delta^0 H^x}{dT}$$

$$= \Delta^0 C_P$$

$$= -bR. \qquad (25)$$

In order to gain a comprehensive view of the data, we shall consider the average of all the b values given in Table VII, which is 24·82, corresponding to an average value of $\Delta^0 C_P$ of $-49·33$. According to the ice-berg hypothesis, this difference between the heat capacity of the activated complex and that of the ground-state solute may be attributed to the complete freezing of n water molecules:

$$\Delta^0 C_P = n(C_{P,\text{ice}} - C_{P,\text{water}}) = -8·1 \, n \, \text{cal/mole-deg.} \qquad (26)$$

The average value of n becomes 6·08, suggesting that 6 molecules of water around the solute molecule have become completely frozen in the activated complex, or that a larger number of solvent molecules has become partially frozen. With limits denoting the range of values rather than the accuracy $(dE_A/dT)_P$ is 54 $\pm$ 5 for the methyl halides (Heppolette and Robertson, 1959), 34 $\pm$ 8 for alkyl and aryl sulphonates (Robertson, Heppolette and J. M. Scott, 1959) 32 $\pm$ 8 for acid anhydrides (Koskikallio, 1954), and 49·5 $\pm$ 0·5 for various aquations (S. C. Chan, 1967).

Little difference results in this interpretation if the break-down frequency term is replaced by kT/h. On the other hand, if we use the relaxation equation (10.61),

$$v_b = kT/4\pi r^3 \eta, \qquad (27)$$

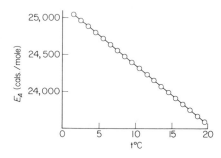

Fig. 2. Apparent energies of activation for the hydrolysis of tertiary butyl chloride. (Moelwyn-Hughes, Robertson and Sugamori, 1965).

some of the apparent energy of activation must be attributed to the viscous energy, B (equation 3.127), since

$$E_A = B + \Delta^0 H^x,\qquad(28)$$

and

$$dE_A/dT = dB/dT + \Delta^0 C_P.\qquad(29)$$

According to Table 3.VIII, $(dB/dT)_p$ varies from -32 at 25°C to $+29$ at 100°C. From the mean value of dE_A/dT obtained from Table 7, the value of $\Delta^0 C_P$ may range from -17 to -78 cal/mole^{-1}.deg^{-1}. That the rate of reaction should vary inversely with respect to the viscosity is consistent with a relaxation process, and with the approximate parallelism found by Laughton and Robertson (1959) between the fluidity and the rate ratio, k_{D_2O}/k_{H_2O}, for deuterolysis and hydrolysis (see Fig. 3).

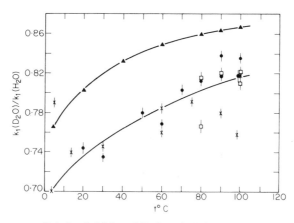

Fig 3. Upper curve. Relative fluidities of D_2O and H_2O. Lower curve. Relative rates of hydrolysis and deuterolysis of various halides. (After Laughton and Robertson, 1959.)

Reasons have been advanced in Chapter 5 for supposing the rate of some reactions in solution to vary in direct proportion to the viscosity. In that case

$$E_A = -B + \Delta^0 H^x,\qquad(30)$$

and

$$dE_A/dT = -dB/dT + \Delta^0 C_P.\qquad(31)$$

The value of $\Delta^0 C_P$ now ranges from -81 at 25° to -20 at 100°. Thus the viscosity correction, whichever way it is applied, does not alter the sign of

$\Delta^0 C_P$. Further discussions of the problem have been given by R. E. Robertson and J. B. Hyne (1955), Tommila, Tiilikamen and Voipio (1955), J. R. Fox and G. Kohnstam (1963) and J. R. Hulett (1964).

The Solvolysis of Tertiary Butyl Chloride in Pure Solvents

We first consider methanol as solvent. After dissolution of *t*-butyl chloride in the dry solvent, the following changes can in principle occur:

$$Bu^tX + MeOH \rightarrow Bu^tOMe + H^+ + X^- \tag{a}$$

$$Bu^tX \rightarrow Me_2C\!:\!CH_2 + H^+ + X^- \tag{b}$$

$$Bu^tX + MeOH \rightarrow Bu^tOH + MeX \tag{c}$$

$$MeX + MeOH \rightarrow Me.O.Me + H^+ + X^- \tag{d}$$

By the method of electrical conductivity, the rate of production of ions is found to be approximately unimolecular with regard to the solute. Analysis of the products of reaction indicates the complete absence of *iso*butene and, within the accuracy of the distillations, a quantitative yield of methyltertiarybutyl ether. The reaction goes to completion. The elimination reaction (b) does not therefore take place, and the production of ions must consequently be due to reaction (a) or reaction (d). Experiments with the methyl halides in pure methanol indicate that reaction (d) is slower by a factor of about 10^{-3} than the rates of increase in conductivity of the t-butyl halide solutions in the same solvent at the same temperature. Reaction (d) can thus be dismissed as contributing insignificantly to the rate-determining process. The chemical change taking place is consequently the simple methanolysis represented by equation (a). Because the equivalent conductance of an electrolyte depends on its concentration, a quantitative relationship between them must be established before the rate of change of electrical conductivity, κ, can be converted into the rate of chemical change, which is what we are primarily interested in. By denoting the true unimolecular constant by k_1 and the apparent unimolecular constant by k_1' we have the general equation (32),

$$k_1 = -d \ln (c_\infty - c)/dt = k_1' f \tag{32}$$

where c is the concentration of halide ion at time t, and c_∞ is the final concentration,

$$k_1' = -d \ln (\kappa_\infty - \kappa)/dt \tag{33}$$

and

$$f = d \ln (c_\infty - c)/d \ln (\kappa_\infty - \kappa). \tag{34}$$

In the region where the square-root conductivity law is obeyed, we have

$$\Lambda = 10^6 \kappa/c = \Lambda_0 - Ac^{1/2}, \tag{35}$$

where κ is the specific conductivity, and c the concentration in mmoles/1. Consequently

$$f = \left[1 - \frac{A}{\Lambda_0} \left(\frac{c_\infty^{3/2} - c^{3/2}}{c_\infty - c} \right) \right] \left[1 - \frac{3}{2} \frac{A}{\Lambda_0} c^{1/2} \right]^{-1}. \tag{36}$$

By means of these equations, accurate unimolecular constants are found throughout the course of each run. The concentrations of solute used seldom exceeded 1 millimole/litre. The results (Biordi and Moelwyn-Hughes, (1962) are summarised in Table VIII.

TABLE VIII

A Summary of the Kinetics of the Methanolysis of Tertiary Butyl Halides

$$\log_{10} k_1 (\text{sec}^{-1}) = a - b \log_{10} T - c/T$$

	Bu^{t}Cl	Bu^{t}Br	Bu^{t}I
a	76·402	69·927	64·998
b	21·905	19·354	17·981
c	8424·7	7899·8	7277·6
E_0 (cal/mole)	38,550	36,150	33,300
$-\partial E_A/\partial T$ (cal/mole-deg)	43·53 $\pm$ 19	38·46 $\pm$ 5·6	35·73 $\pm$ 5·2
E_A at 25°	25,570	24,679	22,647
k_1 (sec^{-1}) at 25°	8·77 $\times$ 10^{-7}	3·47 $\times$ 10^{-5}	1·25 $\times$ 10^{-4}
Relative k_1	1	40	143

The temperature variation of E_A has been found for the solvolysis of tertiary butyl chloride in a variety of alcohols by E. S. Rudakov and Y. A. Kivalin (1964), one instance of whose results is given in the last entry of Table IX. Winstein and Fainberg (1957) give many values of E_A in various

TABLE IX

The Solvolysis of Tertiary Butyl Chloride in Pure Solvents

$$\log_{10} k_1 (\text{sec}^{-1}) = a - b \log_{10} T - c/T$$

Solvent	a	b	c	E_0	$-dE_A/dT$	E_A at 25°C	k_1 at 25°C
H_2O	134·49	40·928	10,359·1	47,400	81·33	24,167	3·12 $\times$ 10^{-2}
CH_3OH	76·40	21·905	8,424·7	38,550	43·53	25,570	8·77 $\times$ 10^{-7}
n-C_4H_9OH	141·40	43·76	12,140	55,550	97·06	29,600	2·27 $\times$ 10^{-7}

solvents at a single temperature. They note that, with the exception of water, the Arrhenius term A is nearly constant, while E_A varies from 21·6 k/cal in formic acid to 26·7 in ethanol. By measuring the gain in free energy attending dissolution, $\Delta^0 G = -RT \ln (p_2/x_2)$, they have been able to refer the free energy of the solute in the ground state to the common reference of the free energy of the liquid solute, and by subtracting a constant term $RT \ln k_1$ to estimate the relative free energy of the activated solute in each solvent.

The Hydrolysis of Tertiary Butyl Chloride in Acetone–Water Mixtures

Tommila (1954) and his collaborators have measured, by the conductimetric method, the rate of hydrolysis of tertiary butyl chloride at temperatures between 10 and 50°C in aqueous solutions of varying compositions

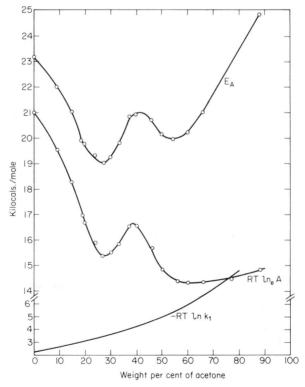

FIG. 4. The hydrolysis of tertiary butyl chloride in acetone-water mixtures (data of Tommila, 1954).

containing up to 88 per cent by weight of acetone. At 25°C, the rate of hydrolysis in the acetone-rich medium is lower by a factor of about 10^{-6} than the rate in pure water. The parameters of the Arrhenius equation at 25°C vary with respect to the solvent composition in the manner shown in Fig. 4. The term $RT \ln_e A$ is, as we have shown, proportional to the entropy of activation, and is seen to pass through two minima and one maximum. E_A behaves similarly. The difference, $-RT \ln k_1 = E_A - RT \ln A$ increases, smoothly, as the acetone content is increased, from 2·2 to 8·4 kcal, revealing neither maximum nor minimum. The term $-RT \ln k_1$, which is proportional to the free energy of activation is represented by the lowest curve in Fig. 4.

The Hydrolysis of the Tertiary Butyl Halides in Methanol–Water Mixtures

The smoothness with which the logarithm of the rate constant varies with solvent composition is also found in the solvolysis of the tertiary butyl halides. In terms of the mole fraction, x_m, of methanol the velocity constants at 25°C are reproduced by the empirical equations:

$$Bu^t Cl: \log_{10} k_1 = \bar{2}.464 - 5.147 x_m + 0.626 x_m^2 \tag{37}$$

$$Bu^t Br: \log_{10} k_1 = \bar{1}.881 - 4.806 x_m + 0.465 x_m^2$$

$$ButI: \log_{10} k_1 = 0.186 - 4.938 x_m + 0.849 x_m^2$$

The data for the chloride are in satisfactory agreement with earlier work (Olson and Halford, 1937; Speith and Olson, 1955; Winstein and Fainberg, 1957; Bunton and Nayak, 1959). From the summary given in Table X, it is

TABLE X

A COMPARISON OF THE RATES OF REACTION OF THE t-BUTYL HALIDES WITH METHANOL AND WATER AT 298·16°K

	Chloride	Bromide	Iodide
Relative rates of methanolysis	1	40	143
Relative rates of hydrolysis	1	26	53
k_1 (hydrolysis)/k_1 (methanolysis)	$3\cdot3 \times 10^4$	$2\cdot2 \times 10^4$	$1\cdot2 \times 10^4$

seen that the relative rates of hydrolysis and methanolysis of the three tertiary butyl halides differ from the relative rates of hydrolysis of the methyl halides (Table III;) $k_1(CH_3Cl): k_1(CH_3Br): k_1(CH_3I):: 1:17:3$.

The Solvolysis of Tertiary Butyl Chloride in Ethanol–Water Mixtures

Winstein and Fainberg (1957) have measured E_A for the solvolysis of tertiary butyl chloride in water–ethanol mixtures of all compositions. After conversion of the composition unit from volume fractions to mole fractions, their results are given by the middle curve in Fig. 5, which shows a sharp minimum at a water mole fraction of 0·835. Arnett, W. G. Bentrude, J. J. Burke and P. McC. Duggleby (1965) have made a very accurate thermometric measurement of the heat gained during the dissolution of this solute in solvents of varying mole-fraction, standardizing their values on Giauque's heat of dissolution of sulphuric acid in water. They find, as shown in the lowest curve of Fig. 5, that the heat absorbed on dissolving passes through an equally marked maximum at a solvent composition of 0·845 mole-fraction of water. Thus the solvent changes E_A and ΔH_{diss} in opposite ways. Their sum gives the smooth curve at the top of the graph. Hence most of the effect of the solvent on the velocity of reaction is due to changes which it brings about in the ground state of the solute. This conclusion is identical with that reached by an analysis of the temperature effect on the velocity of reaction.

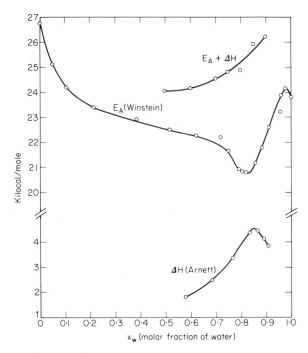

FIG. 5. The solvolysis of tertiary butyl chloride in ethanol-water mixtures.

The velocity constants from which the energies of activation have been derived are composite values, being the sums of the pseudo-unimolecular constants for hydrolysis and ethanolysis. The products of reaction, tertiary butanol and ethyl-tertiarybutyl ether, are found to be formed in the same proportion as the composition of the solvent so that, in the region of the maxima and minima there is no selective attack on the solute.

The free energy, enthalpy and entropy of activation can be obtained from rate constants at different temperatures. Adopting the transition state notation (equation 4.35), Winstein and Fainberg find the values shown graphically in Fig. 6. The free energy of activation ($\Delta F^{\ddagger} \equiv \Delta G^{\ddagger}$), which is

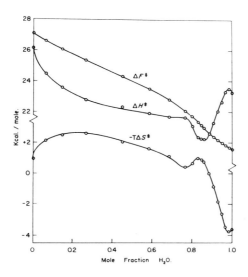

FIG. 6. Kinetic parameters for the solvolysis of tertiary butyl chloride in aqueous ethanolic solutions. (After Winstein and Fainberg.)

proportional to $-RT \ln k_1$, is seen to vary smoothly, and almost linearly, with respect to solvent composition. By definition, it is the difference between the total energy (or enthalpy, at constant pressure) and the bound energy, $T\Delta S$, both of which pass through extrema, due, as we have seen, to changes in the enthalpy of the solute in the ground state.

Arnett (1965) has shown that maxima in the heat of dissolution of most of the solutes examined by him appear in aqueous solutions of high water content, Fig. 7 shows this effect for highly polar solutes. The effect appears to be general, and has been interpreted in terms of the structure of liquid water (F. Franks and D. J. G. Ives, 1966).

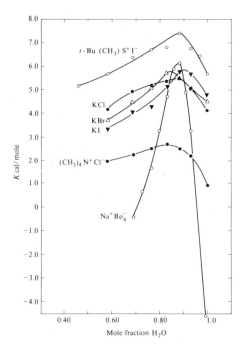

FIG. 7. (After Arnett.)

The Order of Reaction with Respect to Water

Let us consider the methanol–water system, containing a solvolizable solute at a concentration of one millimole per litre of solution. The simplest way of interpreting the effect of solvent composition on the rate of solvolysis is to assume that, in solutions containing relatively little methanol, the rate of change in the concentration, c, of solute consists of the following two components,

$$-dc/dt = k_w c_w^{n_w} c + k_m c_m^{n_m} c \tag{38}$$

where c_w and c_m are the concentrations of water and methanol, respectively, and n_w and n_m are the orders of reaction with respect to molecules of the medium. Concurrent hydrolysis and methanolysis in methanol-rich systems require for their formulation a number of other rate constants and integers lying between n_w and n_m, which are here omitted. Then,

$$k_1 = k_w c_w^{n_w} \left[1 + \left(\frac{k_m c_m^{n_m}}{k_w c_w^{n_w}} \right) \right].$$

From the two rate constants relating to the pure solvents, the second term is known to be small enough to be ignored so that,

$$\ln k_1 = \ln k_w + n_w \ln c_w \tag{39}$$

then

$$d \ln k_1/dx_w = n_w d \ln c_w/dx_w$$

and, c_w being expressed in moles/litre;

$$c_w = \frac{1000x_w}{V_m - (V_m - V_w)x_w} \tag{40}$$

where V_m and V_w are the partial molar volumes. In the limit, as x_w approaches 1, we have

$$\left(\frac{d \ln c_w}{dx_w}\right)_{x_w \to 1} = \frac{V_m - (dV_w/dx_w)x_{w \to 1}}{V_w}. \tag{41}$$

V_w is the ordinary molar volume of water, and V_m the partial molar volume of methanol in an infinitely dilute aqueous solution. Experiments show the second term in the numerator to be negligible. Hence, since $x_w + x_m = 1$,

$$n_w = -\frac{V_w}{V_m}\left(\frac{d \ln k_1}{dx_m}\right)_{x_m \to 0}. \tag{42}$$

At 288·2°K, $V_w = 18.03$ cc/mole, and $V_m = 38.27$. From the empirical equation (37) for the solvolysis of the tertiary butyl halides, we find n to be 5·6 (chloride), 5·2 (bromide) and 5·4 (iodide). Tommila gives 6·7 (chloride) and 5·7 (bromide). From his work on the hydrolysis of ethyl bromide in acetone–water mixtures, he finds $n_w = 2.3$, but on less accurate kinetic data. D. A. Brown and R. F. Hudson (1953) give $n_w = 4 \pm 0.5$ from the effect of dioxan on the rate of hydrolysis of benzoyl chloride (see further references in Chapter 10).

A similar result is found from Winstein and Fainberg's work on the solvolysis of tertiary butyl chloride in aqueous ethanolic solutions (Fig. 6). In the water-rich regions, $-d \ln k_1/dx_w$ is seen to be -9. On applying equation (42) in the form

$$n_w = \frac{V_w}{V_e}\left(\frac{d \ln k_1}{dx_w}\right)_{x_w \to 1} \tag{42a}$$

where V_e is the molar volume of ethanol (57·56 cc), we find that n_w is 4·7.

The Common Ion Effect in Hydrolysis in Aqueous Solution

The pseudo-unimolecular rate constants for the hydrolysis of methyl bromide ($CH_3Br + H_2O \to CH_3OH + H^+ + Br^-$) and methyl iodide

($CH_3I + H_2O \rightarrow CH_3OH + H^+ + I^-$) are unaffected by the presence of added alkali metal bromides and iodides (Moelwyn-Hughes, 1938). On the other hand, the pseudo-unimolecular constant for the hydrolysis of methyl fluoride is decreased by addition of fluoride ion and hydrofluoric acid (Table IV). During a run commencing with pure methyl fluoride in water, the pseudo-unimolecular constant,

$$\bar{k} = \left(\frac{1}{a - x}\right)\frac{dx}{dt},$$

consequently decreases with the progress of reaction.

Following and simplifying the treatments which led to equations (9.18) and (12.14), we formulate the kinetic scheme as follows:

$$\underset{a-y}{RX} \underset{k_2}{\overset{k_1}{\rightleftharpoons}} \underset{x-y}{R^+} + \underset{x}{X^-}$$

$$\downarrow k_5 \text{ (irreversible)}$$

$$ROH + H^+.$$

When the concentration of the intermediate ion, R^+, has reached a steady value, the rate of reaction becomes

$$\frac{dx}{dt} = \frac{k_1 k_5(a - x)}{k_5 + k_2 x}, \tag{43}$$

and

$$k_1 t = \left(1 + \frac{k_2}{k_5}a\right) \ln\left(\frac{a}{a - x}\right) - \frac{k_2}{k_5}x.$$

From the definition of $\bar{k}$, it follows that

$$\frac{1}{\bar{k}} = \frac{1}{k_1} + \left(\frac{k_2}{k_1 k_5}\right)x. \tag{44}$$

The approximately linear plot found for methyl fluoride at 393·1°K is shown in Fig. 8, from which we find $k_1 = 1\cdot94 \times 10^{-5} \text{ sec}^{-1}$, and $k_2/k_5 = 11\cdot26$ litre/mole. If we write $k_5 = k_5'[H_2O]$, we obtain the ratio $k_2/k_5' = 6\cdot3 \times 10^2$, which is greater than can be accounted for simply in terms of the partial molar volumes of the anion and the solvent, but is consistent with the fact that k_2 refers to a reaction between oppositely charged ions while k_5 refers to a reaction between a cation and a polar molecule.

When the initial system contains a concentration, c, of anions of the type generated during hydrolysis, we have

$$k_1 t = \left[1 + \frac{k_2}{k_5}(a + c)\right] \ln\frac{a}{a - x} - \frac{k_2}{k_5}x, \tag{45}$$

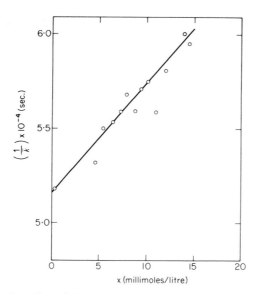

FIG. 8. The retarding effect of the products of reaction on the rate of hydrolysis of methyl fluoride.

and

$$\frac{1}{k} = \frac{1}{k_1} + \frac{k_2}{k_1 k_5}(c + x).$$ (46)

It is clear that the anomalous behaviour of methyl fluoride as compared with the other methyl halides can be interpreted according to equation (15) or equation (44). The inhibition of the rate could thus be due to an activated molecule of the reactant, as first imagined, or to a product ion. Later work, discussed below, suggests that sometimes the active intermediate may be an equilibrated system of an intimate ion-pair and a first-shell ion pair.

Evidence Concerning the Critical Intermediate from Isotopic Analysis and from the Electrolyte Effect

The rate of hydrolysis of benzyl chloride ($PhCH_2Cl + H_2O \rightarrow PhCH_2OH + H^+ + Cl^-$) at 30°C has been accurately measured as a function of pH. The only mechanism capable of explaining the rates of hydrolysis and the rates of isotopic exchange of Cl^- and H^+ is that written above, in neutral or acidic solution, accompanied by the rate-determining step $PhCH_2Cl + OH^- \rightarrow PhCH_2OH + Cl^-$ in alkaline solution. The total evidence eliminates the possibility of an intermediate mechanism involving the carbonium ion $PhCH_2^+$ (Tanabe and Sano, 1962). Benzyl chloride thus behaves exactly like methyl chloride. The rate of hydrolysis of benzal chloride ($PhCHCl_2 + H_2O \rightarrow PhCHO + 2H^+ + 2Cl^-$) decreases with the addition of sodium chloride. The same mechanism as that suggested in the case of methyl fluoride has been firmly established by means of equation (44), the intermediate ion being $PhCHCl^+$ (Tanabe and Ido, 1965).

The effect of certain strong electrolytes on the rate of hydrolysis of tertiary butyl chloride in water at $292 \cdot 90°K$ is summarized in Table XI (Moelwyn-Hughes, 1961), from which we note, in the first place, that there is no evidence

TABLE XI

THE EFFECT OF CERTAIN ELECTROLYTES ON THE RATE OF HYDROLYSIS OF TERTIARY BUTYL CHLORIDE IN WATER AT $292 \cdot 90°K$

Butyl chloride (millimoles/litre)	Electrolyte	[Electrolyte] (millimoles/litre)	$k_1 \times 10^2$ (sec^{-1})	Standard error
4·08	None	0	1·4959	0·0072
4·36	HCl	8·75	1·5173	0·0060
4·46	NaOH	13·5	1·4817	0·0037
2·04	KCl	100·0	1·6749	0·0135

for retardation of the reaction rate by the chloride ion. The term k_2 of equation (43) may thus be omitted, and the rate of reaction is simply $dx/dt = k_1(a - x)$, which is the rate of ionization. This hypothesis, as we have seen, is that postulated for substitution reactions of certain dihalogenoethanes by Slator (1905), of benzhydryl chloride by Ward (1927) and of tertiary butyl chloride by Hughes (1935). We note in the second place that, instead of exerting a retarding influence, the presence of chloride ion causes an increase in the rate of reaction.

More extensive work on the effect of the common ion has been carried out in mixed solvents. Bateman, Hughes and Ingold (1940) measured the rate of hydrolysis of tertiary butyl bromide at 25 and 35°C with and without the addition of lithium bromide in a 90:10 acetone:water mixture, and interpreted their results assuming the inorganic solute to be completely dissociated (Bateman, Church, Hughes, Ingold and Taher, 1940. Olson and Komecny, 1953; and G. R. Nash and C. B. Monk 1955) have shown that in this mixed solvent, the dissociation constants at 25°C are 6·3 (HBr), 12·0 (LiBr) and 13·0 (KBr), each to be multiplied by 10^{-3}. With these results, Nash and Monk have been able to calculate the true ionic strengths, and to show that the logarithm of Ingold's constants at 25°C vary linearly with respect to the ionic strength.

The theoretical treatment proposed by Ingold and his collaborators to account for the electrolyte effect is based on the supposition that the reactive intermediate is a polar molecule with terminal charges of $\varepsilon/2$ and $-\varepsilon/2$, separated by an estimated distance of 2·6 Å. Their resulting equation closely resembles that of an earlier treatment. According to a critical examination by Nash and Monk, both treatments account satisfactorily for the observed increases in hydrolysis rates with increasing ionic strength.

A kinetic investigation of the hydrolysis, in aqueous acetone solution, of *p*-benzhydryl *p*-nitrobenzoate, in the absence and in the presence of the azide ion, using O^{18} and C^{14}, indicates the formation of two intermediates

$$RX \rightleftharpoons [R^+X^-] \rightleftharpoons [R^+][X^-].$$
$$\quad\quad\quad\text{I} \quad\quad\quad\quad \text{II}$$

I denotes an intimate ion-pair, which can reform the original ester while maintaining its configuration. II represents a solvent-separated pair which changes with partial or complete racemization (H. L. Goering and J. F. Levy, 1964; Winstein, Klinedinst and Clippinger, 1961).

Free Energy and Enthalpy Changes Attending the Ionization of a Chemically Reactive Solute

If, as seems probable in certain instances, the pseudo-unimolecular constant, k_1, for hydrolysis is in fact the rate of ionization, we may estimate the ionization constant, K, from the ratio k_1/k_2, accepting Eigen's value of k_2. For methyl fluoride at 393°K, we thus find K to be approximately 10^{-15}, and $\Delta^0 G = 27$ kcal. As far as may be judged by extrapolation from the data of Table 3.VIII, $\Delta^0 G$ for the ionization of water at this temperature is 25 kcal. Within the errors of the estimates, these quantities may be taken as equal, which again reminds us of the possibility that the key rate in hydrolysis is that of the ionization of the water molecule.

By a cyclic process, estimates can be made of the energy required to ionize the methyl halides in water. The accepted ionization potential of the methyl radical is that given by Waldron (1954), and the heats of escape of the ions from aqueous solutions at 25°C are taken from Moelwyn-Hughes (1964). The last line in Table XII gives the energies of activation for the hydrolyses at this temperature, and have been used in estimating the heat of solvation of the methyl ion. The adopted value of -123 kcal/gramme ion may well be in error, but the relative energies of ionization are unaffected by its absolute value.

TABLE XII

Thermodynamic Cycle at 25°C to Estimate the Energy of Ionization of the Methyl Halides in Aqueous Solution

X	F	Cl	Br	I
CH_3X (g) $\rightarrow$ CH_3 (g) + X (g)	116·3	76·0	63·3	47·2
CH_3 (g) $\rightarrow$ CH_3^+ (g) + $\ominus$	233·1	233·1	233·1	233·1
X (g) + $\ominus$ $\rightarrow$ X^- (g)	−97·8	−87·3	−82·0	−75·7
CH_3X (g) $\rightarrow$ CH_3^+ (g) + X^- (g)	251·6	221·8	214·4	204·6
CH_3X (s) $\rightarrow$ CH_3X (g)	4·4	5·7	6·3	6·4
CH_3^+ (g) $\rightarrow$ CH_3^+ (s)	−123·0	−123·0	−123·0	−123·0
X^- (g) $\rightarrow$ X^- (s)	−109	−77	−69	−60
CH_3X (s) − CH_3^+ (s) + X^- (s)	24	27	29	28
Observed E_A for hydrolyses at 25°C	27·9	27·2	25·9	28·0

REFERENCES

Arnett, W. G. Bentrude, J. J. Burke and P. McC. Dugglebey, *J. Amer. Chem. Soc.*, **87**, 1541 (1965).

Arrhenius, *Z. physikal Chem.*, **4**, 226 (1889).

Bateman, Church, Hughes, Ingold and Taher, *Trans. Chem. Soc.*, 979, (1940).

Bateman, Hughes and Ingold, *Trans. Chem. Soc.*, 960 (1940).

Biordi and E. A. Moelwyn-Hughes, *Trans. Chem. Soc.*, 4291 (1962).

Brown, D. A. and R. F. Hudson, *Trans. Chem. Soc.*, 3252 (1953).

Bunton and Nayak, *Trans. Chem. Soc.*, 3843 (1959).

Callendar, *Proc. Roy. Soc.*, **A, 80**, 466 (1908).

Chan, S. C., *Trans. Chem. Soc.*, 291 (1967).

Egan and Kemp, *J. Amer. Chem. Soc.*, **60**, 2097 (1938).

Findlay, *Osmotic Pressure*, Longmans, London (1913).

Fox, J. R. and G. Kohnstam, *Trans. Chem. Soc.*, 1593 (1963).

Franks, F. and D. J. G. Ives, *Quart. Rev.*, **20**, 1 (1966).

Glew, D. N. and E. A. Moelwyn-Hughes, *Proc. Roy. Soc.*, **A, 211**, 254 (1952).

Goering, H. L. and J. F. Levy, *J. Amer. Chem. Soc.*, **86**, 120 (1964).

Heppolette and R. E. Robertson, *Proc. Roy. Soc.*, **A, 252**, 273 (1959).

Hinshelwood and Prichard, *Trans. Chem. Soc.*, **127**, 327 (1925).

Hulett, J. R., *Quart. Rev.*, **18**, 227 (1964).

Hughes, *Trans. Chem. Soc.*, 255 (1935).

Jones, C. M. and Lewis, *Trans. Chem. Soc.*, **117**, 1120 (1920).

Koskikallio, *Ann. Akad. Sci. Fennicae*, II, 57 (1954).

Kullgren, *Z. physikal Chem.*, **41**, 407 (1902).

Lamble and W. C. McC. Lewis, *Trans. Chem. Soc.*, **107**, 233 (1915).

Laughton and Robertson, *Can. J. Chem.*, **37**, 1491 (1959).

Moelwyn-Hughes, E. A., *Kinetics of Reactions in Solution*, p. 152, 1st Ed., Oxford (1933).

Moelwyn-Hughes, E. A., *Physical Chemistry*, 2nd Ed. (1964).
Moelwyn-Hughes, E. A., *Proc. Roy. Soc.*, **A**, **164**, 295 (1938).
Moelwyn-Hughes, E. A., *Proc. Roy. Soc.*, **A**, **220**, 386 (1953).
Moelwyn-Hughes, E. A., *Trans. Chem. Soc.*, 1517 (1961).
Moelwyn-Hughes, E. A., *Trans. Faraday Soc.*, **25**, 81 (1929).
Moelwyn-Hughes, E. A., *Trans. Chem. Soc.*, 779 (1938).
Moelwyn-Hughes, E. A., *Z. physikal Chem.*, **B26**, 281 (1934).
Moelwyn-Hughes, E. A., R. E. Robertson and S. Sugamori, *Trans. Chem. Soc.*, 1965 (1965).
Moran and Lewis, *Trans. Chem. Soc.*, **121**, 1613 (1922).
Nash, G. R. and C. B. Monk, *J. Amer. Chem. Soc.*, 1899 (1955).
Olson and Halford, *J. Amer. Chem. Soc.*, **59**, 2644 (1937).
Olson and Komecny, *J. Amer. Chem. Soc.*, **75**, (1953).
Robertson, Heppolette and J. M. Scott, *Can. J. Chem.*, **37**, 803 (1959).
Robertson and J. B. Hyne, *Can. J. Res.*, **33**, 1544 (1955).
Rudakov, E. S. and T. A. Kivalin, *Organic Reactivity*, *Tartu*, **II**, 114 (1964).
Scatchard, *J. Amer. Chem. Soc.*, **43**, 2387 (1921).
Slator, *Trans. Chem. Soc.*, **87**, 485 (1905).
Smolenski and Kozlowski, *Nature*, p. 771 (1934).
Speith and Olson, *J. Amer. Chem. Soc.*, **77**, 1412 (1955).
Spohr, *Z. physikal Chem.*, **2**, 195 (1888).
Tanabe and Ido, *J. Res. for Catalysis*, Hokkaido Univ., **XII**, **3**, 223 (1965).
Tanabe and Sano, *J. Res. for Catalysis*, Hokkaido Univ., **X**, **2**, 173 (1962).
Tommila, *Acta Chem. Scand.*, **8**, 258 (1954).
Tommila, E., M. Tiilikainen and A. Voipio, *Ann. Acad. Sci. Fennicae, Chem.*, **65** (1955).
Tommila, E., M. Tiilikainen and A. Voipio, *Ann. Acad. Sci. Fennicae*, **A**, **II**, *Chem.*, 64 (1955).
Waldron, *Trans. Faraday Soc.*, **50**, 102 (1954).
Ward, *Trans. Chem. Soc.*, 2285 (1927).
Wilhelmy, *Pogg. Annalen*, **81**, 413 (1850).
Winstein and Fainberg, *J. Amer. Chem. Soc.*, **79**, 5937 (1957).
Winstein, Klinedinst and Clippinger, *J. Amer. Chem. Soc.*, **83**, 4986 (1961).
Urech, *Berichte*, **17**, 2175 (1884).

13

PRESSURE EFFECTS

The pioneer work of the last century, carried out at pressures up to 500 atmospheres, showed that the rate of bimolecular processes in aqueous solution could be favourably or adversely affected by external pressure, and that the logarithm of the velocity constant measured at constant temperature varied linearly with respect to pressure within this region. These two features are illustrated in Fig. 1. Later work, carried out at pressures as high as 40,000 atmospheres, indicates that the validity of the linear logarithmic relationship is limited to regions of relatively low pressures.

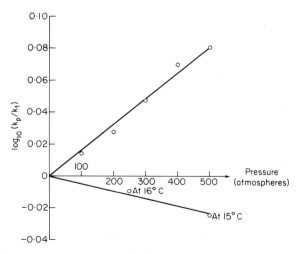

FIG. 1. The influence of pressure on the velocity of catalysed hydrolyses in aqueous solution. Upper curve: The hydrolysis of methyl acetate, catalysed by N-hydrochloric acid at 14°C. Lower curve: The hydrolysis of sucrose, catalysed by N-hydrochloric acid.

The Quasi-thermodynamic Theory of the Effect of Pressure on Rate Constants

Consider the chemical equilibrium represented by the equation

$$A + B \underset{k_1}{\overset{k_2}{\rightleftarrows}} AB.$$

The equilibrium constant, K, is defined by either of the following equations

$$K = \frac{[AB]}{[A][B]} = \frac{k_2}{k_1}, \tag{1}$$

where the bracketed terms denote activities and the k terms velocity coefficients. From the thermodynamic equation

$$\ln K = -\Delta G^0/RT, \tag{2.7}$$

it follows that

$$\left(\frac{\partial \ln K}{\partial P}\right)_T = -\frac{1}{RT}\left(\frac{\partial \Delta G^0}{\partial P}\right)_T = -\frac{\Delta V^0}{RT} \tag{2}$$

where ΔV^0 is the sum of the partial molar volumes of the products less the sum of the partial molar volumes of the reactants, each in some standard state, say, for example, at unit molarity in a solution at atmospheric pressure;

$$\Delta V^0 = \Sigma V_j^0 - \Sigma V_i^0. \tag{3}$$

For brevity let us write

$$\Delta V^0 = V_2 - V_1. \tag{4}$$

Equation 2 is simply a quantitative expression of le Chatelier's principle of mobile equilibrium: when a reaction is attended by a decrease in volume, it is favoured by an increase in pressure, and *vice versa*. Equation (4) may be written as follows

$$\Delta V^0 = (V_2 - V_c) - (V_1 - V_c), \tag{5}$$

in which V_c is an arbitrary volume. On combining equations (1), (2), and (4), we see that

$$\left(\frac{\partial \ln k_2}{\partial P}\right)_T - \left(\frac{\partial \ln k_1}{\partial P}\right)_T = -\frac{(V_c - V_1)}{RT} + \frac{(V_c - V_2)}{RT}. \tag{6}$$

The resolution of this equation into two separate equations, one purporting to hold for the forward and one for the reverse reaction, is equivalent to giving V_c a unique, as distinct from an arbitrary, value, and is a pure assumption which can be justified only *a posteriori*. If it is valid, we arrive at the relations:

$$\left(\frac{\partial \ln k_2}{\partial P}\right)_T = -\frac{(V_c - V_1)}{RT}$$

and

$$\left(\frac{\partial \ln k_1}{\partial P}\right)_T = -\frac{(V_c - V_2)}{RT}.$$

In general, we may therefore write (van't Hoff, 1901; M. G. Evans and M. Polanyi, 1935, 1937)

$$\left(\frac{\partial \ln k}{\partial P}\right)_T = -\frac{(V_c - V)}{RT} = -\frac{\Delta V_c}{RT}. \tag{7}$$

This relation implies that the velocity coefficient of a chemical reaction increases with increasing pressure, when the volume, V_c, of the active molecules is less than the volume, V, of the reactants. If, on reacting, a molecule must pass through a more voluminous state, its velocity of reaction will be adversely affected by pressure. Because the velocities of most of the reactions hitherto studied increase with increasing pressure, we infer that the critical complex is generally smaller than the normal reactants. In other words, complexes in the transition state are usually more compact than those in the ground state.

When the linear relationship between $\ln k$ and P is maintained, ΔV_c is constant, and equation 2 can be integrated to give

$$k_P = k_0 \exp\left(-P\Delta V_c/RT\right), \tag{8}$$

where k_P is the velocity constant when the reacting system is under an external pressure, P, and k_0 is the velocity constant in the absence of external pressure. Effectively, k_0 may be taken as the value of the velocity constant at atmospheric pressure. We obtain for these reactions the values of ΔV_c given in Table I, which also illustrates the applicability of equation (8). From the data of Gibson, Fawcett and Perrin (1935) and of E. G. Williams, Perrin and Gibson (1936), values of ΔV_c may be obtained for a variety of reactions. We find, for example that it is $-11\cdot45$ cc for the reaction between acetic anhydride and ethanol in toluene solution. Equation (8) satisfactorily reproduces the experimental rate constants (Table II) at pressures up to 8,500 atmospheres. Further results obtained by the same investigators are shown in Fig. 2, from which we note that the plot of $\ln k$ against P is sometimes a curve, indicating that ΔV_c is a function of the pressure. For the purpose of comparisons, we shall therefore confine the discussion of some of the data (Tables III and V) to values of ΔV_c relating to relatively low pressures.

TABLE I

THE INFLUENCE OF PRESSURE ON HYDROLYSES CATALYSED BY N-HYDROCHLORIC ACID
IN AQUEOUS SOLUTION

Reaction	Tempera-ture (°C)	Pressure (atmo-spheres)	k_P/k_0	
			Observed	Calculated
The hydrolysis of 5 per cent.	14	1	1·00	1·00
methyl acetate ($\Delta V_c =$	14	100	1·03	1·04
$-8·7$ cc)	14	200	1·07	1·08
	14	300	1·12	1·12
	14	400	1·17	1·16
	14	500	1·21	1·20
The hydrolysis of 20 per cent.	16	1	1·00	1·00
sucrose in water ($\Delta V_c =$	16	250	0·98	0·97
$+2·5$ cc)	15	500	0·94	0·95

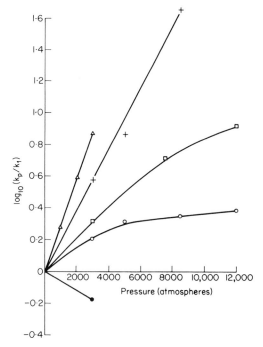

FIG. 2. The influence of pressure on the velocity of various reactions (after Gibson, Fawcett, Perrin and Williams (1935)).

1. △ $(CH_3CO)_2O + C_2H_5OH \rightarrow CH_3COOC_2H_5 + CH_3COOH$ in C_2H_5OH at 20°C.
2. + $(CH_3CO)_2O + C_2H_5OH \rightarrow CH_3COOC_2H_5 + CH_3COOH$ in $C_6H_5CH_3$ at 40°C.
3. ⊡ $CH_2Cl.COONa + NaOH \rightarrow CH_2OH.COONa + HCl$ in H_2O at 40·05 ± 0·05°C.
4. ⊙ $C_2H_5I + C_2H_5ONa \rightarrow C_2H_5OC_2H_5 + NaI$ in C_2H_5OH at 25·1 ± 0·1°C.
5. ◎ $(C_6H_5)(C_6H_5CH_2)(CH_3)(C_3H_5)NBr \rightarrow (C_6H_5)(C_6H_5CH_2)(CH_3)N + C_3H_5Br$ in $CHCl_3$ at 25°C.

TABLE II

THE INFLUENCE OF PRESSURE ON THE
REACTION BETWEEN ACETIC ANHYDRIDE
AND ETHYL ALCOHOL IN TOLUENE SOLUTION
AT 40°C ($\Delta V_c = -11.45$ CC/GRAMME-MOLE)

P	k_P/k_0	k_P/k_0
(atmospheres)	Observed	Calculated
1	1.00	1.00
3,000	3.78	3.81
5,000	7.24	9.28
8,500	44.8	44.1

Reactions in Non-aqueous Media

We recall the classification of bimolecular reactions into normal, fast and slow categories, according as A_2 in the equation $k_2 = A_2 \exp(-E_A/RT)$ is equal to, greater than or less than $10^{10 \pm 0.5}$ litres per mole-second. The reaction labelled 5 in Table III is representative of the normal class, and is seen to be but slightly affected by pressure. Reactions 1, 2 and 6 are slow, and yield large negative values of ΔV_c in each case. The only fast reaction is number 2, which is seen to exhibit a positive value of ΔV_c. The classification based on the values of A_2 is therefore qualitatively reflected in the pressure effects.

The results found for the unimolecular decomposition of methyl-phenyl-benzyl-allyl ammonium bromide and pentaphenyl ethane (reactions 2 and 8) may be profitably compared. Pentaphenylethane, which is only slightly polar, dissociates into two free radicals in a non-polar solvent which is structurally similar to the reactant and products. The magnitude of ΔV_c can be readily understood as arising from the stretching of the Ph_3C-CPh_2H bond, occurring without requiring any radical change in the interaction energy with the solvent molecules. The normal and activated states of the reactant $R_1R_2R_3R_4NBr$ are known to be highly polar, and the relatively small value of ΔV_c may be regarded as the sum of the positive contribution due to the stretching of the solute and the negative contribution due to the stronger interaction energy of the highly polar activated solute with the polar solvent molecules.

Reaction 4, which is pseudo-unimolecular, is distinguished from the other slow reactions in that $[\partial(\Delta V_c)/dT]_P$ is positive. If, as suggested by Moelwyn-Hughes and Rolfe (1932), the rate-determining step is the reaction between

TABLE III

CRITICAL INCREMENTS OF VOLUME FOR CERTAIN REACTIONS IN NON-AQUEOUS MEDIA

Reference	Reaction	Solvent	$t°C$	ΔV_c (cc/gramme-mole)
1	$C_5H_5N + C_2H_5I$	$(CH_3)_2CO$	30·3	− 16·4
			40	− 17·6
			50	− 18·6
			60	− 19·6
2	$(CH_3)(C_6H_5)(C_6H_5CH_2)$	$CHCl_3$	25·0	+ 3·3
	$(C_3H_5)NBr$		35·0	+ 3·3
			44·9	+ 3·9
3	$(CH_3CO)_2O + C_2H_5OH$	$C_6H_5CH_3$	20	− 9·6
			40	− 11·4
			60	− 12·5
			80	− 12·9
4	$(CH_3CO)_2O + C_2H_5OH$	C_2H_5OH	20	− 16·4
			30	− 15·9
			40	− 12·7
5	$C_2H_5I + C_2H_5O^-$	C_2H_5OH	14·9	− 3·7
			25·1	− 4·1
6	2Cyclopentadiene	Pure liquid	0	− 18·6
			20	− 20·8
			30	− 23·1
			40	− 26·1
7	$oCH_3.C_6H_4N(CH_3)_2 + CH_3I$	CH_3OH	40	− 26·6
8	$(C_6H_5)_3CCH(C_6H_5)_2$	$C_6H_5CH_3$	70	+ 13·1

REFERENCES

1–5. Gibson, Fawcett and Perrin, *Proc. Roy. Soc.*, **A, 150**, 223 (1935); Williams, Perrin and Gibson, *ibid.*, **154**, 684 (1936).
6. Newitt and Wassermann, *Trans. Chem. Soc.*, 735 (1940).
7. A. H. Ewald, *Faraday Soc. Dissc.* **22** (1956).
8. K. E. Weals, *Trans. Chem. Soc.*, 2959 (1954); *Faraday Soc. Dissc.*, **22**, 122 (1956).

acetic anhydride and the ethoxide ion, the effect of pressure on the rate may be due in part to its effect on the degree of ionization of ethanol.

The Critical Volume and its Temperature Variation

Just as the energy of a critically activated complex (loosely referred to as the transition 'state') cannot be evaluated until the partial molar energies of the solutes in the ground state are known, so the volume of the critical complex cannot be found until the partial molar volumes of the solutes in the ground state are known. The latter have seldom been measured, and their values are anticipated to differ from the molar volumes of the reactants in their pure states. In the absence of the required information, we shall ignore the difference between the partial molar volumes of ethyl iodide and pyridine in toluene solution and their molar volumes as liquids, and thus construct Table IV. We note that the coefficient of cubical expansion of

TABLE IV

APPROXIMATE ESTIMATES OF THE MOLAR VOLUME (IN CC) AND COEFFICIENT OF THERMAL EXPANSION OF THE CRITICAL COMPLEX IN THE REACTION BETWEEN C_2H_5I AND C_5H_5N

$t°C$	$V(C_2H_5I)$	$V(C_5H_5N)$	ΔV_c	V_c
35	81·91	81·62	−17·0	146·5
45	82·88	82·47	−18·1	147·3
55	83·99	83·35	−19·1	148·2
$\alpha \times 10^3$ (deg^{-1})	1·255	1·049	—	0·577

the complex is about one-half that of either reactant. It approaches, in fact, the coefficient for crystalline salts formed from highly polarizable ions. The molar volume of *cyclo*pentadiene at 20°C is 82·9 cc. From Newitt and Wassermann's results, it follows that the molar volume of the complex dicyclopentadiene is 145·8 cc. Unfortunately, α for the monomer is not known. If it is assumed to lie between the values of α for *cyclo*pentene and benzene, we find the coefficient of expansion of the reactive complex to be nearly zero.

The Volume of Activation at Various Pressures

Two of the curves in Fig. 2 indicate clearly that $(\partial \ln k/\partial P)_T$ for some reactions is far from constant. In these instances, ΔV_c increases, i.e. becomes less negative, with increasing pressure. That this should be so is not surprising for, from equation (4) we see that

$$\left(\frac{\partial \Delta V_c}{\partial P}\right)_T = \left(\frac{\partial V_2}{\partial P}\right)_T - \left(\frac{\partial V_1}{\partial P}\right)_T. \tag{9}$$

ΔV_c can thus be independent of pressure only in the unlikely event of the reactants and the complex responding identically to pressure. The position is analogous to the temperature variation of the apparent energy of activation,

E_A. In dealing with that phenomenon we took, as a first approximation, the constancy of $(\partial E_A/\partial T)_P$. The corresponding approximation here is that $(\partial \Delta V_c/\partial P)_T$ shall be constant; but we shall see that the conditions which justify the adoption of such an assumption are very limited.

When the compressibility, β, is constant and the pressure is low, we have $V_g = V_g^0(1 - \beta P)$ for the reactants in their ground state, and, by analogy, $V_c = V_c^0(1 - \beta_c P)$ for the critically activated complex. Then

$$\Delta V_c = \Delta V_c^0 - (V_c^0\beta_c - V_g^0\beta_g)P, \tag{10}$$

and the volume of activation varies linearly with respect to pressure, as found, for example, by Hyne, Golinkin and Laidlaw (1966) in their work on the solvolysis of benzyl chloride in ethanol-water mixtures at 50.25°C. In a variety of mixed solvents, they find the pseudo-unimolecular constant to be reproduced, up to pressures of 4,000 atmospheres, by equations of the form

$$\ln k = \ln k^0 + AP + BP^2, \tag{11}$$

where the constant A is positive, and B is negative, except in the case of pure water, where B also is positive. It follows that

$$\Delta V_c = -RT(A + 2BP), \tag{12}$$

and that $(\partial \Delta V_c/\partial P)_T$ is positive, except for aqueous solutions. Benson and Berson (1964), basing their treatment on Tait's isotherm for liquids, arrive at the equation

$$\ln k = \ln k^0 + AP + BP^{1.52}, \tag{13}$$

according to which ΔV_c should vary almost linearly with respect to $P^{1/2}$, since

$$\Delta V_c = -RT(A + 1.52BP^{0.52}). \tag{14}$$

While these empirical relationships reproduce with reasonable accuracy some data at fairly low pressures, neither satisfactorily accounts for the high pressure results. The reaction between ethyl bromide and sodium methoxide in methanol solution at 25° has been studied by Hamann (1958), 1957) at pressures up to 40,000 atmospheres. Values of ΔV_c derived from his data are plotted in Fig. 3 as a function of P and of $P^{1/2}$. Both curves tend to be linear at relatively low pressures. The general conclusion to be drawn is that, while pressure influences the reaction rate at all pressures, its effect becomes progressively less as the pressures increases.

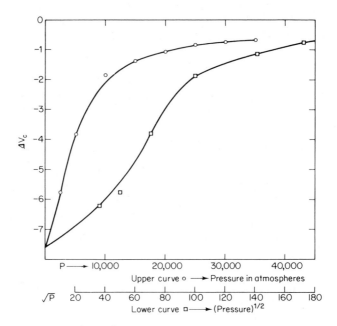

FIG. 3. $\Delta V_c(cc) = -RT\left(\dfrac{\partial \ln k_2}{\partial P}\right)$ for $C_2H_5Br + CH_3ONa$ in CH_3OH at 25°C. (Data of Hamann.)

Reactions in Aqueous Solution

From the data summarized in Table V, we see that the classification of reactions based on the magnitude of the Arrhenius constants A_2 is maintained in aqueous as in non-aqueous solutions. The hydrogen-ion catalysed inversion of sucrose is fast, and ΔV_c is positive. The corresponding hydrolyses of methyl acetate and γ-hydroxybutyrolactone are slow, and ΔV_c is negative. The same regularity is found among the ionic reactions, now to be considered.

Ionic Reactions

The bimolecular velocity constant for ionic reactions at zero ionic strength can be represented by the equation (Christiansen, 1924; Scatchard, 1932)

$$\ln k_2^0 = \ln Z^0 - \frac{E_n}{RT} - \frac{z_A z_B \varepsilon^2}{DrkT}. \tag{7.22}$$

What determine a reaction as normal, fast or slow are the values of the ionic

TABLE V

CRITICAL INCREMENTS OF VOLUME FOR CERTAIN REACTIONS IN AQUEOUS SOLUTIONS

Reference	Reaction	$t°C$	ΔV_c (cc/gramme-mole)
*	$C_{12}H_{22}O_{11} + H_3O^+$	15·5	$+2·5$
†	$CH_3COOCH_3 + H_3O^+$	14	$-8·7$
‡	$\overline{CH_2CH_2CH_2C}\overset{\displaystyle O}{O} + H_3O^+$	35	$-8·4$
‡	$CH_2OH.CH_2CH_2COOH +$ H_3O^+	35	$-2·9$
§	$C_6H_5CH_2Cl + H_2O$	50·25	-8 ± 1
‖	$CH_2BrCOOCH_3 + S_2O_3^{--}$	24·5	$+3·2$
‖	$CH_2BrCOO^- + S_2O_3^{--}$	24.5	-4.8
‖	$Co(NH_3)_5Br^{++} + OH^-$	30	$+8·5$
¶	$CH_2ClCOO^- + OH^-$	40	$-6·1$
		80·3	$-6·2$
**	$NH_4^+ + NCO^-$	60	$+14$
††	$tert.C_4H_9Cl +$ $\begin{cases} 80\%(v)H_2O \\ 20\%(v)C_2H_5OH \end{cases}$	30	-22

REFERENCES

* Röntgen, *Annalen*, **45**, 98 (1892); Cohen and de Boer, *Z. physikal Chem.*, **84**, 41 (1913).
† Rothmund, *Z. physikal Chem.*, **20**, 168 (1896). For the catalysed hydrolyses of other esters and amides, see Laidler and D. Chen., *Trans. Faraday Soc.*, **54**, 1020 (1958).
‡ Osborn and Whalley, *ibid.*, **58**, 2144 (1962).
§ Hyne, Golinkin and Laidlaw, *J. Amer. Chem. Soc.*, **88**, 2104 (1966).
‖ Burriss and Laidler, *Trans. Faraday Soc.*, **51**, 1497 (1955); Laidler, *Trans. Faraday Soc., Disc.* **22**, 88 (1956).
¶ Gibson, Fawcett, Perrin and Williams, *Proc. Roy. Soc.*, A, **150**, 223 (1935); **154**, 684 (1936).
** H. G. David and S. D. Hamann, *Trans. Faraday Soc.*, **50**, 1188 (1954). The value of ΔV_c given here has been calculated from the limiting experimental gradient, $(\partial \ln k_2 / \partial P)_T$, at low pressures.
†† J. Buchanan and S. D. Hamann, *Trans. Faraday Soc.*, **49**, 1425 (1953); Hyne *et al.*, *J. Amer. Chem. Soc.*, **88**, 2104 (1966).

product, $z_A z_B$, and of the constant L which governs the variation, with respect to temperature, of the permittivity, D:

$$\left(\frac{\partial \ln D}{\partial T} \right)_P = -L. \tag{7.26}$$

Provided Z_0, E_n, and r which have been defined in Chapter 7, are not affected by changes in temperature, we have (E. A. Moelwyn-Hughes, 1936)

$$A_2^0 = Z^0 . \exp\left(- z_A z_B \varepsilon^2 L/kDr\right). \tag{7.34}$$

We may now seek to find to what extent equation (7.22) can account for the variation of k_2^0 with respect to pressure. On differentiating equation (7.22) with respect to pressure at constant temperature, we have

$$\left(\frac{\partial \ln k_2^0}{\partial P}\right)_T = \left(\frac{\partial \ln Z^0}{\partial P}\right)_T - \frac{1}{RT}\left(\frac{\partial E_n}{\partial P}\right)_T + \frac{z_A z_B \varepsilon^2}{DrkT}\left[\left(\frac{\partial \ln r}{\partial P}\right)_T + \left(\frac{\partial \ln D}{\partial P}\right)_T\right],$$

and, from equation (7),

$$\Delta V_c = -RT\left(\frac{\partial \ln Z^0}{\partial P}\right)_T + \left(\frac{\partial E_n}{\partial P}\right)_T - \frac{z_A z_B \varepsilon^2 N_0}{Dr}\left[\left(\frac{\partial \ln r}{\partial P}\right)_T + \left(\frac{\partial \ln D}{\partial P}\right)_T\right]. \tag{15}$$

A rough estimate of the first term on the right-hand side may be made by assuming that Z^0 varies directly or inversely with respect to the viscosity. Then, as may be seen from Table VIII, the contribution to ΔV_c arising from this source is $\mp 1\cdot 294$ cc in water at 30°C. Little is known concerning $(\partial E_n/\partial P)_T$, except that it may be considerable. To evaluate the remaining term in equation (15), it is convenient to introduce the compressibility, β, and the density, ρ, of the solvent, so that

$$\left(\frac{\partial \ln D}{\partial P}\right)_T = \left(\frac{\partial \ln D}{\partial V}\right)_T\left(\frac{\partial V}{\partial P}\right)_T = -\beta\left(\frac{\partial \ln D}{\partial \ln V}\right)_T = \beta\left(\frac{\partial \ln D}{\partial \ln \rho}\right)_T. \tag{16}$$

The contribution to ΔV_c ascribable to the effect of pressure on the permittivity is thus

$$\Delta V_c = -\frac{N_0 z_A z_B \varepsilon^2 \beta}{Dr}\left(\frac{\partial \ln D}{\partial \ln \rho}\right)_T. \tag{17}$$

The compressibility of water at 20°C is $4\cdot 68 \times 10^{-5}$ atm^{-1}, and the value of $(\partial \ln D/\partial \ln \rho)_T$ derived by J. S. Jacobs and A. W. Lawson (1952) from the data of Kryopoulos (1926) is $1\cdot 34 \pm 0\cdot 02$. Hence, if ΔV_c is expressed in cc/mole and r in Angstrom units, we have, at this temperature, the relationship

$$\Delta V_c = -\frac{10\cdot 7 z_A z_B}{\mathring{r}}. \tag{18}$$

In testing this equation (Moelwyn-Hughes, 1967), by evaluating $\mathring{r}$, we have used the experimental values of ΔV_c relating to low pressure (Laidler, 1956). The results are given in Table VI, where, unfortunately, only two of the data

TABLE VI

CRITICAL INTERIONIC DISTANCES AFFORDED BY EQUATION (17)

Reaction	$t°C$	ΔV_c	$z_A z_B$	$\mathring{r}(Å)$
$CH_2ClCOO^- + OH^-$	40	$-6\cdot1$	$+1$	$(1\cdot75)$
$NH_4^+ + NCO^-$	60	$+14$	-1	$(0\cdot76)$
$CH_2BrCOO^- + S_2O_3^{--}$	24·5	$-4\cdot8$	$+2$	4.46
$Co(NH_3)_5Br^{++} + OH^-$	30	$+8\cdot5$	-2	$2\cdot52$

refer to temperatures near to that for which equation (18) applies. For these two reactions the critical interionic separations afforded by equation (18) compare favourably with the values found by the numerous other methods described in Chapter 7. It therefore appears that the effect of pressure on the velocity of ionic reactions in aqueous solution is significantly, if not predominantly, to be traced to its effect on the permittivity of the medium.

It is a matter of interest to derive an expression for the pressure effect on the rates of ionic reactions when, contrary to the conditions ordinarily prevailing, the reacting ions and the complex are all spherical, with evenly distributed charges. From the equation

$$\ln k_2^0 = \text{constant} + \frac{\varepsilon^2}{2DkT}\left[\frac{z_A^2}{r_A} + \frac{z_B^2}{r_B} - \frac{(z_A + z_B)^2}{r_C}\right], \qquad (7.50)$$

and equation (13.7), we have

$$\Delta V_c = \frac{N_0\varepsilon^2\beta}{2DkT}\left(\frac{\partial \ln D}{\partial \ln \rho}\right)_T\left[\frac{z_A^2}{r_A} + \frac{z_B^2}{r_B} - \frac{(z_A + z_B)^2}{r_C}\right]$$
$$-\frac{N_0\varepsilon^2}{6DkT}\left[\frac{z_A^2\beta_A}{r_A} + \frac{z_B^2\beta_B}{r_B} - \frac{(z_A + 2_C)^2\beta_C}{r_C}\right]. \qquad (19)$$

No knowledge of r_C is required when $z_A = -z_B$. Let us therefore consider the case where $z_A = +1$ and $z_B = -1$, as is true in the ammonium cyanate reaction. Let us further assume that $r_A = r_B$ ($= r_i$, say), which is not far from the truth. Subject to these limitations, equation (19) becomes

$$\Delta V_c = \frac{N_0\varepsilon^2}{Dr_ikT}\left[\beta\left(\frac{\partial \ln D}{\partial \ln \rho}\right)_T - \frac{1}{3}\beta_i\right]. \qquad (19a)$$

The compressibility, β_i, of either ion can safely be taken as less than that of the solvent, β. We thus recover an equation closely resembling equation (17). The model envisaged in this treatment, however, is highly artificial, and has been shown to be inadequate in other respects (Buchanan and Hamann, 1953).

Reactions between Ions and Polar Molecules

For a bimolecular reaction between an ion of charge e and a polar molecule with dipole of magnitude μ, we have

$$\ln k_2^0 = \ln Z^0 - \frac{E_n}{RT} - \frac{e\mu}{Dr^2kT}. \tag{9.10}$$

On making the same assumptions as in the treatment of ion-ion reactions, we obtain for the electrostatic contribution to the volume of activation:

$$\Delta V_c^e = -\frac{N_0 e\mu}{Dr^2}\left[\left(\frac{\partial \ln D}{\partial P}\right)_T + 2\left(\frac{\partial \ln r}{\partial P}\right)_T\right]. \tag{20}$$

The first term in the square brackets is positive and the second negative, so that positive, negative and zero contributions may be made to ΔV_c. Their magnitudes, however, are less than those discussed above; and there are insufficient data to test them. If the electrostatic component, E_e, of the activation energy is always inversely proportional to the permittivity, D, it follows from equation (17) that the component of ΔV_c due to it is

$$\Delta V_c = -E_e\beta\left(\frac{\partial \ln D}{\partial \ln \rho}\right)_T. \tag{21}$$

Since the function in the brackets is never very different from unity, ΔV_c depends chiefly on the magnitudes of the compressibility and the electrostatic energy.

TABLE VII

GROUND STATE PARTIAL MOLAR VOLUMES (CC) OF SOME SIMPLE SOLUTES IN WATER AT 290·16°C

Solute	Molar volume at atmospheric or orthobaric pressure	Partial molar volume in water at atmospheric pressure	Difference
CH_4	82·7	37	45·7
CH_3F	61·5	35·9	25·6
CH_3Cl	55·2	46·2	9·0
CH_3Br	57·1	53·0	4·1
CH_3I	62·63	63·7	−1·1
CH_3OH	40·23	38·27	1·96

The Partial Molar Volumes of Certain Solutes in the Ground State

Much recent work has been carried out on the partial molar volumes of solutes, particularly those of gases and electrolytes in water; but little is known concerning their pressure dependence (Bridgman, 1949; 1946). At ordinary pressures, the partial molar volume of a solute is frequently not very different from its value in the pure condensed phase, but may vary in a complicated manner with respect to the composition of the solution. Water shows unusually large effects, as the values of methane (Kritchevsky and Ilunskaya, 1945) and the methyl halides (Holland and Moelwyn-Hughes, 1956) show (Table VII).

The Partial Molar Volume in the Ground State, as a Function of Solvent Composition

Hyne and his collaborators (1966) have measured the partial molar volume of benzyl chloride in various mixtures of water and alcohols. Their results for methanol and tertiary butanol at 50.25°C (Fig. 4) reveal maxima, which in

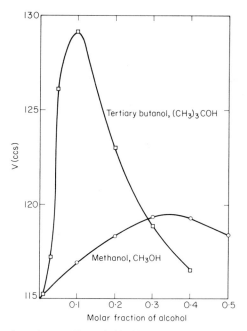

FIG. 4. Partial molar volumes of benzyl chloride, $C_6H_5CH_2Cl$, in aqueous alcoholic solutions at 50.25°C. (Liquid volume = 118.1 cc) (after H. S. Golikin, I. Lee and J. B. Hyne).

methanol occurs at a solvent composition of 0·35 mole fraction of methanol. Their values of ΔV_c for the solvolysis of benzyl chloride pass through a minimum at a composition of 0·43 mole fraction of methanol. Thus the change, with respect to solvent composition, of the activation volume ΔV_c is partly due to the change in the partial volume of the solute in its initial state. The situation is the parallel of that found by Arnett for the change, with respect to composition, of the enthalpy of activation and of the partial molar entropy of the solute in the ground state. A third parallel is provided by our knowledge, in certain instances, of the change, with respect to temperature, of the enthalpy of activation, E_A, and the absolute partial molar enthalpy of the solute in its ground state.

The variables employed experimentally in order to arrive at a reliable knowledge of some of the properties of activated solutes in pure solvents are seen to be temperature, pressure and solvent-composition. Careful analysis of the results afforded in these three fields of enquiry can lead to a sufficient knowledge of the activated complex (loosely, the transition state) to justify a discussion of constitutive influences within the solute molecule, such as electron drifts, induction and resonance.

The Effect of Pressure on the Break-down Frequency, v

That the apparent energy of activation is a measure of the difference between the average energy of the active molecules and the average energy of all the molecules (Tolman, Chapter 10) is true only when the probability of break-down of the active molecules is independent of temperature (La Mer, 1933), which we now know to be seldom true. Similarly, that the apparent volume of activation is a measure of the difference between the partial molar volume of the activated complex and the sum of the partial molar volumes of solutes in the ground state is true only when the probability of breakdown is independent of the pressure.

Henry's law is obeyed by most solutes even at high pressures. This fact and the proportionality between molar fraction and concentration, n, which prevails in dilute solutions entitle us to express the chemical potential of a solute at pressure P in terms of its chemical potential at some reference pressure, P^0, and of its concentration as follows:

$$\mu_g(P) = \mu_g(P^0) + kT \ln n_g + \int_{P^0}^{P} v_g \, dP.$$

The subscript g denotes the ground state. $\mu_g(P^0)$ includes the interaction energy of solute and solvent. v_g is the partial molecular volume. Similar terms are included in the expression of μ for the critically activated complex (or "transition state"), supplemented by its internal energy, ε:

$$\mu_c(P) = \mu_c(P^0) + kT \ln n_c + \int_{P^0}^{P} v_c \, dP + \varepsilon.$$

On equating these expressions, we obtain the equilibrium equation

$$\left(\frac{n_c}{n_g}\right)_P = \left(\frac{n_c}{n_g}\right)_{P^0} \exp\left(-\varepsilon/kT\right) . \exp\left(-\int_{P^0}^{P} \frac{(v_c - v_g)}{kT} \, dP\right). \tag{22}$$

The number of molecules decomposing per cc per second is $n_c(P)v$, and the first-order rate constant, k_1, is $n_c(P)v$ divided by $n_g(P)$. Hence,

$$\ln k_1 = \text{constant} + \ln v - \frac{\varepsilon}{kT} - \int_{P^0}^{P} \frac{(v_c - v_g)}{kT} \, dP,$$

and

$$\left(\frac{\partial \ln k_1}{\partial P}\right)_T = \left(\frac{\partial \ln v}{\partial P}\right)_T - \frac{1}{kT}\left(\frac{\partial \varepsilon}{\partial P}\right)_T - \frac{(v_c - v_g)}{kT}. \tag{23}$$

The volume of activation is seen to be

$$\Delta v_c = v_c - v_g = -kT\left(\frac{\partial \ln k_1}{\partial P}\right)_T + kT\left(\frac{\partial \ln v}{\partial P}\right)_T - \left(\frac{\partial \varepsilon}{\partial P}\right)_T, \tag{24}$$

which contains two terms more than appear in equation (7).

If the internal energy, ε, possessed by the activated complex is electronic, it is doubtful whether moderate pressures will affect it. On the other hand, if ε consists of an elastic energy in Hooke's form $\varepsilon = (1/2)Kx^2$ (K is the restoring force constant, and x the distension), $(\partial \varepsilon/\partial P)_T$ becomes $Kx(\partial x/\partial P)_T$, which may be positive or negative. Such a term may well be relevant to the dissociation of pentaphenylethane, making the true value of Δv_c greater than that given in Table III. We next consider the term $(\partial \ln v/\partial P)_T$.

Equations have been derived in earlier passages for the rate constants of reactions in solution which are directly proportional to the viscosity or inversely proportional to it. In the former case, ΔV_c includes the term $RT(\partial \ln \eta/\partial P)_T$, which is usually positive. Estimates of this volume contribution are given in Table VIII, as derived from Bridgmann's values of η at 1 atmosphere and at 1,000 atmospheres. The contribution to ΔV_c from this source is seen to be considerable, except in the case of water, the most important of all solvents. Moreover, at 25°C, to which most kinetic measurements in aqueous solution refer, the effect of pressure on the viscosity is almost negligible, and ΔV_c due to this source is less than 1 cc. Nevertheless, there is quite a striking parallelism between the catalytic coefficients for the inversion of cane sugar (Cohen and Valeton, 1918) and the viscosity of

TABLE VIII

VISCOSITY CONTRIBUTIONS TO ΔV_c AT 30°C

Solvent	$RT(\partial \ln \eta/\partial P)_T$ (cc/mole)
CS_2	6·66
CCl_4	20·10
$CHCl_3$	12·09
CH_3OH	9·57
C_2H_5OH	11·5
C_4H_9OH	18·4
$isoC_3H_7OH$	19·6
nC_8H_{18}	18·7
$(CH_3)_2CO$	12·9
H_2O	1·294
H_2O (at 10·3°C)	−1·092
H_2O (at 0°C)	−1·844

water even in that complicated region of water structure where high pressures diminish the velocity (Table IX). If, therefore, the velocity coefficient at a constant temperature can be expressed as a function of the viscosity raised to an integral ordinal number, that number is one in the inversion of cane sugar.

When, on the other hand, the rate constant is inversely proportional to the viscosity, as in diffusion-controlled reactions, the contribution to ΔV_c is negative.

At relatively low pressures, the effect of pressure on the viscosity of a liquid at constant temperature is represented empirically by the equation $\eta = \eta^0 e^{KP}$, where K is a constant. At high pressures, other terms are required to reproduce the quite unusual effect of pressure. Bridgman has pointed out that, whereas

TABLE IX

A COMPARISON OF THE RELATIVE VISCOSITIES OF AQUEOUS SOLUTIONS OF SUCROSE UNDER PRESSURE WITH THE RELATIVE VELOCITIES OF INVERSION

P (atmospheres)	At 15°C		At 25°C		At 35°C		At 45°C	
	η_P/η_1	k_P/k_1	η_P/η_1	k_P/k_1	η_P/η_1	k_P/k_1	η_P/η_1	k_P/k_1
1	1·00	1·00	1·00	1·00	1·00	1·00	1·00	1·00
500	0·96	0·94	1·01	1·02	1·03	1·03	1·03	1·04
1,000	—	—	1·04	1·05	1·05	1·06	1·06	1·06
1,500	—	—	1·07	1·05	1·07	1·06	1·08	1·06

for most properties the relative effect of pressure diminishes as the pressure is increased, the reverse is true in the case of viscosity. The viscosity of *iso*propyl alcohol, for example, at 30°C and at 30,000 atmospheres is greater by a factor of about 2×10^7 than its viscosity at unit pressure. This ratio is much larger than that anticipated from the data of Table VIII. Hamann (1958) has measured the rate of the reaction between ethyl bromide and sodium eugenoxide in this solvent at pressures exceeding 30,000 atmospheres. The plot of $\ln k_2$ against P passes through a maximum, beyond which a further increase in pressure lowers the rate of reaction. He has argued that in this region of high viscosity the coefficient of diffusion is so low as to convert the mechanism of reaction from that prevailing at low pressures to a diffusion-controlled process.

Enthalpy and Energy of Activation

It is implied in the differential form of the Arrhenius equation $E_A = RT^2(d \ln k/dT)$ that the temperature change is carried out in a system at constant pressure. For greater consistence with thermodynamic usage we refer to E_A as the enthalpy of activation, and define it as

$$\Delta H_c = RT^2(\partial \ln k/\partial T)_P. \tag{25}$$

The corresponding definition of the energy of activation is

$$\Delta E_c = RT^2(\partial \ln k/\partial T)_V. \tag{26}$$

The relationship between ΔH_c and ΔE_c may be obtained by regarding the velocity constant as being, in the first place, a function of temperature and pressure, so that

$$d \ln k = \left(\frac{d \ln k}{dP}\right)_T dP + \left(\frac{d \ln k}{dT}\right)_P dT,$$

and in the second place as a function of temperature and volume, so that

$$d \ln k = \left(\frac{d \ln k}{dV}\right)_T dV + \left(\frac{d \ln k}{dT}\right)_V dT.$$

By eliminating $d \ln k$, and rearranging the terms, we have

$$\left(\frac{d \ln k}{dT}\right)_V = \left(\frac{d \ln k}{dT}\right)_P - \left(\frac{dV}{dT}\right)_P \left(\frac{dP}{dV}\right)_T \left(\frac{d \ln k}{dP}\right)_T.$$

Reference to equations (25) and (26) then enables us to write

$$\Delta E_c = \Delta H_c - RT^2 \left(\frac{dV}{dT}\right)_P \left(\frac{dP}{dV}\right)_T \left(\frac{d \ln k}{dP}\right)_T. \tag{27}$$

But the coefficients of isobaric expansion and isothermal compression are defined by the well-known relations

$$\alpha = \frac{1}{V}\left(\frac{dV}{dT}\right)_P$$

and

$$\beta = -\frac{1}{V}\left(\frac{dV}{dP}\right)_T.$$

Hence

$$\Delta E_c = \Delta H_c + \frac{RT^2\alpha}{\beta}\left(\frac{d\ln k}{dP}\right)_T. \tag{28}$$

This equation (M. G. Evans and M. Polanyi, 1937) has been applied by Newitt and Wassermann (1940) to the kinetics of the dimerization of *cyclo*-pentadiene in the liquid phase, with results which are shown in Table X.

TABLE X

ARRHENIUS CONSTANTS, A_P, ΔH_c AND A_V, ΔE_c FOR THE DIMERIZATION OF LIQUID CYCLOPENTADIENE

P (atmospheres)	ΔH_c	$\frac{RT^2\alpha}{\beta}\left(\frac{d\ln k}{dP}\right)_T$	ΔE_c	$\log_{10} A_P$	$\log_{10} A_V$
	(Kcal/gramme-mole)			(A in litres/gramme mole-sec.)	
1	$16\cdot6 \pm 0\cdot4$	$4\cdot1$	$20\cdot7 \pm 0\cdot7$	$6\cdot1 \pm 0\cdot3$	$9\cdot2 \pm 0\cdot5$
1,000	$17\cdot7 \pm 0\cdot4$	$1\cdot4$	$19\cdot1 \pm 0\cdot8$	$7\cdot4 \pm 0\cdot3$	$8\cdot1 \pm 0\cdot5$
3,000	$18\cdot6 \pm 0\cdot4$	$1\cdot1$	$19\cdot7 \pm 0\cdot7$	$8\cdot8 \pm 0\cdot3$	$8\cdot8 \pm 0\cdot5$

The authors conclude "that, in contrast to ΔH_c and A_P, which increase markedly with increasing pressure, the pressure dependence of the activation energy ΔE_c and A_V is smaller than the experimental error". The A terms are defined as usual in the integrated forms of the Arrhenius equation:

$$k = A_P \cdot \exp\left(-\Delta H_c/RT\right) = A_V \cdot \exp\left(-\Delta E_c/RT\right), \tag{29}$$

which are obtained from equations (25) and (26) on the assumption that ΔH_c at constant pressure and ΔE_c at constant volume are independent of temperature.

By combining equations (28) and (7), we have

$$\Delta E_c = \Delta H_c - \frac{T\alpha\Delta V_c}{\beta}, \tag{30}$$

and, since the kinetic pressure, P_k, is $T\alpha/\beta$,

$$\Delta E_c = \Delta H_c - P_k \Delta V_c. \tag{31}$$

An application of equation (30) to a reaction in acetone solution is given in Table XI. In aqueous solution the difference between ΔE_c and ΔH_c is naturally less than the differences given here. It is, for example, 207 cal for the hydrolysis of methyl acetate at 15°C, and -59 cal for the hydrolysis of sucrose.

TABLE XI

An Estimate of the Difference between E_P and E_V for the Reaction between Pyridine and Ethyl Iodide in Acetone Solution

T (°K)	$\alpha \times 10^3$ (deg^{-1})	ΔV_c (cc/gramme-mole)	$\beta \times 10^{10}$ (cm^2/dyne)	$\Delta E_c - \Delta H_c$ (cal./gramme-mole)
303·1	1·5085	−16·4	0·990	1,809
313·1	1·5315	−17·6	1·035	1,944
323·1	1·5545	−18·6	1·080	2,066
333·1	1·5775	−19·6	1·125	2,188

The extensive work of Shu-lin P'eng, Sapiro, Linstead and Newitt (1938) on the kinetics of the reactions between acetic acid and equimolar quantities of various alcohols under pressure provides yet another example of the approximately linear relationship between the terms A and E_A (or ΔH_c) of the Arrhenius equation. The rate of formation of ethyl acetate increases by a factor of 24 when the pressure is raised from one atmosphere to 4,000 atmospheres, and the term A_P increases by a factor of 35. Pressure thus affects the rate more through a change in A_P than a change in E_A. Gibson, Fawcett and Perrin (1935) who found the following constants for the Hecht reaction in ethylalcoholic solution:

$$\text{at 1 atmosphere:} \quad k_2 = 2\cdot13 \times 10^{11} \times \exp\left(-20740/RT\right);$$

$$\text{at 2,980 atmospheres:} \quad k_2 = 3\cdot72 \times 10^{11} \times \exp\left(-20800/RT\right),$$

have concluded that, up to the pressure measured, the effect of increasing the pressure is almost entirely reflected in an increase of the collision number.

REFERENCES

Benson and Berson, *J. Amer. Chem. Soc.*, **86**, 259 (1964).

Bridgman, *The Physics of High Pressure*, p. 146, Bell, London (1949); *Rev. Mod. Phys.*, **18**, 38 (1946).

Buchanan and Hamann, *Trans. Faraday Soc.*, **49**, 1425 (1953).

Christiansen, *Z. physikal Chem.*, **113**, 35 (1924).

Cohen and Valeton, *Z. physikal Chem.*, **92**, 433 (1918).

Evans, M. G. and M. Polanyi, *Trans. Faraday Soc.*, **31**, 875 (1935); **33**, 449 (1937).

Gibson, Fawcett and Perrin, *Proc. Roy. Soc.*, **A**, **150**, 223 (1935).

Hamann, *Trans. Faraday Soc.*, **54**, 507 (1958); *Physicochemical Effects of Pressure*, Butterwoth, London (1957).

Holland and E. A. Moelwyn-Hughes, *Trans. Faraday Soc.*, **52**, 297 (1956).

Hyne, Golinkin and Laidlaw, *J. Amer. Chem. Soc.*, **88**, 2104 (1966).

Jacobs, J. S. and A. W. Lawson, *J. Chem. Physics*, **20**, 1161 (1952).

Kritchevsky and Ilunskaya, *Acta Physicochim.*, U.S.S.R., **20**, 327 (1945).

Kryopoulos, *Z. physikal Chem.*, **40**, 507 (1926).

Laidler, *Trans. Faraday Diss.*, **22**, 88 (1956).

Moelwyn-Hughes, E. A., *J. Phys. Chem.*, **71**, 4120 (1967).

Moelwyn-Hughes, E. A., *Proc. Roy. Soc.*, **A**, **155**, 308 (1936).

Moelwyn-Hughes, E. A. and Rolfe, *Trans. Chem. Soc.*, 241 (1932).

Newitt and Wassermann, *Trans. Chem. Soc.*, 735 (1940).

Scratchard, *Chem. Rev.*, **10**, 229 (1932).

Shu-lin P'eng, Sapiro, Linstead and Newitt, *Trans. Chem. Soc.*, 784 (1938).

Williams, E. G., Perrin and Gibson, *Proc. Roy. Soc.*, **A**, **154**, 684 (1936).

van't Hoff, *Vorlesungen über theoretische und physikalische Chemie*, **1**, p. 236, Braunschweig (1901).

14

FAST REACTIONS AND RELAXATION PHENOMENA

In chemical kinetics, as elsewhere, the adjective fast is a comparative one which cannot, therefore, be precisely defined. Loosely we may call a reaction rapid when it is half over in a few seconds, which is the utmost that unaided manipulative dexterity can accomplish. By indirect means, however, and with the aid of modern techniques, it is possible to measure the rate of a reaction with a half-life as low as 10^{-11} sec. The various new approaches have fully confirmed and greatly extended our knowledge of fast reactions. Let us first consider what had been established before the advent of recent methods.

Determination of the Rate of Some Fast Reactions by Classical Methods

(i) The first-order constant governing the mutarotation of most of the reducing sugars in aqueous solutions of strong acids or bases is reproduced by the equation

$$k = k_0 + k_{H^+}[H^+] + k_{OH^-}[OH^-] \tag{1}$$

or, since the ionization of water is rapidly established,

$$k = k_0 + k_{H^+}[H^+] + \frac{Kk_{OH^-}}{[H^+]}, \tag{2}$$

where K is the ionic product of water. The first-order constant should therefore have a minimum value when

$$k_{OH^-} = k_{H^+}[H^+]_{min}^2/K. \tag{3}$$

The catalytic coefficient k_{H^+} can be readily measured, and the pH at the minimum point determined electrometrically. Since the ionic product, K, is known, k_{OH^-} can be determined. In the catalytic decomposition of nitramide at 24·84°C, it is found to be $5·10 \times 10^9$ litre/mole-sec. (La Mer and Marlies, 1935). The half-life of this fast reaction is $1·36 \times 10^{-10}$ sec in normal alkaline solution. This method is due to Wijs (1893) who successfully used it to measure the ionic product of water from catalytic coefficients.

(ii) One of the rate-determining steps in the reaction between iodine and hydrogen peroxide in aqueous solution is given by the expression

$$\text{rate} = k_2[H_2O_2]\,[OI^-].$$

The hyperiodite ion is produced in the equilibrated reaction

$$I_2 + H_2O \rightleftharpoons 2H^+ + I^- + OI^-;\qquad K = [H_+]^2[OI^-]^2/[I_2].$$

In neutral solutions, this component of the rate equation should therefore vary inversely as the concentration of hydrogen ion, since

$$\text{rate} = k_2[H_2O_2]\frac{\{K[I_2]\}^{1/2}}{[H^+]}. \tag{4}$$

From the known hydrolysis constant of iodine in water k_2 is found to be 2×10^{11} litre/mole-sec at 25°C (Liebhafsky, 1932). Liebhafsky emphasized that the magnitude of this rate constant coincided with the limiting value according to the collision theory.

(iii) The electrical conductivity of ethanol containing a trace of ammonia increases slowly during storage. Ogston (1936) attributed the effect to a pseudo-unimolecular rate-determining ionization

$$C_2H_5OH \rightarrow C_2H_5O^- + H^+;\qquad k_1 = 3 \times 10^{-8}\ \text{sec}^{-1}\ \text{at 25°C}$$

which is followed by the rapid reaction of the hydrogen ion with ammonia. The ionic product of pure ethanol is known from electrometric studies (Macfarlane and H. Hartley, 1932).

$$C_2H_5OH \underset{k_2}{\overset{k_1}{\rightleftharpoons}} C_2H_5O^- + H^+ : K_{25°} = k_1/k_2$$

$$= 1·9 \times 10^{-19}\ (\text{gramme-ions/litre})^2. \tag{5}$$

Hence the bimolecular constant k_2 governing the union of the ions is about $1·6 \times 10^{11}$ litre/mole-sec.

(iv) In a system containing equal initial concentrations, a, of reactants, the bimolecular velocity constant is given by equation (6.39) in terms of the half-life, $t_{1/2}$, $(k_2 = 1/t_{1/2}a)$, so that very fast reactions can be studied by working with small concentrations. With colloidal particles in solution and in the gas phase, the actual number of particles per unit volume can be counted, and rates can be measured at a concentration of 3.6×10^{11} particles per unit volume, which corresponds to an initial concentration, a, of 6×10^{-13} moles per litre. As explained in Chapter 5, k_2 is about 3.8×10^9 litres/mole-sec. The principle of using low concentrations applies also to certain reactions involving radioactive solutes, the molar concentrations of which can be accurately measured when as low as 10^{-8}.

(v) The bimolecular constant governing the deactivation of excited molecules in solution has the same order of magnitude. Its measurement is made possible because the Stern–Volmer equation (5.20) allows an evaluation of k_2/k_1, where k_1 is known from the optical properties of the fluoroescent solute to have the order of magnitude of 10^{-7} to 10^{-9} sec^{-1} (Forster, 1953; E. J. Bowen, 1953; G. G. Guilbault, 1967).

(vi) Most rate constants decrease as the temperature is lowered. A reaction which is very fast at room temperatures can therefore be measured at low temperatures (see Table 9.XIII and G. N. Lewis and G. T. Seaborg, 1939).

Determination of the Rates of Fast Reactions by Recent Methods

Numerous relatively recent methods have been employed to measure the rates of very fast reactions (see E. F. Caldin, 1964; S. Claesson, 1968). Some of them do not require the introduction of new physicochemical principles but rely on ingenious experimentation rendered possible by such instruments as photomultipliers and oscilloscopes. Others, such as those based on the measurement of dielectric loss and the absorption of ultrasonic waves, are based on important new principles. We shall deal briefly with instances of both kinds.

(i) FLOW METHODS

The principle of the flow method is a simple one. Instead of following the course of reaction by measuring some property of a static solution, the two reactant solutions are forced separately through narrow orifices by means of plungers into a chamber where, still flowing, they are thoroughly mixed and allowed to flow out in a common stream through a capillary tube in which the chemical reaction continues. At different positions along the tube, the composition of the solution is estimated, for example, by recording its refractive index. The difference between the concentrations found at two given positions along the tube, divided by the very short time taken by the solution to cover this distance yields the instantaneous, dc/dt, rate of chemical change. The method is due to Hartridge and Roughton (1923) who applied it chiefly to reactions of biological interest, such as $CO_2 + OH^-$ $\rightarrow HCO_3^-$, for which k_2 in water at 25°C is $3.44 \times 10^{+3}$ litre/mole-sec.

Von Halban and Eisner (1936) have measured the rates of some fast analytical reactions, such as the starch–iodine test by this means. Improved adaptations employ intermittent flows (the stopped-flow method).

(ii) PHOTOCHEMICAL METHODS

In a system which absorbs light, its intensity, I, at a distance d cm from its source is related as follows to its initial intensity, I_0; $I = I_0 e^{-\alpha n d}$, where n is the number of absorbing molecules per cc, and α is the absorption coefficient. The dimensions of α are those of cm^2 per molecule. Experimentalists generally use the decadic equation $I = I_0 . 10^{-\varepsilon c d}$, where c is the concentration in gramme-moles per litre, and ε is termed the extinction coefficient. The molar optical density is defined as $\varepsilon c d = \log_{10}(I_0/I)$. The relationship between the extinction and absorption coefficients is clearly

$$\varepsilon = \left(\frac{N_0}{1{,}000 \times 2{\cdot}30}\right)\alpha.$$

Both coefficients provide an optical method for determining the concentration of the absorbing molecules. The fraction of light absorbed is

$$(I_0 - I)/I_0 = 1 - I/I_0 = 1 - e^{-\alpha n d}.$$

When there is only slight absorption, we have the approximate equation

$$(I_0 - I)/I_0 = \alpha n d.$$

The absorption coefficient can be measured by determining I at a fixed value of d, using systems of various concentrations, or, as is more usual, by measured I at various values of d in a system with a fixed concentration. The use of a pair of compensating photoelectric cells makes it relatively easy to measure the intensity of the incident light, I_0, and that of the emergent light, $I_0 - I$.

The optical arrangement used by Rabinowitch and W. C. Wood (1936) in their study of solutions of the halogens in carbon tetrachloride and other inert solvents is sketched in Fig. 1. The two photocells PC1 and PC2 are

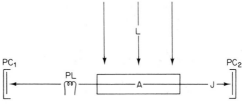

FIG. 1. Apparatus for the study of photostationary state of iodine (after Rabinowitch and Wood, 1936).

placed at opposite ends of the reaction vessel, which is 10 cm long. The proportionality between the extinction coefficient towards light of wavelength 5,500 Å and the concentration of halogen molecules is first established. In the particular experiment to which we shall refer, the halogen X_2 was iodine in CCl_4 at 20°C and at a concentration of $2\cdot2 \times 10^{-5}$ mole/litre = $1\cdot33 \times 10^{16}$ molecules/cc. The symbol X is used so as not to confuse the symbol for intensity with that for the iodine atom. Vertically above the reaction vessel is placed a powerful search-light, L, whose rays are concentrated downwards on to the reaction vessel by means of a parabolic mirror. In the presence of this intense light, the absorption coefficient was found to be less than in its absence (see Fig. 2). The decrease in the concentration

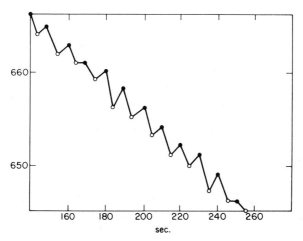

FIG. 2. The dissociation-effect in iodine solution. Points show the extinction of the solution in the dark; circles that during illumination.

of iodine molecules is attributed to their photolytic decomposition into atoms. The light intensities and concentrations in the presence and absence of the illumination are evidently related as follows:

$$\frac{I_0 - I_{\text{dark}}}{I_0 - I_{\text{light}}} = \frac{n_{\text{dark}}}{n_{\text{light}}}.$$

The fractional decrease in the concentration of iodine molecules,

$$\frac{n_{\text{dark}} - n_{\text{light}}}{n_{\text{dark}}} = \frac{I_{\text{light}} - I_{\text{dark}}}{I_0 - I_{\text{dark}}}$$

in the experiment referred to is found to be $8\cdot7 \times 10^{-5}$. The photostationary concentration of iodine atoms is thus $2 \times 1.33 \times 10^{16} \times 8.7 \times 10^{-5} = 2.32 \times 10^{12}$ atoms/cc.

The rate at which iodine molecules are decomposed and formed in unit volume is given by the equation

$$-\frac{dn_{X_2}}{dt} = \gamma \frac{dn_{hv}}{dt} - k_2 n_X^2, \tag{6}$$

where n_{X_2} is the concentration of iodine molecules, γ is the quantum efficiency, (dn_{hv}/dt) is the number of quanta of radiation absorbed per second in 1 cc, and k_2 is the bimolecular coefficient governing the combination of iodine atoms. Under photostationary conditions, these rates are equal. The rate of absorption of quanta is found to be 4.2×10^{14} quanta per cc per sec. Assuming γ to be unity, we thus have

$$k_2 = 4.2 \times 10^{14}/n_X^2 = 7.8 \times 10^{-11} \text{ cc/atom-sec}$$

$$= 4.7 \times 10^{10} \text{ litre/gramme-atom-sec}.$$

According to the kinetic theory of gases, the number of collisions per cc per second between iodine atoms at a molecular concentration, n, is $2n^2\sigma^2(\pi kT/m)^{1/2}$ where m is the atomic mass. From the viscosity of gaseous iodine, $\sigma = 3.89$ Å. If union takes place at each encounter, the calculated value of k_2 at 20°C is 7.43×10^{-11} cc/molecule-sec $= 4.48 \times 10^{10}$ litre/mole-sec, in satisfactory agreement with experiment. The formation of iodine molecules from iodine atoms in the gas phase does not take place except in the presence of a third particle. The formation of iodine molecules from iodine atoms in solution takes place at each encounter, because the solvent molecules act as the third partners.

(iii) FLASH METHODS

In flash photolysis, the system to be examined is held in a long tube beneath, and parallel with, another tube through which an intense flash of light, lasting about 10^{-5} sec, is discharged (see Fig. 3). The chemical changes induced by the light include the formation of free radicals, atoms, ions and various electronically excited species whose fate is examined by subjecting the system to a second flash of white light. The interval between the two flashes is measured in microseconds (10^{-6} sec). After passage through the reacting system, this beam of light is passed through a spectrograph and its spectrum recorded rapidly at intervals which are also measured in microseconds. Examination of the lines in the spectra obtained after known time intervals reveals the story of the transient species generated by the first flash, and allows of the measurement of the rates of the extremely fast reactions

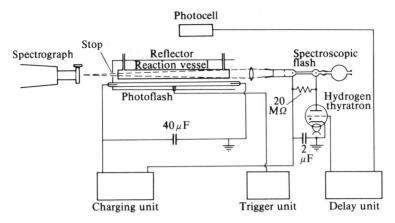

FIG. 3. Schematic diagram of apparatus for flash photolysis and kinetic spectroscopy (after R. G. W. Norrish, 1965).

taking place. The method is due to Norrish and G. Porter (1954), and has been extensively employed (Norrish and Thrush, 1956), yielding bimolecular rate constants which are as high as 10^{+10} litre/mole-sec (see also Osugi, Kusuhara and Hirayama, 1967).

(iv) PULSED RADIOLYSIS

Instead of the intense radiant energy used in flash photolysis to initiate decomposition, a beam of hard X-rays or high energy electrons (10^6 volts) may be used. McCarthy and MacLachlan (1961) irradiated liquid cyclohexanol at 25° with a beam of high energy (6 MeV) electrons which imparts to the liquid molecules sufficient energy to produce free radicals at concentrations of about 0·2 millimole/litre. The concentration is estimated from the energy absorption. At right angles to the high energy electric beam, light is passed, and is analysed by a photomultiplier. Absorption, which occurs in the region of 3,000 Å, is found to fall off in intensity to half its initial value in about 3×10^{-5} sec (see Fig. 4). The change is found to be bimolecular with $k_2 = 3\cdot4 \times 10^8$ litre/mole-sec. The mechanism proposed is

$$\underset{CH_2}{\overset{CH_2-CH_2}{\diagdown}}\underset{CH_2-CH_2}{\overset{}{\diagup}}\underset{OH}{\overset{H}{C}} \rightarrow \underset{CH_2}{\overset{CH_2-CH_2}{\diagdown}}\underset{CH_2-CH_2}{\overset{}{\diagup}}C-OH + H$$

followed by the bimolecular union of the free radicals to form the diol. The bimolecular constant for the union of the radicals $C_6H_{17}O$ is very rapid, with $k_2 = 2 \cdot 7 \times 10^{10}$, and E_A in the diffusion controlled range of $1 \cdot 7$ to $5 \cdot 5$ kcal/mole.

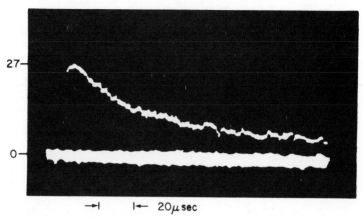

FIG. 4. Oscilloscope trace showing the percentage absorption of energy in liquid cyclohexanol after irradiation by a high energy electron beam (after McCarthy and MacLachlan, 1961).

Relaxation Methods

When the composition, x_e, of a system at equilibrium is changed to the composition x, the disturbed system reverts to its equilibrium state at a rate which is proportional to the displacement:

$$-\frac{d(x - x_e)}{dt} = \frac{1}{\tau}(x - x_e),$$

or

$$-\frac{d\Delta x}{dt} = \frac{\Delta x}{\tau}. \tag{6.23}$$

This, as we have seen, is the basic law of relaxation phenomena. When applied to reversible unimolecular reactions, the relaxation time, τ, is given by the equation

$$1/\tau = k_A + k_B, \tag{6.21}$$

When a forward unimolecular reaction is opposed by a bimolecular reaction, the relaxation time is given by the equation

$$1/\tau = k_A + k_B(n_A^0 + n_B^0). \tag{6.71}$$

Clearly, the experimental determination of relaxation times provides a powerful means of measuring rate constants. In particular, the measurement of short relaxation times leads to the evaluation of fast reaction rate constants.

The agencies which have been used to disturb equilibrated systems include pressure, heat, light, static and alternating electric fields, γ-rays, electron beams and ultrasonic waves. As it is impossible to describe all the methods adopted we shall confine ourselves to the effects of ultrasonic waves, after first dealing briefly with viscosity, which is the first relaxation phenomenon to have been studied.

Maxwell's Theory

The general relationship between the strain, s, produced in a body by displacement, under a stress f is

$$\frac{df}{dt} = \frac{1}{\beta}\frac{ds}{dt} - \frac{f}{\tau}. \tag{7}$$

In the case of a solid, τ is so large that the last term may be neglected and the strain produced is at all times proportional to the stress. If, however, the body is viscous, f does not remain constant, but tends to disappear at a rate proportional to f; that is, the body relaxes. In a simple compression, the stress, f, is the force per unit area, or the additional pressure, $P - P^0$, and the strain, s, is $-d \ln V$, where V is the volume, so that the relaxation equation equation becomes

$$\frac{dP}{dt} = -\frac{1}{\beta}\frac{d \ln V}{dt} - \frac{P - P^0}{\tau}. \tag{8}$$

If the rate of change of the strain is constant, we may write

$$B = -\frac{1}{\beta}\frac{d \ln V}{dt}, \tag{9}$$

and integrate the equation

$$\frac{dP}{dt} = B - \frac{P - P^0}{\tau}, \tag{10}$$

obtaining

$$P - P^0 = B\tau[1 - \exp(-t/\tau)]. \tag{11}$$

From the form of this pressure–time relationship (Fig. 5) it is seen that the

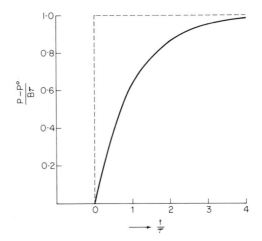

Fig. 5. Real and hypothetical attainment of the stationary state (full line and dotted line respectively).

stress is initially zero, and increases exponentially to the limiting and steady value $B\tau$. The relaxation time is the reciprocal of the first-order constant governing the process which leads to the steady state. In the absence of the relaxation, the pressure–time curve would follow the dotted lines, indicating an instantaneous response of the stress to the strain. We may therefore regard the relaxation time as an approximate measure of the time lapse between the application of the stress and the response of the system to it.

Let us next consider the pressure–time relationship which would result when, at a certain instant $t = T$, the strain were to become constant. Then $B = 0$, and

$$\frac{dP}{dt} = -\frac{P - P^0}{\tau},$$

and

$$\ln (P - P^0) = -t/\tau + a \text{ constant}.$$

Since T is a time during which the steady state has been reached, $P - P^0 = B\tau$ at $t = T$. Then

$$\ln B\tau = -T/\tau + a \text{ constant}.$$

Hence

$$P - P^0 = B\tau \exp\left[-(t - T)/\tau\right]. \tag{12}$$

The stress thus diminishes exponentially from $B\tau$ to zero, as may be seen from the second part of Fig. 6.

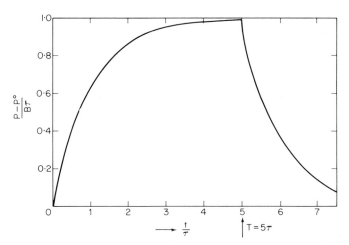

F$_{IG}$. 6. Attainment of a stationary state, followed by a unimolecular relapse after removal of the strain.

Let us suppose that a system first reaches the steady state, and that after a time T the strain becomes constant. At a time $2T$, let the strain again vary linearly with respect to time. If these processes were to be repeated, the pressure–time curve would consist of a linked series of curves of the type shown in Fig. 6. Such an artificially interrupted system vaguely resembles what is found in actual systems through which sound waves are transmitted. The alternate compressions and dilatations, however, vary smoothly and continuously with respect to time. As we shall see, both f and s are proportional to $\sin 2\pi\nu t$, where ν is the frequency of the sound wave.

Let us next apply the steady state equation $P - P^0 = B\tau$ to a liquid consisting of spherical molecules of diameter a each. The force exerted on each molecule is

$$X = \pi a^2 (P - P^0)$$

$$= \pi a^2 \tau B$$

$$= -\pi a^2 \frac{\tau}{\beta} \frac{d \ln V}{dt}$$

$$= -\pi a^2 \frac{\tau}{\beta} \frac{3 d \ln a}{dt}$$

$$= -\frac{3\pi a\tau}{\beta} \frac{da}{dt}.$$

But, according to Stokes' (1856) law for the steady linear motion of a sphere of diameter, a, in a medium of viscosity, η,

$$X = -3\pi\eta a \frac{da}{dt}.$$

Hence

$$\tau = \eta\beta, \tag{13}$$

which is Maxwell's (1868) fundamental equation.

Applied to an ideal gas, for which $\eta = (1/2)mn\bar{c}\lambda$, $\beta = 1/nkT$ and $\bar{c} = (8kT/\pi m)^{1/2}$, we see that

$$\tau = \frac{4}{\pi}\frac{\lambda}{\bar{c}} = \frac{4}{\pi}\cdot\tau_c. \tag{14}$$

The relaxation time is thus very nearly the ratio of the mean free path, λ, to the average velocity, $\bar{c}$. This ratio is evidently the average time, τ_c, between two successive collisions.

The value of τ_c afforded by this equation for chlorine gas at 0°C and 1 atm is 9.44×10^{-11} sec. By a method to be described later, the experimental value of the relaxation time under the same conditions is 4.7×10^{-6} sec (A. Eucken and R. Becker, 1934), which is greater than the calculated value by about 5×10^4. Thus there occurs in this gas a much slower relaxation process than can be explained by Maxwell's theory. In other words, a relatively long time elapses between the application of a stress and the establishment of a steady response to it. We shall see that this time lapse can be traced to the difficulty of converting the translational energy of the molecules into internal vibrational energy. Very roughly it may be said that 5×10^4 collisions between chlorine molecules in the gas phase are required before such a conversion of energy can take place. The determination of the relaxation times in a gas thus leads to the evaluation of the probability of excitation of vibrational quanta, which is a significant factor in the kinetics of chemical change. Before proceeding to our main task, which is an understanding of the ultrasonic method of determining the rates of very fast chemical reactions, we must digress to deal with the basic aspects of the theory of sound which, though mathematically somewhat difficult, is essential. (See K. F. Herzfeld and T. A. Litovitz, 1959; L. Bergmann, 1949; M. Eigen, 1959; R. O. Davies and J. Lamb, 1957; J. Lamb, 1960; M. Eigen, 1954; T. L. Cottrell and J. C. McCourbrey, 1961.)

The Passage of Sound through an Elastic Medium

Sound consists of the movement of molecules vibrating harmonically with respect to space and time. It resembles light in its harmonic nature, but differs from it in three respects, (1) the velocity of sound is much less than that of light, (2) its propagation requires material media, and (3) particle displacements are longitudinal, i.e. in the direction of propagation. The molecular motions in the medium through which sound travels set up regions of alternate compression and rarefaction at distances apart of $(1/2)\lambda$, where λ is the wave length. Sound of a constant frequency, v, can be readily generated in various ways, and its wave length, like that of light, can be accurately measured interferometrically. Consequently, the velocity $(c = \lambda v)$ of sound, and of ultrasonic waves, may be determined under a variety of conditions. It is found to be greater than the average molecular velocity. In fact, the cyclical changes in compression and in many other properties which accompany the passage of sound waves take place more quickly than the system can absorb or lose heat. The disturbance is consequently adiabatic. When there are no viscous forces at work, no transitions between molecules in different energy states and no chemical changes, the passage of sound is not attended by energy loss or change in propagation velocity; and it is with systems conforming to these requirements that we must first be concerned.

Let us consider a medium, containing n molecules per cc., to be enclosed in a cylinder of unit cross-sectional area, and confine attention to a small element of the cylinder, of length dx (Fig. 7). Let the pressure on the left-hand

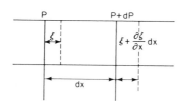

FIG. 7. Relationship between pressure change, dP, and change, $\partial\xi$ in particle displacement during the passage of sound.

side be P and that on the right hand side be $P + dP$. The force acting on the medium within the volume dx in the positive direction is thus

$$P - \left[P + \left(\frac{\partial P}{\partial x}\right) dx \right] = -\frac{\partial P}{\partial x} dx.$$

The number of molecules on which the force acts is $n\,dx$. Hence the average force, X, acting on one molecule is

$$X = -\frac{1}{n} \frac{\partial P}{\partial x}. \tag{15}$$

By Newton's second law, this force is given in terms of the mass, m, and the velocity, u, as $m(\partial u/\partial t)$. Hence

$$m\frac{\partial u}{dt} = -\frac{1}{n}\frac{\partial P}{\partial x}.$$

The density of the medium is $\rho = mn$, so that

$$\frac{\partial u}{\partial t} = -\frac{1}{\rho}\frac{\partial P}{\partial x}. \tag{16}$$

To derive the equation of continuity, we note that the number of molecules entering the section from the left-hand side per second is nu. The number leaving from the right hand side per second is $n[u + (\partial u/\partial x)\,dx]$. The gain in the number of molecules per second is thus $-n(\partial u/\partial x)\,dx$. As the volume concerned is dx, the increase in concentration per second is $-n(\partial u/\partial x)$. This can clearly be written as $\partial n/\partial t$. Therefore

$$\frac{\partial n}{\partial t} = -n\frac{\partial u}{\partial x}$$

and consequently

$$\frac{\partial \ln n}{\partial t} = \frac{\partial \ln \rho}{\partial t} = -\frac{\partial u}{\partial x}. \tag{17}$$

Finally, we must consider the volume change accompanying the passage of sound. Let ξ denote the displacement, at any instant, of a molecule from its position of minimum energy on the left hand side of the section. Because the transference of pressure is neither instantaneous nor independent of the position of the molecule, the molecular displacement on the right hand side is $\xi + (\partial \xi/\partial x)\,dx$. The original volume of the section is dx, and its volume during the passage of sound $dx + (\partial \xi/\partial x)\,dx$. Hence the relative increase in volume is

$$\frac{dV}{V} = d \ln V = \frac{\partial \xi}{\partial x}. \tag{18}$$

This important and dimensionless variable, which is a function of x and t, is sometimes denoted by d, the "dilatation", or by $-s$, where s is the "condensation". Maxwell called s the strain or distortion. It can be conveniently defined in a slightly different way by the equation

$$\rho = \rho^0(1 + s), \tag{19}$$

where ρ^0 is the density of the medium in the absence of sound. Then

$$\frac{\partial \ln \rho}{\partial t} = \frac{1}{1 + s}\cdot\frac{\partial s}{\partial t}, \tag{20}$$

and, from equations (17) and (20),

$$\frac{\partial s}{\partial t} = -(1 + s)\frac{\partial u}{\partial x}. \tag{21}$$

By combining equations (16) and (19), we have

$$\frac{\partial u}{\partial t} = -\frac{1}{\rho}\frac{\partial P}{\partial x} = -\frac{1}{\rho}\frac{\partial P}{\partial s}\frac{\partial s}{\partial x} = -\frac{1}{\rho^0(1 + s)}\frac{\partial P}{\partial s}\frac{\partial s}{\partial x}. \tag{22}$$

An approximate relationship between P and ρ may be obtained from the general expression for the compressibility,

$$\beta = \partial \ln \rho/\partial P \tag{23}$$

by integrating on the assumption that β is a constant (denoted by β_S) independent of the pressure. Then

$$P = P^0 + (1/\beta_S) \ln (\rho/\rho^0)$$
$$= P^0 + (1/\beta_S) \ln (1 + s), \tag{24}$$

where P^0 is the constant external pressure, and β_S is the compressibility of the system at this pressure. The subscript s is to remind us of the isentropic condition. Then

$$\frac{\partial P}{\partial s} = \frac{1}{(1 + s)\beta_S}, \tag{25}$$

and equation (22) becomes

$$\frac{\partial u}{\partial t} = -\frac{1}{\rho^0\beta_S(1 + s)^2}\frac{\partial s}{\partial x}. \tag{26}$$

Experiments show that s is usually very much smaller than unity, so that equations (21), (24) and (26) are effectively

$$\partial s/\partial t = -\partial u/\partial x, \tag{27}$$

$$P = P^0 + s/\beta_S, \tag{28}$$

and

$$\frac{\partial u}{\partial t} = -\frac{1}{\rho^0\beta_S}\frac{\partial s}{\partial x}. \tag{29}$$

Equation (27) may be differentiated as follows:

$$\frac{\partial^2 s}{\partial t^2} = -\frac{\partial}{\partial t}\left(\frac{\partial u}{\partial x}\right) = -\frac{\partial^2 u}{\partial t\,\partial x} = -\frac{\partial^2 u}{\partial x\,\partial t} = -\frac{\partial}{\partial x}\left(\frac{\partial u}{\partial t}\right).$$

On using equation (29), we have

$$\frac{\partial^2 s}{\partial t^2} = \frac{1}{\rho^0 \beta_S} \frac{\partial^2 s}{\partial x^2}. \tag{30}$$

Since

$$\frac{\partial s}{\partial t} = \frac{\partial s}{\partial P} \frac{\partial P}{\partial t} \quad \text{and} \quad \frac{\partial S}{\partial x} = \frac{\partial s}{\partial P} \frac{\partial P}{\partial x}$$

we may also write

$$\frac{\partial^2 P}{\partial t^2} = \frac{1}{\rho^0 \beta_S} \frac{\partial^2 P}{\partial x^2}. \tag{31}$$

These differential equations are in a well known form, giving, for example, the solutions

$$s = s^0 \sin \omega(t - x/c) \tag{32}$$

where the velocity of propagation, c, is $\lambda v = \lambda(\omega/2\pi)$, and

$$c^2 = 1/\rho^0 \beta_S. \tag{33}$$

By means of equation (23), we have

$$c^2 = (\partial P/\partial \rho)_S, \tag{34}$$

and from the thermodynamic relationship between the isothermal and adiabatic compressibilities,

$$\frac{\beta_T}{\beta_S} = \frac{C_P}{C_V} = \gamma,$$

we have

$$c^2 = \gamma/\rho^0 \beta_T, \tag{35}$$

which is Laplace's amended form of Newton's equation for the velocity of sound in an elastic medium. In an ideal gas, $\beta_T = 1/P = 1/nkT$. Hence

$$c = (\gamma kT/m)^{1/2}. \tag{36}$$

By combining equation (32) with equation (19), (28) and (29) respectively, we see that

$$\rho - \rho^0 = \rho^0 s = \rho^0 s^0 \sin \omega\left(t - \frac{x}{c}\right), \tag{37}$$

$$P - P^0 = \frac{s}{\beta_S} = \frac{s^0}{\beta_S} \sin \omega\left(t - \frac{x}{c}\right), \tag{38}$$

and

$$u = cs = cs^0 \sin \omega\left(t - \frac{x}{c}\right). \tag{39}$$

Thus the change in density, the change in pressure, and the particle velocity are all in phase. Their average values are zero, and their maximum values are the coefficients of $\sin \omega(t - x/c)$. Moreover, s is seen to be the ratio of the velocity with which the molecules are displaced to the velocity with which the sound is propagated. The latter is always positive, but the former assumes positive, zero and negative values. Since $u = \partial\xi/\partial t$, the displacement

$$\xi = -\frac{s^0 c}{\omega} \cos \omega\left(t - \frac{x}{c}\right) \tag{40}$$

is seen to be in phase with the force

$$X = m\frac{\partial u}{\partial t} = mcs^0\omega \cos \omega\left(t - \frac{x}{c}\right) = -m\omega^2\xi. \tag{41}$$

The kinetic energy per molecule is

$$\varepsilon_{\text{kin}} = (1/2)mu^2 = (1/2)m(cs)^2 = (1/2)m(cs^0)^2 \sin^2 \omega\left(t - \frac{x}{c}\right) \tag{42}$$

and the potential energy per molecule is

$$\varepsilon_{\text{pot}} = -\int X\partial\xi = m\omega^2 \int \xi\partial\xi = (1/2)m\omega^2\xi^2$$

$$= (1/2)m(cs^0)^2 \cos^2 \omega\left(t - \frac{x}{c}\right). \tag{43}$$

The total energy of the molecule due to the transmission of sound is thus a constant, independent of x and t, as is the energy density

$$E = n(\varepsilon_{\text{kin}} + \varepsilon_{\text{pot}}) = (1/2)\rho(cs^0)^2. \tag{44}$$

The solution of equation (16) can, of course, be written in the alternative form:

$$s = A \exp\left[i\omega\left(t - \frac{x}{c}\right)\right],$$

where A is a constant, and i is $(-1)^{1/2}$.

The Classical Theory of Sound Absorption

When sound is propagated through a viscous medium, there is exerted on the molecules a viscous force, in addition to the pressure force given by equation (15). Such a force is, as Newton showed, proportional to the viscosity η, and to the velocity gradient $\partial u/\partial z$ measured in a direction perpendicular to that of the flow. The viscous force exerted over unit area on the left of the section shown in Fig. 7 is $-\eta(\partial u/\partial z)$, and on the right of the section is $-\eta[(\partial u/\partial z) + (\partial^2 u/\partial z^2)\,dz]$. The net viscous force on the matter contained in the volume dz is thus $+\eta(\partial^2 u/\partial z^2)\,dz$, and on the matter in unit volume is $\eta(\partial^2 u/\partial z^2)$. Hydrodynamical theory allows us to equate this to $\frac{4}{3}\eta(\partial^2 u/\partial x^2)$. We thus obtain Stokes' equation

$$\rho_0 \frac{\partial u}{\partial t} = -\frac{\partial P}{\partial x} + \frac{4}{3}\eta \frac{\partial^2 u}{\partial x^2}. \tag{45}$$

Its solution can be formed by assuming that u, $P - P^0$ and s are all proportional to the expression

$$\exp\left[-\alpha x + i\omega\left(t - \frac{x}{c}\right)\right], \tag{46}$$

where α, the absorption coefficient, is the reciprocal of a distance at which the amplitude has fallen to a fraction e^{-1} of its value at the origin. A dimensionless and more convenient absorption coefficient, μ, is defined as follows:

$$\mu = \alpha\lambda = \alpha c/v = \alpha 2\pi c/\omega. \tag{47}$$

The exponential expression can now be written in the form

$$\left[i\omega t - \frac{i\omega x}{c}\left(1 - \frac{i\mu}{2\pi}\right)\right]. \tag{48}$$

When u and s are separately made proportional to this expression, and equation (27) is again used, we find

$$s = -\frac{u}{c}\left(1 - \frac{i\mu}{2\pi}\right). \tag{49}$$

In contrast with equation (39), u and c are no longer in phase—a result which is typical of most absorptive processes. When $P - P^0$ is set proportional to the exponential expression (48), we find, after substituting the various results in equation (45), that

$$\rho_0 i\omega u = \frac{i\omega}{c}\left(1 - \frac{i\mu}{2\pi}\right)(P - P^0) - \frac{4}{3}\eta\left[\frac{\omega}{c}\left(1 - \frac{i\mu}{2\pi}\right)\right]^2 u. \tag{50}$$

Elimination of the term u from equations (49) and (50) yields the result

$$\frac{s}{P - P^0} = \frac{1}{\rho^0 c^2}\left(1 - \frac{i\mu}{2\pi}\right)^2\left[1 - \frac{4}{3}\frac{i\omega\eta}{\rho^0 c^2}\left(1 - \frac{i\mu}{2\pi}\right)^2\right]^{-1}. \tag{51}$$

According to equations (28) and (33), however,

$$\frac{s}{P - P^0} = \frac{1}{\beta_S} = \frac{1}{\rho^0 c_0^2} \tag{52}$$

where c_0 is the velocity of propagation in the absence of the viscous force. Hence

$$\left(\frac{c}{c_0}\right)^2 = \left(1 - \frac{i\mu}{2\pi}\right)^2 \left[1 - \frac{4}{3} \frac{i\omega\eta}{\rho^0 c^2} \left(1 - \frac{i\mu}{2\pi}\right)^2\right]^{-1}. \tag{53}$$

During the absorption of sound at low frequencies, its velocity changes only slightly, so that c/c_0 may be taken as unity. Hence, approximately,

$$\left(1 - \frac{i\mu}{2\pi}\right)^2 = \left(1 + \frac{4i\omega\eta}{3\rho_0 c^2}\right)^{-1}. \tag{54}$$

On expanding both sides of this equation as far as the second term, we have

$$\mu = \frac{4\pi\omega\eta}{3\rho^0 c^2} \tag{55}$$

which is G. G. Stokes' (1845) result. By means of equation (33), we can also write

$$\mu = (4/3)\pi\omega\eta\beta_S. \tag{56}$$

The absorption coefficient of sound at low frequencies is thus directly proportional to its frequency and to the viscosity and adiabatic compressibility of the medium. From equation (47) we see that the amplitude absorption coefficient is

$$\alpha = \frac{2\omega^2\eta}{3\rho^0 c^3}. \tag{57}$$

According to Maxwell's fundamental theory, the relaxation time, τ, for compressional changes in a fluid is the product of its viscosity and its compressibility (equation 13). If an approximate allowance is made for the persistence of velocity after collision by introducing the factor 4/3, we have

$$\tau = (4/3)\eta\beta_S, \tag{58}$$

and consequently

$$\mu = \pi\omega\tau. \tag{59}$$

By means of equations (52) and (58), equations 53 may now be written as follows:

$$\left(\frac{c_0}{c}\right)^2 \left(1 - \frac{i\mu}{2\pi}\right)^2 = 1 - i\omega\tau\left(\frac{c_0}{c}\right)^2\left(1 - \frac{i\mu}{2\pi}\right)^2, \tag{60}$$

from which we see that

$$\left(\frac{c_0}{c}\right)^2 \left(1 - \frac{i\mu}{2\pi}\right)^2 = \frac{1}{1 + i\omega\tau}. \tag{61}$$

By squaring out the left hand side and multiplying numerator and denominator on the right hand side by $1 - i\omega\tau$, we have

$$\left(\frac{c_0}{c}\right)^2 \left[1 - \frac{i\mu}{\pi} - \left(\frac{\mu}{2\pi}\right)^2\right] = \frac{1 - i\omega\tau}{1 + \omega^2\tau^2} = \frac{1}{1 + \omega^2\tau^2} - \frac{i\omega\tau}{1 + \omega^2\tau^2},$$

which contains real and imaginary quantities on both sides. We may therefore equate the coefficients of i on both sides, and obtain for the absorption coefficient the expression

$$\mu = \pi \left(\frac{c}{c_0}\right)^2 \frac{\omega\tau}{1 + \omega^2\tau^2}. \tag{62}$$

On equating the real terms on both sides of the equation we find

$$1 - \left(\frac{\mu}{2\pi}\right)^2 = \left(\frac{c}{c_0}\right)^2 \frac{1}{1 + \omega^2\tau^2}. \tag{63}$$

By eliminating $(c/c_0)^2$, we obtain the following quadratic equation

$$\left(\frac{\mu}{2\pi}\right)^2 + \frac{2}{\omega\tau}\left(\frac{\mu}{2\pi}\right) - 1 = 0,$$

of which only the positive root has any meaning. Hence

$$\frac{\mu}{2\pi} = \frac{1}{\omega\tau}[\sqrt{1 + \omega^2\tau^2} - 1]. \tag{64}$$

Expansion of the square-root term, which is valid for low frequencies, leads again to equation (59). On eliminating μ from equations (62) and (64) we have the dispersion equation

$$\left(\frac{c}{c_0}\right)^2 = \frac{2(1 + \omega^2\tau^2)}{(1 + \omega^2\tau^2)^{1/2} + 1}. \tag{65}$$

Equations (62) and (65) are shown in Fig. 8 where the absorption coefficient and the sound velocity are plotted as functions of $\omega\tau$. Since τ is a constant for any given system, these plots show how absorption and velocity vary with the frequency of the sound wave. As ω increases, μ at first also increases, and passes through a maximum value of $\pi/2$ when $\omega\tau = 1$. Equation (62) can therefore be written as follows:

$$\frac{\mu}{\mu_{max}} = 2 \frac{\omega\tau}{1 + \omega^2\tau^2}. \tag{66}$$

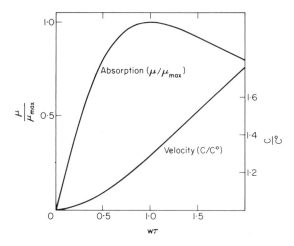

Fig. 8. The theoretical absorption coefficient and the velocity of sound as a function of the frequency.

On the other hand, c/c_0 increases continuously as ω is increased. μ is much more sensitive than c to changes in ω. Thus, for example, when $\omega\tau$ is raised from 0.1 to 0.5, μ is increased by a factor of 3.5, while c/c_0 is increased by only 9 per cent. We therefore see that, as a first approximation, the term c/c_0 in the absorption equation (62) may be taken as unity, so that

$$\mu = \pi \frac{\omega\tau}{1 + \omega^2\tau^2}, \tag{67}$$

with a maximum value of $\pi/2$ when $\omega\tau = 1$.

From measurements of the coefficients of absorption at various sound frequencies, the relaxation time, τ, may be found in various ways. For instance, if ω_1 and ω_2 are the values of ω found when $\mu/\mu_{max} = \frac{1}{2}$, then $\tau = (2 - \sqrt{3})/\omega_1 = (2 + \sqrt{3})/\omega_2$. Another method depends on the direct proportionality between the coefficient of absorption and the frequency which prevails at relatively low frequencies, when equation (62) reduces to equation (59).

Experimental results on the absorption of sound are often given, not as μ, but as α/v^2, which, according to equations (47) and (67), is

$$\frac{\alpha}{v^2} = \frac{2\pi^2\tau}{c} \frac{1}{1 + \omega^2\tau^2}. \tag{68}$$

Thus, unless $\omega\tau$ becomes significant in comparison with 1,

$$\frac{\alpha}{v^2} = \frac{2\pi^2\tau}{c}, \tag{69}$$

indicating that the amplitude absorption coefficient should vary directly as the square of the frequency. By means of equations (58) and (33), we have the equivalent expressions

$$\frac{\alpha}{v^2} = \frac{8\pi^2\eta\beta_S}{3c} = (8/3)\pi^2(\beta_S)^{3/2}\rho_0^{1/2}, \tag{70}$$

as derived by G. G. Stokes (1845). The absorption coefficient thus estimated is frequently referred to as the classical absorption coefficient.

Numerical Results for Water and Carbon Disulphide

Let us apply equations (33) and (58) to water at 40°C and to carbon tetra-chloride at -29°C, using the densities, compressibilities and viscosities listed in Table 1. We note that the calculated values of the sound velocities

TABLE I

ACOUSTICAL CONSTANTS FOR WATER AND CARBON DISULPHIDE

Liquid	H_2O	CS_2
t°C	40	-29
ρ_0 (gramme/cc)	0·9922	1·3373
β_S (cm²/dyne)	$4·278 \times 10^{-11}$	$3·92 \times 10^{-11}$
η (gramme/cm-sec.)	$6·551 \times 10^{-3}$	$5·68 \times 10^{-3}$
c (cm/sec)	$1·535 \times 10^5$	$1·38 \times 10^5$
τ (sec), calculated (equation 58)	$3·736 \times 10^{-13}$	$2·98 \times 10^{-13}$
τ (sec), observed (equation 67)	$1·135 \times 10^{-12}$	$2·60 \times 10^{-9}$
τ (observed)/τ (calculated)	3·04	$8·75 \times 10^3$

and the relaxation times are fairly similar for the two liquids, but that the ratio τ (observed)/τ (calculated) is about 3 for water and about 4,000 for carbon disulphide. It has not proved difficult, by the inclusion of a subsidiary viscous energy in equation (45) to account for the factor of 3 for water (L. Tisza, 1942; F. E. Fox and G. D. Rock, 1946). The large disparity between the observed and classically computed values of the relaxation time for carbon disulphide is attributed, as in the case of gaseous chlorine, to the slowness of the energy exchange between the external and internal modes of motion of the molecules (H. O. Kneser, 1938).

Vibrational Relaxation in Liquids

The heat capacity of any system may be resolved into external and internal components, so that we may write

$$C = C^{\text{ext}} + C^{\text{int}}. \tag{71}$$

The external component of the heat capacity is the temperature coefficient of the energy associated with the motion of the centre of gravity of the molecule and with its bulk rotation about the centre of gravity. The internal component is the temperature coefficient of the energy due to internal vibrations and rotations. When heat capacities are measured by static methods, ample time is allowed for the unimpeded interchange of energies, external and internal. It is known, however, that the interconversion of, for example, translational and vibrational energies is a process which requires time for its establishment. A diatomic molecule possessing one quantum of vibrational energy may have to collide several times with another molecule or atom before losing it or gaining another quantum. Consequently, if a rapid method, rather than a static method, is used to measure heat capacities, it may happen that insufficient time is allowed for the internal component to contribute fully, or even at all, to the observed values. The instantaneous value of the internal energy under such circumstances differs from the value it would have if there were uninhibited exchange of the energies associated with the internal and external motions. According to the general law of relaxation, the rate at which this difference decreases is proportional to the difference itself. Moreover, since energy is directly proportional to temperature, provided the heat capacity is constant, we may state the relaxation law in terms of temperatures, as follows:

$$-dT^{\text{int}}/dt = (1/\tau)(T^{\text{int}} - T^{\text{ext}}) \tag{72}$$

where τ is the relaxation time for the process of energy exchange, T^{int} is directly proportional to the instantaneous value of the internal energy, and T^{ext} is directly proportional to value of the internal energy when it exchanges freely with its surroundings. In the absence of slow energy exchanges, these two temperatures are equal to the constant temperature, T, of the system. Because T does not vary with time, we may replace the term on the left of equation (72) by $-d(T^{\text{int}} - T)$. On adding and subtracting T from the term within the brackets, and rearranging, we have

$$(T^{\text{int}} - T) + \tau \frac{d(T^{\text{int}} - T)}{dt} = T^{\text{ext}} - T. \tag{73}$$

It is now assumed that

$$T^{\text{int}} - T = A \exp(i\omega t),$$

where A is a constant, and $\omega/2\pi$ is the frequency of the disturbance responsible for the temperature difference. Then

$$\frac{d(T^{\text{int}} - T)}{dt} = i\omega A^{i\omega t} = i\omega(T^{\text{int}} - T).$$

Consequently,

$$\frac{T^{\text{int}} - T}{T^{\text{ext}} - T} = \frac{1}{1 + i\omega\tau}, \tag{74}$$

and

$$\frac{dT^{\text{int}}}{dT^{\text{ext}}} = \frac{1}{1 + i\omega\tau}. \tag{75}$$

The total energy intake for a small rise in temperature is

$$dE = C^{\text{ext}} dT^{\text{ext}} + C^{\text{int}} dT^{\text{int}}$$

$$= \left(C^{\text{ext}} + C^{\text{int}} \frac{dT^{\text{int}}}{dT^{\text{ext}}}\right) dT^{\text{ext}}$$

$$= \left(C^{\text{ext}} + \frac{C^{\text{int}}}{1 + i\omega\tau}\right) dT^{\text{ext}}.$$

The physical assumption is now made that the behaviour of the system is effectively governed by T^{ext}, so that the effective heat capacity is

$$C_{\text{eff}} = C^{\text{ext}} + \frac{C^{\text{int}}}{1 + i\omega\tau} = C^{\text{ext}} - C^{\text{int}}\left(\frac{i\omega\tau}{1 + i\omega\tau}\right). \tag{76}$$

As the internal heat capacity is not greatly affected by pressure, we may write

$$(C_P)_{\text{eff}} = C_P^{\text{ext}} + \frac{C^{\text{int}}}{1 + i\omega\tau} = C_P - C^{\text{int}}\left(\frac{i\omega\tau}{1 + i\omega\tau}\right), \tag{77}$$

and

$$(C_V)_{\text{eff}} = C_V^{\text{ext}} + \frac{C^{\text{int}}}{1 + i\omega\tau} = C_V - C^{\text{int}}\left(\frac{i\omega\tau}{1 + i\omega\tau}\right). \tag{78}$$

Hence

$$(C_P - C_V)_{\text{eff}} = C_P^{\text{ext}} - C_V^{\text{ext}} = C_P - C_V. \tag{79}$$

The ratio of the effective heat capacities is seen to be

$$\gamma_{\text{eff}} = \frac{C_P^{\text{ext}} + i\omega\tau C_P^{\text{ext}} + C^{\text{int}}}{C_V^{\text{ext}} + i\omega\tau C_V^{\text{ext}} + C^{\text{int}}} = \frac{C_P + C_P^{\text{ext}}}{C_V + C_V^{\text{ext}}}. \tag{80}$$

The static value of the ratio is

$$\gamma_0 = C_P / C_V.$$

Hence

$$\frac{\gamma_0}{\gamma_{\text{eff}}} = \left(1 + \frac{C_V^{\text{ext}}}{C_V} \cdot i\omega\tau\right)\left(1 + \frac{C_P^{\text{ext}}}{C_P} \cdot i\omega\tau\right)^{-1}. \tag{81}$$

We shall next apply equation 51 to a system absorbing energy, but free from viscous forces. Remembering (equation 28) that $ds/dP = \beta_S$, we have

$$\beta_S = \frac{1}{\rho_0 c^2}\left(1 - \frac{i\mu}{2\pi}\right)^2.$$

The corresponding expression when there is no absorption is

$$\beta_S^0 = \frac{1}{\rho_0 c_0^2}.$$

Hence

$$\left(\frac{c_0}{c}\right)^2\left(1 - \frac{i\mu}{2\pi}\right)^2 = \frac{\beta_S}{\beta_S^0}. \tag{82}$$

Now $\beta_S = \beta_T/\gamma_{\text{eff}}$, and $\beta_S^0 = \beta_T/\gamma_0$, where β_T is the isothermal compressibility. The ratio of the compressibilities in equation (82) can therefore be replaced by $\gamma_0/\gamma_{\text{eff}}$. On combining equations (81) and (82), we have

$$\left(\frac{c_0}{c}\right)^2\left(1 - \frac{i\mu}{2\pi}\right)^2 = \left(1 + \frac{C_V^{\text{ext}}}{C_V} \cdot i\omega\tau\right)\left(1 + \frac{C_P^{\text{ext}}}{C_P} \cdot i\omega\tau\right)^{-1}. \tag{83}$$

After equating the imaginary terms on both sides of this equation, we find the following expression for the absorption coefficient:

$$\mu = \pi\left(\frac{c}{c^0}\right)^2 \frac{\left(\dfrac{C_P^{\text{ext}}}{C_P} - \dfrac{C_V^{\text{ext}}}{C_V}\right)\omega\tau}{1 + \left(\dfrac{C_P^{\text{ext}}}{C_P}\right)^2 \omega^2\tau^2}, \tag{84}$$

which is more complicated than equation (62), but has the same form. Let us define a derived relaxation time γ_1 as follows:

$$\tau_1 = \left(\frac{C_P^{\text{ext}}}{C_P}\right)\tau, \tag{85}$$

which enables us to write:

$$\mu = \pi\left(\frac{c}{c_0}\right)^2 \cdot \frac{(C_P - C_V)C^{\text{int}}}{C_V(C_P - C^{\text{int}})} \cdot \frac{\omega\tau_1}{1 + \omega^2\tau_1^2}. \tag{86}$$

Let the term involving the heat capacities be denoted by r:

$$r = \frac{(C_P - C_V)C^{\text{int}}}{C_V(C_P - C^{\text{int}})},$$ (87)

so that equation (86) may be written as follows:

$$\mu = \pi r \left(\frac{c}{c_0}\right)^2 \frac{\omega \tau_1}{1 + \omega^2 \tau_1^2},$$ (88)

or, if we ignore the difference between c and c_0,

$$\mu = \pi r \frac{\omega \tau_1}{1 + \omega^2 \tau_1^2},$$ (89)

which has the maximum value of

$$\mu_{\max} = \pi r/2.$$ (90)

After equating the real terms on both sides of equation (83), we find

$$\left(\frac{c_0}{c}\right)^2 \left[1 - \left(\frac{\mu}{2\pi}\right)^2\right] = 1 - r\frac{\omega^2 \tau_1^2}{1 + \omega^2 \tau_1^2}.$$ (91)

Now $(\mu/2\pi)^2$ seldom exceeds 0·001, and can be ignored in comparison with unity, so that effectively

$$\left(\frac{c_0}{c}\right)^2 = 1 - r\frac{\omega^2 \tau_1^2}{1 + \omega^2 \tau_1^2}.$$ (92)

At the frequency corresponding to maximum absorption, therefore,

$$\left(\frac{c_0}{c_{\max}}\right)^2 = 1 - (r/2).$$ (93)

On combining equations (88) and (91), we see that

$$\mu = \pi r \frac{\omega \tau_1}{1 + \omega^2 \tau_1^2 (1 - r)}$$ (94)

giving a maximum absorption

$$\mu_{\max} = \pi \left(\frac{r}{2 - r}\right).$$ (95)

The factor r is sometimes expressed in terms of the compressibilities. We have for the static compressibility

$$\beta_S^0 = \beta_T(C_V/C_P),$$

and for the effective compressibility

$$\beta_S = \beta_T(C_V^{\text{eff}}/C_P^{\text{eff}}).$$

Hence

$$\frac{\beta_S^0 - \beta_S}{\beta_S^0} = 1 - \left(\frac{C_V^{\text{eff}}}{C_V}\right) \Big/ \left(\frac{C_P^{\text{eff}}}{C_P}\right).$$

On using equations (77) and (78), we obtain the expression

$$\frac{\beta_S^0 - \beta_S}{\beta_S^0} = \frac{i\omega\tau(C_P - C_V)C^{\text{int}}}{C_V[C_P + i\omega\tau(C_P - C^{\text{int}})]}.$$

We now multiply across and equate the coefficients of i on both sides of the equation, obtaining

$$\frac{\beta_S^0 - \beta_S}{\beta_S^0} = \frac{(C_P - C_V)C^{\text{int}}}{C_V(C_P - C^{\text{int}})} \qquad (96)$$

Comparison with equation (87) shows that

$$r = \frac{\beta_S^0 - \beta_S}{\beta_S^0}. \qquad (97)$$

Then

$$\mu = \pi\left(\frac{c}{c^0}\right)\left(\frac{\beta_S^0 - \beta_S}{\beta_S^0}\right)\frac{\omega\tau_1}{1 + \omega^2\tau_1^2}. \qquad (98)$$

J. H. Andrea, E. L. Heasell and J. Lamb (1956) find the absorption of carbon disulphide in the ultrasonic frequency range 2×10^6 to 200×10^6 sec^{-1} to obey equation (94), with a maximum value of μ at a frequency of $78 \times 10^6 \text{ sec}^{-1}$. This corresponds to $\omega = 4.90 \times 10^8$ radians per second, and, since $\omega\tau_1 = 1$ at this point, $\tau_1 = 2.04 \times 10^{-9}$ sec. From calorimetric and spectroscopic sources respectively, we have, in calories per mole-degree, $C_P = 18.11$, $C_V = 11.74$, and $C^{\text{int}} = 3.93$. Hence, from equations (85) and (87) respectively, $\tau = 2.60 \times 10^{-9}$ sec, and $r = 0.150$. The theoretical value of μ_{max}, according to equation (95), becomes 0.255, in satisfactory agreement with the observed value of 0.262. Finally, c_0/c_{max}, from equation (93), is 1.04.

The Influence of Pressure on the Relaxation Time

The absorption coefficient, α, for carbon disulphide at 25° varies as follows with respect to the frequency, f of the sound wave absorbed (J. Lamb, 1956;

Herzfeld and Litovitz, 1959.):

$$10^{17} \frac{\alpha}{f^2} = \frac{c_0}{c} \cdot \frac{5867}{1 + (2\pi\tau f)^2}$$

c is the velocity of propagation of the sound wave during absorption, and c_0 its velocity in the static system. For the absorption of sound of very low frequencies, only the leading term, i.e. the numerator, need be considered. Equations (68) and (70) provide the following expressions:

$$\frac{\alpha}{v^2} = \frac{2\pi^2\tau}{c} = \frac{8\pi^2\eta\beta_s}{3c}.$$

The effect of increasing the pressure on these variables is shown in the following table:

P (kgramme/cm^2)	$(\alpha/f^2) \times 10^{17}$ (cm^{-1}/sec^2)	$\tau \times 10^9$ (sec)	$c_0 \times 10^{-5}$ (cm-sec^{-1})
1	5,880	3·44	1·314
978	2,940	2·30	1·534

Increasing the pressure a thousand-fold thus causes an increase of 53 per cent in the vibration frequency. Only a small fraction of this increase is due to the effect of pressure on c. The same increase in pressure causes a decrease of about 25 per cent in the compressibility, and an increase of 59 per cent in the viscosity. Thus the greatest single factor is that of viscosity to which the collision numbers in this system, as in those discussed in Chapter 5 (equations 53–58), seems to be directly proportional.

Relaxations due to the Displacement of Chemical Equilibrium

Hitherto we have dealt with only two kinds of relaxations, namely, the classical relaxation caused by the time lag between the application of stress and the attainment of steady motion in liquids and gases, and the more pronounced relaxation due to the reluctance of molecules in either state to convert translational into vibrational energy. We have next to deal with relaxations caused by the displacement of chemical equilibrium occasioned by the passage of ultrasonic waves through the system.

Reversible Unimolecular Reactions; $A \rightleftarrows B$

Because most ultrasonic investigations on equilibria of this type have been carried out in pure liquids rather than in dilute solutions, it will be convenient to express the rate constants and the equilibrium constants in terms of molar fractions x_A amd x_B rather than in terms of concentrations:

$$A \underset{k_B}{\overset{k_A}{\rightleftarrows}} B$$

$$x_A \quad x_B.$$

Let $x_A = x_A^0 - \Delta x$ and $x_B = x_B^0 + \Delta x$, where the superscripts denote equilibrium values, and Δx represents the extent of the displacement of the system from these values. The rate of reaction is

$$dx_B/dt = k_A(x_A^0 - \Delta x) - k_B(x_B^0 + \Delta x)$$

$$= k_A x_A^0 - k_B x_B^0 - (k_A + k_B)\Delta x.$$

To allow for the effect of the displacement on the rate constants, write $k_A = k_A^0 + \Delta k_A$, and $k_B = k_B^0 + \Delta k_B$. Then, ignoring the products of two differentials, we have

$$dx_B/dt = k_A^0 x_A^0 - k_B^0 x_B^0 + x_A^0 \Delta k_A - x_B^0 \Delta k_B - (k_A^0 + k_B^0)\Delta x$$

$$= k_A^0 x_A^0 - k_B^0 x_B^0 + x_A^0 k_A^0 \Delta \ln k_A - x_B^0 k_B^0 \Delta \ln k_B - (k_A^0 + k_B^0)\Delta x.$$

Now

$$d\Delta x_B/dt = dx_B/dt - dx_B^0/dt$$

$$= dx_B/dt - (k_A^0 x_A^0 - k_B^0 x_B^0)$$

$$= -(k_A^0 + k_B^0)\Delta x + x_A^0 k_A^0 \Delta \ln k_A - x_B^0 k_B^0 \Delta \ln k_B.$$

Since $x_A^0 k_A^0 = x_B^0 k_B^0$ and $K = k_A/k_B$,

$$d\Delta x_B/dt = -(k_A^0 + k_B^0)\Delta x + x_A^0 k_A^0 \Delta \ln K$$

$$= (k_A^0 + k_B^0)\left[-\Delta x + \left(\frac{x_A^0 k_A^0}{k_A^0 + k_B^0} \right) \Delta \ln K \right].$$

The coefficient of $\Delta \ln K$ is seen to be the product of the mole fractions $x_A^0 x_B^0$. Hence

$$-d\Delta x_B/dt = (k_A^0 + k_B^0)(\Delta x_B - x_A^0 x_B^0 \Delta \ln K)$$

$$= (1/\tau)(\Delta x_B - x_A^0 x_B^0 \Delta \ln K). \tag{99}$$

where τ has been defined in equation (6.21) as $1/(k_A^0 + k_B^0)$.

If the cause of the displacement from the equilibrium is the presence of a sound wave of frequency $v(= \omega/2\pi)$, it may be assumed that $\Delta x_B = Ae^{i\omega t}$, where A is a constant. Then

$$\Delta x_B = \frac{x_A^0 x_B^0 \Delta \ln K}{1 + i\omega \tau},$$

of which the real part is

$$\Delta x_B = x_A^0 x_B^0 \Delta \ln K \left(\frac{1}{1 + \omega^2 \tau^2} \right). \tag{100}$$

Now K is a function of the independent variables, T and P, so that

$$\Delta \ln K = (\partial \ln K / \partial T)_P \, \Delta T + (\partial \ln K / \partial P)_T \, \Delta P,$$

or, since

$$\Delta H = RT^2 (\partial \ln K / \partial T)_P,$$

and

$$\Delta V = -RT(\partial \ln K / \partial P)_T,$$

$$\Delta x_B = x_A^0 x_B^0 \left[\frac{\Delta H}{RT^2}(T - T_0) - \frac{\Delta V}{RT}(P - P_0) \right] \left(\frac{1}{1 + \omega^2 \tau^2} \right). \tag{101}$$

Now ΔH is the heat absorbed when one gram-mole of A is converted into B; therefore, the heat absorbed during the displacement of the equilibrium to the extent Δx_B is $q = \Delta x_B \Delta H$. This quantity divided by the increase in temperature corresponds to an internal heat capacity:

$$\frac{C^{\text{int}}}{R} = \frac{q}{R(T - T_0)} = \frac{\Delta x_B \Delta H}{R(T - T_0)} = x_A^0 x_B^0 \left[\left(\frac{\Delta H}{RT} \right)^2 - \frac{\Delta H \Delta V}{R^2 T} \frac{(P - P_0)}{(T - T_0)} \right]$$

$$\times \left(\frac{1}{1 + \omega^2 \tau^2} \right)$$

$$= x_A^0 x_B^0 \left(\frac{\Delta H}{RT} \right)^2 \left[1 - \frac{T \Delta V}{\Delta H} \frac{(P - P_0)}{(T - T_0)} \right] \left(\frac{1}{1 + \omega^2 \tau^2} \right).$$

In experiments at low frequencies, the last bracketed term is nearly unity. Moreover, isomeric changes are generally attended by small or even zero volume changes. In such cases, we have

$$\frac{C^{\text{int}}}{R} = x_A^0 x_B^0 \left(\frac{\Delta H}{RT} \right)^2. \tag{102}$$

If, in addition, the change in entropy attending the reaction can be neglected, we may identify ΔH with the standard free energy change, ΔG^0. Then

$$K = x_B^0 / x_A^0 = \exp(-\Delta H / RT), \tag{103}$$

and since $x_A^0 + x_B^0 = 1$, we see that

$$x_A^0 = [1 + \exp(-\Delta H / RT)]^{-1},$$

and

$$x_B^0 = \exp(-\Delta H/RT)[1 + \exp(-\Delta H/RT)]^{-1}.$$

Hence

$$x_A^0 x_B^0 = \exp(-\Delta H/RT)[1 + \exp(-\Delta H/RT)]^{-2}$$
$$= [\exp(\Delta H/2RT) + \exp(-\Delta H/2RT)]^{-2}$$
$$= \left[2\cosh\left(\frac{\Delta H}{2RT}\right)\right]^{-2}$$

and

$$\frac{C^{\text{int}}}{R} = \left(\frac{\Delta H}{2RT}\operatorname{sech}\frac{\Delta H}{2RT}\right)^2, \tag{104}$$

which is W. Schottky's equation for isentropic isomerism. It shows a maximum value of C^{int}/R of 0·439, when $\Delta H/2RT$ is 1·20 (Fig. 9).

In order to apply these considerations to the determination of the rates and energies of activation of the opposing unimolecular reactions, we begin with the ultrasonic determination of the factor r as given by the equation

$$\mu_{\text{max}} = \pi \frac{r}{2-r}. \tag{95}$$

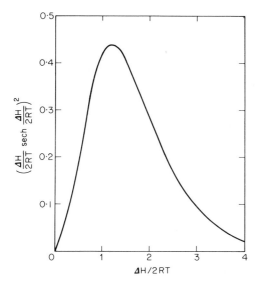

Fig. 9. Schottky's equation.

We next make use of equation (87), remembering that C^{int}/R cannot here exceed 0.439, and is therefore small. Equation (87) therefore becomes

$$C^{int} = r\left(\frac{C_V C_P}{C_P - C_V}\right) \tag{105}$$

From the value of C^{int} so obtained, ΔH may be estimated from equation (104). The relaxation time is given by equation (67). Let us apply these equations to the isomerism of acraldehyde at 25°C, (Fig. 10) (J. Lamb, 1960), for which ΔH is found to be 2·1 kcal/mole, and

$$\frac{1}{\tau} = \omega = 2\pi v = k_A^0 + k_B^0 = 1·59 \times 10^9 \text{ sec}^{-1}.$$

On combining this equation with the equilibrium equation 103, we have

$$k_A^0 = \frac{1}{\tau} \cdot \frac{\exp(-\Delta H/RT)}{1 + \exp(-\Delta H/RT)} = 4·43 \times 10^7 \text{ sec}^{-1}$$

and

$$k_B^0 = \frac{1}{\tau} \cdot \frac{1}{1 + \exp(-\Delta H/RT)} = 1·53 \times 10^9 \text{ sec}^{-1}.$$

The apparent energies of activation at 25°C, not here distinguished from the corresponding enthalpies, are seen to be given by the equations

$$\Delta H_A = -RT^2\left(\frac{d \ln \tau}{dT}\right)_P + \frac{\Delta H}{1 + \exp(-\Delta H/RT)}$$

$$= -RT^2\left(\frac{d \ln \tau}{dT}\right)_P + 1·73 \text{ kcal}$$

and

$$\Delta H_B = -RT^2\left(\frac{d \ln \tau}{dT}\right)_P - \frac{\Delta H . \exp(-\Delta H/RT)}{1 + \exp(-\Delta H/RT)}$$

$$= -RT^2\left(\frac{d \ln \tau}{dT}\right)_P - 0·37 \text{ kcal}.$$

It is seen that $\Delta H = \Delta H_A - \Delta H_B$. Lamb* has measured the temperature coefficients of the relaxation times for various reactions, and has thus been able to evaluate the energies of activation of the direct and reverse reactions

---

* Fuller accounts are given by Mason, *Physical Acoustics* (Academic Press, London (1964, 1965)).

in numerous isomerisms, such as that of acraldehyde, which may be re-presented as follows:

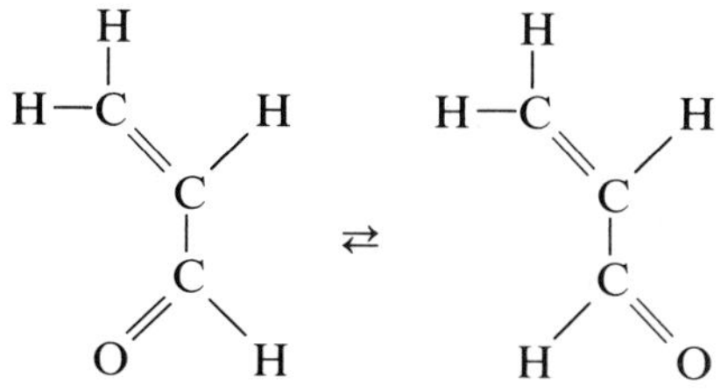

Fig. 10. *Isomerism* of acraldehyde.

When the hydrogen atoms in the aldehydic and methylenic groups are replaced by methyl radicals, giving mesityl oxide, no relaxation is observed, indicating that isomerism does not take place under the conditions of experiment. Steric hindrance in that instance raises energy barriers which are too high to be surmounted. The barriers opposing free rotation about the carbon–carbon axis in 1-bromopropane are considerably lower. Other instances are given in Table II.

TABLE II

<small>ENTHALPIES OF ACTIVATION FOR REVERSIBLE ISOMERISMS FROM ULTRASONIC RELAXATION MEASUREMENTS AT 25° (AFTER J. LAMB, 1960)</small>

Liquid	kcal/mole ΔH^0	ΔH_B	ΔH_A
1,1,2-trichloroethane	2·2	5·8	8·0
1,1,2-tribromoethane	1·6	6·4	8·0
1-bromopropane	1·3	3·6	4·9
1,2-dichloropropane	1·1	4·7	5·8
1,2-dibromopropane	0·9	4·9	5·8
sym-tetrabromoethane	0·9	4·3	5·2
2-methylbutane	0·9	4·7	5·6
2,3-dimethylbutane	1·0	2·8	3·8
2-methylpentane	0·9	3·9	4·8
3-methylpentane	0·9	4·1	5·0
acrolein	2·1	5·0	7·1
crotonaldehyde	1·9	5·5	7·4
cinnamaldehyde	1·5	5·6	7·1
methacrolein	3·1	5·3	8·4
furacrolein	1·2	5·1	6·3

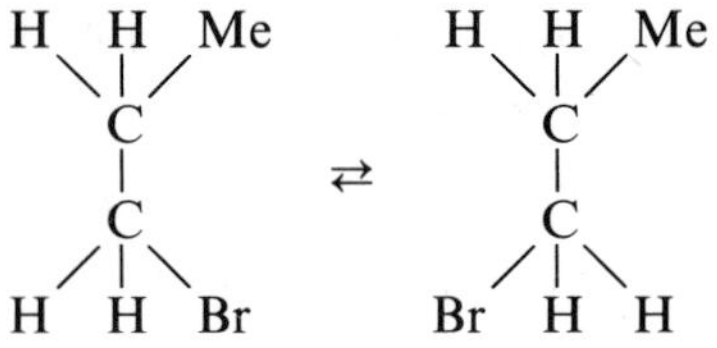

FIG. 11. Isomerism of 1-bromopropane.

Reversible Dissociations or Ionizations: $AB \rightleftarrows A + B$

The ultrasonic method for the study of the kinetics of reversible dissociations in the gaseous phase was applied by W. T. Richards and J. A. Reid (1933) to the well-known system $N_2O_4 \rightleftarrows 2NO_2$. We shall here consider the application of the same method to the ionization of solutes in dilute aqueous solution (M. Eigen, 1954; M. Eigen, K. Kustin and H. Maas, 1961; J. Struer, E. Yeager, T. Sachs and F. Hovorka, 1963; M. Eigen and L. de Maeyer, 1955).

We have seen that the bimolecular constant for the recombination of the ions, and the equilibrium constant, $K = k_1/k_2$, are related as follows to the relaxation time:

$$1/\tau = k_2(K + n_A^0 + n_B^0), \tag{6.71}$$

where the n^0 terms denote equilibrium concentrations. Equation (98), when c/c^0 is taken as unity, gives for the absorption coefficient

$$\mu = \frac{\beta_S^0 - \beta_S}{\beta_S^0} \pi \frac{\omega\tau}{1 + \omega^2\tau^2}. \tag{98a}$$

To evaluate the compressibility ratio, we note that the volume of the system is a function of pressure, temperature and the degree of ionization:

$$V = f(P, T, \alpha),$$

so that

$$dV = \left(\frac{\partial V}{\partial P}\right)_{T,\alpha} dP + \left(\frac{\partial V}{\partial T}\right)_{P,\alpha} dT + \left(\frac{\partial V}{\partial \alpha}\right)_{T,P} d\alpha.$$

At constant temperature, we have

$$\left(\frac{dV}{dP}\right)_T = \left(\frac{\partial V}{\partial P}\right)_{T,\alpha} + \left(\frac{\partial V}{\partial \alpha}\right)_{P,T}\left(\frac{\partial \alpha}{\partial P}\right)_T.$$

On dividing each term by V and multiplying by -1,

$$\beta_T^0 = \beta_T - \frac{1}{V}\left(\frac{\partial V}{\partial \alpha}\right)_{P,T}\left(\frac{\partial \alpha}{\partial P}\right)_T.$$

Since $\beta_T/\beta_S = C_P/C_V$, we may replace the ratio of the adiabatic compressibilities by that of the isothermal compressibilities, and obtain

$$\frac{\beta_T^0 - \beta_T}{\beta_T^0} = -\frac{1}{\beta_T^0 V}\left(\frac{\partial V}{\partial \alpha}\right)_{P,T}\left(\frac{\partial \alpha}{\partial P}\right)_T. \tag{106}$$

If c is the total concentration of electrolyte in moles per litre, we have

$$V = c(1 - \alpha)V_{AB} + c\alpha V_A + c\alpha V_B + V_{\text{solvent}} = 1{,}000 \text{ cc}, \tag{107}$$

where the volume terms with subscripts are partial molar quantities. Then

$$(\partial V/\partial \alpha)_{P,T} = c(V_A + V_B - V_{AB}) = c\Delta V,$$

where ΔV is the increase in volume attending the ionization of one mole of electrolyte. Consequently

$$\frac{\beta_T^0 - \beta_T}{\beta_T^0} = -\frac{c\Delta V}{1{,}000\beta_T^0}\left(\frac{\partial \alpha}{\partial P}\right)_T.$$

By means of equation (102), we may write

$$\left(\frac{\partial \alpha}{\partial P}\right)_T = \left(\frac{\partial \alpha}{\partial \ln K}\right)_T\left(\frac{\partial \ln K}{\partial P}\right)_T = -\frac{\Delta V}{RT}\left(\frac{\partial \alpha}{\partial \ln K}\right)_T.$$

Hence

$$\frac{\beta_T^0 - \beta_T}{\beta_T^0} = \frac{c(\Delta V)^2}{1{,}000\beta_T^0 RT}\left(\frac{\partial \alpha}{\partial \ln K}\right)_T.$$

Finally, since $K = c\alpha^2/(1 - \alpha)$,

$$\ln K = \ln c + 2 \ln \alpha - \ln (1 - \alpha),$$

$$\frac{\partial \ln K}{\partial \alpha} = \frac{2}{\alpha} + \frac{1}{1 - \alpha} = \frac{2 - \alpha}{\alpha(1 - \alpha)}.$$

The absorption coefficient is thus

$$\mu = \pi \cdot \frac{(\Delta V)^2}{1{,}000\beta_T^0 RT} \cdot \frac{c\alpha(1 - \alpha)}{(2 - \alpha)} \cdot \frac{\omega\tau}{1 + \omega^2\tau^2}. \tag{108}$$

This equation has been verified for various electrolytes in aqueous solution at 25°C, when β_T^0 is 4.62×10^{-5} atm. Thus, for a given electrolyte at a given concentration, μ passes through a maximum when $\omega = 1/\tau$. The value of μ at the maximum is increased as the total concentration, c, is increased, but, on account of α, the relationship is not a linear one.

Let us consider the ionization of ammonia in aqueous solution, for which $K = k_1/k_2 = 1.81 \times 10^{-5}$ gramme-ions/litre at this temperature, and for which

$$\Delta V = V(\text{NH}_4^+) + V(\text{OH}^-) - V(\text{NH}_3) - V(\text{H}_2\text{O})$$
$$= 18\cdot0 \quad\;\; + 5\cdot2 \quad\;\; - 17\cdot9 \quad\;\; - 18\cdot0 \quad\;\; = -23\cdot1 \text{ cc/mole.}$$

We see that

$$\mu = 1\cdot465 \frac{c\alpha(1 - \alpha)}{2 - \alpha} \frac{\omega\tau}{1 + \omega^2\tau^2},$$

and, since α is small and $\omega\tau$ is unity at the maximum,

$$\mu_{\text{max}} = 0\cdot366c\alpha.$$

Two sets of data taken from Eigen's work are summarised in Table III.

TABLE III

AQUEOUS AMMONIA AT 25°C

c (moles/ litre)	α	$(n_A^0 + n_B^0) = 2c\alpha N_0$ (ions/cc)	v_{max} (sec^{-1})	$(1/\tau = \omega_{\text{max}} = 2\pi v_{\text{max}}$ (sec^{-1})
0·5	$6\cdot15 \times 10^{-3}$	$3\cdot70 \times 10^{18}$	$2\cdot5 \times 10^7$	$1\cdot6 \times 10^8$
1·0	$4\cdot25 \times 10^{-3}$	$5\cdot12 \times 10^{18}$	$3\cdot5 \times 10^7$	$2\cdot2 \times 10^8$

By solving the equation

$$1/\tau = k_1 + (n_A^0 + n_B^0)k_2, \tag{6.71}$$

we find $k_2 = 4\cdot2 \times 10^{-11}$ cc/molecule-sec $= 2\cdot6 \times 10^{10}$ litre/mole-sec. The value of k_1 from these data alone is somewhat high, but is the difference of two large quantities. When the complete data are similarly analysed, k_2 from the gradient becomes 3×10^{10} litres per mole-second, and k_1 from the intercept becomes $4\cdot6 \times 10^{+5}$, in agreement with the equation $k_1 = Kk_2$. Further instances, not all obtained in this way, are summarized in Table IV. It is seen that, with few exceptions, the bimolecular rate constants for the recombination of univalent ions of opposite sign have the order of magnitude expected if the process were governed by diffusion. The high rate constant for the union of the hydrogen and hydroxyl ions has already been discussed (equation 5.29). The unimolecular rate constant for the ionization process, on the other hand, varies widely. M. Eigen and H. Eyring (1962) have found that $k_2 \times 10^{-10}$ for the combination of the hydrogen ion with the amino-benzoate ions has the values 5·8 (*ortho*), 4·6 (*meta*) and 3·7 (*para*).

TABLE IV

Rate Constants for Ionizations and Ionic Recombinations in Aqueous Solutions at 25°C

Reaction	k_2 (litre/mole-sec)	k_1 (sec^{-1})
$H_2O \rightleftarrows H^+ + OH^-$	$(1.3 \pm 0.2) \times 10^{11}$	2.3×10^{-5}
$H_2S \rightleftarrows H^+ + SH^-$	$(7.5 \pm 1) \times 10^{10}$	4.3×10^3
$HF \rightleftarrows H^+ + F^-$	1×10^{11}	6.7×10^7
$H_2SO_3 \rightleftarrows H^+ + HSO_3^-$	2.5×10^8	1.0×10^7
$CH_3COOH \rightleftarrows H^+ + CH_3COO^-$	4.5×10^{10}	8.1×10^5
$C_6H_5COOH \rightleftarrows H^+ + C_6H_5COO^-$	3.7×10^{10}	2.4×10^6
$NH_4OH \rightleftarrows NH_4^+ + OH^-$	3×10^{10}	4.6×10^5
$(CH_3)_4NOH \rightleftarrows (CH_3)_4N^+ + OH^-$	1×10^{10}	6×10^5
$H_2O + CN^- \rightleftarrows HCN + OH^-$	3.7×10^9	5.2×10^4

The absorption coefficients for di-divalent electrolytes in water, when plotted against the frequency of sound, show two distinct maxima indicating two relaxation processes. These have been attributed by Eyring and his collaborators to the successive removal of water molecules from rapidly formed ion-pairs:

$$\left[M^{++}O \overset{H}{\underset{H}{<}} \quad O \overset{H}{\underset{H}{<}} X^{--} \right]_{aq.} \overset{k_\alpha}{\underset{k_{-\alpha}}{\rightleftarrows}} H_2O + \left[M^{++}O \overset{H}{\underset{H}{<}} X^{--} \right]_{aq.} \overset{k_\beta}{\underset{k_{-\beta}}{\rightleftarrows}} H_2O$$

$$+ [M^{++}, X^{--}]_{aq}$$

The first-order rate constants k_α and $k_{-\alpha}$ have the order of magnitude 10^9 sec^{-1}, and are virtually independent of the nature of the cation M^{++}. k_β and $k_{-\beta}$, on the other hand, are highly sensitive to the nature of the cation, and have rate constants varying from 10^2 to 10^8 sec^{-1}. From their temperature variation, the energies of partial de-solvation have been estimated, and found to be in agreement with electrostatic estimates, as modified by ligand field effects. The processes are the conversion of completely solvated ion pairs into inner shell ion pairs and finally into intimate ion pairs.

Dimerisation Equilibria in Dilute Solution

In order to formulate the kinetic expression for dimerisation equilibria in dilute solution, represented by the equation

$$A_2 \overset{k_1}{\underset{k_2}{\rightleftarrows}} 2A,$$

we follow the treatment which led to equation (6.69), with the necessary modification

$$n_D = n_D^0 + \Delta n,$$

$$n_M = n_M^0 - 2\Delta n,$$

where D denotes the dimer and M the monomer. We then find

$$\frac{d\Delta n}{dt} = -[k_1 + 4k_2(n_M^0)^2]\Delta n,$$

$$= -k_1[1 + 4K(n_M^0)^2]\Delta n,$$

where

$$K = k_1/k_2.$$

The relaxation time τ is, as before, the reciprocal of the coefficient of Δn in these equations, and is obtained from the absorption coefficient

$$\mu = \pi r \frac{\omega\tau_1}{1 + \omega^2\tau_1^2}, \qquad (89)$$

where

$$r = \frac{(C_P - C_V)C^{\text{int}}}{C_V(C_P - C^{\text{int}})}, \qquad (87)$$

and

$$\tau_1 = \left|\frac{C_P - C^{\text{int}}}{C_P}\right|\tau. \qquad (85)$$

Compared with the complete heat capacity, C_P, the internal heat capacity, C^{int}, is known to be small in this instance, so that

$$\mu = \pi r \frac{\omega\tau}{1 + \omega^2\tau^2}$$

and

$$r = \frac{(C_P - C_V)C^{\text{int}}}{C_P C_V}. \qquad (105)$$

The relaxation time may be obtained from the proportionality between the absorption coefficient at low frequencies and the frequencies themselves:

$$\mu = \pi r \omega \tau = (2\pi^2 r \tau)\nu.$$

Finally,

$$\mu_{\max} = (\pi/2)r.$$

The ratio r of equation (105) is sometimes expressed in another way, by making use of the thermodynamic equations $\beta_T/\beta_S = C_P/C_V$ and

$$C_P - C_V = \frac{\alpha^2 V T}{\beta_T}.$$

Then

$$r = \frac{\alpha^2 V T}{\beta_S C_P^2} . C^{\text{int}}, \tag{109}$$

where α is the coefficient of isobaric expansion and V is the molar volume.

W. Maier and H. D. Rudolph (1957) have used the ultrasonic method to measure the rates of dimerisation of benzoic acid in various solvents at different temperatures. Their results at two temperatures, using carbon tetrachloride as solvent are given in Table V, and may be summarized as follows:

$$k_2 = 9\cdot2 \times 10^{12} . \exp[-(3,080 \pm 860)RT]$$

$$k_1 = 5\cdot1 \times 10^{15} . \exp[-(13,430 \pm 80)RT].$$

TABLE V

KINETICS OF FORMATION AND
DECOMPOSITION OF THE BENZOIC
ACID DIMER IN CARBON
TETRACHLORIDE SOLUTION
(after Maier)

$t°C$	$k_2 \times 10^{-10}$ (litre/mole-sec)	$k_1 \times 10^{-5}$ (sec^{-1})
25	$5\cdot1 \pm 0\cdot3$	$7\cdot3 \pm 0\cdot1$
55	$8\cdot2 \pm 0\cdot6$	58

The gain in heat content attending the decomposition of the dimer is thus about 10 kcal, in reasonable agreement with values found for other carboxylic acids in this solvent. The magnitude of k_2 suggests a diffusion-controlled mechanism for the union of the monomers, and this view is supported by the magnitude of $E_{A,2}$, for, according to this mechanism

$$E_{A,2} = RT + B. \tag{6.31}$$

At 40°C, the calculated value of this apparent energy of activation is

$$E_{A,2} = 551 + 2,375 = 2,926 \text{ cal},$$

which is near to the experimental value. According to equation (5.36), the pre-exponential term becomes

$$A_2 = \frac{N_0}{1,000} \frac{8kTe}{3b} = 1{\cdot}08 \times 10^{12}, \tag{6.36}$$

which is less than the experimental value by a factor of 9. The adoption of Eyring's equation (5.65) leads to the theoretical expression

$$A_2 = \frac{V_m}{1,000} \frac{8kTe}{3h} = 4{\cdot}36 \times 10^{12}, \tag{110}$$

which agrees with experiment within the limits of experimental error. The magnitude of the term A_1 cannot be understood without supposing that internal degrees of freedom are involved.

Change in Liquid Structure

Let us consider a chemically pure liquid the molecules of which are capable of existing, at constant temperature and pressure, in two structurally distinct patterns, A and B, with different co-ordination numbers and molar volumes. There is nothing in the derivation of equation (101) to limit its application to isomerisms as ordinarily understood or to invalidate its use in the study of liquid structure. For this purpose, we consider changes which do not involve changes in temperature, so that, from equation (101), omitting the relaxation factor, we have

$$\Delta x_B = -x_A^0 x_B^0 \frac{\Delta V}{RT}(P - P^0). \tag{111}$$

Quite generally for a binary system, the molar volume may be expressed as follows in terms of the molar fractions, x, and the partial molar volumes:

$$V = x_A V_A + x_B V_B,$$

so that

$$\frac{dV}{dP} = x_A \frac{dV_A}{dP} + x_B \frac{\partial V_B}{\partial P} + V_A \frac{\partial x_A}{\partial P} + V_B \frac{\partial x_B}{\partial P}$$

$$= x_A \frac{\partial V_A}{\partial P} + x_B \frac{\partial V_B}{\partial P} + (V_B - V_A) \frac{\partial x_B}{\partial P}.$$

The isothermal compressibility of the system is found by multiplying each term by $-(1/V)$:

$$\beta_T = -\frac{1}{V}\left(\frac{dV}{dP}\right)_T = -\frac{1}{V}\left(\frac{x_A \, \partial V_A + x_B \, \partial V_B}{\partial P}\right)_T - \frac{\Delta V}{V}\left(\frac{\partial x_B}{\partial P}\right).$$

The first term on the right hand side is recognized as the compressibility of the undisturbed or frozen system, which Herzfeld and Litovitz denote by β_∞:

$$\beta_T = \beta_\infty - (\Delta V/V)(\partial x_B/\partial P).$$

The last term, which is the additional contribution to the compressibility arising from the disturbance in the system, is obtained from equation (111):

$$\beta_T = \beta_\infty + x_A^0 x_B^0 \frac{(\Delta V)^2}{VRT}.$$

The fractional change in compressibility is

$$\frac{\beta_T - \beta_\infty}{\beta_T} = \frac{x_A^0 x_B^0}{\beta_T} \frac{(\Delta V)^2}{VRT}.$$

We now proceed as in previous passages by assuming that Δx_B obeys the general relaxation law, with a relaxation time τ, and that the parameters x and P vary in proportion to $\exp(i\omega t)$. We obtain finally the following expression for the absorption coefficient

$$\mu = \pi \left[\frac{x_A^0 x_B^0}{\beta_S^0} \frac{(\Delta V)^2}{VRT} \right] \frac{\omega \tau}{1 + \omega^2 \tau^2} \tag{112}$$

which is seen to be entirely analogous to equation (89), with the square-bracketed term taking the place of r.

Experiments on the measurement of the absorption of ultrasonic waves by such a system at various sound frequencies thus affords a value of $\omega = 1/\tau = k_A^0 + k_B^0$, and, in principle, of the volume increase attending the conversion of liquid A into liquid B. Supplementary assumptions, however, are required before the unknown quantities k_A^0, k_B^0 and ΔV can be separately evaluated (L. Hall, 1948; K. F. Herzfeld and T. A. Litovitz, 1959). It is first assumed that the relaxation time of liquid water in the absence of any disturbance is given in terms of its molar volume, V and viscosity η by the equation afforded by the kinetic theory for ideal gases:

$$\tau = \frac{\pi}{4} \cdot \frac{V}{RT} \cdot \eta. \tag{113}$$

As usual in reversible processes, $1/\tau$ is the sum of two rate constants which are here taken as those governing the changes A to B and B to A. These processes are taken to occur *via* the formation of an activated state, using Herzfeld's frequency term kT/h as the average probability of transformation per second. The process held responsible for the additional compressibility of water over the classical value is the change characterised by k_A. This introduces into the expression for the absorption coefficient the factor $k_A^0/(k_A^0 + k_B^0)$, which, according to equation (103), is found to be x_B^0. The revised expression for μ is thus

$$\mu = \frac{(1 - x_B^0)(x_B^0)^2}{\beta_S^0} \frac{(V_2 - V_1)^2}{VRT} \frac{\omega\tau}{1 + \omega^2\tau^2}. \tag{114}$$

in which there are still a number of unknown quantities.

Application to Liquid Water

Water has long been regarded as a solution containing molecules of monohydrol, H_2O, and various polymers, denoted by $(H_2O)_s$ where the integer s has been ascribed values ranging from 2 (van Laar, 1899; W. Sutherland, 1900; Bousfield and Lowry, 1905) to 9 (Ducloux, 1912). A. Eucken (1948) has treated water as consisting of four species in equilibrium, namely, H_2O, $(H_2O)_2$, $(H_2O)_4$, and $(H_2O)_8$. Röntgen (1891) regarded water as a solution of a species with the density of ice dissolved in monohydrol, and Bernal and Fowler (1933) have produced evidence suggesting that at least two structural types, quartz and tridymite, coexist in liquid water. Shifts in the equilibria established in liquid water with rising temperature have been held responsible for the anomalous temperature variation of its density (For recent data, see Steckel and Szapiro, 1963).

L. Hall (1948) has succeeded in interpreting the absorption coefficient of liquid water using the very simple idea that the liquid consists of two species only, one with a molar volume higher than, and the other with a molar volume lower than that of ordinary water. Then, since

$$V = x_A V_A + x_B V_B = V_A + x_B \Delta V,$$

the value of x_B can be found provided the molar volume of species A and the volume change ΔV are known. Hall, on the evidence provided by X-ray spacings and compressibility data, has argued that the ratio

$$(V_B - V_A)/V = 0.47, \tag{115}$$

is independent of temperature. Herzfeld and Litovitz (1959), accepting this figure conclude that the molar fraction x_B at 4°C is 0.26. Combining equation (115) with the resulting equation

$$V = 0.74V_A + 0.26V_B = 18.0 \text{ cc per mole},$$

yields the values $V_A = 15.6$, and $V_B = 24.1$ cc per mole. The more detailed estimates of the molar fraction x_B at various temperatures given by Herzfeld and Litovitz are shown in column 2 of Table VI, along with the molar volumes

TABLE VI

MOLAR VOLUMES OF ORDINARY WATER AND OF THE TWO ISOMERS (CCS)

$t°C$	x_B	V	V_A (quartz)	V_B (tridymite)	$V_B - V_A$
0	0·281	18·0180	15·64	24·11	8·470
3·98	0·275	18·0156	15·69	24·16	8·465
25	0·256	18·0685	15·90	24·39	8·492
60	0·263	18·3228	16·06	24·67	8·613

of ordinary water and of the two postulated varieties. It will be observed that, in the region in which water exhibits its maximum density, $(V_B - V_A)$ is very nearly constant, so that, from the equation $V = V_A + (V_B - V_A)x_B$, the expansion with respect to temperature at constant pressure is nearly

$$\frac{dV}{dT} = \frac{dV_A}{dT} + (V_B - V_A)\frac{dx_B}{dT}$$

and the coefficient of expansion becomes

$$\alpha = \frac{V_A}{V}\alpha_A + \frac{(V_B - V_A)}{V} \cdot \frac{dx_B}{dT}. \tag{116}$$

The explanation provided by J. D. Bernal and R. H. Fowler for the zero coefficient of expansion of ordinary water is that, at 4°C, the positive contribution of the first term in equation (116) is offset by the negative contribution of the second term.

From the values of V_A listed in Table V, the coefficient of cubical expansion is seen to be $\alpha_A = 2·86 \times 10^{-4}$ deg^{-1}. This lies in the region of normal values for liquid, such as mercury and carbon tetrachloride, for which $\alpha \times 10^4$ at 0°C are 1·8197 and 11·5 respectively.

From the values of x_B the equilibrium constants for the reaction $A \rightleftarrows B$ may be calculated:

$$K = x_B^0/x_A^0 = \exp(-\Delta G^0/RT). \tag{103}$$

To within 1 per cent the average value of ΔG^0 is RT. Hence

$$\Delta S^0 = -d\Delta G^0/dT = S_B^0 - S_A^0 = -R$$

$$S_B^0 = S_A^0 - R, \tag{117}$$

as found independently by L. Hill and by K. F. Herzfeld and T. A. Litovitz, who suggest that the quartz-like arrangement may have the full communal entropy, like the diamond, while the tridymite or open structure has zero entropy, i.e. is completely ordered.

Relaxation Phenomena and Slow Reactions

The experimental measurement of relaxation times and their theoretical interpretation have, as in the numerous examples discussed in this Chapter, enlarged our knowledge of fast reactions. It is possible that they may also assist in the interpretation of slow reactions. Before exploring such a possibility (Chapter 16), we must derive a simple, but by no means obvious theorem.

The Kinetics of Vibrational Excitation

In an equilibriated system of harmonic oscillators under quantal conditions, let N_0 molecules in the ground state have no energy; N_1 molecules in the first quantum state have energy hv each; N_2 molecules in the second quantum state have energy $2hv$ each, and so on. We next consider a gaseous system of diatomic oscillators, not necessarily in a state of thermal equilibrium, in order to find the rate at which it attains or reverts to the equilibrium state. Whether the system is or is not at thermal equilibrium, we can retain the following expressions for the total number of oscillators and the total vibrational energy respectively:

$$N = N_0 + N_1 + N_2 + \ldots \tag{118}$$

and

$$E = O + N_1 hv + 2N_2 hv + 3N_3 hv$$
$$= hv(N_1 + 2N_2 + 3N_3 + \ldots). \tag{119}$$

The gain in the vibrational energy of the molecules is due entirely to transitions from given levels to levels immediately above them, because only transitions between neighbouring levels are possible. The rate of gain of oscillatory energy due to asceding transitions is thus

$$dE/dt = hv(k_{01} N_0 + k_{12} N_1 + k_{23} N_2 + \ldots), \tag{120}$$

where the k's are unimolecular constants. Now the probability of transition from one state to another is proportional to the quantum number of the higher state. This rule, while being a logical deduction from wave mechanics, may also be reached by regarding interactions between molecules and vibrational quanta as essentially bimolecular. Since

$$k_{01} : k_{12} : k_{23} : : 1 : 2 : 3, \tag{121}$$

we can write

$$dE/dt = hvk_{01}(N_0 + 2N_1 + 3N_2 + 4N_3 + \ldots)$$
$$= hvk_{01}(N_0 + N_1 + N_2 + N_3 + \ldots + N_1 + 2N_2 + 3N_3 + \ldots). \quad (122)$$

On using equations (118) and (119), we can therefore write

$$dE/dt = k_{01}Nhv + k_{01}E. \quad (123)$$

The loss in vibrational energy of the molecules is due entirely to transitions from given levels to levels immediately beneath them so that

$$dE/dt = -hv(k_{10}N_1 + k_{21}N_2 + k_{32}N_3 + \ldots). \quad (124)$$

There are, of course, no descending transitions from the ground level. In this series also, the probability of transition is proportional to the quantum number of the higher state, i.e.,

$$k_{10}:k_{21}:k_{32}::1:2:3. \quad (125)$$

so that $2k_{10}$ may be written for k_{21}, and $3k_{10}$ for k_{32}, and so on. The rate of change of the vibrational energy due to descending transitions is accordingly

$$dE/dt = -hvk_{10}(N_1 + 2N_2 + 3N_3 + \ldots)$$
$$= -k_{10}E. \quad (126)$$

On addition, the total rate of loss of vibrational energy becomes:

$$-dE/dt = k_{10}E - k_{01}E - k_{01}Nhv$$
$$= (k_{10} - k_{01})\left[E - \left(\frac{k_{01}}{k_{10} - k_{01}}\right)Nhv \right]. \quad (127)$$

If thermal equilibrium is maintained at a certain temperature, T, we may write

$$k_{01}/k_{10} = N_1/N_0 = \exp(-hv/kT). \quad (128)$$

Then at any other temperature,

$$-\frac{dE}{dt} = (k_{10} - k_{01})\left[E - \frac{Nhv}{\exp(hv/kT) - 1} \right]. \quad (129)$$

The second term in the second brackets is the total vibrational energy, E_v, of a system on N quantized, linear oscillators under equilibrium conditions. We thus arrive at the relaxation equation

$$-dE/dt = (k_{10} - k_{01})(E - E_v),$$

or

$$-dE/dt = (1/\tau)(E - E_v),$$

where

$$1/\tau = k_{10} - k_{01} = k_{10}[1 - \exp(-h\nu/kT)]. \tag{130}$$

By combining equations (128) and (130) we see that

$$k_{10} = \frac{1}{\tau[1 - \exp(-h\nu/kT)]} \quad \text{and} \quad k_{01} = \frac{\exp(-h\nu/kT)}{\tau[1 - \exp(-h\nu/kT)]}. \tag{131}$$

Now a unimolecular constant, k, may be defined, as explained in Chapter 6, in two ways, (i) it is the fraction to the total number of particles undergoing change in unit time, and (ii) it is the probability, per unit time, that any particle, taken at random from a large population, shall undergo the change. The second of these definitions is what is wanted here. k_{10}, for example, is the probability that, within one second, any molecule in the first quantum state shall lose a quantum of energy and thus descend to the ground state. In the absence of radiation the only means whereby a molecule may gain or lose a quantum of internal energy is by collision with another molecule or atom. The unimolecular constants must therefore be proportional to the number, Z, of collisions suffered by one molecule in one second, and must be equal to this number multiplied by the average probability, p, that a collision shall be, from the transitional aspect, effective. Thus, for example,

$$k_{10} = Zp_{10}$$

and

$$k_{01} = Zp_{01}.$$

According to the kinetic theory of gases (Chapter 2), the number of collisions made per second by one molecule of an ideal gas is given in terms of the average velocity, $\bar{c}$, and mean free path, λ, as

$$Z = \frac{\bar{c}}{\lambda} = \frac{\bar{c}mn\bar{c}}{2\eta} = \frac{4nkT}{\pi\eta} = \frac{4}{\pi} \cdot \frac{P}{\eta}$$

where P is the pressure and η the viscosity. τ_c is defined as the reciprocal of Z. We thus have for the average probability that a collision shall be effective in decreasing or increasing the vibrational energy by one quantum the respective equations:

$$p_{10} = \frac{k_{10}}{Z} = \frac{\tau_c}{\tau[1 - \exp(-h\nu/kT)]}; p_{01} = \frac{k_{01}}{Z} = \frac{\tau_c \exp(-h\nu/kT)}{\tau[1 - \exp(-h\nu/kT)]}. \tag{132}$$

It is to be observed that, for a given value of τ at a given temperature, p_{10} is greater than p_{01}. The difference gets less as the temperature is raised.

REFERENCES

Andrea, J. H., E. L. Heasell and J. Lamb, *Proc. Phys. Soc.*, **B, 69**, 625 (1956).

Bergmann, L., *Der Ultraschall*, Herzel, Zurich (1949).

Bernal, J. D. and R. H. Fowler, *J. Chem. Physics*, **1**, 515 (1933).

Bousfield and Lowry, *Phil. Trans.*, **204A**, 283 (1905).

Bowen, E. J., *Fluorescence of Solutions*, Longmans, London (1953).

Caldin, E. F., *Fast Reactions in Solution*, Blackwell, Oxford (1964).

Claesson, S., (Ed.), *Fast Reactions and Primary Processes in Chemical Kinetics*, Wiley, Chichester (1968).

Cottrell, T. L. and J. C. McCourbrey, *Molecular Energy Transfer in Gases*, Butterworth, London (1961).

Davies, R. O. and J. Lamb, *Quart. Rev.*, **11**, 134 (1957).

Ducloux, *J. Chim. Phys.*, **10**, 73 (1912).

Eigen, M., *Internationales Kolloquium über Schnelle Reaktionen in Lösungen*, Max Planck Gesellschaft (1959).

Eigen, M., *Z. physikal. Chem.*, *Neue Folge*, **1**, 176 (1954).

Eigen, M. and H. Eyring, *J. Amer. Chem. Soc.*, **84**, 3254 (1962).

Eigen, M., K. Kustin and G. Maas, *Z. physical. Chem.*, **30**, 130 (1961).

Eigen, M. and L. de Maeyer, *Z. Elektrochem*, **59**, 986 (1955).

Eucken, A., *Z. Elektrochem*, **52**, 255 (1948).

Eucken, A. and R. Becker, *Z. physikal. Chem.*, **B,27**, 219 (1934).

Förster, *Fluoreszenz organischer Verbindungen*, Vanden–Hoeck and Ruprecht, Göttingen (1951).

Fox, F. E. and G. D. Rock, *Phys. Rev.*, **70**, 68 (1946).

Guilbault, *Fluorescence*, Arnold, London (1967).

Hall, L., *Phys. Rev.*, **73**, 772 (1948).

Hartridge and Raughton, *Proc. Roy. Soc.*, **104**, 376 (1923).

Herzfeld, K. F. and T. A. Litovitz, *Absorption and Dispersion of Ultrasonic Waves*, Academic Press, New York and London (1959).

Kneser, H. O., *Physikal. Z.*, **39**, 800 (1938).

Lamb, J., *Z. Elektrochem.*, **64**, 135 (1960).

La Mer and Morlies, *J. Am. Chem. Soc.*, **57**, 1812 (1935).

Liebhafsky, *J. Amer. Chem. Soc.*, **54**, 1792, 3499 (1932).

Macfarlane and H. Hartley, *Phil. Mag.*, **13**, 425 (1932).

Maier, W. and H. D. Rudolph, *Z. physikal. Chem.*, **10**, 83 (1957).

Maxwell, *Phil. Mag.*, IV, **35**, 129, 185 (1868).

McCarthy and MacLachlan, *Trans. Faraday Soc.*, **57**, 1107 (1961).

Norrish, R. G. W., *Science*, **149**, 1470 (1965).

Norrish and G. Porter, *Faraday Soc. Diss.*, **17**, 40 (1954).

Norrish and Thrush, *Quart. Rev.*, **10**, 149 (1956).

Ogston, *Trans. Chem. Soc.*, 1023 (1936).

Osugi, Kusuhara and Hirayama, *Res. Phys. Chem.*, Japan, 37, 94 (1967).

Rabinowitch and W. C. Wood, *Trans. Faraday Soc.*, **32**, 547 (1936).

Richards, W. T. and J. A. Reid, *J. Chem. Physics*, **1**, 114 (1933).

Röntgen, *Wied. Ann.*, **45**, 91 (1891).

Steckel and Szapiro, *Trans. Faraday Soc.*, **59**, 331 (1963).

Stokes, G. C., *Proc. Camb. Phil. Soc.*, **8**, 287 (1845).

Stokes, G. G., *Proc. Camb. Phil. Soc.*, **9**, 5 (1856).

Struer, J., E. Jeager, T. Sachs and F. Hovorka, *J. Chem. Physics*, **38**, 587 (1963).

Sutherland, W., *Phil. Mag.*, **50**, 460 (1900).

Tisza, *Phys. Rev.*, **61**, 531 (1942).
van Laar, *Z. physikal. Chem.*, **31**, 12 (1899).
von Halban and Eisner, *Helv. Chim Acta*, **19**, 915 (1936).
Wijs, *Z. physikal Chem.*, **12**, 1514 (1893).

15

CORRELATIONS

Correlations

The study of the statics and kinetics of reactions in solution has revealed a number of empirical relationships between pairs of experimental variables, such as $\Delta^0 H$ and $T\Delta^0 S$ in equilibrium systems, and E_A and $RT \ln A$ in kinetic systems. Though the meaning of many of these correlations is only partly understood, they are useful adjuncts to knowledge, and have inspired a wealth of theoretical speculation. We shall here consider a few examples of these empirical correlations, beginning in the field of chemical statics, although kinetic correlations take historical precedence.

We note in the first place that no reputable correlation between the thermodynamic variables $T\Delta^0 S$ and $\Delta^0 H$ in simple and comparable gaseous systems has been recorded. It is clear from Table I, for instance, that the enormous range in the magnitude of the equilibrium constants is determined principally by changes in $\Delta^0 H$. Moreover, the absolute values of $\Delta^0 S$ have been interpreted completely in terms of measured values of moments of

TABLE I

THERMODYNAMIC CONSTANTS FOR A SET OF COMPARABLE GASEOUS REACTIONS
AT 25°C

Reaction	K	$\Delta^0 S$ cal-mole^{-1}-deg^{-1}	$\Delta^0 H$ cal/mole^{-1}
$2HCl \rightleftarrows H_2 + Cl_2$	$5{\cdot}50 \times 10^{-34}$	$-4{\cdot}74$	43,960
$2HBr \rightleftarrows H_2 + Br_2$	$1{\cdot}05 \times 10^{-19}$	$-5{\cdot}07$	24,380
$2HI \rightleftarrows H_2 + I_2$	$5{\cdot}01 \times 10^{-4}$	$-5{\cdot}20$	2,950

inertia, masses and vibration frequencies. In these systems, correlation between $\Delta^0 H$ and $T\Delta^0 S$ is neither found nor expected. That empirical relationships between $\Delta^0 H$ and $T\Delta^0 S$ frequently appear when these variables refer to reactions in solution suggests that their origin may be traced to the solvent, or, more strictly, to its interactions with the solutes.

Static Correlations between Solubility Constants

We shall here examine some results on the solubility of certain gases and liquids in various solvents at different temperatures. In order to work with positive values of $\Delta^0 H$ and $\Delta^0 S$, we shall have in mind the process of escape of the solute X from solution to the gas phase:

X (at unit concentration in solution) $\rightarrow$

X (at unit concentration in the gaseous phase).

The ratio of the equilibrium concentrations of the solute in the two phases is denoted by Ostwald's absorption coefficient,

$$s = \frac{c_s}{c_g}. \tag{1}$$

The molar gain in standard free energy attending the escape is thus

$$\Delta^0 G = RT \ln s. \tag{2}$$

It follows that

$$\Delta^0 H = -RT^2 \frac{d \ln s}{dT}, \tag{3}$$

and

$$\Delta^0 S = -R \ln s + \frac{\Delta^0 H}{T}. \tag{4}$$

Table II summarizes some experimental results on the solubility of the 1899; Winkler, 1906; Ramstedt, 1911; E. Rutherford, 1911; Schulze, 1920; Valentiner, 1927; Lannung, 1930). We shall not elaborate on the values of the standard entropy of these solutes in water (S_s^0), or on the parallelism elements of group VIII of the periodic classification in water (Estreicher, between the heat of escape, $\Delta^0 H$ and the latent heat, L, of vaporization of the pure liquid solute. These are matters which have been discussed in Chapter 3. We simply note that, empirically, the family of thermodynamic variables governing the escape of these elements from water at 25°C are

TABLE II

$$s = c_s/c_g$$

$$\Delta^0 G = RT \ln s.$$

Gas	$s \times 10^3$	ΔG^0 (cal/gramme-m)	ΔH^0 (cal/gramme-m)	ΔS^0 (cal/gramme-mole-deg)	S_g^0 (cal/gramme-mole-deg)	S_s^0 (cal/gramme-mole-deg)	L (cal/gramme-mole)
He	9·47	2,760	0	9·3	23·8	14·5	23
Ne	11·1	2,660	320	10·0	28·6	18·6	413
A	34·7	2,000	2,110	13·8	30·6	16·8	1,500
Kr	39·6	1,910	2,870	16·0	32·8	16·8	2,314
X	120	1,260	3,460	15·8	34·2	18·4	3,210
Rn	218	900	4,030	16·6	35·8	19·2	4,340

related as follows:

$$T\Delta^0 S = 2,772 + 0.636\Delta^0 H. \tag{5}$$

Lannung (1930) extended his measurements to the solubilities of gases other than the inert elements, and derived, as follows, an equation of this form. He provisionally regarded the term $\Delta^0 G$ of equation (2) as being independent of temperature, so that $\Delta^0 S = 0$, and $\Delta^0 G = \Delta^0 H$. Then s should be equal to $\exp(\Delta^0 H/RT)$. On closer inspection, however, he found it empirically necessary to write $s = a \exp(\Delta^0 H)RT$, where, for the complete series of solutes, $-\log_{10} a = 0.3 + \Delta^0 H/3,000$ approximately. We may identify a with $\exp(-\Delta^0 S/R)$, so that, at 25°C,

$$T\Delta^0 S = 410 + 0.46\Delta^0 H. \tag{6}$$

The analogous relationship afforded by the data of Table 3.III on the escape of the methyl halides from aqueous solutions at the same temperature is

$$T\Delta^0 S = 648 + 0.685\Delta^0 H. \tag{7}$$

Correlations of this kind are not confined to aqueous solutions. The thermodynamic variables governing the escape of various alcohols from benzene solution (J. A. V. Butler and P. Harrower, 1937) are related as follows:

$$T\Delta^0 S = 7,938 + 0.397\Delta^0 H. \tag{8}$$

The escaping process here is from a hypothetical solution, in which the mole fraction of the solute is unity, to the gaseous phase where its partial pressure is 1 mm. This change in the standard state merely shifts the first constant term in equation (8) without significantly affecting the coefficient of $\Delta^0 H$. Similarly, for the escape of a given solute (benzoyl chloride) from a variety of aromatic solvents (M. G. Evans and M. Polanyi, 1936),

$$T\Delta^0 S = 3 + 0.474\Delta^0 H. \tag{9}$$

In this series, the transfer of the solute is from solution to the pure liquid. Further correlations of this kind have been observed and discussed by J. A. V. Butler (1937), R. P. Bell (1937) and by I. M. Barclay and J. A. V. Butler (1938). For the escape of various non-polar gases from carbon tetrachloride solution, for example, Bell finds, from the Ostwald absorption coefficients:

$$T\Delta^0 S = -645 + 0.364\Delta^0 H. \tag{10}$$

Discussion of Entropy–Enthalpy Relationships based on Solubility Equilibria

It is evident that linear relationships between standard changes in entropy and heat content for sets of ostensibly similar systems are fairly general, holding for the escape of different solutes from a common solvent, and for the escape of a given solute from a variety of solvents, including water, some special features of which have been discussed in Chapters 3 and 14. We shall here deal briefly with some heterogeneous equilibria between the gaseous phase and a solution so dilute that the probability of two solutes being in contact may be ignored. The gain in free energy, ΔG_{esc}, attending the removal of a gram-mole of solute from dilute solution to the gaseous phase is then $-N_0 c\phi$, where c is the number of solvent molecules immediately surrounding a single solute molecule, and ϕ is the interaction energy associated with one solute-solvent contact. ϕ is a function of the permittivity, D, of the solvent, the equilibrium distance, a, between the effective centres of the solute and solvent molecules, and θ, the angle between the direction of the radius and that of the line of centres. In its dependence on a, ϕ may be assumed to vary inversely as a raised to an integral power.

If, in dealing with a set of equilibria, the most relevant variable were D, while c, a and θ remained independent of temperature, we would derive the following relationship:

$$T\Delta^0 S/\Delta^0 H = LT/(LT - 1) \tag{11}$$

which is small and negative for non-polar liquids like toluene but larger and positive for polar liquids like water (3.63) and acetophenone (5.50). It is however, inherently unlikely that changes in the permittivity alone can account for the facts, or that all the other variables, especially a and c, should be independent of temperature.

Let us next allow for changes in a with respect to temperature, while regarding c and the other variables as constant. The simplest set of reactions to consider in this category is the escape of non-polar solutes from non-polar solvents. Let the interaction energy of a single solute-solvent contact be expressed as $\phi = -B/a^6$, where B is London's attraction constant, and a is the equilibrium separation of solute and neighbouring solvent molecules. The gain in free energy attending the escape of a gram mole of solute from an infinitely dilute solution is $\Delta^0 G_{esc} = N_0 cB/a^6$. If c and B are independent of temperature, it follows that

$$T\Delta^0 S/\Delta^0 H = \frac{6(d \ln a/dT)_P}{1 + 6(d \ln a/dT)_P}. \tag{12}$$

We can identify $3(d \ln a/dT)_P$ with an effective coefficient of cubical expansion, α, which is 3 times the coefficient of expansion of the solute–solvent link. Then

$$T\Delta^\circ S/\Delta^\circ H = 2\alpha T/(1 + 2\alpha T). \tag{13}$$

From the empirical equation (10) governing the escape of non-polar molecules from carbon tetrachloride solution, the experimental ratio is 0·364, which requires α at 25°C to be $9\cdot5 \times 10^{-4}$ deg^{-1}. The coefficient of expansion of the bond would thus be only slightly smaller than that for the solvent–solvent bond, for which $\alpha = 12\cdot05 \times 10^{-4}$ deg^{-1}.

When the solute molecule is polar, greater difficulties arise in expressing its average energy of interaction with the surrounding solvent molecules. If it may be assumed that the polar molecule, of moment μ_B, rotates freely, the average energy of its interaction with one solvent molecule, of polarizability, α_A, is $-\alpha_A \mu_B^2/a^6$. The argument of the preceding paragraph again holds. On combining the data of Butler and Harrower on the solubility of alcohols in benzene (equation 8) with the theoretical equation (13), the effective coefficient of cubical expansion, α, becomes $8\cdot31 \times 10^{-4}$, which again is comparable with but slightly less than the value of $\alpha = 12\cdot09 \times 10^{-4}$ for the pure solvent at 25°.

Further difficulties are found in attempting to express the interaction energy of a solute–solvent pair when each partner is polar, unless we make still more drastic assumptions. For a pair of freely rotating dipoles, the average energy of interaction is $\phi = -2\mu_A^2\mu_S^2/3kTa^6$. Following the same treatment we now have,

$$\frac{T\Delta^0 S}{\Delta^0 H} = \frac{1 + 2\alpha T}{2(1 + \alpha T)}. \tag{14}$$

Aqueous solutions of the methyl halides are not entirely suitable systems on which to test this equation, because according to spectroscopic and calorimetric evidence, free rotation of these solute molecules is probably confined to the single rotation about the carbon–halogen axis. Using the value of $T\Delta^0 S/\Delta^0 H$ given by equation (7) α becomes $1\cdot97 \times 10^{-3}$, which is somewhat greater than for pure methyl iodide, $\alpha = 1\cdot24 \times 10^{-3}$.

The other extreme case is when neither polar molecule is free to rotate. The electrostatic contribution to the free energy of escape is then

$$\Delta G_{esc} = \frac{2N_0 c \mu_A \mu_B}{Da^3},$$

where c is the number of near neighbours, the μ's are dipole moments, and a is the average separation of the centres of the dipoles in their equilibrium state. We then find that

$$\frac{T\Delta S_{esc}}{\Delta H_{esc}} = -\frac{(LT - 3\alpha T)}{1 - LT + 3\alpha T}. \tag{15}$$

When α is zero, we recover equation (11), and when L is zero we have

$$\frac{T\Delta S_{esc}}{\Delta H_{esc}} = \frac{3\alpha T}{1 + 3\alpha T}, \tag{16}$$

which resembles equation (13). From the experimental value of L for water and the experimental value of $T\Delta S_e/\Delta H_e$ found for the escape of the methyl halides from water (equation 7), αT is thus found to be $0\cdot49$, which greatly exceeds the value for pure water ($0\cdot033$) but lies near to that for methyl iodide ($0\cdot37$). It follows that the coefficient of expansion of the halide–water bond more closely resembles that of the liquid halide than that of the solvent.

The significant result is that, despite the approximations made, the empirical relationship between the entropy and the enthalpy of escape can be roughly interpreted in terms of the effect of temperature, at constant pressure, on the permittivity of the medium and the expansivity of the solute–solvent bond.

There is much evidence for supposing that the average co-ordination number, c, of a liquid molecule decreases as the temperature is raised at constant pressure, and it is reasonable to anticipate a similar behaviour for the co-ordination number of a solute in dilute solution. The quantitative relationship between c and T is, however, extremely complicated, and cannot be formulated without making a greater number of assumptions than we are here allowing ourselves.

Static Correlations based on the Distribution of Energy

Many of the static correlations doubtless arise from the simple fact that, when a series of solutes of increasing complexity in a common solvent is considered, the increase in entropy, the decrease in heat capacity and the increase in enthalpy attending escape from solution all increase roughly in the direction of increasing size of the solute molecule. The larger the solute, the greater is the number of solute–solvent interactions to be reckoned with, and the greater is the number of oscillators among which the total energy may be distributed. Such a hypothesis would account for the empirical proportionality (D. N. Glew and E. A. Moelwyn-Hughes, 1953) between $\Delta^0 H_O$ and $T\Delta^0 C_P$ measured for the escape of the methyl halides from water. The average value of this ratio at 25°C is 1·45. If we ignore the difference between energy and enthalpy, equations (3.36) and (3.37) become

$$\Delta^0 H = \Delta^0 H_O - (s - \tfrac{1}{2})RT. \tag{17}$$

and

$$\Delta^0 S = R\left[\ln 3 + \left(s - \frac{1}{2}\right)\ln\frac{\Delta^0 H_O}{(s - \tfrac{1}{2})RT}\right] \tag{18}$$

On using the empirical relationship $\Delta^0 H_O/(s - \tfrac{1}{2})RT = 1·45$, we see that

$$\Delta^0 H = 0·45(s - \tfrac{1}{2})RT \tag{19}$$

and

$$T\Delta^0 S = 1·098RT + 0·83\Delta^0 H. \tag{20}$$

The more accurate value of 1·73 given by D. N. Glew (1962) for the ratio $\Delta^0 H_O/T\Delta^0 C_P$ yields 0·75 as the coefficient of $\Delta^0 H$.

Static Correlations in Hydrogen Bond Formation

Thiocyanic acid forms hydrogen-bonded complexes in carbon tetra-chloride solution, with numerous bases, according to the scheme

$$B + HNCS \rightleftarrows B \cdot HNCS; \qquad K = [\text{complex}]/[B][HNCS]. \tag{21}$$

The equilibrium constants at 25°C, measured by determining the shift in the NH absorption frequency ($3,469 \text{ cm}^{-1}$), are found to vary in the direction of the electronegativity of the proton acceptor (Barakat, J. Nelson, S. M. Nelson and Pullin, 1968). The set of data are summarized by the equation

$$T\Delta S^0 = -930 + 0·55\Delta H^0. \tag{22}$$

Static Correlations between Ionization Constants

Let the ionization of benzoic acid be denoted by K_u, and the standard gain in free energy attending its ionization by $\Delta^0 G_u = -RT \ln K_u$. The partial molecular free energy of benzoic acid containing a dipole of moment μ_X in the para position is greater than that of the unsubstituted acid by $\varepsilon\mu_X/Dr^2$, where r is a distance—not precisely defined—between the centre of the dipole and that of the ionizable hydrogen ion. If we ignore the difference between the free energies of the anions, we thus have (Waters, 1929; Wolf, 1929) $\Delta^0 G_s = \Delta^0 G_u - \varepsilon\mu_X/Dr^2$, where the subscript s denotes the substituted acid. Consequently

$$RT \ln (K_s/K_u) = -\Delta^0 G_s + \Delta^0 G_u = \varepsilon\mu_X/Dr^2. \qquad (23)$$

This equation has been applied by A. Eucken (1932) to the experimental ionization constants of many para-substituted benzoic acids in aqueous solution at 25°C. Using reasonable estimates of r, ranging from 5·9 to 7·0 Å, he found the average value of the effective permittivity to be 5·6, indicating that most of the electrostatic effect due to the dipole is exerted through the benzene ring rather than through the solvent medium (Schwarzenbach, 1936). Eucken found that the effective permittivity varied in the same direction as the macroscopic permittivity. Waters (1929) has rightly pointed out that μ_X should be reduced by the moment, μ_{C-H}, of the carbon–hydrogen bond in the unsubstituted acid.

Hammett (1940) denotes the difference between the decadic logarithms of the ionization constants of the substituted and unsubstituted acids by the symbol σ,

$$\sigma = \log_{10} (K_s/K_u), \qquad (24)$$

and refers to it as the substitution constant. Clearly, according to the Waters–Eucken theory

$$\sigma = \frac{\varepsilon\mu_X}{2\cdot303 kTDr^2}. \qquad (25)$$

Values of σ obtained by Hammett are seen from Table III to vary in the same direction as the dipole moment of the para-substituent. That different values of σ should be required for the nitro group in benzoic and phenolic systems indicates the need for some extension to this simple formulation.

The basic assumption underlying equation (23) is that the difference between the standard free energy change attending the ionization of the substituted and unsubstituted acids is due entirely to the difference in the chemical potentials of the unionized acids. No allowance has been made for the difference in the chemical potentials of the anions produced.

From the ionization constants of benzoic acid and of several di- and tri-substituted methylbenzoic acids measured in aqueous solution (J. M. Wilson, N. E. Gore, J. E. Sawbridge and F. Cardenas–Cruz (*Trans. Chem. Soc.*, **B**, 852 (1967)) over a range of temperatures, the following approximate

TABLE III

Para-substituent	σ (Hammett)	$\mu \times 10^{18}$ e.s.u.
NH_2	-0.660	-1.55
CH_3	-0.170	-0.45
H	0	0
F	$+0.062$	$+1.57$
Cl	$+0.227$	$+1.70$
NO_2	$+0.778$ (benzoic)	$+4.19$
	$+1.27$ (phenolic)	

relationship emerges at 25°C:

$$T\Delta^0 S = 5{,}575 + 0.771\Delta^0 H. \tag{26}$$

We can naturally summarise the results in the equivalent form

$$\Delta^0 G = -5{,}575 + 0.229\Delta^0 H, \tag{27}$$

which indicates that the standard gain in free energy attending ionization consists of a term which is common to all the acids in the series, plus a term which varies directly as the standard gain in enthalpy. Different correlation constants are found when groups other than the methyl group are substituted into benzoic acid, and when a given group is substituted in various positions.

If the difference between the standard free energies of the ionizations of substituted and unsubstituted acids were due solely to the simple electro-static effect transmitted through the benzene ring, the sign of the coefficient of $\Delta^0 H$ would be negative. The omission of the role of the anion and of all the non-electrostatic factors doubtless accounts for the difference in sign.

Electrostatic principles have been applied by numerous workers (Bjerrum, 1923; Ebert, 1925; Gane and Ingold, 1928; Kirkwood and Westheimer, 1938) to the problem of the relative ionization constants of monobasic and dibasic acids. The difficulties encountered and the degree of success attained are similar to those discussed above (see D. J. G. Ives and P. D. Marsden, 1965).

Static Correlations between Hydration Constants

The reversible hydration of the carbonyl group, which may be written as $R_1 R_2 CO + H_2 O \rightleftarrows R_1 R_2 C(OH)_2$ has been extensively studied by means of nuclear magnetic resonance and ultraviolet spectroscopy (R. P. Bell and

J. C. Clunie, 1952) and relationships between various thermodynamic parameters established (R. P. Bell and A. O. McDougall, 1960). The following equation

$$T\Delta^0 S = -920 + 0.69\Delta^0 H \qquad (28)$$

fits (with one exception) all the data provided by P. Greenzaid, Zvi Rappoport and D. Samuel (1967) at 25°C. They refer to the hydration of various aldehydes in water, acetonitrile, chloroform and carbon tetrachloride.

Some Kinetic Correlations

From the equivalent equations

$$k_n = v \exp\left(-\Delta^0 G^x / RT\right) = A \cdot \exp\left(-E_A / RT\right) \qquad (4.22)$$

we see that

$$RT \ln A = RT \ln v + E_A - \Delta^0 G^x,$$

and, if v is independent of E_A,

$$RT \frac{d \ln A}{dE_A} = 1 - \frac{d(\Delta^0 G^*)}{dE_A}. \qquad (29)$$

The ratio on the left-hand side is clearly the kinetic analogue of the thermodynamic ratio $T\Delta^0 S / \Delta^0 H$ dealt with in static systems. If, therefore, in a series of comparable reactions, the standard free energies of activation were equal to some generic constant plus the apparent energy of activation, there would be no correlation between E_A and $\ln A$; this we may regard as normal behaviour for reactions of all kinetic orders.

The claim (Hinshelwood and Fairclough, 1957; Fairclough, 1938; Hegan and Wolfenden, 1939) that $\ln A$ frequently varies as $E_A^{-1/2}$ cannot be substantiated.

Let us apply equation (4.22) to a set of unimolecular or pseudo-unimolecular reactions, assuming that v is independent of temperature. It follows that $E_A = \Delta^0 H^*$ for each member, and therefore that

$$k_1 = v \cdot \exp\left(-\Delta^0 G^* / RT\right)$$
$$= v \cdot \exp\left(\Delta^0 S^* / R\right) \cdot \exp\left(-E_A / RT\right). \qquad (4.22)$$

An empirical equation frequently found for sets of reactions of varying kinetic orders may be summarized as follows:

$$k_1 = B \cdot \exp\left(E_A / R\theta\right) \cdot \exp\left(-E_A / RT\right) \qquad (30)$$

where θ has the dimensions of temperature, and has a common value for the series of reactions. It follows that

$$\frac{\Delta^0 S^*}{R} = \ln\left(\frac{B}{v}\right) + \frac{E_A}{R\theta}. \tag{31}$$

For a set of carboxylic acids decomposing unimolecularly in aqueous solution, for example, θ is found to be 458°K. (Boeseken, 1927; Gapon, 1930; G. M. Schwab, 1928; Holzschmidt 1931; E. A. Moelwyn-Hughes, 1928).

The term entropy of activation has, as we have seen, been variously defined. Equation (4.23) can be written as follows:

$$\Delta^0 S^* = R\left[\ln\left(\frac{A_n}{v}\right) - \frac{d \ln v}{d \ln T}\right], \tag{32}$$

where A_n is the pre-exponential term of the Arrhenius equation, and v is a specific vibration frequency. If we adopt Herzfeld' expression kT/h for v, we obtain the entropy of activation as defined by Eyring:

$$\Delta^0 S^{\ddagger} = R\left[\ln\frac{A_n}{(kT/h)} - 1\right]. \tag{4.37}$$

Mansel Davies and A. Edwards (1967) and D. L. Levi (1946) have measured dielectrically the unimolecular constants governing the rotation of large rigid molecules, such as tetraphenyl cyclopentadienone, in a polystyrene matrix. They find a linear relationship between $\Delta^0 S^{\ddagger}$ and E_A for a series of reactions with E_A values ranging from about zero to 130 kcal (Fig. 1). We note that $dE_A/d(\Delta^0 S^{\ddagger}) = 417°$K, and that, expressed in kilo calories per mole,

$$T\Delta^0 S^{\ddagger} = -2\cdot5 + 0\cdot715 E_A. \tag{33}$$

Since

$$\Delta^0 S^* = \Delta^0 S^{\ddagger} + R\left[\ln\frac{hv}{kTe} + \left(\frac{d \ln v}{d \ln T}\right)_P\right], \tag{34}$$

it is quite possible that the correct plot passes through the origin, with

$$T\Delta^0 S^* = 0\cdot715 E_A. \tag{35}$$

The entropy of activation is thus directly proportional to the apparent energy of activation. It is interesting to compare the numerical term with that derived in equation (20). The relation between the kinetic or quasi-thermodynamic variables for the rotation of various solutes in solution seems to resemble that between the static or genuine thermodynamic variables for escapes from solution. It would appear that the rotation of a solute molecule is a step towards its escape.

R. J. Meakins (1959) has studied molecular rotation in crystalline long-chain esters and ethers. E_A is again found to vary linearly with respect to $\Delta^0 S^{\ddagger}$.

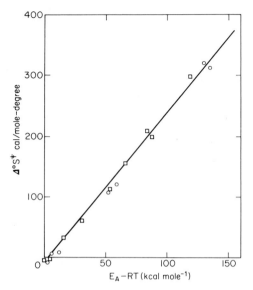

F IG. 1. Correlation between entropy and energy of activation in dipole relaxations (Levi, 1946; Davies and Edwards, 1967).

Correlations between Static and Kinetic Constants: Brönsted's Equation

Up to the present, we have dealt with empirical correlations involving equilibrium processes, such as dissolution, ionization and hydration, and those involving kinetic processes, such as hydrolysis and decarboxylation. One of the earliest and most fruitful of the empirical relationships is that established by Brönsted (1928) between the catalytic coefficients of acid and basic solutes which promote general catalysis and the ionization constants of the parent acids and bases.

It is impossible to predict which reactions are catalysed only by hydrogen ions and which are catalysed by hydrogen ion, ammonium ion, water, and any other undissociated acid. For reactions catalysed by anything which can yield a proton, Brönsted (1928) found that the catalytic coefficient is related as follows to the dissociation constant of the proton-container:

$$k_A = aK_A^\alpha. \tag{36}$$

a and α are specific isothermal constants for the system. A corresponding relation

$$k_B = bK_A^{-\beta} \tag{37}$$

holds for basic catalysis, where K_A is the dissociation constant of the acid

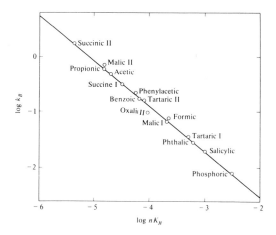

FIG. 2. General basic catalysis of the nitramide decomposition.

from which the base is formed. This empirical relation is illustrated logarithmically in Fig. 2, where n is an integer (Brönsted and Pedersen, 1924).

To translate these relationships into the realm of energy we must assume, though without direct experimental support, that a and α in acid catalysis, and b and β in basic catalysis, are independent of temperature. Then

$$E_A = E_a + \alpha H_A \tag{38}$$

and

$$E_B = E_b - \beta H_A, \tag{39}$$

where E_A, E_B stand for the Arrhenius energies of activation of the catalysed reactions, H_A for the increase in heat content attending the ionization of the acid, and E_a, E_b are generic constants.

On applying relation (38) to catalysis by hydrogen ion, for which H_A is obviously zero, we see that E_A is in fact E_H, so that

$$E_A = E_H + \alpha H_A. \tag{40}$$

In words, the energy of activation for a reaction catalysed by an undissociated acid exceeds the energy of activation of the same reaction catalysed by hydrogen ion by an amount which is proportional to the energy of dissociation of the acid. This fact is one that is very readily interpreted in terms of the general theory of the forces holding the proton to the water molecule and to other bases (Horiuti and Polanyi, 1935; E. A. Moelwyn-Hughes, 1936).*

On applying relation (39) to catalysis by water molecules, for which H_A is again obviously zero, we see that $E_b = E_W$, so that

$$E_B = E_W - \beta H_A, \tag{41}$$

which is illustrated in Table 6 from data (Pedersen, 1934) on the enolization of ethyl acetoacetate.

On combining equation (82) as applied to catalysis by hydrogen ion and water molecules, with equation (83) as applied to catalysis by hydroxyl ion and water molecules, we have the evident result:

$$\frac{E_H - E_{OH}}{H_W} = \beta - \alpha. \tag{42}$$

But (Table XI, 10) the ratio on the left is 0·44 for a number of reactions; hence, very roughly,

$$(\beta - \alpha) \sim \tfrac{1}{2}. \tag{43}$$

There are not sufficient data to test this relation, which depends on a number of assumptions still requiring substantiation. In general, however, β is known to be greater than α. Two cases of interest arise. When $\alpha = \tfrac{1}{2}$, $\beta = 1$, that is $E_{OH} = E_b - H_W$. Applied to the specific case examined by Pedersen,

TABLE IV

ENERGETICS OF THE BASE-CATALYSED ENOLIZATION
OF AN ACETOACETIC ESTER

Catalyst	E_B	$E_W - \beta H_A$
CH_2COO^-	13,400	13,500
CH_2OHCOO^-	13,100	13,600
CH_2ClCOO^-	13,800	(13,800)
$H_2PO_4^-$	14,000	13,000
SO_4^-	17,500	13,700
H_2O	14,200	14,200

---

* The numerical constants obtained by the late Dr. A. Sherman for the quantal relation cited in the latter work are:

$$E = 0.8D_1 - (0.75 - 0.00825\Delta H)D_2,$$

where E is the energy of activation, D_1 is the dissociation energy of the bond broken, and D_2 that of the bond formed, in kcal/gramme mole. For fuller accounts see Adam, *The Physics and Chemistry of Surfaces*, 3rd ed., p. 333, Oxford, (1941); Bell, *Acid-Base Catalysis*, Oxford, (1941).

this leads to a value of almost zero for the reaction as catalysed by hydroxyl ions. Secondly, when $\beta = \frac{1}{2}$, $\alpha = 0$, that is, there is no acid catalysis. We may therefore predict that β is not far off 0·5 for reactions which, like the catalytic hydrolysis of trichloracetic esters, are insensitive to hydrogen ions.

Hammett's Equation

Hammett observed that $\log(k_s/k_u)$ for a series of reactions at a constant temperature varied nearly linearly with respect to $\log(K_s/K_u)$, where the K terms refer to equilibrium constants for the parent acid or base to which the reactant esters or other derivatives are related. We thus have $\log(k_s/k_u) = \rho \log(K_s/K_u) + A$, where ρ is the experimental gradient and A is an integration constant. From the definition (equation 47) of the substituent constant, σ, we may write $\log(k_s/k_u) = \rho\sigma + A$. But, when σ is zero, k_s/k_u is unity, so that A is zero. Then

$$\log(k_s/k_u) = \rho\sigma \qquad (44)$$

or

$$\log k_s = \log k_u + \rho\sigma.$$

The empirical constant ρ is termed the reaction constant. It is constant for all substituents, depending only on the series of reactions. Extensive tables have been published for both ρ and σ. Combined with values of k_u, they summarize conveniently a large volume of experimental data (Hammett, 1940; Streitwiesser, 1962; I. Lee and Y. J. Park, 1963). Some values of σ have already been given. Values of ρ are found to be positive or negative, and often near unity.

Some idea of the physical meaning to be attached to ρ and σ may be gained by the simple electrostatic treatment of the relative rates of the alkaline saponification of para-substituted ethyl benzoates and the relative ionization constants of the corresponding para-substituted benzoic acids. We found

$$\log\left(\frac{k_s}{k_u}\right) = \frac{-z_A \varepsilon \mu_x \cos\theta}{2·303 k T r_2^2} \cdot \frac{1}{D_2} = \rho\sigma \qquad (9.23a)$$

and

$$\log\left(\frac{K_s}{K_u}\right) = \frac{\varepsilon\mu_X \cos\theta_i}{2·303 k T r_i^2} \cdot \frac{1}{D_i} = \sigma \qquad (45)$$

where the angles, distances and permittivities are not identical in the two series. Their ratio has the order of magnitude unity, and may be positive or negative.

The Equations of Grunwald and Winstein, Swain, Taft and Edwards

There have been many variations and extensions of Hammett's equation, which can itself, be regarded as an extension of Brönsted's formula. Grunwald and Winstein (1948) dealing primarily with solvolytic reactions, express the difference between the logarithm of the rate constant, k, of a pseudo-unimolecular reaction in any solvent at a temperature, T, and the logarithm of the rate constant, k^0, of the same reaction in 80 per cent ethanol–water at the same temperature as the product of two terms, m and Y:

$$\log(k/k^0) = mY. \tag{46}$$

The product mY is readily found, and m is taken as unity for tertiary butyl chloride. The values of Y thus obtained are regarded as measuring the relative ionizing power of the solvents. C. G. Swain, R. B. Moseley and D. E. Brown (1955) resolve a similar difference into the sum of two products:

$$\log(k/k^0) = sn + s'e, \tag{47}$$

where s and s' are substrate constants based on the value $s = 1$ for the hydrolysis of methyl bromide, and the nucleophilic and electrophilic constants n and e are both normalized to zero for water. R. W. Taft (1957) adds to Hammett's equation two dimensionless terms, S and R, to allow empirically for the effects of steric inhibition and resonance:

$$\log(k/k^0) = \sigma\rho + S + R. \tag{48}$$

J. O. Edwards (1954) has proposed the equation

$$\ln(k/k^0) = \alpha E_n + \beta H,$$

in which E_n is a nucleophilic constant, H denotes the relative basicity of electron donors, and α and β are substrate constants. Since these and the many similar equations advanced are based on the logarithm of the isothermal velocity constants they are seen to represent attempts at the difficult task of resolving the total free energy of activation into its various components. The degree of success attending these efforts may be judged from the admirable reviews of S. W. Benson (1960), E. S. Amis (1966) and Leffler and Grunwald (1963).

Discussion

It is one of the major tasks of chemistry to understand the effect of molecular structure on the rate of chemical change, and it is being tackled with ever-increasing diligence by chemists of all disciplines. There is naturally a wide variety of objectives, emphases and methods of approach. Some investigators are primarily concerned with comparisons of the rates of reactions of comparable compounds in a common medium at constant temperature and pressure. Others prefer to concentrate on a single reaction in various media and under as great a range of external conditions as is possible. There are difficulties and dangers in both fields.

(i) Even if a reaction is kinetically simple, its rate constant may still be a composite quantity. Thus, for example, an ostensible bimolecular velocity constant is often the product of a unimolecular constant and an equilibrium constant, as in equation (1.4), and a change in molecular structure, medium, temperature and pressure may effect either or both.

(ii) The ratio, k_β/k_α, of the catalytic coefficients for the hydrolysis of β- and α-methylglucosides under identical conditions of temperature, pressure and activity coefficient is greater than 1 below 115°C, and less than 1 above this temperature. Theories based on standard free energies of activation for such pairs must be such that $\Delta^0 G_\beta^* - \Delta^0 G_\alpha^*$ (sometimes written as $\Delta\Delta^0 G^*$) shows a change of sign. Reflection on the advances made in the kinetics of reaction in solution leaves no doubt that the energy of activation—even the apparent value—is a more reliable guide to chemical reactivity than the velocity constant itself or its logarithm multiplied by RT.

(iii) When a pair of similarly constituted reactants undergo chemical change at different rates under identical conditions, it is fair to assume that the energies of activation are different. It is, however, false to infer that the energies of the two activated reactants are different. They may be identical; and the difference in energies of activation can sometimes be traced to the difference between the energies of the two reactants in the ground state. Experimental evidence dealt with in earlier chapters indicates that the effect of temperature and medium on the rates of certain reactions are due more to changes in the ground state energies than to changes in the activated state energies.

(iv) Consistent with the lines of thought delineated in earlier passages of this book, stress has been laid on the electrostatic influences on rates of reaction (Moelwyn-Hughes and Sherman, 1936). Some of the variables considered here are the number, s, of oscillators, the number, c, of solute–solvent contacts the distances apart, r_{ij}, of relevant partners, and the variations of these with respect to temperature, pressure and permittivity. There remains ample scope for exploration of what can be termed the internal factors affecting the stability of the activated molecule or complex. Of these, the distribution of electron density is probably the most important, as instanced in the wave-mechanical treatment by Coulson and Chandra of the effect of polar substituents on reaction rate, and in the empirical estimates which are amply discussed in the numerous texts on physical-organic chemistry.

(v) Sound empirical observations contain the germ of truth, but because they can be interpreted in many ways, which truth is often difficult to find. In the search for it, most investigators agree that the experimental conditions of reactants and solvents must be kept as constant as is possible, with the exception of one variable. In this way, the proportionalities between energy of activation and heat of ionic solution (Table 6.XII) and between energy of activation and deformation constants (Table 6.XXI) have been established; and, from Hine's work (Table 6.XX) a quantitative explanation for the effect of steric hindrance can be given in terms of intermolecular forces.

REFERENCES

Amis, E. S., *Solvent Effects on Reaction Rates and Mechanism*, Chapter VI, Academic Press, New York (1966).

Barakat, J. Nelson, S. M. Nelson and Pullin, *Trans. Faraday Soc.*, **64**, 41 (1968).

Barclay, I. M. and J. A. V. Butler, *Trans. Faraday Soc.*, **34**, 1445 (1938).

Bell, R. P., *Trans. Faraday Soc.*, **33**, 496 (1937).

Bell, R. P. and J. C. Clunie, *Trans. Faraday Soc.*, **48**, 439 (1952).

Bell, R. P. and A. O. McDougall, *Trans. Faraday Soc.*, **56**, 1281 (1960).

Benson, S. W., *Foundations of Chemical Kinetics*, Chapter XVI, McGraw-Hill, New York (1960).

Bjerrum, *Z. physikal. Chem.*, **106**, 219 (1923).

Boeseken, *Rec. trav. chim. Pays-bas*, **46**, 574 (1927).

Brönsted, *Chem. Rev.*, **5**, 231 (1928).

Brönsted and Pedersen, *Z. physikal. Chem.*, **108**, 195 (1924).

Butler, J. A. V., *Trans. Faraday Soc.*, **33**, 229 (1937).

Butler, J. A. V. and P. Harrower, *Trans. Faraday Soc.*, **33**, 71 (1937).

Davies, Mansel and A. Edwards, *Trans. Faraday Soc.*, **63**, 2163 (1967).

Ebert, *Ber.*, **58B**, 175 (1925).

Edwards, J. O., *J. Amer. Chem. Soc.*, **76**, 1540 (1954).

Estreicher, *Z. physikal. Chem.*, **31**, 176 (1899).

Eucken, A., *Z. angew. Chem.*, **45**, 203 (1932).

Evans, M. G. and M. Polanyi, *Trans. Faraday Soc.*, **33**, 1333 (1936).

Fairclough, *Trans. Chem. Soc.*, 1186 (1938).

Gane and Ingold, *Trans. Chem. Soc.*, 1594 (1928).

Gapon, *Ukraine Chem. J.*, **5**, 169 (1930).

Glew, D. N., *J. Phys. Chem.*, **66**, 605 (1962).

Glew, D. N. and E. A. Moelwyn-Hughes, *Discussions Faraday Soc.*, **15**, 150 (1953).

Greenzaid, P., Zvi Rappoport and D. Samuel, *Trans. Faraday Soc.*, **63**, 2131 (1967).

Grunwald and Winstein, *J. Amer. Chem. Soc.*, **70**, 846 (1948).

Hammett, *Physical Organic Chemistry*, p. 188, McGraw-Hill, New York (1940).

Hegan and Wolfenden, *Trans. Chem. Soc.*, 508 (1939).

Hinshelwood and Fairclough, 'The functional relation between the constants of the Arrhenius equation', *Trans. Chem. Soc.*, 1575 (1957).

Holzschmidt, *Z. Anorg. Chem.*, **200**, 82 (1931).

Horiuti and M. Polanyi, *Acta Physicochim, U.S.S.R.*, **2**, 505 (1935).

Ives, D. J. G. and P. D. Marsden, *Trans. Chem. Soc.*, 649 (1965).

Kirkwood and Westheimer, *J. Chem. Physics*, **6**, 513 (1938).

Lannung, *J. Amer. Chem. Soc.*, **52**, 68 (1930).

Lee, I. and T. J. Perk, *J. Korean Chem. Soc.*, **7**, 238 (1963).

Levi, D. L., *Trans. Faraday Soc.*, **42**, 152 (1946).

Leffler and Grunwald, *Rates and Equilibria of Organic Reactions*, Wiley, New York (1963).

Meakins, R. J., *Trans. Faraday Soc.*, **55**, 1694 (1959).

Moelwyn-Hughes, E. A., *Acta Physicochim*, **4**, 173 (1936).

Moelwyn-Hughes, E. A., *Trans. Faraday Soc.*, **24**, 309 (1928).

Moelwyn-Hughes, E. A. and Sherman, *Trans. Chem. Soc.*, 101 (1936).

Pedersen, *J. Physical Chem.*, **38**, 1019 (1934).

Ramstedt, *Le Radium*, **8**, 253 (1911).

Rutherford, E., *Radioactive Substances and their Radiations*, p. 377, Cambridge (1911).

Schulze, *Z. physikal. Chem.*, **95**, 257 (1920).

Schwab, G. M., *Z. physikal. Chem.*, **B,5**, 406 (1928).

Schwarzenbach, *Z. physikal. Chem.*, **A,176**, 133 (1936).

Streitwieser, *Solvolytic Displacement Reactions*, McGraw-Hill, New York (1962).

Swain, C. G., R. B. Moseley and D. E. Brown, *J. Amer. Chem. Soc.*, **77**, 3731 (1955).

Taft, R. W., *J. Chem. Physics*, **26**, 93 (1957).

Valentiner, *Z. physikal Chem.*, **42**, 253 (1927).

Waters, *Phil. Mag.*, **8**, 436 (1929).

Wilson, J. M., N. E. Gore, J. E. Sawbridge and F. Cardenus-Cruz, *Trans. Chem. Soc.*, **B,** 852 (1967).

Winkler, *Z. physikal Chem.*, **55**, 344 (1906).

Wolf, *Z. physikal. Chem.*, **B**, **3**, 128 (1929).

16

REACTIONS BETWEEN POLAR MOLECULES

The kinetics of the formation of quaternary ammonium salts is the subject which has received much attention. The rate of reaction between various amines $R_1R_2R_3N$ and halides R_4X, measured extensively in a variety of solvents at different temperatures and pressures, is found to be predominantly bimolecular but sensitive to impurities. The reactions are exothermic, and the rate constants are lower by several powers of ten than those expected according to any theory which identifies the apparent with the true energy of activation. Despite the large volume of experimental data available and the numerous discussions based upon them, the mechanism of these reactions remains obscure.

The Influence of the Solvent: Experimental Results

The solvent effect was first systematically investigated by Menschutkin (1890) who measured, at 100°C, the rate at which ethyl iodide and triethylamine,

$$C_2H_5I + (C_2H_5)_3N \rightarrow (C_2H_5)_4NI,$$

combine in various media. His relative constants (Table I) show that the reaction is always faster in an aromatic solvent than in the corresponding aliphatic solvent. The "slow" solvents are the hydrocarbons and ethers; the "fast" solvents are the alcohols and ketones. The positive catalytic influence of the solvent thus runs roughly parallel with its general chemical reactivity—a result which justified Menschutkin in classifying the solvent influence among specific chemical effects. The fast solvents are those which favour ionization of solutes, i.e. have high permittivities. Trimethylamine with alkyl hiodide behave similarly (M. A. Abraham, 1969).

More recent results have been given in terms of the parameters A and E_A of the Arrhenius equation. Whether we examine a set of reactions in a given solvent, or a given reaction in different solvents, E_A and $\log A$ generally change in the same direction (Table II), with a gradient $dE_A/Rd \ln_e A$ in the neighbourhood of 250°.

TABLE I

Solvent		Relative k
Hydrocarbons	Hexane	0·13
	Heptane	0·17
	Xylene	2·2
	Benzene	2·4
Halogen Derivatives	Propyl chloride	4·0
	Chlorobenzene	17·4
	Bromobenzene	20·3
	α-Bromonaphthalene	84·9
Ethers	Ethyl-*iso*-amyl ether	0·47
	Diethyl ether	0·57
	Phenetole	16·0
	Anisole	30·3
Esters	*Iso*-butyl acetate	4·3
	Ethyl acetate	16·7
	Ethyl benzoate	19·4
Alcohols	*Iso*-butyl alcohol	19·4
	Ethyl alcohol	27·4
	Allyl alcohol	32·5
	Methyl alcohol	38·0
	Benzyl alcohol	100·0
Ketones	Acetone	45·7
	Acetophenone	97·3

The reactions listed in Table II, with the exception of the last one, have been examined also in mixed solvents, such as binary mixtures of hydrocarbons and ketones. When the parameters E_A and $\log A$ are plotted as functions of the solvent composition at constant temperature and pressure, the curves may show maxima, or minima, or double maxima or a maximum and minimum. To discuss these phenomena without a knowledge of the properties of the reactants in their ground states would be futile, as work on the statics of dissolution (Chapter 10) makes clear.

TABLE II

REACTIONS OF AMINES AND HALIDES IN VARIOUS SOLVENTS (E_A IN kcal; A IN LITRES/GRAM-MOLE-SEC)

Reference	Reaction	C_6H_6 ($D = 2.3$)		C_2H_5OH ($D = 25.0$)		$(CH_3)_2CO$ ($D = 31.0$)		$C_6H_5NO_2$ ($D = 35.8$)	
	Solvent:	E_A	$\log_{10} A$	E_A	$\log_{10} A$	E_A	$\log_{10} A$	E_A	$\log_{10} A$
1	$(C_2H_5)_3N + C_2H_5I$	11.4	3.26	—	—	—	—	11.6	4.91
2	$(C_2H_5)_3N + C_2H_5Br$	11.7	3.96	—	—	—	—	11.2	2.47
3	$(C_2H_5)_3N + C_6H_5CH_2Cl$	10.8	3.62	14.1	6.99	—	—	9.2	4.07
3	$C_5H_5N + C_6H_5CH_2Cl$	13.4	4.81	14.3	6.80	—	—	8.4	2.81
4	$C_5H_5N + CH_2{:}CHCH_2Br$	13.95	4.81	14.65	6.25	13.30	5.55	13.0	5.74
5	$C_5H_5N + CH_3I$	14.1	5.41	18.0	8.32	14.0	6.49	13.7	6.59
6	$C_6H_5NH_2 + CH_3COC_6H_4Br$	8.1	0.92	13.9	6.83	11.1	4.46	13.5	5.84

REFERENCES

1. Grim, Ruf, and Wolff, Z. *physikal Chem.*, **B, 13**, 301 (1931). Error of ± 0.3 in E_A assessed by the investigators.
2. Hemptinne and Bekaert, *ibid*, **28**, 225 (1899). Error in E_A difficult to assess, but may be as high as ± 2.0. Estimated dE_A/dT is as high as $+155 \pm 25$.
3. Muchin, Ginsberg, and Moissejeva, *Ukraine Chem.*, J., **2**, 136 (1926).
4. Hawkins, *Trans. Chem. Soc.*, **121**, 1170 (1922). Errors in E_A estimated, successively, as ± 0.02, ± 0.23, ± 0.16, and ± 0.62.
5. Pickles and Hinshelwood, *ibid*, 1353 (1936); Fairclough and Hinshelwood, *ibid*, 1573 (1937).
6. Cox, *ibid*, **119**, 142 (1921).

The Influence of the Permittivity of the Solvent

If the total energy of activation, E, of reactions between polar molecules includes an electrostatic contribution, E_e, due to dipolar interaction, we have

$$E = E_n + E_e$$

$$= E_n + \frac{\mu_A \mu_B}{Dr^3} f(\theta_A, \theta_B), \tag{1}$$

where the μ terms are the dipole moments of the solutes, D is the permittivity of the solvent, r is the distance apart of the dipole centres in the reactive complex and f is a function of the angles, θ_A and θ_B, of inclination of the dipolar axes to the line of centres. Let us examine some consequences of this simple treatment in the light of the data of reactions 1 and 6 of Table II.

(i) For a given reaction in a variety of solvents at constant temperature and pressure, we expect a linear relationship between the logarithm of the bimolecular rate constant and the reciprocal of the permittivity:

$$\left[\frac{d \ln k_2}{d(1/D)}\right]_{T,P} = -\frac{\mu_A \mu_B f(\theta_A \theta_B)}{r^3 kT}. \tag{2}$$

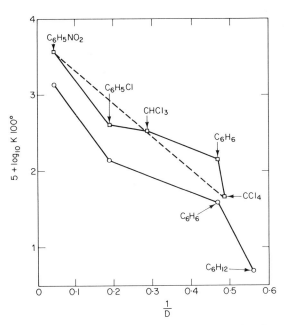

FIG. 1. The influence of the permittivity on the velocity of the reactions between ethyl iodide and triethylamine (⊙) and between methyl iodide and pyridine (⊡).

The extent to which this is obeyed is shown in Fig. 1, from which, after allowing f its maximum value of 2, r is found to be 1·78 Å.

(ii) For a series of homologous reactions in a given solvent at constant temperature, the apparent energy of activation, E_A, may be expected to vary linearly with respect to the product of the dipole moments of the solutes. We have

$$E_A = E_n + \frac{\mu_A\mu_B f(\theta_A, \theta_B)(1 - LT)}{Dr^3}. \tag{3}$$

The results of Winkler and Hinshelwood (1935) and of E. G. Williams and Hinshelwood (1934) on certain reactions in benzene solution (Table III) may be summarized in the following linear equations, giving the energies in calories per mole and expressing the magnitude of the dipole moments in Debye units:

$$E_A = 5,600 + 2,840\mu_A\mu_B,$$

$$E_A = 5,800 + 390\mu_A\mu_B,$$

$$E_A = 5,500 + 400\mu_A\mu_B.$$

When the experimental gradients $dE_A/d\mu_A\mu_B$ are compared with those given by equation (3), again taking $f(\theta_A, \theta_B)$ as 2, we find the following respective values of $\mathring{r}$: 1·49, 2·67 and 2·86.

TABLE III

DIPOLE MOMENTS AND APPARENT ENERGIES OF ACTIVATION (CALORIES) OF CERTAIN BIMOLECULAR REACTIONS BETWEEN UNDISSOCIATED MOLECULES IN BENZENE SOLUTION

Reactants		$\mu_{base} \times 10^{18}$	$\mu_{halide} \times 10^{18}$	$\mu_A\mu_B \times 10^{36}$	E_A
Base	Halide				
$(CH_3)_3N + CH_3I$		0·82	1·41	1·16	8,790
$(C_2H_5)_3N + CH_3I$		0·90	1·41	1·27	9,300
$C_5H_5N + CH_3I$		2·16	1·41	3·05	14,250
$C_9H_7N + CH_3I$		2·19	1·41	3·09	14,350
$C_5H_5N + CH_3I$		2·16	1·41	3·05	14,250
$C_5H_5N + C_2H_5I$		2·16	1·78	3·76	15,760
$C_5H_5N + C_3H_7^nI$		2·16	1·84	3·98	16,100
$C_5H_5N + C_3H_7^\beta I$		2·16	1·95	4·21	18,000
$p\text{-}CH_3.C_6H_4.NH_2 + C_6H_5COCl$		1·31	2·5	3·3	6,800
$H.C_6H_5.NH_2 + C_6H_5COCl$		1·53	2·5	3·8	7,350
$p\text{-}Cl.C_6H_5.NH_2 + C_6H_5COCl$		2·95	2·5	7·4	7,600
$p\text{-}NO_2.C_6H_5.NH_2 + C_6H_5COCl$		6·2	2·5	15·4	11,800

(iii) For a given reaction in a variety of solvents, the apparent energy of activation may be expected, from equation (3), to vary linearly with respect to the variable $(1 - LT)/D$. Figure 2 shows this to be sometimes the case.

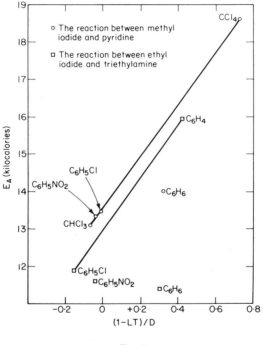

Fɪɢ. 2

The difference between chloroform and carbon tetrachloride as solvents is striking. Although the rates of reaction in the two solvents differ by a factor of 7·5 only, the difference in E_A is 5,500 cal.

Methyl iodide in carbon tetrachloride solution at 25°C reacts with trimethylamine about 3×10^4 times as fast as it reacts with pyridine. The increase is due in nearly equal measure to an increase in the A and E_A parameters.

Deviations from Strict Bimolecularity

Grimm, Ruf and Wolff (1931) found a drift in the second-order rate constant governing the reaction between ethyl iodide and trimethylamine in benzene and in *p*-dichlorbenzene. Moelwyn-Hughes and Hinshelwood (1932) detected a similar behaviour for the same reaction in carbon tetrachloride solution. The effect has been investigated in detail for the slightly simpler reaction between methyl iodide and trimethylamine in carbon tetrachloride solution (Fahim and Moelwyn-Hughes, 1956). The activity coefficients of the separate reactants were first measured. Those for methyl iodide at 25°C are given by the equation $\log_{10} \gamma_2 = -0.0306c + 0.00149c^2$ where c is the concentration in moles per litre of solution. The activity coefficient for the amine is found to decrease rapidly with increasing concentration in the region of kinetic interest, and to pass through a minimum value at a mole-fraction of about 0.09. During the course of reaction, therefore, the activity coefficient of the amine rises continuously. Attempts to determine the activity coefficient of the pure salt, tetramethylammonium iodide, in the stringently purified solvent proved abortive. Nevertheless, analysis of the end product of the reaction between methyl iodide and the amine indicated the presence of the salt, which, therefore, though insoluble in the pure solvent is soluble in solutions containing either of the reactants. This conclusion is confirmed

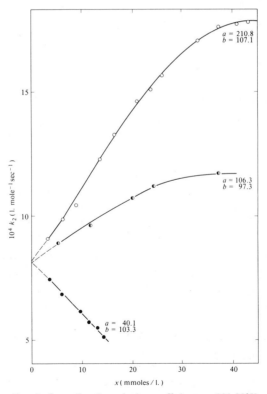

FIG. 3. Second-order velocity coefficients at 298.23°K.

by the work of Vernon and Sheard (1948) who estimated the solubility of this salt in pure benzene to be 6.92×10^{-6} mole/litre. These investigations bring to light a difficulty which is inherent in the study of the kinetics of reactions in solution. It is not the thermodynamic properties of solute A or of solute B in the solvent that is required, but the corresponding properties of each in the presence of the other—which is, on account of the chemical change, not measurable. The present reaction is thus much more complicated than cursory work would indicate. The bimolecular velocity constant, obtained in the usual way, is found to increase with the progress of reaction when the concentration (a) of the halide is in excess (Fig. 3), suggesting an essentially termolecular mechanism (R. F. Hudson and I. Stelzer, 1966). When the initial concentration of the amine (b) is in excess, however, the rate constant decreases with the progress of reaction. In view of these unexplained facts, it is difficult to determine the activation energy precisely. By combining the results of Tommila and Kauranen (1954) with those of Grimm, Ruf and Wolff and of Fahim, it is estimated that $E_A = (18,250 \pm 3250) - (9.35 \pm 4.75)RT$. It follows that dE_A/dT lies between the limits -9 and -28cal/mole-deg. The negative sign is consistent with the formation of a salt-like activated state, and with the effect of pressure on reactions of this type (Table 12.III). In methylene dichloride, however, dE_A/dT appears to be $+4 \pm 0.4$ in the same units (J. H. Beard and P. H. Plesch, 1964).

Preliminary Discussion of the Kinetics of Slow Reactions in Solution

Not all reactions between polar molecules in solution are slow, in the sense of the collision theory. The reaction between ethyl sulphide and diethylamine, for example, proceeds at a rate somewhat greater than that computed on the simplest assumptions of that theory. Most reactions of amines, aliphatic and aromatic, with alkyl and aryl halides, particularly in non-polar solvents are slow. The bimolecular velocity constant for the reaction between methyl iodide and trimethylamine in carbon tetrachloride solution, for examples, may be expressed as follows:

$$k_2 \text{ (litre/mole-sec)} = 1.33 \times 10^4 . \exp(-9,760/RT). \tag{4.51}$$

The slowness is directly expressed as the quotient of the Arrhenius parameter, and the standard collision frequency, Z^0, which has the order of magnitude of 10^{11} in these units. Roughly, then, $P = A_2/Z^0$ is about 7.5×10^{-6} in this instance.

Slowness can also be expressed in terms of entropy of activation. According to one of the definitions of this variable

$$\Delta^0 S^* = R \left[\ln \left(\frac{A_2}{v} \right) - \frac{d \ln v}{d \ln T} \right]. \tag{4.26}$$

When applied to bimolecular reactions, we naturally replace v by Z^0, and $d \ln v / d \ln T$ by $\frac{1}{2}$, so that

$$\Delta^0 S^* = R\left[\ln\left(\frac{A_2}{Z^0}\right) - \frac{1}{2}\right].\tag{4.52}$$

For reactions which proceed with the normal rate according to the collision theory, the entropy of activation thus defined is $-R \ln 2 = -0.139$ cal/mole-deg, which is negligible. In the present instance, $\Delta^0 S^* = -32.5$ in the same units. If we accept the definition of entropy of activation proposed by Eyring

$$\Delta S^{\ddagger} = R\left[\ln\left(\frac{A_n}{kT/h}\right) - 1\right].\tag{4.37}$$

In the present instance, $\Delta S^{\ddagger} = -41.7$. The proximity of the two estimated entropies arises from the *numerical* similarity of Z^0 (litre/mole-sec) to kT/h (sec^{-1}).

Let us consider a reaction for which $\Delta^0 S$ has been measured from the temperature variation of the equilibrium constant: $CH_3I + (CH_3)_2NC_6H_5 \rightarrow (CH_3)_3C_6H_5NI$ in nitrobenzene solution. $\Delta^0 S$ for the formation of phenyltrimethylammonium iodide is -36.9 cal/mole-deg. From the kinetic results of Essex and Gelormini, we have $\Delta^0 S^* = -18.2$, and $\Delta^0 S^{\ddagger} = -27.4$. In this instance, about half the total change in entropy attending the complete reaction has been reached in forming the critical complex.

Comparisons of this kind are based on the resolution of the total entropy change into components kinetically attributable to entropy changes associated with the formation of the critical complex, in exactly the same way as the total enthalpy change has been resolved (equation 4.13). They allow certain inferences to be drawn concerning the nature of the critical complex.

Denoting the initial, critical and final states by the subscripts i, c and j respectively, the standard gain in entropy of the reaction may be resolved as follows:

$$\Delta^0 S = S_j^0 - S_i^0$$
$$= (S_c^0 - S_i^0) - (S_c^0 - S_j^0)$$
$$= \Delta^0 S_1 - \Delta^0 S_2.$$

Thus, for example, if the standard gain in the thermodynamic entropy attending the reaction, $\Delta^0 S$, were equal to the entropy of activation, $\Delta^0 S_1$, for the forward reaction, the entropy of activation for the reverse reaction, $\Delta^0 S_2$, would be zero, and therefore the entropy of the active complex would equal that of the product (Eyring and Wynne-Jones, 1935; Kohnstamm and Scheffer, 1911; Trautz, 1910, 1915, 1935; R. E. Roberts, 1933). In the example just considered, the amine and the halide form a complex with a decrease in entropy; the salt forms the complex with an increase in entropy. These conclusions are consistent with association in the first step and with a complicated decomposition in the reverse step. The reliability of the conclusions depends on the assumptions that v is independent of T or directly proportional to it.

To express, as is done here, the slowness of reaction in terms which have the dimensions of entropy is not to solve the problem but to restate it.

A Possible Explanation for the Slowness of Solvolytic Reactions

Some reactions in solution, notably solvolyses, which ostensibly take place between molecules of solute and solvent may in fact be reactions between the solute and an ion formed by the solvent. The velocity would then appear to be slow because the concentration of ions is small. To test the hypothesis, we require a knowledge of the ionization constant of the solvent at various temperatures, which is seldom available.

The first quantitative application of this hypothesis was to the reaction

$$(CH_3CO)_2O + C_2H_5OH \rightarrow CH_3COOC_2H_5 + CH_3COOH,$$

which has been shown to be a slow reaction. In two typically inert solvents the bimolecular velocity coefficients are

$$k_2 = 9{\cdot}16 \times 10^4 \times \exp(-12,400/RT) \quad \text{(hexane)}$$

and

$$k_2 = 4{\cdot}14 \times 10^5 \times \exp(-13,400/RT) \quad \text{(carbon tetrachloride)}.$$

The reaction has also been measured in ethanol solution (E. A. Moelwyn-Hughes and Rolfe, 1932; Williams, Perrin and Gibson, 1936), under which conditions the rate—now naturally a pseudounimolecular one—is found to be given by the equation

$$k_1 = 5{\cdot}87 \times 10^7 \times \exp(-17,550/RT).$$

If the reaction takes place only between acetic anhydride molecules and the ethoxide ion, we have

$$k_2 = -\frac{d[(CH_3CO)_2O]}{dt} \frac{1}{[(CH_3CO)_2O][C_2H_5O^-]}$$

$$= k_1/[C_2H_5O^-]$$

$$= k_1 K^{-1/2},$$

where

$$K = [C_2H_5O^-][H^+].$$

In this particular instance electromotive and thermal data are available

(Macfarlane and H. Hartley, 1932) which can be combined to give the ionic concentration at various temperatures:

$$K^{1/2} = 8 \cdot 1 \times 10^{-6} \times \exp{(5{,}820/RT)}.$$

On combining the kinetic with the static information we have

$$k_2 = 7 \cdot 2 \times 10^{12} \times \exp{(-11{,}730/RT)}.$$

The rate of reaction, superficially regarded as slow, is thus actually somewhat greater than can be accounted for by assuming that each activating collision between the acetic anhydride molecule and the ethoxide ion is effective.

How far the ionization hypothesis can extend to other reactions, or, in fact, to this reaction in other solvents, cannot be judged until further thermodynamic constants are measured. In the meantime we must guard against assuming that a molecule in solution is un-ionized simply because we do not know its ionization constant.

The reverse operation has sometimes been tried. The ionization constant of a sparingly ionized solute may be computed by assuming that the rate of reaction attributed to the parent molecule is in fact due to an ion produced by it. The degree of ionization of methyl iodide in nitrobenzene solution would thus appear to have the order of magnitude 10^{-20}.[*]

A Tentative Interpretation of Slow Reactions

It has been argued (Moelwyn-Hughes, 1932) that, in reactions between molecules, "the observed critical increments (E_A) are false, in the sense that—by some unknown mechanism—they have been depressed from their normal values". Such a depression can occur if the transfer of energy from the translational to the vibrational form—or *vice versa*—were to be a difficult process, depending in a complicated manner on the temperature. Why such a difficulty should arise in reactions between unchanged solute molecules while being apparently absent from reactions involving a charged solute is not clear. The principle underlying the idea can be illustrated with reference to such energy transfers in gaseous ammonia. We have seen that activation in a complicated molecule may consist of direct accumulation of vibrational energy in one mode of vibration, which is that with the lowest frequency, and that the spread of energy to other modes can be relatively rapid (P. G. T. Fogg, P. A. Hanks and J. D. Lambert, 1953). The lowest internal vibration frequency in the ammonia molecule is $950 \, \text{cm}^{-1}$. We have also seen (equation 13.109) that the probability of a 01 energy acquisition

---

[*] *Acta Physicochim. U.S.S.R.*, 4, 201 (1936).

during a collision is given by the equation

$$p_{01} = \frac{\tau_c}{\tau} \cdot \frac{\exp(-hv/kT)}{1 - \exp(-hv/kT)} = \frac{\tau_c}{\tau} \frac{1}{\exp(hv/kT) - 1}, \qquad (14.132)$$

where τ_c is the time between successive collisions, and τ is the relaxation time. The two relaxation times detected in gaeous ammonia (W. Griffith, 1950; Buschmann and K. Schäfer, 1941; Herzfeld and Litovitz, 1959) are given in Table IV, along with the values of τ_c calculated from the viscosities given by

TABLE IV

Transition Probabilities and Relaxation Times for Gaseous Ammonia

$t°C$	(obs, τ sec)	(calc, τ_c sec)	hv/kT	p_{01}
20	1.2×10^{-7}	7.86×10^{-11}	4.663	6.9×10^{-2}
40	1.8×10^{-7}	8.47×10^{-11}	4.366	3.6×10^{-2}
40	1.45×10^{-6}	8.47×10^{-11}	4.366	4.44×10^{-2}

Chapman and Cowling (1929) using Maxwell's equation $\tau_c = (\pi/4)(\eta/P)$. From the values of p_{01} given in the last column, it can be seen that, if bimolecular constants involving ammonia were proportional to them, they would contribute to $RT^2(d \ln k_2/dT)_P$ amounts varying from -6 to -25 kcal/mole. The smaller of these terms would suffice to account for most of the apparent slowness of the reactions between amines and halides in non-polar solvents. It is to be stressed that it is not the actual value of p_{01} but its temperature variation that is important.

This explanation of the slowness of reaction between polar molecules is admittedly only one of many explanations that have been advanced. Reorganization of solvent molecules depends on the differing polarities of normal and active molecules, and it is possible that, as in certain proton-transfer reactions, the net rate is governed by the rate of such solvent reorganisation (G. J. Hills, P. J. Ovenden and D. R. Whitehouse, 1965). Before the problems raised in this Chapter can be solved, much more work is clearly to be done. In the meantime, closer attention must be paid to the observation that the second-order rate "constant" between amines and halides in non-polar media is not a true constant.

REFERENCES

Abraham, M. A., *Chem. Comm.*, 1307 (1969).
Beard, J. H. and P. H. Plesch, *Trans. Chem. Soc.*, 3682 (1964).
Buschmann and K. Schäfer, *Z. physikal. Chem.*, **B, 50**, 73 (1941).

Chapman and Cowling, *The Mathematical Theory of Non-Uniform Gases*, Cambridge (1929).

Eyring, H. and Wynne-Jones, *J. Chem. Physica*, **3**, 492 (1935).

Fahim and Moelwyn-Hughes, E. A., *Trans. Chem. Soc.*, 1034 (1956).

Fogg, P. G. T., P. A. Hanks and J. D. Lambert, *Proc. Roy. Soc.*, A, **219**, 490 (1953).

Griffith, W., *J. Appl. Physics*, **21**, 1319 (1950).

Grimm, Ruf and Wolff, *Z. physikal Chem.*, **B13**, 301 (1931).

Herzfeld and Litovitz, *Absorption and Dispersion of Ultrasonic Waves*, Academic Press, New York and London (1959).

Hills, G. J., P. J. Ovenden and D. R. Whitehouse, *Faraday Soc. Diss.*, **39**, 207 (1965).

Hudson, R. F. and I. Stelzer, *Trans. Chem. Soc.*, 775 (1966).

Kohnstamm and Scheffer, *Verlag Akad. Wetensch. Amsterdam*, **19**, 878 (1911).

Macfarlane and H. Hartley, *Phil. Mag.*, **13**, 425 (1932).

Menschutkin, *Z. physikal. Chem.*, **6**, 41 (1890).

Moelwyn-Hughes, E. A., *Chem. Rev.*, **10**, 260 (1932).

Moelwyn-Hughes, E. A. and Hinshelwood, *Trans. Chem. Soc.*, 230 (1932).

Moelwyn-Hughes, E. A. and Rolfe, *Trans. Chem. Soc.*, 241 (1932).

Roberts, R. E. and Soper, *Proc. Roy. Soc.*, A, **140**, 71 (1933).

Tommila and Kawranen, *Acta. Chem. Scand.*, **8**, 1152 (1954).

Trautz, *Z. physikal. Chem.*, **68**, 637 (1910); *Z. Elektrochem*, **25**, 4 (1915).

Vernon and Sheard, *J. Amer. Chem. Soc.*, **70**, 2035 (1948).

Williams, E. G. and Hinshelwood, *Trans. Chem. Soc.*, 1079 (1934).

Williams, Perrin and Gibson, *Proc. Roy. Soc.*, A, **154**, 684 (1936).

Winkler and Hinshelwood, *Trans. Chem. Soc.*, 1147 (1935).

APPENDIX 1

THE ENTHALPY OF CAVITY FORMATION

In general, $H = E + PV$, and changes in H at constant temperature therefore consist of three parts:

$$dH = dE + P\,dV + V\,dP.$$

Because the energy, E, can be expressed as a function of temperature and volume, we may write

$$dE = (\partial E/\partial T)_V\,dT + (\partial E/\partial V)_T\,dV,$$

so that

$$dH = (\partial E/\partial T)_V\,dT + (\partial E/\partial V)_T\,dV + P\,dV + V\,dP.$$

It follows that

$$(\partial H/\partial V)_{T,P} = (\partial E/\partial V)_T + P.$$

By a well known thermodynamic theorem,

$$(\partial E/\partial V)_T + P = T(\partial P/\partial T)_V = \alpha T/\beta,$$

where α is the coefficient of isobaric expansion, and β is the isothermal compressibility. Hence

$$(\partial H/\partial V)_{T,P} = \alpha T/\beta.$$

Both α and β are functions of the volume. If their quotient remains constant while the volume of the system is increased by the molar volume, V, the increase in enthalpy attending the formation of a hole of molar size is

$$\Delta H_{T,P} = \alpha T V/\beta.$$

If the argument can be extended to the formation of a cavity sufficient to house a molecule of volume v_2 in solvent 1,

$$\Delta H_{T,P} = \alpha_1 T V_2 \beta_1,$$

where α_1 and β_1 refer to the solvent, and V_2 is the molar volume of the solute (Eley, 1939).

APPENDIX 2

HEAT OF EXPANSION OF THE SOLVENT

A correction, usually slight, is required in determining E_A so as to allow for the fraction of the total enthalpy increase due to expansion of the solution. In dilute solutions, this may be taken as the enthalpy of expansion of the solvent, which is readily obtained from the coefficient of isobaric expansion, particularly in the case of water, for which a simple empirical relationship is found.

Coefficient of Expansion of Water at Atmospheric Pressure

The specific volume of water, v_1, at atmospheric pressure as given by Dorsey, *Properties of the Ordinary Water Substance*, p. 203, Reinhold, New York (1940) is accurately reproduced by the equation (Glew, private communication)

$$\ln_e v_1 = A + B \ln_e T + C/T$$

where

$$A = -6.70781$$
$$B = 1.012566$$

and

$$C = 280.633.$$

It follows that the coefficient of expansion is

$$\alpha = \left(\frac{d \ln_e v_1}{dT} \right)_P = \frac{B}{T} - \frac{C}{T^2}.$$

At the temperature of maximum density (4·0°C), α is zero and

$$T_{max} = C/B.$$

The heat of expansion, in cal/mole^{-1}, is

$$\Delta H = RT^2\alpha = BRT - RC$$

$$= 2 \cdot 012136T - 557 \cdot 6634.$$

The standard error in $d(\Delta H)/dT$ is $7 \cdot 257 \times 10^{-3}$ cal/mole-deg at all temperatures. The standard error in ΔH varies from $8 \cdot 344 \times 10^{-2}$ to $3 \cdot 282 \times 10^{-1}$ cal/mole.

A less accurate equation results if we ignore the difference between B and unity. Then

$$\ln_e v_1 = -6 \cdot 604 + \ln_e T + \frac{551}{RT}$$

or

$$v_1 = 1 \cdot 358 \times 10^{-3}T \cdot \exp(551/RT).$$

The approximate expression for the heat of expansion is therefore $\Delta H = RT^2\alpha = RT - 551$, which is sufficiently accurate for most kinetic purposes.

APPENDIX 3

THE TEMPERATURE VARIATION OF THE VISCOSITY OF WATER

Collision frequencies in solution may be independent of the viscosity or may vary, directly or inversely with respect to it, and it is essential to know how this property varies with respect to temperature. Otherwise, misleading values will be obtained for the true energy of activation, i.e. the difference between the energy of the reacting molecules and the average energy of all the molecules.

The simplest equation which we have found to reproduce empirically the temperature variation of the viscosity of water is

$$\log_{10} \eta = \overline{355} . 035 + \frac{275 \cdot 5}{R} \log_{10} T - \frac{0 \cdot 408 T}{2 \cdot 303 R} + \frac{49,883}{2 \cdot 303 R T},$$

or

$$\ln_e \eta = \overline{818} . 6 + \frac{275 \cdot 5}{R} \ln_e T - \frac{0 \cdot 408 T}{R} + \frac{49,883}{R T}. \qquad (137)$$

It follows that the viscous energy is

$$B = -R T^2 \left(\frac{d \ln_e \eta}{d T} \right)_P = 49,883 - 275 \cdot 5 T + 0 \cdot 408 T^2, \qquad (138)$$

and its temperature coefficient is

$$d B / d T = -275 \cdot 5 + 0 \cdot 816 T. \qquad (139)$$

Numerical values at rounded temperatures are given in Table VIII. The coefficient dB/dT, which changes its sign at 64·4°C, may radically affect estimates of the heat content and heat capacity of reactive complexes in aqueous solution.

TABLE VIII

THE VISCOSITY OF WATER, THE VISCOUS ENERGY AND
ITS TEMPERATURE COEFFICIENT

$t°C$	$\eta \times 10^3$ (gramme/cm-sec)	B (cal)	dB/dT
0	17·89	5,083	− 52·6
25	8·94	4,013	− 32·2
50	5·50	3,473	− 11·8
75	3·80	3,423	+ 8·6
100	2·84	3,903	+ 29·0

APPENDIX 4

KINETIC FORMULATION OF ISOTOPIC EXCHANGE REACTIONS, WITH SIMULTANEOUS ALLOWANCE FOR SOLVOLYSIS AND RADIOACTIVE DECAY

We shall adopt the same procedure as in Chapter 6, but will begin with the bimolecular exchanges and the pseudo-unimolecular hydrolyses, without allowing for radioactive decay.

If x denotes, in moles per litre, the extent of chemical change, at time t, due to the reversible bimolecular substitution, and y denotes the extent of chemical change due to the pseudo-unimolecular hydrolysis of the methyl iodide isomers, we have the following reaction scheme:

$$\underset{a-(x+y)}{CH_3I} + \underset{b-x+z}{*I^-} \underset{k_2}{\overset{k_2}{\rightleftarrows}} \underset{x-z}{CH_3^*I} \quad \underset{c+x+y}{I^-}$$

$$\underset{a-(x+y)}{CH_3I} + H_2O \overset{k_1}{\to} CH_3OH + \underset{y+z}{H^+} + \underset{c+x+y}{I^-}$$

$$\underset{x-z}{CH_3^*I} + H_2O \overset{k_1}{\to} CH_3OH + \underset{y+z}{H^+} + \underset{b-x+z}{*I^-} .$$

Here z stands for the decrease in the concentration of radioactive methyl iodide due to hydrolysis. In terms of the new variables, defined as follows

$$p = x + y,$$

$$q = x - z,$$

515

the various concentrations at time t are:

$$[CH_3I] = a - p,$$

$$[CH_3^*I] = q,$$

$$[I^-] = c + p,$$

$$[^*I^-] = b - q,$$

$$[H^+] = p - q.$$

The rate of production of hydrogen ions is given by the equation

$$\frac{d[H^+]}{dt} = \frac{d(p - q)}{dt} = k_1[CH_3I] + k_1[CH_3^*I]$$

$$= k_1(a - p) + k_1 q$$

$$= k_1[a - (p - q)].$$

Since $[H^+]_0$ is zero, we have, on integration

$$[H^+] = p - q = a[1 - \exp(-k_1 t)]. \tag{1}$$

The rate of production of radioactive methyl iodide is given by the equation

$$\frac{d[CH_3^*I]}{dt} = \frac{dq}{dt} = k_2[CH_3I][^*I^-] - k_2[CH_3^*I][I^-] - k_1[CH_3^*I]$$

$$= k_2(a - p)(b - q) - k_2 q(c + p) - k_1 q$$

$$= k_2[ab - (a + c)q - bp] - k_1 q. \tag{2}$$

On eliminating p from equations (118) and (119), we have

$$\frac{dq}{dt} = -[k_2(a + b + c) + k_1]q + k_2 ab \exp(-k_1 t)$$

$$= -Aq + B \exp(-k_1 t), \tag{3}$$

where

$$A = k_2(a + b + c) + k_1, \tag{4}$$

and

$$B = k_2 ab. \tag{5}$$

Equation (3) may be rearranged to give $dq + Aq\, dt = B \exp(-k_1 t)\, dt$. Multiply throughout by $\exp(At)$:

$$\exp(At)\, dq + qA \exp(At)\, dt = B \exp[(A - k_1)t]\, dt.$$

The integral on the left hand side is $q \exp(At)$. Hence

$$q \exp(At) = \frac{B}{A - k_1} . \exp[(A - k_1)t] + C,$$

where C is a constant. Multiply throughout by $\exp(-At)$:

$$q = \frac{B}{A - k_1} . \exp(-k_1 t) + C \exp(-At). \qquad (6)$$

On differentiating, we have

$$\frac{dq}{dt} = -\frac{Bk_1}{a - k_1} . \exp(-k_1 t) - CA . \exp(-At).$$

Now

$$-\frac{Bk_1}{A - k_1} = B - \frac{AB}{A - k_1}.$$

Therefore

$$\frac{dq}{dt} = B \exp(-k_1 t) - \frac{AB}{A - k_1} . \exp(-k_1 t) - CA . \exp(-At)$$

$$= B \exp(-k_1 t) - A\left[\frac{B}{A - k_1} . \exp(-k_1 t) + C \exp(-At)\right]$$

$$= B . \exp(-k_1 t) - Aq,$$

which is equation (120), as required. Since q is zero when t is zero,

$$C = -\frac{B}{(A - k_1)},$$

and

$$q = \frac{B}{A - k_1}[\exp(-k_1 t) - \exp(-At)].$$

On substituting expressions (121) and (122) for A and B, we have

$$q = \frac{ab}{a + b + c}\{\exp(-k_1 t) - \exp(-[k_2(a + b + c) + k_1]t)\},$$

or

$$[CH_3^*I] = \frac{ab}{a + b + c} . \exp(-k_1 t)\{1 - \exp[-k_2(a + b + c)t]\}, \qquad (7)$$

which is the desired solution. The ratio of the inorganic activity to its initial value is

$$R = \frac{[*I^-]}{[*I^-]_0} = \frac{b-q}{b} = 1 - \frac{q}{b}$$

$$= 1 - \left(\frac{a}{a+b+c}\right) \exp\left(-k_1 t\right)\{1 - \exp\left[-k_2(a+b+c)t\right]\}. \tag{8}$$

After an infinite lapse of time, R is seen to be unity, because, due to the irreversible hydrolysis of the two forms of the methyl halide, all the iodide is then in the inorganic state.

No allowance has yet been made for the spontaneous decay of radioactive iodine, whether in the inorganic form of $*I^-$ or in the organic form of CH_3^*I. The decay constant, λ, of I^{131} is 9.97×10^{-7} sec^{-1}, corresponding to a half-life of 8·05 days. The ratio, R, as defined above, is the ratio of the number of Geiger counts which would be found for a sample of inorganic iodide to the initial number if there were no radioactive decay. The ratio of the observed number of counters is less by a factor of $e^{-\lambda t}$, so that

$$\frac{\text{Observed counts per minute at time } t}{\text{Observed counts per minute at zero time}} = \frac{C_t}{C_0} = R \exp\left(-\lambda t\right). \tag{9}$$

On eliminating R from equations (125) and (126), we have

$$\frac{C_t}{C_0} \cdot \exp\left(+\lambda t\right) = 1 - \left(\frac{a}{a+b+c}\right) \exp\left(-k_1 t\right)\{1 - \exp\left[-k_2(a+b+c)t\right]\}. \tag{10}$$

To evaluate the bimolecular constant, this equation may be written in the form

$$k_2 = -\frac{1}{t}\left(\frac{1}{a+b+c}\right) \ln\left[1 - \left(\frac{a+b+c}{a}\right)\left[1 - \frac{C_t}{C_0}\exp\left(+\lambda t\right)\right]\exp\left(+k_1 t\right)\right].$$

The term on the left-hand side of equation (127) is the ratio $[*I^-]/[*I^-]_0$ when allowance has been made for the decay, and is the ratio required in the study of the chemical changes:

$$\left(\frac{[*I^-]}{[*I^-]_0}\right)_{\text{corrected}} = 1 - \left(\frac{a}{a+b+c}\right) \exp\left(-k_1 t\right)\{1 - \exp\left[-k_2(a+b+c)t\right]\}.$$

After an infinite lapse of time, this corrected ratio is also unity, because all the iodide is in the ionic form. When hydrolysis can be ignored, on the other hand, the ratio of the final to the initial concentration is

$$\left(\frac{[*I^-]_\infty}{[*I^-]_0}\right)_{\text{corrected}} = \frac{b+c}{a+b+c}. \tag{11}$$

Equation (127) is the exact kinetic expression for reversible, symmetrical bimolecular, isotopic exchanges accompanied by unimolecular solvolyses and corrected for radioactive decay. It has been successfully applied by Swart and Le Roux (1957, 1956) to the system discussed here. Its derivation has been given here in full because some of the equations appearing in their paper have, unfortunately, been printed in error and uncritically adopted by later workers.*

REFERENCES

Dorsey, *Properties of the Ordinary Water Substances*, p. 203, Reinhold, New York (1940).
Eley, *Trans. Faraday Soc.*, **35**, 1281 (1939).
Swart and Le Roux, *Trans. Chem. Soc.*, 406 (1957); 2110 (1956).

---

* Drs. Swart and Le Roux, in private communications with the Author, agree with the version given here, which is the same as that appearing in their original manuscripts. The errors referred to are doubtless editorial or typographic.

AUTHOR INDEX

Numbers in parentheses are reference numbers and are included to assist in locating references in which authors' names are not mentioned in the text. Numbers in italics refer to pages on which the references are listed.

A

Abraham, M. A. 497, *507*
Acree, S. E. 168, *217*
Adam, K. R., 375, *378*
Albery, W. J., 355, 356, 363, 374, *378, 379*
Alley, S. K., 37, *69*
Amis, E. S., 271, *296*, 492, *494*
Anbar, M., 32, *32*, 173, *216*
Anderson, D. L., 37, *69*
Andrea, J. H., 455, *475*
Arnett, W. G., 398, *406*
Arrhenius, S., 77, *97*
Ashdown, A. A., 149, *156*
Ashmore, P. G., 344, *379*
Averbach, I., 307, *337*

B

Bach, F., 366, *379*
Barclay, I. M., 480, *494*
Basolo, F., 197, 202, *216, 216*
Bauer, G., 361, *380*
Bauer, S. H., 37, *71*
Baxendale, J. H., 173, 186, *215*
Beard, J. H., 503, *507*
Becker, R., 440, *475*
Bell, R. P., 344, 349, 355, 359, 363, 369, 370, 373, 374, *378, 379*, 480, 486, *494*
Bender, M. L., *296*
Bennett, G. M., 172, *215*, 310, *337*
Benson, S. W., 492, *494*
Bentrude, J. J. 398, *406*
Berger, K., 35, *70*
Bergmann, L., 440, *475*
Bernal, J. D., 50, *69*, 470, *475*
Biegeleisen, J., 363, *379*
Bishop, D. M., 280, 290, *296*
Bjerrum, N., 63, *69*, 76, *97*, 160, 176, *215*

Bodenseh, H. K., 64, 65, *70*
Bodenstein, M., 93, *97*
Boer, J. H., 35, *70*
Bonhoeffer, K. F., 366, *379*
Bowen, E. J., 93, *97*, 299, *337*, 431, *475*
Boyd, R. N., 122, *123*
Bradley, R. S., 5, *17*, 100, *122*
Bradley, W. E., 75, *98*
Brand, J. C. D., 172, *215*
Branton, G. R., 304, *337*
Bray, 31
Briegleb, G., 35, *70*
Brinkman, M., 290, *296*
Britton, D., 366, *379*
Brockway, L. O., 35, *71*
Brönsted, J. N., 160, *215*
Brookfield, K. F., 136, *156*
Brown, D. A., 401, *406*
Brown, D. E., 492, *495*
Brown, H. F., 233, *259*
Brugger, W., 310, *338*
Buchachenko, A. L., 336, *337*
Bunton, C. A., 375, *379*
Butler, J. A. V., 258, *259*, 479, 480, *494*
Butler, P. J. R., 105, *122*
Buxton, G. V., 185, *215*

C

Caldin, E. F., 431, *475*
Calvin, M., 194, *216*
Cardenus-Cruz, F., 484, *495*
Chalk, A. J., 378, *379*
Chan, S. C., 197, 199, 201, *215*, 391, 392, *406*
Chapman, A. W., 327, *337*
Chapman, D. L., 187, *215*, 298, *337*
Christiansen, J. A., 174, 175, *215*
Claesson, S., 431, *475*

SUBJECT INDEX